Y0-AAA-608

Mathematical Sciences Research Institute Publications

21

Mathematical Sciences Research Institute Publications

Volume 1	Freed and Uhlenbeck: Instantons and Four-Manifolds Second Edition
Volume 2	Chern (ed.): Seminar on Nonlinear Partial Differential Equations
Volume 3	Lepowsky, Mandelstam, and Singer (eds.): Vertex Operators in Mathematics and Physics
Volume 4	Kac (ed.): Infinite Dimensional Groups with Applications
Volume 5	Blackadar: K-Theory for Operator Algebras
Volume 6	Moore (ed.): Group Representations, Ergodic Theory, Operator Algebras, and Mathematical Physics
Volume 7	Chorin and Majda (eds.): Wave Motion: Theory, Modelling, and Computation
Volume 8	Gersten (ed.): Essays in Group Theory
Volume 9	Moore and Schochet: Global Analysis on Foliated Spaces
Volume 10	Drasin, Earle, Gehring, Kra, and Marden (eds.): Holomorphic Functions and Moduli I
Volume 11	Drasin, Earle, Gehring, Kra, and Marden (eds.): Holomorphic Functions and Moduli II
Volume 12	Ni, Peletier, and Serrin (eds.): Nonlinear Diffusion Equations and their Equilibrium States I
Volume 13	Ni, Peletier, and Serrin (eds.): Nonlinear Diffusion Equations and their Equilibrium States II
Volume 14	Goodman, de la Harpe, and Jones: Coxeter Graphs and Towers of Algebras
Volume 15	Hochster, Huneke, and Sally (eds.): Commutative Algebra
Volume 16	Ihara, Ribet, and Serre (eds.): Galois Groups over Q
Volume 17	Concus, Finn, Hoffman (eds): Geometric Analysis and Computer Graphics

(continued)

Y.N. Moschovakis
Editor

Logic from Computer Science

Proceedings of a Workshop held November 13–17, 1989

Springer-Verlag
New York Berlin Heidelberg London Paris
Tokyo Hong Kong Barcelona Budapest

Yiannis N. Moschovakis
Department of Mathematics
University of California at Los Angeles
Los Angeles, CA 90024-1766
USA

Mathematics Subject Classifications: 03-06, 68-06, 68QXX, 03D05

With 10 Figures.

Library of Congress Cataloging-in-Publication Data
Logic from computer science: proceedings of a workshop held November 13–17, 1989/Yiannis Moschovakis, editor.
p. cm. – (Mathematical Sciences Research Institute publications; 27)
Includes bibliographical references.
ISBN 0-387-97667-1
1. Logic, Symbolic and mathematical – Congresses. 2. Computer science – Congresses. I. Moschovakis, Yiannis N. II. Series.
QA9.A1L64 1991
511.3 – dc20 91-28180

Printed on acid-free paper.

Production managed by Francine Sikorski; manufacturing supervised by Robert Paella.
Photocomposed copy provided by MSRI using AMS-TEX and LaTEX.
Printed and bound by Book-mart, North Bergen, NJ.
Printed in the United States of America.

9 8 7 6 5 4 3 2 1

ISBN 0-387-97667-1 Springer-Verlag New York Berlin Heidelberg
ISBN 3-540-97667-1 Springer-Verlag Berlin Heidelberg New York

PREFACE

The volume is the outgrowth of a workshop with the same title held at MSRI in the week of November 13–17, 1989, and for those who did not get it, *Logic from Computer Science* is the converse of *Logic in Computer Science*, the full name of the highly successful annual LICS conferences. We meant to have a conference which would bring together the LICS community with some of the more traditional "mathematical logicians" and where the emphasis would be on the flow of ideas *from* computer science *to* logic rather than the other way around. In a LICS talk, sometimes, the speaker presents a perfectly good theorem about (say) the λ-calculus or finite model theory in terms of its potential applications rather than its (often more obvious) intrinsic, foundational interest and intricate proof. This is not meant to be a criticism; the LICS meetings are, after all, organized by the IEEE Computer Society. We thought, for once, it would be fun to see what we would get if we asked the speakers to emphasize the relevance of their work for logic rather than computer science and to point out what is involved in the proofs. I think, mostly, it worked. In any case, the group of people represented as broad a selection of logicians as I have seen in recent years, and the quality of the talks was (in my view) exceptionally, unusually high. I learned a lot and (I think) others did too.

Many of the talks—including some excellent ones—were expository and the speakers did not wish to write them up. I know we missed some good ones, but this is already one of the largest proceedings volumes of an MSRI workshop. In addition to papers by invited speakers, the final list of contents includes a paper co-authored by G. Longo, who was scheduled to speak but was unable to come, and a contributed paper by J. Lipton.

The following is the schedule of talks at the conference:

Monday

D. S. Scott	Intrinsic topology
P. Martin-Löf	Mathematics of infinity
S. Feferman	Logics for termination and correctness of functional programs
P. Kanellakis	Unification problems and type inference
V. Breazu-Tannen	Comparing the equational theories of typed disciplines
J. C. Mitchell	PCF considered as a programming language
A. R. Meyer	Completeness for flat, simply typed $\lambda\Omega$-Calculus

Tuesday

N. Immerman	Theory of finite models: definitions, results, directions
P. Kolaitis	$0-1$ Laws for higher order logic
Y. Gurevich	Finite compactness theorem
K. J. Compton	A deductive system for existential least fixed point logic
M. Y. Vardi	Global optimization problems for database logic programs
A. Nerode	Concurrent computation as game playing
H. P. Barendregt	Logics, type systems and incompleteness
B. Trakhtenbrot	Synchronization, fixed points and strong conjunction
S. Abramsky	Topological aspects of non-well-founded sets

Wednesday

A. Scedrov	Bounded linear logic
A. Ehrenfeucht	Relational structures, graphs and metric spaces
P. Freyd	Recursive types reduced to inductive types
D. Leivant	Computing and the impredicativity of Peano's arithmetic
S. Buss	The undecidability of k-provability for the sequent calculus
G. Takeuti	Bounded arithmetic and computational complexity
J. Krajíček	Propositional logic relevant to bounded arithmetic
A. Urquhart	Relative complexity of propositional proof systems
P. Pudlak	Some relations between subsystems of arithmetic and the complexity of computations

Thursday

J.-Y. Girard	Linear logic and geometry of interaction
S. A. Cook	Complexity theory for functionals of finite type
M. Hennessy	A proof system for communicating processes with value passing
R. Statman	Diophantine and word problems in the λ-calculus and combinatory algebra
L. Moss	Logics of set-valued feature structures
W. Maass	On the complexity of learning a logic definition
M. Magidor	On the logic of non-monotonic reasoning
R. J. Parikh	Some recent developments in reasoning about knowledge
R. Fagin	A model theory of knowledge

Friday

P. G. Clote	Alogtime and a conjecture of S.A. Cook
J. H. Gallier	A "pot pouri" of unification procedures in automated theorem processing
H. Barringer	The imperative future
E. Dahlhaus	On interpretations
V. Pratt	Two-dimensional logic: a ubiquitous phenomenon
E. Börger	A logical framework for operational semantics of full PROLOG
M. C. Fitting	Bilattices in logic and in logic programming
K. Fine	Unique readability

We are grateful to the *National Science Foundation*, the *Mathematical Sciences Institute*, the *Center for the Study of Language and Information* and (most of all) *MSRI* for the financial support which made the workshop possible. Personally, I cannot thank enough Irving Kaplansky, Arlene Baxter and their gracious Staff who freed me from all administrative hassles during the workshop. The proceedings were expertly typed by Margaret Pattison, who also managed very efficiently and without complaint the interminable exchange of electronic, telephone, and primitive, mail communications with the multiple authors of 22 papers, spread out from Canada to Czechoslovakia. Finally, the task of organizing such a large conference and bringing to it good people representing a broad spectrum of mathematical and computer science logic was quite formidable. I imposed on many of my friends and some whom I knew only vaguely, for advice and for persuading their colleagues and friends to come. Most particularly I want to thank Jon Barwise, Sam Buss, Steve Cook, Martin Davis, Melvin Fitting,

Yuri Gurevich, Neil Immerman, Phokion Kolaitis, Ken Kunen (who did not attend in the end), Albert Meyer, Anil Nerode, Rohit Parikh, Vaughn Pratt and Dana Scott, without whose advice and help in the initial stages this workshop would not have materialized.

Yiannis N. Moschovakis

CONTENTS

THE IMPERATIVE FUTURE:
PAST SUCCESSES $\Rightarrow$ FUTURE ACTIONS

HOWARD BARRINGER AND DOV GABBAY

ABSTRACT. We present a new paradigm of reading future temporal formulae imperatively. A future formula is understood as a command to take action to make it true. With this point of view, temporal logic can be used as an imperative language for actions and control.

1. INTRODUCTION

We distinguish two views of logic, the declarative and the imperative. The declarative view is the traditional one, and it manifests itself both syntactically and semantically. Syntactically a logical system is taken as being characterised by its set of theorems. It is not important how these theorems are generated. Two different algorithmic systems generating the same set of theorems are considered as producing the same logic. Semantically a logic is considered as a set of formulae which hold in all models. The model $\mathcal{M}$ is a static semantic object. We evaluate a formula φ in a model and, if the result of the evaluation is positive (notation $\mathcal{M} \models \varphi$), the formula holds. Thus the logic obtained is the set of all valid formulae in some class $\mathcal{K}$ of models.

Applications of logic in computer science have mainly concentrated on the exploitation of its declarative features. Logic is taken as a language for describing properties of models. The formula φ is evaluated in a model $\mathcal{M}$. If φ holds in $\mathcal{M}$ (evaluation successful) then $\mathcal{M}$ has property φ. This view of logic is, for example, most suitably and most successfully exploited in the areas of databases and in program specification and verification. One can present the database as a deductive logical theory and query it using logical formulae. The logical evaluation process corresponds to the computational querying process. In program verification, for example, one can describe in logic the properties of the programs to be studied. The description plays the rôle of a model $\mathcal{M}$. One can now describe one's specification as a logical formula φ, and the query whether φ holds in $\mathcal{M}$ (denoted $\mathcal{M} \vdash \varphi$) amounts to verifying that the program satisfies the specification. These methodologies rely solely on the declarative nature of logic.

Logic programming as a discipline is also declarative. In fact it advertises itself as such. It is most successful in areas where the declarative component is dominant, for example deductive databases. Its procedural features are computational. In the course of evaluating whether $\mathcal{M} \vdash \varphi$, a procedural reading of $\mathcal{M}$ and φ is used. φ does not imperatively act on $\mathcal{M}$, the declarative logical features are used to guide a procedure, that of taking steps for finding whether φ is true. What does not happen as part of the logic is the reading of φ imperatively, resulting in some action. In logic programming such actions (e.g. assert) are obtained as side-effects through special non-logical imperative predicates and are considered undesirable. There is certainly no conceptual framework within logic programming for properly identifying only those actions which have logical meaning.

The declarative view of logic allows for a variety of logical systems. Given a set of data Δ and a formula φ, we can denote by $\Delta\ ?\ \phi$ the query of φ from Δ. Different procedures or logics will give different answers, e.g. $\Delta \vdash_{L1} \varphi$ or $\Delta \nvdash_{L2} \varphi$ depending on the logic. Temporal and modal logics allow for systems of databases as data, so the basic query/data situation becomes

$$\{t_i : \Delta_i\} ? t : \phi$$

if Δ_i hold at t_i then we query whether ϕ holds at t? The t_i may be related somehow, e.g. being time points. The basic situation is still the same. We have (distributed) data and we ask a specific query. The different logical systems (modal, temporal) have to do with the representation and reasoning with the distributed data. There are various approaches. Some use specialised connectives to describe the relations between $t_i : \Delta_i$. Some use a metalogic to describe it (for example classical logic). Some reason directly on labelled data. They all share the common theme that they answer queries from given databases, however complex they are and whatever the procedure for finding the answer is.

Implicit in the data/query approach is our view of the system from the outside. We are looking from the outside at data, at a logic and at a query, and we can check from the outside whether the query follows from the data in that logic. Temporal data can also be viewed this way. However, in the temporal case, we may not necessarily be outside the system. We may be in it, moving along with it. When we view the past, we can be outside, collect the data and query it in whatever logic we choose. This is the declarative approach. When we view the future, from where we are, we have an option. We can either be god-like, wait for events to happen and query them or we can be human and influence events as they occur, or even make them occur if we can. The result of this realization is the imperative view of the future.

A temporal statement like "John will wait for two hours", can only be read declaratively as something evaluated at time t, by an observer at a later time (more than two hours). However, an observer at time t itself has the further option of reading it imperatively, resulting in the action of forcing John to wait for two hours. In effect what we are doing is dynamic model construction (see section 4). To summarize, we distinguish three readings of logical formulae: the traditional declarative reading, finding answers from data; the logic programming procedural reading, supporting the declarative and giving procedures for finding the answer; the imperative reading which involves direct actions.

Reading the future imperatively in this manner is the basis of the imperative use of temporal logic as a paradigm for a new executable logic language. This idea is new. It is the theme this paper wants to introduce; the idea that a future statement can be read as commands. In traditional declarative uses of logic there are some aspects of its use. In logic programming and deductive databases the handling of integrity constraints borders on the use of logic imperatively. Integrity constraints have to be maintained. Thus one can either reject an update or do some corrections. Maintaining integrity constraints is a form of executing logic, but it is logically ad-hoc and has to do with the local problem at hand. Truth maintenance is another form. In fact, under a suitable interpretation, one may view any resolution mechanism as model building which is a form of execution. In temporal logic, model construction can be interpreted as execution. Generating the model, i.e. finding the truth values of the atomic predicates in the various moments of time, can be taken as a sequence of execution.

As the need for the imperative executable features of logic is widespread in computer science, it is not surprising that various researchers have touched upon it in the course of their activity. However there has been no conceptual methodological recognition of the imperative paradigm in the community, nor has there been a systematic attempt to develop and bring this paradigm forward as a new and powerful logical approach in computing.

The logic USF, defined by Gabbay [Gab89], was the first attempt at promoting the imperative view as a methodology, with a proposal for its use as a language for controlling processes. The papers [BFG+90, GH89] give initial steps towards the realisation of this imperative view of temporal logic as a viable programming paradigm.

2. Executing the Future

Consider a temporal sentence of the form:

antecedent about the past $\Rightarrow$ consequent about the present and future

this can, in general, be interpreted as if the "antecedent (about the past)" is true then **do** the "consequent (about the present and future)". Adopting this imperative reading, in fact, yields us with an execution mechanism for temporal logics. We take this as the basis of our approach, called METATEM. This name, in fact, captures two key aspects of our developing framework:

- the use of TEMporal logic as a vehicle for specification and modelling, via direct execution of the logic, of reactive systems;
- the direct embodiment of META-level reasoning and execution, in particular, identifying metalanguage and language as one and the same, providing a highly reflective system.

Generally, the behaviour of a (reactive) component is given by specifying the interactions that occur between that component and the context, or environment, in which it is placed. In particular, a distinction needs to be made between actions made by the component and those made by the environment. In METATEM, we perpetuate such distinctions. The behaviour of a component is described by a collection of temporal rules, in the form mentioned above. The occurrence, or otherwise, of an action is denoted by the truth, or falsity, of a proposition. The mechanism by which propositions are linked to actual action occurrences is not an issue here and is left to the reader's intuition. However, the temporal logic used below does distinguish between component and environment propositions.

A METATEM program for controlling a process is presented as a collection of temporal rules. The rules apply universally in time and determine how the process progresses from one moment to the next. A temporal rule is given in the following clausal form

past time antecedent $\Rightarrow$ future time consequent.

This is not an unnatural rule form and occurs in some guise in most programming languages. For example, in imperative programming languages it corresponds to the conditional statement. In declarative logic programming, we have the Horn clause rule form of Prolog and other similar languages. In METATEM, the "past time antecedent" is a temporal formula referring strictly to the past; the "future time consequent" is a temporal formula referring to the present and future. Although we can adopt a declarative interpretation of the rules, for programming and execution purposes we take an imperative reading following the natural way we ourselves tend to behave and operate, namely,

on the basis of the past **do** the future.

Let us first illustrate this with a simple merging problem. We have two queues, let us say at Heathrow Airport, queue 1 for British Citizens, and queue 2 for other nationalities. The merge box is immigration. There is a man in charge, calling passengers from each queue to go through immigration. This is a merge problem. We take "snapshots" each time a person moves through. The propositions we are interested in are, "from which queue we are merging" and "whether the British Citizens queue is longer than the other queue". We must not upset the left queue; these people vote, the others don't. We of course get a time model, as we go ahead and merge.

We want to specify that if queue 1 is longer, we should next merge from queue 1. The desired specification is written in temporal logic. Let $m1$ be "merge from queue 1", $m2$ be "merge from queue 2" and b be "queuel is longer than queue2". Note that $m1$ and $m2$ are controlled by the program, and b is controlled by the environment. The specification is the conjunction of formulae:

$$\Box(m1 \vee m2)$$

$$\Box\neg(m1 \wedge m2)$$

$$\Box(b \Rightarrow \bigcirc m1).$$

Here $\Box$ is the always operator and $\bigcirc$ stands for next, or tomorrow.

We want to use the *specification* to control the merge while it is going on. To explain our idea look at the specification $b \Rightarrow \bigcirc m1$. If we are communicating with the merge process and at time n we see that b is true, i.e. the left queue 1 is longer than queue 2, we know that for the specification to be true, $\bigcirc m1$ must be true, i.e. $m1$ must be true at time $n+1$. Since we are at time n and time $n+1$ has not come yet, we can make $m1$ true at time $n+1$. In other words, we are reading $\bigcirc m1$ as an imperative statement. Execute $m1$ next.

The specification, $b \Rightarrow \bigcirc m1$, which is a declarative statement of temporal logic, is now perceived as an executable imperative statement of the form:

If b then **next do** $m1$.

In practice we simply tell the merge process to merge from queue 1. Our point of view explains the words "the imperative future". Of course when adopting this point of view, we must know what it means to **do** or **execute** $m1$ or indeed any other atom or formula. To **execute** mi we merge from *queuei*. How do we **execute** b? We cannot make the queue longer under the conditions of the example at Heathrow. In this case we just helplessly wait to see if more passengers come.

More specifically, given a specification expressed by a formula φ in temporal logic using connectives, we use the separation theorem (we assume we have a fully expressive set of connectives) to rewrite the specification φ in the form

$$\bigwedge_i (\varphi^i \Rightarrow \psi_1^i \vee \varphi_2^i)$$

where φ^i is a pure past formula, ψ_1^i is a boolean expression of atoms and ψ_2^i is a pure future formula.

As we have seen above, there may be various ways of separating the original formula φ. The exact rewrite formulation requires an understanding of the application area. This is where the ingenuity of the programmer is needed. Having put φ in separated form we read each conjunct

$$\varphi^i \Rightarrow \psi_1^i \vee \psi_2^i$$

as:

If φ^i is true at time n **then do** $(\psi_1^i \vee \psi_2^i)$

More formally

Hold $(\varphi^i) \Rightarrow$ **Execute** $(\psi_1^i \vee \psi_2^i)$

We must make it clear though that there is a long way to go, if want to build a practical system.

In the general case the METATEM framework works as follows. Given a program consisting of a set of rules R_i, this imperative reading results in the construction of a model for the formula $\Box \bigwedge_i R_i$, which we refer to as the program formula. The execution of a METATEM program proceeds, informally, in the following manner. Given some initial history of execution:

(1) determine which rules currently apply, i.e. find those rules whose past time antecedents evaluate to true in the current history;
(2) "jointly execute" the consequents of the applicable rules together with any commitments carried forward from previous times — this will result in the current state being completed and the construction of a set of commitments to be carried into the future;
(3) repeat the execution process for the next moment in the context of the new commitments and the new history resulting from 2 above.

The "joint execution" of step 2 above relies upon a separation result for non-strict future-time temporal formulae. Given such a formula ψ, it can be written in a logically equivalent form as

$$\bigvee_{i=1}^{n} f_i$$

where each f_i is either of the form $\varphi_i \wedge \bigcirc\psi_i$ or of the form φ_i where each φ_i is a conjunction of literals, i.e. i.e. a present time formula, and ψ_i is a future (non-strict) time formula. Thus φ_i can be used to build the current state, and recursive application of the execute mechanism on ψ_i (with the program rules) will construct the future states.

It is worth mentioning that although the METATEM program rule form of "past implies future" is a syntactic restriction on temporal formulae, it is not a semantic restriction (see [Gab89] for proof details).

3. EXAMPLES

To further demonstrate the execution process we offer two simple examples. The first is a resource manager, the second is the ubiquitous "dining philosophers".

3.1. A Resource Manager. Consider a resource being shared between several (distributed) processes, for example a database lock. We require a "resource manager" that satisfies the constraints given in Figure 1.

(1) If the resource is requested by a process then it must eventually be allocated to that process.
(2) If the resource is not requested then it should not be allocated.
(3) At any one time, the resource should be allocated to at most one process.

FIGURE 1. Resource Manager Constraints

To simplify the exposition of the execution process, we restrict the example to just two processes. Let us use propositions r_1 and r_2 to name the occurrence of a request for the resource from process 1 and process 2 respectively. Similarly, let propositions a_1 and a_2 name appropriate resource allocations. It is important to note that the difference between propositions r_i and a_i. The request propositions are those controlled by the environment of the resource manager, whereas the allocation propositions are under direct control of the resource manager and can not be effected by the environment. Writing the given informal specification in the desired rule form, i.e. "pure past formula implies present and future formula", results in the rules of Figure 2. The symbols ⊙ and ● both denote "yesterday"; however, ⊙φ is false at the beginning of time, whereas ●φ is true at the beginning. The formula

$\neg r_1 \mathcal{Z} (a_1 \wedge \neg r_1)$ denotes that $\neg r_1$ has been true either since the beginnning of time or since $(a_1 \wedge \neg r_1)$ was true; it thus specifies that there is no outstanding request. Given particular settings for the environment propositions, i.e.

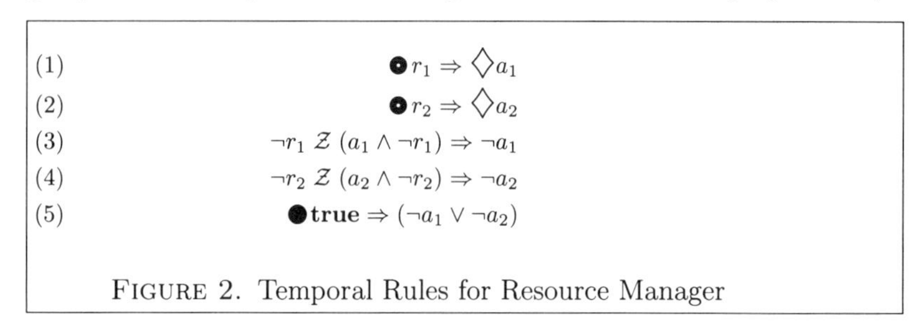

$$(1) \quad \odot r_1 \Rightarrow \Diamond a_1$$
$$(2) \quad \odot r_2 \Rightarrow \Diamond a_2$$
$$(3) \quad \neg r_1 \mathcal{Z} (a_1 \wedge \neg r_1) \Rightarrow \neg a_1$$
$$(4) \quad \neg r_2 \mathcal{Z} (a_2 \wedge \neg r_2) \Rightarrow \neg a_2$$
$$(5) \quad \bullet \mathbf{true} \Rightarrow (\neg a_1 \vee \neg a_2)$$

FIGURE 2. Temporal Rules for Resource Manager

r_1 and r_2, the execution will determine appropriate values for the allocation propositions. Figure 3 gives the first seven steps of a trace obtained for the given environment inputs (note that other behaviours are possible).

Requests		*Allocations*		*Time*	*Commitments*
r_1	r_2	a_1	a_2	*Step*	
y	n	n	n	0	
n	n	y	n	1	
y	y	n	n	2	
n	n	y	n	3	$\Diamond a_2$
y	n	n	y	4	
n	y	y	n	5	
n	n	n	y	6	

FIGURE 3. Sample Execution of Resource Manager

Step 0: The environment has requested the resource for process 1. To see how the current state is completed, we must find which rules from Figure 2 apply. Clearly rules 1 and 2 do not apply; we are currently at the beginning of time and hence $\odot \phi$, for any ϕ, is false. The other rules do apply and require that both a_1 and a_2 are made false. No extra commitments are to be carried forward. Execution proceeds to the next time step.

Step 1: First, there are no new requests from the environment. However, in examining which rules apply, we note that hypothesis of rule 1 is true (there was a request in the previous moment), hence we must "execute" $\Diamond a_1$. Also, rule 4 and rule 5 applies, in fact, the latter rule applies at every step because its hypothesis is always true. To execute $\Diamond a_1$, we execute $a_1 \vee (\neg a_1 \wedge \bigcirc\Diamond a_1)$. We have a choice; however, our mechanism will prioritise such disjunctions and attempt to execute them in a left to right order. One reason for this is that to satisfy an eventuality such as $\Diamond\phi$, we must eventually (in some future time step) satisfy ϕ; so we try to satisfy it immediately. [1] If we fail then we must carry forward the commitment to satisfy $\Diamond\phi$. Here we can make a_1 true. Rule 4 requires that a_2 is false, leaving rule 5 satisfied.

Step 2: The environment makes a request for both process 1 and process 2. The hypotheses of rules 1 and 2 are false, whereas those of rules 3 and 4 are true. So a_1 and a_2 are made false. No commitments are carried forward.

Step 3: No new requests from the environment, but there are two outstanding requests, i.e. both rule 1 and 2 apply. However, rule 5 requires that only one allocation may be made at any one time. The execution mechanism "chose" to postpone allocation to process 2. Thus, the eventuality from rule 1 is satisfied, but the eventuality for rule 2 is postponed, shown in the figure by carrying a commitment $\Diamond a_2$.

Step 4: A further request from the environment occurs, however, of interest here is the fact that $\Diamond a_2$ must be satisfied in addition to the commitments from the applicable rules. Note that this time, rule 4 does not apply as there is an outstanding request. Fortunately, the execution mechanism has no need to further postpone allocation to process 2 and a_2 is made true.

Steps 5 – 6: Similar to before.

3.2. Dining Philosophers. Our second example illustrates a rather simple solution to the dining philosophers problem. First, consider, if you can, a dining philosopher, Philosopher_A, alone in a closed environment. What is it that we wish to express about this philosopher? Basically that he never goes hungry, i.e. that he eats sufficiently often. So in our abstraction we use the proposition eaten_A to denote that Philosopher_A has just eaten. Hence a rather high level description of the behaviour of this chap might be given by the temporal logic formula

$$\Box\Diamond\text{eaten}_A.$$

However, since we know that he will eventually be rather more sociable and want to dine with colleagues, we'll give him a couple of forks and get him

[1] This particular choice of execution is described in more detail in the paper [BFG+90]

used not to using his fingers. In order to have eaten, we pretend that he needs to have possessed a fork in his left hand and one in his right hand for at least two moments; we denote possession of forks in left and right hands by the propositions fork_{AL} and fork_{AR}, respectively. Thus we should add to the above the formula

$$\Box(\text{eaten}_A \Rightarrow \odot(\odot(\text{fork}_{AL} \wedge \text{fork}_{AR}) \wedge (\text{fork}_{AL} \wedge \text{fork}_{AR}))).$$

Writing this as a METATEM process in rule form, we get the following

$$\bullet\,\textbf{true} \Rightarrow \Diamond\text{eaten}_A$$

$$\neg\odot(\odot(\text{fork}_{AL} \wedge \text{fork}_{AR}) \wedge (\text{fork}_{AL} \wedge \text{fork}_{AR})) \Rightarrow \neg\text{eaten}_A$$

which we will refer to as the Philosopher_A program. Let us execute, i.e. run, not kill, Philosopher_A. Remember that he is alone in his own closed world, with, therefore, everything under his own control. The interpretation process in fact will produce a run of Philosopher_A that has him eating infinitely often, that is if it were given enough time. Although he doesn't necessarily eat continuously, one possible execution trace will satisfy the temporal formula $\Diamond\Box\text{eaten}_A$. For example, all moments have fork_{AL} and fork_{AR} true, but the first two moments have eaten_A false. The first few steps of a trace produced by a prototype METATEM interpreter [Fis88] are given in Figure 4. In the first step, the interpreter sets fork_{AL} true but fork_{AR} false. Since both forks have not been held by the philosopher for the last two moments, eaten_A must be made false and the satisfaction of $\Diamond\text{eaten}_A$ must be postponed. A commitment to satisfy $\Diamond\text{eaten}_A$ is carried forward to step 1. The birthdate of the commitment tags the formula. The next two steps have fork_{AL} and fork_{AR} set true, but again postponement of satisfaction of $\Diamond\text{eaten}_A$ has to occur. These new commitments are absorbed into the commitment to satisfy $\Diamond\text{eaten}_A$ from step 0. In step 3 the philosopher can eat, therefore no commitment is carried forward. In step 4, we have a repeat of the situation from step 0. It is more interesting to create a process with two, or more, philosophers. In such a situation, the philosophers need to share forks, i.e. there should only be two real forks for two philosophers, three forks for three philosophers, etc. Consider just two philosophers. Assuming fork_{BL} and fork_{BR} are the propositions indicating that Philosopher_B has left and right forks, the following interfacing constraint needs to be placed.

$$\Box(\neg(\text{fork}_{AL} \wedge \text{fork}_{BR}) \wedge \neg(\text{fork}_{AR} \wedge \text{fork}_{BL}))$$

Propositions			*Time*	*Commitments*
fork_{AL}	fork_{AR}	eaten_A	*Step*	
y	n	n	0	$\Diamond\text{eaten}_A(0)$
y	y	n	1	$\Diamond\text{eaten}_A(0)$
y	y	n	2	$\Diamond\text{eaten}_A(0)$
n	n	y	3	
y	n	n	4	$\Diamond\text{eaten}_A(4)$
y	y	n	5	$\Diamond\text{eaten}_A(4)$
y	y	n	6	$\Diamond\text{eaten}_A(4)$
n	n	y	7	
y	n	n	8	$\Diamond\text{eaten}_A(8)$

FIGURE 4. Sample Trace of Single Philosopher

In rule form we thus have:

$$\begin{aligned}
\bullet\mathbf{true} &\Rightarrow \Diamond\text{eaten}_A \\
\neg \odot(\odot(\text{fork}_{AL} \wedge \text{fork}_{AR}) \wedge (\text{fork}_{AL} \wedge \text{fork}_{AR})) &\Rightarrow \neg\text{eaten}_A \\
\bullet\mathbf{true} &\Rightarrow \Diamond\text{eaten}_B \\
\neg \odot(\odot(\text{fork}_{BL} \wedge \text{fork}_{BR}) \wedge (\text{fork}_{BL} \wedge \text{fork}_{BR})) &\Rightarrow \neg\text{eaten}_B \\
\bullet\mathbf{true} &\Rightarrow \neg(\text{fork}_{AL} \wedge \text{fork}_{BR}) \\
\bullet\mathbf{true} &\Rightarrow \neg(\text{fork}_{AR} \wedge \text{fork}_{BL})
\end{aligned}$$

If we now consider execution of this set of rules, we should observe that both philosophers do eventually eat. Imagine the following potentially disastrous execution, e.g. Figure 5, from Philosopher_A's point of view.

Propositions						*Time*	*Commitments*
fork_{AL}	fork_{AR}	eaten_A	fork_{BL}	fork_{BR}	eaten_B	*Step*	
n	n	n	y	y	n	0	$\Diamond\text{eaten}_A(0)$, $\Diamond\text{eaten}_B(0)$
n	n	n	y	y	n	1	$\Diamond\text{eaten}_A(0)$, $\Diamond\text{eaten}_B(0)$
n	n	n	y	y	y	2	$\Diamond\text{eaten}_A(0)$
n	n	n	y	y	y	3	$\Diamond\text{eaten}_A(0)$
n	n	n	y	y	y	4	$\Diamond\text{eaten}_A(0)$
n	n	n	y	y	y	5	$\Diamond\text{eaten}_A(0)$
n	n	n	y	y	y	6	$\Diamond\text{eaten}_A(0)$

FIGURE 5. Bad Trace for Two Philosophers

The interpreter decides to allocate both forks to Philosopher_B forever, i.e.

it attempts to maintain $\neg \text{fork}_{AL}$, $\neg \text{fork}_{AR}$, fork_{BL} and fork_{BR} true. Thus the interpreter is not forced to make eaten_B false and hence will always be able to satisfy the constraint $\Diamond \text{eaten}_B$. However, the interpreter is forced to make eaten_A false, because the appropriate fork propositions are not true, and hence the satisfaction of $\Diamond \text{eaten}_A$ is continually postponed. Eventually the loop checking mechanism of the interpreter recognises this potential continual postponement of $\Diamond \text{eaten}_A$ since step 0 and backtracks to make a different choice. This loop checking mechanism will eventually cause both forks to be "allocated" to Philosopher_A. So the disaster, in fact, does not occur and both philosophers eat infinitely often.

Figure 6 gives part of a trace that can be obtained with the METATEM interpreter for this example.

Propositions						*Time*	*Commitments*
fork_{AL}	fork_{AR}	eaten_A	fork_{BL}	fork_{BR}	eaten_B	*Step*	
y	n	n	n	n	n	0	$\Diamond \text{eaten}_A(0)$, $\Diamond \text{eaten}_B(0)$
y	y	n	n	n	n	1	$\Diamond \text{eaten}_A(0)$, $\Diamond \text{eaten}_B(0)$
y	y	n	n	n	n	2	$\Diamond \text{eaten}_A(0)$, $\Diamond \text{eaten}_B(0)$
n	n	y	y	y	n	3	$\Diamond \text{eaten}_B(0)$
n	n	n	y	y	n	4	$\Diamond \text{eaten}_A(4)$, $\Diamond \text{eaten}_B(0)$
y	y	n	n	n	y	5	$\Diamond \text{eaten}_A(4)$
y	y	n	n	n	n	6	$\Diamond \text{eaten}_A(4)$, $\Diamond \text{eaten}_B(6)$
n	n	y	y	y	n	7	$\Diamond \text{eaten}_B(6)$
n	n	n	y	y	n	8	$\Diamond \text{eaten}_A(8)$, $\Diamond \text{eaten}_B(6)$

FIGURE 6. Sample Trace of Two Philosophers

4. A CLASSICAL VIEW

Executing the future can be considered as a special case of a more general computational problem involving quantifiers in Herbrand universes. To explain what we mean, consider the following example.

We are given a database with the following items of data:

(1) $Bad(O)$ "Oliver is bad"

(2) $\forall x[Bad(x) \Rightarrow Bad(S(x))]$ "Any son of a bad person is also bad"

(3) $\exists x Tall(x)$ "Someone is tall"

our query is

(4) $\exists x[Tall(x) \wedge Bad(x)]$ "Is there someone both tall and bad?"

The answer to this query is negative. It does not follow from the database in first order classical logic. If we try to use a Horn Clause procedural computation we get:

Data

(1) $Bad(O)$

(2) $Bad(x) \Rightarrow Bad(S(x))$

(3) $Tall(c)$

Query

(4) $Tall(x) \wedge Bad(x)$

The c is a Skolem constant. The query will fail or loop.

Suppose we add the *extra assumption* 5 to the database, where:

> (5) The existential quantifiers involved (i.e. the $\exists x$ in (3)) refer to the Herbrand universe generated by the function symbol $S(x)$.

This means that the constant c must be an element of the set $\{O, S(O), S^2(O), \ldots\}$. In this case, the query should succeed. There does exist an x which is both tall and bad. How can we logically put this extra input (5) into the database? Are we doomed to reasoning in such cases in the meta-language only?

The answer is that temporal logic does it for us. Let the "flow of time" be the set

$$\{O, S(O), S^2(O), \ldots\}$$

i.e. Oliver and his descendants are considered to be points of time. We let $S^m(O) < S^n(O)$ iff $m < n$.

We can read $Tall(x)$ (respectively $Bad(x)$) to mean *Tall* (respectively *Bad*) is true at "time" x. The data and query now become

Data

(1*)	Bad	*Bad* is true at time O, which is "now"
(2*)	$\Box(Bad \Rightarrow \bigcirc Bad)$	Always if *Bad* is true then *Bad* is also true next
(3*)	$\Diamond Tall$	Sometime *Tall* is true

Query

(4*) $\Diamond(Tall \wedge Bad)$

We see that $\Diamond$ is read exactly as we want it and the simple database (1), (2), (3), (5) is properly logically represented by (1*), (2*), (3*).

We can now see what is the general framework involved. First we are dealing with first order logic with function symbols. Second we are dealing with special quantifiers, which we call γ-*quantifiers* (γ for "generated" universe quantifiers). These quantifiers have the form $\exists_{\{f_i\}}$, $\forall_{\{f_i\}}$, where $\exists_{\{f_i\}}A(x)$, is true iff there exists an element y generated in the Herbrand universe of the function symbols $\{f_i\}$, such that $A(y)$ is true. $\forall_{\{f_i\}}A(x)$ reads for all y in the Herbrand universe of the function symbols $\{f_i\}$, $A(y)$ is true.

The data and query can now be written in the new γ-quantifiers language as follows:

Data

(1**) $Bad(O)$

(2**) $\forall_{\{O,S\}}Bad(x)$

(3**) $\exists_{\{O,S\}}Tall(x)$

Query

(4**) $\exists_{\{O,S\}}(Tall(x) \wedge Bad(x))$

Our MetateM interpreter is nothing more than a model generating method for first order logic with γ-quantifiers for unary Skolem functions. See [BFG+90] for further details.

5. Related Work in Temporal Logic Programming

Currently, the most developed implementation of an executable temporal logic is that of Moszkowski and Hale. Moszkowski [Mos86] has implemented a subset of a temporal logic, known as Tempura, based on interval models [MM84]. In fact, that particular interval temporal logic was first introduced for describing hardware [Mos83] and has now been put to the test with some examples of hardware simulation within Tempura [Hal87]. However, although Tempura is, in the computational sense, expressive enough, Tempura is limited in the sense of logic programming as in Prolog. Concurrent with the development of Moszkowski's work, Tang [Zhi83] defined and implemented a low-level temporal (sequential) programming system XYZ/E. His logic is based upon a linear, discrete, future-time temporal logic. Recent developments in that area have been developments of the logic (to obtain XYZ/D) to handle limited forms of distributed programming. Again, the drawback of this work, from the declarative and logic programming perspective, is the very traditional low-level state machine style. Steps towards the development of richer and more expressive executable subsets have since occurred. In particular, Gabbay [Gab87, Gab89] who describes general execution models, Abadi et al. [AM85, AM86, Aba87b, Aba87a, Bau89] who investigate temporal resolution mechanisms, Sakuragawa [Sak86] for a similar resolution-based "Temporal Prolog" and Hattori et al. [HNNT86] describe a modal logic based programming language for real-time process control, RACCO, based on Temporal Prolog.

REFERENCES

[Aba87a] M. Abadi, *Temporal-Logic Theorem Proving*, PhD thesis, Department of Computer Science, Stanford University, March 1987.

[Aba87b] M. Abadi, *The Power of Temporal Proofs*, In "Proc. Symp. on Logic in Computer Science", pages 123–130, Ithaca, June 1987.

[AM85] M. Abadi and Z. Manna, *Non-clausal Temporal Deduction*, Lecture Notes in Computer Science, **193**, 1–15, June 1985.

[AM86] M. Abadi and Z. Manna, *A Timely Resolution*, In "Proc. Symp. on Logic in Computer Science", pages 176–186, Boston, June 1986.

[Bau89] M. Baudinet, *Temporal Logic Programming is Complete and Expressive*, In "Proceedings of ACM Symposium on Principles of Programming Languages", Austin, Texas, January 1989.

[BFG+90] H. Barringer, M. Fisher, D. Gabbay, G. Gough, and P. Owens, *MetateM: A Framework for Programming in Temporal Logic*, In "Proc. of REX Workshop: Stepwise Refinement of Distributed Systems – Models, Formalisms and Correctness", pages 94–129, LNCS Vol. 430, Springer-Verlag, May 1990.

[Fis88] M. D. Fisher, *Implementing a Prototype* METATEM *Interpreter*, Temple Group Report, Department of Computer Science, University of Manchester, November 1988.

[Gab87] D. Gabbay, *Modal and Temporal Logic Programming*, In A. Galton, editor, "Temporal Logics and their Applications", Academic Press, London, December 1987.

[Gab89] D. Gabbay, *Declarative Past and Imperative Future: Executable Temporal Logic for Interactive Systems*, In B. Banieqbal, H. Barringer, and A. Pnueli, editors, "Proceedings of Colloquium on Temporal Logic in Specification", Altrincham, 1987, LNCS Vol. 398, pages 402–450, Springer-Verlag, 1989.

[GH89] D.M. Gabbay and I.M. Hodkinson, *Nonmonotonic coding of the declarative past*, Draft Report, October 1989.

[Hal87] R. Hale, *Temporal Logic Programming*, In A. Galton, editor, 'Temporal Logics and their Applications", chapter 3, pages 91–119, Academic Press, London, December 1987.

[HNNT86] T. Hattori, R. Nakajima, N. Niide, and K. Takenaka, *RACCO: A Modal-Logic Programming Language for Writing Models of Real-time Process-Control Systems*, Technical report, Research Institute for Mathematical Sciences, Kyoto University, November 1986.

[MM84] B. Moszkowski and Z. Manna, *Reasoning in Interval Temporal Logic*, In "AMC/NSF/ONR Workshop on Logics of Programs", pages 371–383, Berlin, 1984, LNCS Vol. 164, Springer-Verlag.

[Mos83] B. Moszkowski, *Resoning about Digital Circuits*, PhD thesis, Department of Computer Science, Stanford University, July 1983.

[Mos86] B. Moszkowski, *Executing Temporal Logic Programs*, Cambridge University Press, Cambridge, England, 1986.

[Sak86] T. Sakuragawa, *Temporal Prolog*, Technical report, Research Institute for Mathematical Sciences, Kyoto University, 1986, to appear in Computer Software.

[Zhi83] Tang Zhisong, *Toward a Unified Logic Basis for Programming Languages*, In R. E. A. Mason, editor, "Information Processing 83", pages 425–429, IFIP, Elsevier Science Publishers B.V. (North Holland), Amsterdam, 1983.

DEPARTMENT OF COMPUTER SCIENCE, UNIVERSITY OF MANCHESTER, OXFORD ROAD, MANCHESTER M13 9PL, UK.

E-mail: howard@uk.ac.man.cs

A LOGICAL OPERATIONAL SEMANTICS OF FULL PROLOG: PART III. BUILT-IN PREDICATES FOR FILES, TERMS, ARITHMETIC AND INPUT-OUTPUT

EGON BÖRGER

ABSTRACT. Y. Gurevich recently proposed a framework for semantics of programming concepts which directly reflects the *dynamic* and *resource-bounded* aspects of computation. This approach is based on (essentially first-order) structures that evolve over time and are finite in the same way as real computers are (so-called "dynamic algebras").

We use dynamic algebras to give an *operational* semantics for Prolog which, far from being hopelessly complicated, unnatural or machine-dependent, is *simple, natural* and *abstract* and in particular supports the process oriented understanding of programs by programmers. In spite of its abstractness, our semantics is machine executable. It is designed for extensibility and as a result of the inherent extensibility of dynamic algebra semantics, we are able to proceed by stepwise refinement.

We give this semantics for the full language of Prolog including all the usual non-logical built-in predicates. We hope to contribute in this way to reducing the "mismatch between theory and practice ... that much of the theory of logic programming only apply to pure subsets of Prolog, whereas the extra-logical facilities of the language appear essential for it to be practical" (Lloyd 1989). Our specific aim is to provide a mathematically precise but simple logical framework in which *standards* can be defined rigorously and in which different implementations may be compared and judged. This work is in applied logic and suggests several interesting new classes of problems in model theory.

Part I deals with the core of Prolog which governs the selection mechanism of clauses for goal satisfaction including backtracking and cut and closely related built-in control predicates. In Part II the database built-in predicates are treated. In the present Part III we deal with the remaining built-in predicates for manipulation of files, terms, input/output and arithmetic.

1. INTRODUCTION

Y. Gurevich recently proposed a framework for semantics of programming concepts which directly reflects the *dynamic* and *resource-bounded* aspects of computations. This approach is based on (essentially first-order) structures that evolve over time and are finite in the same way as real computers are (so-called "**dynamic algebras**"). These dynamic algebras allow to give an *operational semantics* for real programming languages which, far from being hopelessly complicated, unnatural, or machine-dependent, is:

- **simple**, read: easily comprehensible, mathematically tractable and universally implementable. Mathematical tractability means being precise and appropriate for conventional mathematical techniques to prove properties of programs and program transformations, by which the approach becomes useful as a foundation for program verification and transformation and similar areas. By universally implementable we mean that the operational semantics essentially constitutes an interpreter for the language under consideration which easily can be implemented on arbitrary general purpose machines.
- **natural**, read: it is a direct formalization of the usual informal process oriented descriptions of Prolog which one finds in manuals or textbooks.
- **abstract**, read: independent of any particular machine model but embodying all of the fundamental intuitive concepts of users of the language under consideration.

Historically speaking, the idea of dynamic algebra grew out of general considerations about relations between logic and computer science (see Gurevich 1988) and has since then been applied to give a complete operational semantics for Modula 2 (see Gurevich & Morris 1988) and Smalltalk (see Blakeley 1991) as characteristic sequential languages and for Occam (see Gurevich & Moss 1990) as an example of a parallel and distributed language.

We want to give a complete operational semantics, using dynamic algebras, for the full language of PROLOG, i.e. Prolog including all the usual non-logical built-in predicates for control, arithmetic, input/output and manipulations of terms, files and databases. Our system can be extended without difficulty to handle debugging features as well, although we do not enter into this subject here. Since it is easy to extend dynamic algebras when new features come into the picture, we are able to produce our system by stepwise refinement. Other refinements could be defined which have more sophisticated error handling procedures than the one we consider here. It should also not be too difficult to extend our Prolog algebras to the case of concurrent versions of Prolog, although we do not enter into this subject in this paper.

Our specific goal is to develop the logical framework up to the point where it:

(1) allows the rigorous definition of *standards* by which the fidelity of an implementation with respect to the designer's intentions (correctness) and the relations between different implementations may be judged. The rigor in our definition may help to obtain a certain uniformity of different implementations and provide a tool for correct implementation design.
(2) may be used as a tool for writing an unambiguous *manual* of reasonable size and clarity for Prolog;
(3) may be taken as a precise mathematical basis for writing an introduction to Prolog which is suitable as a *textbook for teaching* the language to students.

In particular, our description of all the basic (logical and non-logical) features of Prolog will be given entirely by first-order structures. This approach is thereby open to the application of conventional mathematical techniques for proving properties of programs and program transformations. The algorithmic part of our system (read: the system of transition rules) is very simple; it facilitates considerably the task of implementing the whole system as interpreter for Prolog. An implementation has been started and is ongoing (see Kappel 1990). We hope to contribute in this way to reducing the "mismatch between theory and practice... that much of the theory of logic programming only applies to pure subsets of Prolog, whereas the extra-logical facilities of the language appear essential for it to be practical" (Lloyd 1989).

The paper is organized as follows: in section 1 we will repeat the notion of dynamic algebra adapted from Gurevich 1988 and fix our language. In section 2 we will review some Prolog algebra components and rules of Part II which will be referred to in this paper. In this way, this paper can be read independently of Parts I and II although for sake of brevity we do not repeat here many explanations, examples and motivating discussion, for which the interested reader is referred to Parts I and II. In section 3 we describe the extensions needed for the usual built-in predicates for file manipulation. Section 4 is devoted to the built-in predicates for arithmetic, manipulation of terms and input-output. In section 5 we will present some interesting new classes of problems in model theory which are suggested by the introduction of dynamic algebras as models of real programming languages. In section 6 we will give references to related work in the literature.

2. Dynamic Algebras

The basic idea of the operational approach to semantics is to give the semantics of a programming language by an *abstract machine* for the execution of the commands of the language. Following Gurevich 1988 we consider an abstract machine to be a finite *mathematical structure*—which embodies all of the basic intuitions possessed by users of the language—together with a set of *transition rules* which reflect the execution of language commands. For the purpose of a semantics of Prolog it is sufficient to consider only first-order structures. The structures are many-sorted and partial. The latter means that the universes may be empty (at given moments in time) and that all functions are partial.

For the sake of simplicity we assume that the universe BOOL = $\{0, 1\}$ of Boolean values is always present. This allows the consideration of functions only, predicates being represented by their characteristic functions.

The **transition rules** we need are all of the form

$$\text{IF } b(x) \text{ THEN } \begin{array}{l} U_1(x) \\ \vdots \\ U_k(x) \end{array}$$

where $b(x)$ is a Boolean expression of the signature of the algebra under consideration, k is a natural number and $U_1(x), \ldots, U_k(x)$ are **updates** of the three types defined below. x is a set of variables ranging over some of the universes of the algebra under consideration. If x occurs in a rule then this rule is treated as a schema (i.e. it can be applied with concrete elements of the relevant universes in place of the variables). To execute such a transition rule means to perform simultaneously all the updates $U_1(x), \ldots, U_k(x)$ in the given algebra if $b(x)$ is true. As a result of such a rule application the given algebra is "updated" to another algebra (of the same signature). For notational convenience we will use nestings of if-then-else.

As update U we allow function updates and extensions of universes. A

function update is an equation of the form

$$f(t_1, \ldots, t_n) := t$$

where $t_1, \ldots, t_n, t$ are terms and f a function of the signature under consideration. To execute such a function update in a given algebra A means to evaluate the terms $t_1, \ldots, t_n, t$ in A, say with resulting elements $e_1, \ldots, e_n, e$ of some of the universes of A, and then to assign e as the new value of f in the argument combination $(e_1, \ldots, e_n)$. As a result we obtain a new Algebra A' of the same signature as A.

A **universe extension** has the form:

$$\text{TAKE } temp(i) = NEW(U) \text{ where } i = (1 \ldots t) \text{ IN } F \text{ ENDTAKE}$$

where U is a universe of the given algebra, F is a sequence of function updates and t a term of the given signature. To apply such a universe extension to an algebra A means to evaluate the term t in A, say to a number n, to add n new elements $temp(1), \ldots, temp(n)$ to the universe U and then to execute simultaneously for each i from 1 to n each function update in the list F. If t always evaluates to 1 we write simply

$$\text{TAKE } temp = NEW(U) \text{ IN } F \text{ ENDTAKE.}$$

To summarize we define following Gurevich 1988:

Definition. . *A* **dynamic algebra** *is a pair (A, T) consisting of a finite, many-sorted, partial first-order algebra A of some signature and a finite set T of transition rules of the same signature.*

3. Selection Core of Prolog

In this section we review a part of "Prolog algebras" and corresponding transition rules which have to do with the basic mechanism of clause selection directed towards satisfaction or failure and assisted by backtracking and cut. We first describe the "world" of Prolog, i.e. we list the main universes and the functions defined on them, and then formulate the transition rules for *stop, selection, success, backtracking, unify* and the control predicate *cut.* We follow the formalization in Part II, Börger (1990) to which we refer the reader for motivation and explanations.

3.1. Basic Universes and Related Functions

(a) The principal universe of Prolog algebras is a set RESSTATE which comes with a *successor* function, called *choicepoint*, and two distinguished elements (individual constants, i.e. 0-place functions) denoted by *currstate* and *nil*:

$$\begin{array}{l} choicepoint : \text{RESSTATE} \rightarrow \text{RESSTATE} \\ currstate,\ nil \in \text{RESSTATE} \end{array}$$

The basic logical unit which is formalized by an element of RESSTATE is an instantaneous description (intermediate *resolution state*) of a Prolog computation which has been started from an initially given query and will terminate either by failure or by finding an answer substitution. The individual constants serve to denote the current resolution state and the empty resolution state respectively. The (partial) successor function *choicepoint* imposes an order on the set of resolution states and is thought of as associating to each resolution state its next *choicepoint* in terms of the Prolog computation.

Each resolution state encodes the sequence of *goals* (coming together with their *cutpoint information*) which still have to be satisfied to terminate the computation successfully. This motivates the introduction of the following **two additional universes**:

(i) GOAL. The elements of this universe are called *activators.* The following function distinguishes between user-defined and built-in activators:

$$is\_user\_defined : \text{GOAL} \rightarrow \text{BOOL}$$

(ii) DECGOAL. The elements of this universe are called *decorated goals.* A decorated goal is thought to represent a goal together with its cutpoint information.

The association of a decorated goal sequence to each resolution state is formalized by a function:

$$decglseq : \text{RESSTATE} \rightarrow \text{DECGOAL}^* .$$

Since Prolog works on each element of DECGOAL* "from left to right," at each moment we need to consider only the first element of the given sequence (actually its activator and its cutpoint information) and its continuation (*restsequence*). This is formalized by introducing the **three functions** *activator, cutpoint, cont* as follows:

(1) activator : $\text{DECGOAL}^* \to \text{GOAL}$
This function associates with each decorated goal sequence the **current activator**, i.e. the activator of the first sequence element.

(2) cutpoint : $\text{DECGOAL}^* \to \text{RESSTATE}$
This function associates with each decorated goal sequence the cut-information of the current activator.

(3) cont : $\text{DECGOAL}^* \to \text{DECGOAL}^*$
This function associates with each decorated goal sequence its continuation, i.e. the remaining sequence after deletion of the current decorated goal.

(b) Two functions are needed which describe **unification** and **substitution**. For this purpose we introduce two new universes which we call SUBST (of *substitutions*) and TERM (of *terms*). The following two functions describe unifying substitutions between terms and the result of terms under substitutions:

$$unify : \text{TERM} \times \text{TERM} \to \text{SUBST} \cup \{\text{nil}\}.$$

This function assigns to two terms either a substitution (which represents "a **unifier**" of these terms (possibly the most general one) if it exists) or *nil*; consequently we will have to assume that *nil* is not an element of SUBST and is to be distinguished from the empty substitution.

$$subres : \text{DECGOAL}^* \times \text{SUBST} \to \text{DECGOAL}^*.$$

This function yields the result of applying the given substitution to the activators (goals) in the given list of decorated goals.

To assure correct operation of such an abstract unify function we have to solve the problem of **renaming** variables of a clause which has to be executed to avoid clashing with existing variables. To this purpose we introduce a renaming level by introducing a new universe INDEX with a "successor" function

$$succ : \text{INDEX} \to \text{INDEX}$$

and with an individual constant

$$varindex \in \text{INDEX}$$

(which we think of as indicating the successor of the highest so far used variable renaming index).

Renaming has to take place when a clause is selected to unify its head with the current activator and to substitute the latter by the clause body. We introduce therefore a universe CLAUSE (of user-defined *clauses*) with individual constant

$$database \in \text{CLAUSE}^*$$

which is thought of as representing the current database and with the following function which intuitively speaking yields for a predicate a fresh, i.e. renamed copy of its definition in the given database:

$$procdef : \text{CLAUSE}^* \times \text{INDEX} \times \text{TERM} \to \text{CLAUSE}^* .$$

procdef yields a copy (at the given renaming index) of the definition (in the given list of clauses) of that predicate whose functor is the same as the main functor of the given term.

In connection with the universe CLAUSE we need **four standard auxiliary functions** whose names speak for themselves: projection function (which yields for a sequence of clauses the component with the given number), length function (which yields the length of a given clause sequence) and functions which to a clause associate its head or body respectively:

$$clcomponent : \text{INTEGER} \times \text{CLAUSE}^* \to \text{CLAUSE}$$
$$length : \text{CLAUSE}^* \to \text{INTEGER}$$

Here INTEGER stands for a universe (of "integers").

$$clhead,\ clbody : \text{TERM} \to \text{TERM}$$

where we assume that CLAUSE is a subset of TERM.

(c) In accordance with our introductory remarks, we assume that our Prolog algebras are equipped with the standard **list operations**. We introduce two individual constants into BOOL:

$$error,\ stop.$$

As long as no error occurs the constant *error* will have value 0, otherwise it will be set to 1. The reader will see that it is easy to introduce into our system more sophisticated error handling if wanted. The constant *stop* during execution will have value 0 and will be switched to 1 in case the computation should be stopped.

3.2. The Transition Rules

In this section we review our rules from Part II, Börger (1990) for the SLD selection mechanism of PROLOG with backtracking and cut. The reader who is not much familiar with Prolog may have advantage to look at the pictorial representation of these rules given in Part I, Börger (1990).

The **stop rule** applies if all possible answers to the initially given goal have been produced already or if the current resolution state is a successfully terminating one. The former condition is verified when the (universe) RESSTATE has become "empty", formally expressed by $currstate = nil$. The second condition under which the stop rule applies is verified when the decorated goal sequence of the current resolution state is the empty sequence, formally:

$$decglseq(currstate) = [\quad].$$

This explains the following stop rule:

$$\begin{array}{l} \text{If } error = 0 \& stop = 0 \\ \quad \& (currstate = nil \text{ or } decglseq(currstate) = [\quad]) \\ \qquad \text{then } stop := 1 \end{array}$$

The **goal success rule** applies if the current activator has been successfully computed, formally expressed by $activator(decglseq(currstate)) = true$. In this situation, Prolog systems ("cancel" the satisfied activator and) start working on the next activator of the current resolution state, read:

$$decglseq(currstate) := cont(decglseq(currstate)),$$

what we will abbreviate by "succeed". We will also use the abbreviation *curractivator* for *activator(decglseq(currstate))*. This explains the following goal success rule:

$$\begin{array}{l} \text{If } error = 0 \& stop = 0 \& curractivator = true \\ \quad \text{then } succeed \end{array}$$

Note that we write *true* for the empty element of GOAL.

The **backtracking rule for user defined predicates** applies if the current activator is user-defined but there is no clause in the database to be tried upon it, formally expressed by

$$procdef(database, varindex, curractivator) = [\quad].$$

Prolog systems then ("remove" the current resolution state as failed and) attempt to satisfy the next alternative by going to *choicepoint(currstate)*. This is expressed by the following backtracking rule for user defined predicates:

$$\begin{array}{l}\text{If } error = 0 \& stop = 0 \& is\_user\_defined(curractivator) = 1 \\ \quad \& procdef(database, varindex, curractivator) = [\ \ ] \\ \qquad \text{then } backtrack\end{array}$$

Note that for the sake of easy understanding and remembering we write "backtrack" for the update $currstate = choicepoint(currstate)$.

The **selection rule** applies when the current activator is user-defined and its definition D in the database is not empty, formally expressed by

$$NOT(procdef(database, varindex, curractivator) = [\ \ ]).$$

Those clauses of D whose head might match the current activator constitute the (currently) possible alternatives which one may have to try in order to satisfy the current activator. For reasons of simplicity of exposition we choose to collect all the clauses in D and to check for matching on backtracking. Therefore our rule expresses that fresh (renamed) copies of the clause bodies of all elements in D are placed in the given order "on top" of the current resolution state; each of them is preceded by the goal to unify the current activator with the relevant clause head and is followed by the current continuation. The cutpoint for each alternative is $choicepoint(currstate)$. This explains the following selection rule:

$$\begin{array}{ll}\text{If} & error = 0 \& stop = 0 \& is\_user\_defined(curractivator) = 1 \\ & \& NOT(procdef(database, varindex, curractivator) = [\,]) \\ \text{then} & \text{LET } Cont\ cont(decglseq(currstate)) \\ & \text{LET } ClList\ procdef(database, varindex, curractivator) \\ & \text{LET } CutPt\ choicepoint(currstate) \\ & \text{TAKE } temp(i) = New\ \ (\text{RESSTATE}) \\ & \qquad\qquad \text{WHERE } i = (1 \ldots length(ClList)) \text{ IN} \\ & \quad \text{put new } resstates\ temp(i) \text{ on top} \\ & \quad decglseq(temp(i)) := \\ & \qquad [\langle curractivator = clhead(clcomponent(i, ClList)), \bullet\rangle, \\ & \qquad \langle clbody(clcomponent(i, ClList)), CutPt\rangle \mid Cont] \\ & \text{ENDTAKE} \\ & varindex := succ(varindex)\end{array}$$

Note.

(1) The abbreviation "put new *resstates temp*(i) on top" stands for the following three updates:

$$currstate := temp(1)$$
$$choicepoint(temp(i)) := temp(i+1)$$
$$choicepoint(temp(length(ClList))) := choicepoint(currstate)$$

(2) "*Let Variable Functioncall*" is used for a short and transparent formulation of rule conclusions in the following usual meaning:

Variable stands (in the updates) for the computed value of *Functioncall*.

(3) • indicates that the function is defined arbitrarily on this argument and that its value will never be needed.

A **unify-rule** applies when $curractivator = (Term1 = Term2)$. Which one of the following two rules applies then depends on the result of the function *unify* (success or failure):

If $error = 0 \& stop = 0 \& curractivator = (Term1 = Term2)$
$\& unify(Term1, Term2) \in$ SUBST
then *succeed with substitution update by* $unify(Term1, Term2)$

If $error = 0 \& stop = 0 \& curractivator = (Term1 = Term2)$
$\& unify(Term1, Term2) = nil$
then *backtrack*

By *succeed with substitution update by Subst* we abbreviate the update:

$$decglseq(currstate) := subres(cont(decglseq(currstate)),\ Subst).$$

The same abbreviation will be used in the sequel whith the respective substitution which will always be clear from the rule premisse.

The **cut rule** applies when $curractivator = !$. Then the cut succeeds and the next choicepoint is updated by the cutpoint (of *decglseq* of *currstate*). This is formally described by:

If $error = 0 \& stop = 0 \& curractivator = !$
then *succeed*
$choicepoint(currstate) := cutpoint(decglseq(currstate))$

4. Built-In Predicates for File Manipulation

In this section we will describe the extension of the algebra of section 2 and the rules which are needed to formalize the main file manipulating built-in predicates *see, seeing, seen, tell, append, telling, told* which are usually found in Prolog systems.

In different Prolog systems one can find a great variety of file management systems depending on whether or not multiple file access is allowed, whether or not (and which) files can be simultaneously in read and write mode etc. The question of which system is most suitable does not belong specifically to *logic* programming; also the ISO Prolog standardization committee at the time of writing this paper is still working on a standard proposal for file manipulating built-in predicates in Prolog. In the following we therefore define rules for a simple file management system which allows only one file access at a time; we also present the modifications which are needed to assure that only the *user* file is allowed to be open simultaneously for reading and writing mode. The essential ingredients of a dynamic algebra formalization of file manipulating built-in predicates in Prolog should become clear from such a system and the reader is invited to refine this description to make it fit his favorite file management system. In particular it should be easy to rewrite this section for the forthcoming ISO Prolog standard for file handling.

4.1. Additional Universes and Related Functions

The two principal new universes which have to be introduced are FILENAME (which we think of as set of (external) names of *existing* files) and FILE (which we think of as set of *existing* files or internal names of existing files). The relation between FILENAME and FILE is realized by a function:

$$denfile : \text{FILENAME} \rightarrow \text{FILE}$$

which associates to each existing file name the file denoted by it (or if one prefers to each external file name an internal one). FILE comes with a distinguished element (individual constant) denoted by *empty* which we think of as formalizing the empty file:

$$empty \in \text{FILE}\,.$$

The universe FILENAME (which in most systems is supposed to be a subset of a set of atoms) comes with three distinguished elements which represent the current input file, the current output file and the standard input and output file (screen) respectively:

$$currinfile,\ curroutfile,\ user \in \text{FILENAME}.$$

Furthermore, a function is needed which tells whether files are "open" ("active") or not:

$$activefile : \text{FILENAME} \to \text{BOOL}$$

where $activefile(File) = 1$ denotes that $File$ is "open".

Related to FILENAME and FILE we need a universe POSITION (of *pointer positions* in files) with a successor function *nextpos* and three functions which we think of as associating to a given file name its initial, current and final pointer position respectively:

$$nextpos : \text{POSITION} \to \text{POSITION}$$
$$firstpos,\ currpos,\ lastpos : \text{FILENAME} \to \text{POSITION}$$

A typical integrity constraint for these functions would be that with respect to the partial order imposed on POSITION by *nextpos* the value of $currpos(File)$ is always in the interval between $firstpos(File)$ and $lastpos(File)$. One can also easily determine at this level of abstraction the role of the pointer positions for empty files and what happens when an attempt is made to read at the end of a file; to concentrate on the essentials we assume this to be done. For the role of such integrity constraints see section 2.1 in Part I, Börger (1990).

4.2. The Transition Rules

We now give the rules which formalize in the extended algebra of section 3.1 the above mentioned file manipulating built-in predicates.

The *see*-**rule** applies when $curractivator = see(File)$. This activator fails and passes control to an error handling procedure together with an error message if $File$ is not the name of an existing file; otherwise it succeeds, updates $currinfile$ to $File$ and in case the latter is not $user$ and not open also opens $File$ with its current pointer position at the beginning of the file. It is assumed that the standard input file is always open and therefore needs no current pointer position update in the *see*-rule. This is formalized by:

$$\begin{array}{l}
\text{If } error = 0 \& stop = 0 \& curractivator = see(File) \text{ then} \\
\quad \text{If } File \notin \text{FILENAME then } error := 1 \\
\qquad\qquad\qquad errormessage := typeerror \\
\qquad\qquad\qquad backtrack \\
\quad \text{If } File \in \text{FILENAME} \\
\qquad \text{then } succeed \\
\qquad\quad currrinfile := File \\
\qquad\quad \text{If NOT}(File = user) \& activefile(File) = 0 \\
\qquad\qquad \text{then } activefile(File) := 1 \\
\qquad\qquad\qquad currpos(File) := firstpos(File)
\end{array}$$

Remark on Error Handling. In the sequel we will also write

$$call\ error\ handler\ with\ errormessage := \ldots$$

for the two updates $error := 1$ and $errormessage := \ldots$. The former passes control to an error handling procedure; in fact once it is executed none of our rules defined so far is applicable any more but only the rules for the error handling procedure—which we do not define here because they belong to the specifics of particular systems. The error handler will give back control to the main system by updating *error* to 0 (and *errormessage* to $undefined$). The same remark applies to all rules where an error detection is included.

A **seeing-rule** applies when $curractivator = seeing(File)$. Which one of the following two rules applies then depends on whether $File$ unifies with the name of the current input file (success) or not (backtracking):

$$\begin{array}{l}
\text{If } error = 0 \& stop = 0 \& curractivator = seeing(File) \\
\qquad \& unify(File, currinfile) \in \text{SUBST} \\
\text{then } succeed\ with\ substitution\ update\ by\ unify(File, currinfile)
\end{array}$$

$$\begin{array}{l}
\text{If } error = 0 \& stop = 0 \& curractivator = seeing(File) \\
\qquad \& unify(File, currinfile) = nil \\
\text{then } backtrack
\end{array}$$

The *seen*-**rule** applies when $curractivator = seen$. This activator succeeds, updates the current input file by the standard input file and makes the former inactive in case it was not the standard input file. This is formalized in the following rule (where one should remember that, as in the *see*-rule, the standard input file needs no activation or current pointer position update):

If $error = 0 \& stop = 0 \& curractivator = seen$ then
  $succeed$
  $currinfile := standinfile$
  If NOT$(currinfile = standinfile)$ then $activefile(currinfile) := 0$

The *tell*-**rule** applies when $curractivator = tell(File)$. This activator succeeds. If $File$ is not an existing file name, then it is made the name of a new, active, empty file to which the current output file is updated; if $File = user$, then the current output file is updated to the standard output file; if $File$ is the name of an existing file which is not the standard output file, then again the current output file is updated to $File$ and if the latter is not open, then it is opened and receives the empty file as denotation. This is formalized by the following rule (where for simplicity of exposition we skip the obvious distinctions for error handling):

If $error = 0 \& stop = 0 \& curractivator = tell(File)$ then
  $succeed$
  If $File \notin$ FILENAME then TAKE $temp =$ NEW(FILENAME) IN
      $activefile(temp) := 1$
      $denfile(temp) := empty$
      $curroutfile := temp$
      $temp := File$
    ENDTAKE
  If $File \in$ FILENAME then $curroutfile := File$
    If NOT$(File = user) \& activefile(File) = 0$
      then $activefile(File) := 1$
        $denfile(File) := empty$

Note. Remember that we assume the pointer positions for the empty file to be predefined. The update $temp := File$ in the preceding rule is not an official update of the kind allowed in our definition of dynamic algebras because $temp$ formally is not a term of the signature. We use this update as an abbreviation for the obvious official update which would use yet another intermediate function which associates to each member of the universe FILENAME its corresponding atom (name of a file in the sense of the syntax of Prolog). Another way to handle this kind of problem could be to also officially allow updates of the kind $temp := t$ in the update list of universe extensions with the meaning that the computed value of t is added to the universe in question ("And t to universe"). See Gurevich 1991.

The *append*-**rule** applies when $curractivator = append(File)$. In contraste to the tell-rule, it allows writing at the end of an existing file without erasing its previous content, independently of its being active or not. Therefore this activator succeeds. If $File$ is not an existing file name, then our rule provokes an error message; if $File$ is the name of an existing file, the current output file is updated to $File$, the current pointer position of $File$ is updated to its last position, and if $File$ is not open, it is then opened. This is formalized by:

If $error = 0 \& stop = 0 \& curractivator = append(File)$ then
  succeed
  If $File \notin$ FILENAME
    then *call error handler*
      $errormessage :=$ *no file with name File exists*
  If $File \in$ FILENAME
    then $curroutfile := File$
      $currpos(File) := lastpos(File)$
      If $activefile(File) = 0$ then $activefile(File) := 1$

The **telling-rules** are literally the same as the seeing-rules with *seeing*, $currinfile$ replaced by *telling*, $curroutfile$ respectively. They apply when $curractivator = telling(File)$. Which one of the following two rules applies then depends on whether $File$ unifies with the current output file (success) or not (backtracking):

If $error = 0 \& stop = 0 \& curractivator = telling(File)$
  $\& unify(File, curroutfile) \in$ SUBST
then *Succeed with substitution update by* $unify(File, curroutfile)$

If $error = 0 \& stop = 0 \& curractivator = telling(File)$
  $\& unify(File, curroutfile) = nil$
then *backtrack*

Also the *told*-**rules** are the same as the seen-rules with *seen*, $currinfile$, $standinfile$ replaced by *told*, $curroutfile$, $standoutfile$ respectively. They apply when

$$curractivator = told.$$

This activator succeeds, updates the current output file by the standard output file and makes the former inactive in case it was not the standard output file. Remember that as for the standard input file so also the standard output file needs no activation or current pointer position update.

If $error = 0 \& stop = 0 \& curractivator = told$ then
  $succeed$
  $curroutfile := standoutfile$
  If NOT$(curroutfile = standoutfile)$ then $activefile(curroutfile) := 0$

Remark. In the preceding formalization of file manipulating built-in predicates an open file can be used without discrimination for reading and writing; as a matter of fact, a file can simultaneously be the value of $currinfile$ and of $curroutfile$. To restrict this to the standard input and output file $user$, it suffices to replace the two file modes 0 (not active) and 1 (active) by four file modes 0, $read$, $write$, $both$. Only $user$ is supposed to be always in mode $both$. In the see-rule one has only to replace the $activefile$-update by

If Not$(activefile(File) = read)$ then $activefile(File) := read$
$currpos(File) := firstpos(File)$

The same modification applies to the $tell$-rule with $write$ instead of $read$; in addition, in its $activefile$-update in the FILENAME-extension one has to replaced 1 by $write$. Similarly, in the $append$-rule one has to replace the $activefile$-update by

If NOT$(activefile(File) = write)$ then $activefile(File) := write.$

The other rules are not affected.

5. Built-In Predicates for Arithmetic, Terms, Input-Output

In this section we describe the different forms of built-in predicates for arithmetical operations, manipulation of terms and input/output among those which have been selected for the standard draft proposal (see DFPS 1989). Most of these predicates are of a purely procedural nature and are not resatisfiable; here again there is not so much of typical for *logic* programming and therefore we treat only typical candidates to clarify the principles from which other similar cases can be worked out by routine.

Remark. When this paper was almost completed I received DFPS (1990) which updates DFPS (1989). The former allows an interesting concrete comparison between the dynamic algebra approach to a formal semantics for full Prolog advocated in this paper and Parts I and II of Börger (1990) and the approach presented by Deransart & Ferrand (1987) which is the basis of the formal presentation in DFPS (1990). In those cases of built-in predicates where DFPS (1990) contains a complete intuitive procedural description of their effect our rules exactly parallel those verbal procedural descriptions and are considerably shorter and technically less involved than the formal substitutes offered in DFPS (1990); compare for ex. the following rules for *functor*, *arg*, *univ*, *name* etc. That our exact rules, formulated in an attempt to extract from implementations and reference manual descriptions the procedural content of the real effect of the built-in predicates in working systems, parallel the attempts in DFPS (1990) to give a complete verbal description and can be considered as their direct formalization shows more than anything else the naturalness and the advantage of the dynamic algebra approach.

a) For the description of **arithmetical built-in predicates** we introduce two new universes NUMBER (of numbers which usually include integers and real numbers) and CONAREXPR (of constant—i.e. variable free—arithmetical expressions which are constructed starting from numbers by applying the available arithmetical functions like -, +, *, /, mod, abs, sqrt, sin, cos etc.). Obviously for each type of numbers one wants to describe, say for a standard, we will include a correponding universe in our algebras such as INTEGER or REAL, and for each arithmetical function one wishes to allow for the construction of arithmetical expressions we will introduce a corresponding function among the corresponding universes: in a natural way, abstract data types are subalgebras of dynamic algebras.

For each of the usual binary arithmetical built-in predicates *arbip* (like $<, >, =:=$, Not=:=, $>=, =<$) we introduce into the algebra a corresponding Boolean valued function

$$Den[arbip] : \text{NUMBER} \times \text{NUMBER} \rightarrow \text{BOOL}.$$

When these built-in predicates are executed for arguments, say, *Expression*1, *Expression*2, then Prolog systems first try to evaluate the arguments *Expression*1, *Expression*2 to numbers and if this succeeds then the computed numbers will be compared by the built-in predicate in question. To describe this evaluation of constant arithmetical expressions to numbers we introduce a function

$$Value : \text{CONAREXPR} \rightarrow \text{NUMBER}.$$

Based on these new universes and functions it is easy to formalize the intended meaning of the usual arithmetical built-in predicates by rules. The *is*-**rule** applies when $curractivator = Result\ is\ Expression$. This goal succeeds if $Expression$ evaluates to a number which unifies with $Result$. If $Expression$ is not a variable-free arithmetical term, then an error message is reported. If the number value of $Expression$ does not unify with $Result$, then this goal fails. This explains the following *is*-rule:

If $error = 0 \& stop = 0 \& curractivator = Result\ is\ Expression$
$\quad \& Expression \in$ CONAREXPR
$\quad \& unify(Result,\ value(Expression)) \in$ SUBST
then *Succeed with substitution update by* $unify(Result,\ value(Expres-$
$sion))$

If $error = 0 \& stop = 0 \& curractivator = Result\ is\ Expression$
$\quad \& Expression \in$ CONAREXPR
$\quad \& unify(Result,\ value(Expression)) = nil$
then *backtrack*

If $error = 0 \& stop = 0 \& curractivator = Result\ is\ Expression$
$\quad \& Expression \notin$ CONAREXPR
then $errormessage :=$ *Expression is not a constant arithmetical term*
succeed (or *backtrack* or *stop* or *call error handler*)

Remark. Different Prolog systems differ in the way they react in case of an error message. The reactions we discovered in analysing the implementations which were available to us range from success over error handling and failure to stop of the whole system. A similar remark applies to other rules.

The *number comparison*-**rules** for any of the usual built-in comparison predicates *cp* among $<, >, =:=, \ldots$ are similar in structure to the *is*-rule: success is obtained when both arguments of *cp* evaluate to numbers and the corresponding comparison of these numbers in the algebra gives a positive result. If one of the two arguments does not evaluate to a number, then an error message is reported. If $Den[cp]$ on the evaluated arguments yields value 0, then the goal fails. This explains the following rule for any built-in comparison predicate cp:

If $error = 0 \& stop = 0 \& curractivator = cp(Expression1, Expression2)$
$\& Expression1 \in$ CONAREXPR $\& Expression2 \in$ CONAREXPR
then If $Den[cp]$ $(Value(Expression1), Value(Expression2)) = 1$ then
succeed
else *backtrack*

If $error = 0 \& stop = 0 \& curractivator = cp(Expression1, Expression2)$
$\& Expression1 \notin$ CONAREXPR or $Expression2 \notin$ CONAREXPR
then $errormessage :=$ *one expression is not a constant arithmetical term*
succeed (or *backtrack* or *stop* or *call error handler*)

Remark. If one wants to further distinguish different types of error to be reported when executing an arithmetical built-in predicate, the only thing we have to do is to refine our rules (and possibly to extend our algebra by new universes corresponding to the new kind of objects to be distinguished, like VAR (for variables) or STRING (for strings) etc.). This kind of Prolog algebra modification is routine work. Also the introduction of CONAREXPR as universe can be avoided if its usual inductive definition is included into the *is*-rule (as is done in DFPS 1990).

b) For the description of **built-in predicates for term comparison, type testing and manipulation** the main universe we have to enrich is TERM. We have to introduce new subuniverses and functions in correspondence with the syntactical subcategories and operations on terms which are the content of the built-in predicates to be formalized. In particular, TERM will contain the following auxiliary universes:

VAR (of variables) CONSTANT (of constant terms)
ATOM (of atoms) NUMBER (of numbers)
STRING (of symbol sequences)

where NUMBER as already mentioned above is usually supposed to contain the subuniverses INTEGER (which contains the constant 0), NATURAL (of all positive integers) and REAL.

LIST (of lists of terms) STRUCTURE (of compound terms).

For LIST we assumed already the usual list operations *Listhead*, *Listtail* and a constant [ ] for the empty list in our algebras. We need furthermore an injective function

$$componentlist : \text{STRUCTURE} \to \text{LIST}$$

which yields for every structure $f(t1, \ldots, tn)$ the list $[f, t1, \ldots, tn]$ of its immediate components. The injectivity of *componentlist* assures that we can use the inverse function which we denote by *inverse*(*componentlist*).

We also need a function

$$variablelist : \text{NATURAL} \rightarrow \text{LIST}$$

which associates to each non negative number n the list of length n each of which components is the variable "_" (which therefore formally has to be assumed to be a distinguished element in VAR. We do not want to enter here into the details of the special status of "_" in connection with substitutions. This has to be carefully considered once SUBST and *unify* are specified in further detail.

The universe STRUCTURE of compound terms comes with three functions:

$$functor : \text{STRUCTURE} \rightarrow \text{ATOM}$$

This function yields the name (the functor symbol) of a compound term.

$$arity : \text{STRUCTURE} \rightarrow \text{NATURAL}$$

This function yields the number of arguments of a compound term.

$$argument : \text{NATURAL} \times \text{STRUCTURE} \rightarrow \text{TERM}$$

This function associates to a pair $(i, Term)$ the i-th argument of the given compound term if it exists.

As in the case of arithmetic built-in predicates we also need for each of the usual binary term comparing built-in predicates *termcp* (like ==, <, >, =<, >= etc.) a corresponding function *Den*[*termcp*] from TERM × TERM into BOOL in our algebras (this expresses the standard ordering of terms found in implementations).

In this extension of Prolog algebras we can formulate the rules for the usual built-in predicates for term comparison, type testing and manipulation. The *unify*-rule has been given already in section 2. The symmetric ***notunify*-rule** is given just for the sake of completeness:

$$\begin{array}{l}\text{If } error = 0 \,\&\, stop = 0 \,\&\, curractivator = \text{NOT}(Term1 = Term2) \\ \text{then If } unify(Term1, Term2) \in \text{SUBST} \text{ then } backtrack \\ \qquad\qquad\qquad\qquad\qquad\qquad\qquad\qquad\qquad \text{else } succeed\end{array}$$

The *term comparison*-**rules** are defined similarly to the number comparison rules as follows, where *cp* is any of the built-in predicates for term comparison ($\equiv$, *not identical*, *less than*, *less than or equal*, *greater than*, *greater than or equal*, ...):

$$\begin{array}{l} \text{If } error = 0 \& stop = 0 \& curractivator = cp(Term1, Term2) \\ \text{then If } Den[cp](Term1, Term2) = 1 \text{ then } succeed \\ \qquad\qquad\qquad\qquad\qquad\qquad\qquad \text{else } backtrack \end{array}$$

Remark. Depending on the ordering of terms one may wish to introduce the detection of errors, for ex. insufficient instantiation of arguments making the ordering indeterminate. It is a routine matter to include such an error detection into our rules (see the number comparison rules).

The *type testing*-**rules**—if applicable—always succeed or fail without further effect and are of the following simple pattern, where *category* stands for any of the syntactical categories *var*, *atom*, *atomic*, *number*, *integer*, *real*, *string*, *structure* for which there is a corresponding universe CATEGORY in our algebras:

$$\begin{array}{l} \text{If } error = 0 \& stop = 0 \& curractivator = category(Term) \\ \text{then If } Term \in \text{CATEGORY then } succeed \\ \qquad\qquad\qquad\qquad\qquad\quad \text{else } backtrack \end{array}$$

DFPS (1989) do not consider any error message for type testing.

For *term manipulation* we describe the rules for $functor(T, F, N)$, $arg(N, T, A)$, $T\ =\ ..L$ (*univ*) and $name(A, L)$. The $functor(Term, Functor, Arity)$-**rule** checks whether $Term$ is a variable or whether it is a constant or compound term whose functor unifies with $Functor$ and whose arity with $Arity$ (arity 0 in case $Term$ is a constant). In the latter case the predicate simply succeeds with the corresponding unifying substitution; in the former case it succeeds by unifying $Term$ with the term $Functor(\_, \ldots, \_)$ where $Arity$ is the number of argument variables. This is formalized by:

If $error = 0 \& stop = 0 \& curractivator = functor(Term, Functor, Arity)$
then
  If $Term \in$ STRUCTURE
    then If $unify([Functor, Arity], [functor(Term), arity(Term)]) \in$ SUBST
      then *succeed with substitution update by* $unify(\dots)$
      else *backtrack*
  If $Term \in$ CONSTANT
    then If $unify([Functor, Arity], [Term, 0]) \in$ SUBST
      then *succeed with substitution update by* $unify(\dots)$
      else *backtrack*
  If $Term \in$ VAR
    then If $Arity = 0 \& Functor \in$ CONSTANT
      then *succeed with substitution update by* $unify(Term, Functor)$
      If $Arity \in$ NATURAL $- \{0\} \& Functor \in$ ATOM
      then *succeed with substitution update by* $Subst$
      else *backtrack*
  Else *error handling*

where $Subst$ stands for the substitution

$$unify(Term,\ inverse(componentlist)([Functor \mid variablelist(Arity))))].$$

For the error handling the following seems to be agreed upon by the standardization committe (see DFPS 1989):

succeed (or *backtrack* or *stop* or *call error handler*)
If $Term \in$ VAR $\&(Functor \in$ VAR or $Arity \in$ VAR)
  then $errormessage :=$ *instantiation error*
If $Arity \notin$ VAR $\cup$ INTEGER
  then $errormessage :=$ *type error*
If $Arity \in$ INTEGER $-$ NATURAL
  then $errormessage :=$ *range error*
If $Arity \in$ INTEGER $\& Arity > max\_arity$
  then $errormessage :=$ *implementation error*

The **arg(N, Term, Arg)-rule** following DFPS (1990), succeeds by unifying Arg with the N-th argument of the structure $Term$ and otherwise fails with possibly some error message. As error messages, instantiation error (if N or $Term$ is a variable), type error (if N is not an integer or $Term$ is a constant), and range error (if N is an integer greater than the arity of $Term$ or a non-natural integer) are considered. The $(N, Term, Arg)$-rule therefore reads as follows:

If $error = 0 \& stop = 0 \& curractivator = arg(N, Term, Arg)$
then If $Term \in$ STRUCTURE $\& N \in$ NATURAL $\& 0 < N \leq arity(Term)$
$\& unify(Arg, argument(N, Term)) \in$ SUBST
then *succeed with substitution update by* $unify(Arg, argument(N, Term))$
else *backtrack*
If $N \in$ VAR or $Term \in$ VAR then $errormessage :=$ *instantiation error*
If $N \notin$ INTEGER then $errormessage :=$ *type error*
If $Term \in$ CONSTANT then $errormessage :=$ *type error*
If $N \in$ INTEGER $-$(NATURAL $- \{0\}$) or $N > arity(Term)$ then
$errormessage :=$ *range error*

Note that this rule formalizes the behaviour of many of the running systems which we tested, namely that in case of an error message the main system reacts by failure. If one wants to take another decision then the error handling part of this rule has to be adapted. Also in our formulation the different error cases are not mutually exclusive.

The **Term = ..List - rule**, following DFPS (1989), succeeds in case $Term$ is a structure or constant by unifying $List$ with the list of components of $Term$ and in case $Term$ is a variable by unifying this variable with the term constituted by the immediate components of $List$. Otherwise it fails. The following are considered as error messages: instantiation error when both $Term$ and either $List$ or the head of $List$ are variables, type error when $List$ is neither a variable nor a list or when the head of $List$ is a compound term or when the head of $List$ is a number or string and the tail of $List$ is not the empty list, implementation error if the tail of $List$ has a length greater than the constant max_arity. This explains the following

$Term = ..List$-rule:

If $error = 0 \& stop = 0 \& curractivator = Term = ..List$ then
    If $Term \in$ STRUCTURE $\& unify(List, componentlist(Term)) \in$ SUBST
        then *succeed with substitution update by* $unify(List, componentlist(Term))$
    If $Term \in$ CONST $\& unify(List, [Term]) \in$ SUBST
        then *succeed with substitution update by* $unify(List, [Term])$
    If $Term \in$ VAR $\& unify(Term, inverse(componentlist)(List)) \in$ SUBST
        then *succeed with substitution update by*
            $unify(Term, inverse(componentlist)(List))$
    else *backtrack*
        If $Term \in$ VAR $\&(List \in$ Var or $Listhead(List) \in$ Var or $ispartial(List))$
            then $errormessage :=$ *instantiation error*)
        If $List \notin$ VAR $\cup$ LIST or $Listhead(List) \in$ STRUCTURE
            or ($Listhead(List) \in$ NUMBER $\cup$ STRING $\& Listtail(List) \neq [\,]$) or *improper List*
            then $errormessage :=$ *type error*
        If $length(Listtail(List)) > max\_arity$
            then $errormessage :=$ *implementation error*

Note that here again we have formalized failure of the main system in case of an error message.

Remark added in proof. (March 1991): In the meantime the ISO Prolog standardization group has discussed to change the third case of the preceding rule what we can express easily by replacing it as follows:

If $Term \in$ VAR $\& Head(List) \in$ CONSTANT
    $\& Tail(List) = empty \& unify(Term, Head(List)) \in$ SUBST
    then *succeed with substitution update by ...*

If $Term \in$ VAR $\& Head(List) \notin$ CONSTANT
    $\& Unify(Term, inverse(componentlist)(List)) \in$ SUBST
    then *succeed with substitution update by ...*

For the **name(Atom, List)-rule** we introduce two more universes into our algebras:

SYMBOL - the alphabet of the basic symbols
ASCIILIST - the set of lists of ASCII-numbers

ASCIILIST may be considered to be a subset of LIST. It comes with an injective function

$$symbollist : \text{ATOM} \cup \text{NUMBER} \rightarrow \text{ASCIILIST}$$

which associates to each atom or number the list of the ASCII-numbers of its symbols.

The built-in predicate $name(Atom, List)$ succeeds in two not necessarily disjoint cases: a) if $Atom$ is an atom or a number it succeeds by unifying $List$ with the list of ASCII-numbers of the symbols of $Atom$, b) if $List$ is a list of ASCII-numbers it succeeds by unifying $Atom$ with the atom which has $List$ as symbollist. Otherwise $name(Atom, List)$ fails with an error message in the following cases: both $Atom$ and $List$ are variables (instantiation error), $List$ is neither a variable nor a list or $Atom$ is neither a variable nor an atom (type error) or $List$ contains the ASCII-number of a control symbol (i.e. of an element of a special subset of SYMBOL) (range error) or $List$ contains the ASCII-number of a symbol which is not allowed in atoms (implementation error). This is formalized in the following $name(Atom, List)$-rule:

If $error = 0 \& stop = 0 \& curractivator = name(Atom, List)$ then
  If $Atom \in \text{ATOM} \cup \text{NUMBER} \& unify(List, symbollist(Atom)) \in \text{SUBST}$
    then *succeed with substitution update by* $unify(List, symbollist(Atom))$
  If $List \in \text{ASCIILIST} \& unify(Atom, inverse(symbollist)(List)) \in \text{SUBST}$
    then *succeed with substitution update by* $unify(Atom, inverse(symbollist)(List))$
  else *backtrack*
    If $Atom, List \in \text{VAR}$
      then $errormessage := instantiation\ error$
    If $List \notin \text{VAR} \cup \text{LIST}$ or $Atom \notin \text{VAR} \cup \text{ATOM} \cup \text{NUMBER}$
      then $errormessage := type\ error$
    If *List contains the* ASCII- *number of a control symbol*
      then $errormessage := range\ error$
    If *List contains the* ASCII- *number of a symbol which is not allowed in an atom*
      then $errormessage := implementation\ error$

Note that here again we have formalized failure of the main system in case of an error message.

c) For the **input built-in predicates** in Prolog there is nothing particular for *logic* programming except the fact that the main data structure *term* is a structure which in classical logic is the main structure for representation of objects. The question arises whether for a safe understanding of purely procedural input and output predicates one really needs a *formal* semantic description. Whatever the answer to this question may be it should nevertheless be clear from the procedural character of the dynamic algebra rules that the input/output procedures defined by the built-in predicates in Prolog can be described in a natural way by transitions in Prolog algebras. Also at the time of writing of this paper the ISO Prolog standardization committee seems not yet to have decided a standard for the input/output Prolog built-in predicates. For the sake of illustration we therefore formalize in Prolog algebras some characteristic examples of such predicates as they appear in running Prolog systems. From these examples the principles should become clear and in particular it should be easy to rewrite this section for the forthcoming Prolog ISO standard.

For the built-in predicates which manage the **character input** we introduce the new universe

$$\text{ASCIINUMBER(read: the set of numbers } 1, 2, \ldots, 126)$$

of ASCII-code numbers of the elements of SYMBOL which comes with an injective function

$$ascii : \text{SYMBOL} \rightarrow \text{ASCIINUMBER}$$

which associates to each symbol its ASCII-code number. Furthermore we introduce a function

$$filesymbcont : \text{FILENAME} \times \text{POSITION} \rightarrow \text{SYMBOL}$$

which yields the symbol which is contained in the given position of the given file.

This extension of Prolog algebras is sufficient to describe the usual character input built-in predicates. The unary **getchar(Char)-rule** advances the current position of the current input file (side effect) and succeeds by unifying *Char* with the ASCII-code number of the next character from the current input file; it gives an error message when *Char* is neither a variable nor the ASCII-number of a symbol (type error) and when the end of the file is reached (range error). This is formalized by the following $getchar(Char)$-rule, where we abbreviate $currpos(currinfile)$ by $currentpos$:

If $error = 0 \& stop = 0 \& curractivator = getchar(Char)$ then
  $currentpos := nextpos(currentpos)$
  If $unify(ascii(filesymbcont(currinfile, currentpos)), Char) \in$ SUBST
    then $succeed\ with\ substitution\ update\ by\ unify(\dots)$
  else $backtrack$ (or $succeed$ or $call\ error\ handler$)
    If $Char \notin$ VAR $\cup ascii$(SYMBOL)
      then $errormessage := type\ error$
    If $currentpos = lastpos(currinfile)$
      then $errormessage := range\ error$

The binary **getchar(Char, File)-rule** is literally the same as the unary $get(Char)$-rule with $currinfile$ replaced by $File$, with the additional premise (success test condition)

$$activefile(File) = read$$

(to assure that $File$ is an existing input file which is open for reading) and with the following additional error handling lines:

If $File \in$ VAR
  then $errormessage := instantiation\ error$
If $File \notin$ VAR $\cup$ ATOM
  then $errormessage := type\ error$
If NOT $activefile(File) = read$
  then $errormessage :=$ IO $-control\ error$

The $get(Char)$-rule which is usually found in current Prolog systems is the same as the unary $getchar(Char)$-rule described above with the additional premise (success test condition)

$$ascii(filesymbcont(currinfile, currentpos)) > 32$$

which assures that only printing symbols—those with ASCII-code number greater than 32—are considered for reading.

The $skip(Nat)$-**rule** has the effect of continuous reading until an input symbol with ASCII-code number Nat is reached in which case it succeeds by reading that symbol. This is formalized by:

If $error = 0 \& stop = 0 \& curractivator = skip(Nat) \& Nat \in \text{NATURAL}$
$\quad \& currentpos \neq lastpos(currinfile)$ then
$\quad currentpos := nextpos(currentpos)$
$\quad$ If $ascii(filesymbcont(currinfile, currentpos)) = Nat$ then $succeed$

If $error = 0 \& stop = 0 \& curractivator = skip(Nat) \& (Nat \notin \text{NATURAL}$
$\quad$ or $currentpos = lastpos(currinfile))$ then
$\quad$ *call error handler*
$\quad errormessage :=$ *Nat is not a natural number or range error*

The rules for the **term input built-in predicates** are similar in structure to the rules for the character input. The function *filesymbcont* is replaced by the corresponding function at the term level:

$$filetermcont : \text{FILENAME} \times \text{POSITION} \rightarrow \text{TERM} \cup \text{SYNTAX\_ERROR\_MESSAGE}$$

which yields as value the first term which appears in the given file starting from the given position.

Instead of the successor function *nextpos* on POSITION at the symbol level a successor function at the term level is used:

$$nexttermpos : \text{FILENAME} \rightarrow \text{POSITION}$$

yielding the first position after the end of the value of *filetermcont* for the given file at its current position. The *read*(*Term*)-**rule** is now easily derived from the *get*(*Char*)-rule. For a more complete description one has also to specify the role of the last position of a file with respect to the syntx error message *end of file is reached* when an attempt is made to read a term starting from *lastpos*(*File*); for the standard input file one has to specify some convention which is used to indicate the end of a term input (usually the dot followed by a non printing symbol). It is a matter of routine to write down this refined version of the *read*(*Term*)-rule with corresponding error handling.

From these examples it should be clear how to write the rules for other term input built-in predicates like *readf* (a ternary built-in predicate for reading a term from a file according to a given format specification), *sreadf* (a built-in predicate with four arguments for reading the next term from a string according to a given format specification which keeps track explicitely of the rest of the string which has not yet been read) and also for operator defining or searching built-in predicates (like $op, currentop$) etc.

(d) For the **output built-in predicates** the same introductory remark applies that was made for the input built-in predicates. To illustrate the principles of a dynamic algebra formalization—if such a formalization should be felt to be needed or useful—we give the rule for the built-in predicates $put(Char)$ and $put(Char, File)$. These predicates have the effect of creating a new position after the last position in the current output file or in $File$ respectively and to assign the given character $Char$ as the symbol content of this new file position. For this purpose we introduce a new function

$$appendsymb : \text{FILE} \times \text{SYMBOL} \to \text{FILE}$$

with the obvious intended meaning. The $put(Char)$-**rule** is then formalized by:

If $error = 0 \& stop = 0 \& curractivator = put(Char)$ then
 If $Char \in$ ASCIINUMBER
  then $succeed$
   TAKE $temp = New$(POSITION) IN
    $put\ temp\ at\ bottom\ of\ curroutfile\ positions$
    $filesymbcont(curroutfile, temp) := inverse(ascii)(Char)$
    $denfile(curroutfile) := appendsymb(denfile(curroutfile), inverse(ascii)(Char))$
   ENDTAKE
 If $Char \notin$ ASCIINUMBER
  then If $Char \in$ VAR then $errormessage := instantiation\ error$
   If $Char \notin$ VAR then $errormessage := type\ error$

The abbreviation $put\ temp\ at\ bottom\ of\ curroutfile\ positions$ is used for the following updates:

$$lastpos(curroutfile) := temp$$
$$nextpos(lastpos(curroutfile)) := temp$$

If in connection with the $end\ of\ file$-condition $lastpos(File)$ is used as end of $File$ (which does not carry any more any file content information), then it suffices to replace in the above $filesymbcont$-update $temp$ by $lastpos(curroutfile)$; intuitively that means that the old last file position will contain the new appended symbol and that the new position will serve as new last position.

The binary $put(Char, File)$-rule is similar with $curroutfile$ replaced by $File$, with the additional premise $activefile(File) := write$ to assure that $File$ is an open output file and with the following additional error handling cases:

If $File \in$ VAR then $errormessage := instantiation\ error$
If $File \notin$ VAR $\cup$ ATOM then $errormessage := type\ error$
If NOT$(activefile(File) = write)$ then $errormessage :=$ IO $-error$

It is a routine matter to adapt the put-rules to a formalization of built-in predicates like $tab(N)$ (to output N blanks), nl (to proceed to the next line on the current output file) and to output predicates at the term level like $write(Term)$ and its variants with format specifications or operator declarations (like $writef$, $writeeq$, $swritef$, $display$, $currentop$ etc.).

6. Some New Problems for Logic

The dynamic algebra approach to semantics of real programming languages suggests to develop a model theory of dynamic algebras which in return would be useful for applications of these algebras to computer science pehomena. This model theory should take into account in an explicit way the finiteness and the dynamic character of dynamic algebras. In particular we see the following questions of interest:

(a) Can something general be said about the relations between (types of) programming languages and (types of) classes of dynamic algebras? The experience made with the dynamic algebra description of Modula-2 (Gurevich & Morris 1988), Occam (Gurevich & Moss 1990), Smalltalk (Blakley 1991) and Prolog indicates that there are intimate relations between the basic object structures and the basic operations of a programming language on one side and on the other side the signature and the form of the rules of corresponding dynamic algebras. For the Modula-2 description in Gurevich & Morris (1989) it was useful for the parameter handling of procedures to allow free variables in the rules which represent some form of universal quantification in the metalanguage (here: in the place where is defined what it means to apply a rule). For the description of selection core and control of Prolog in Part I, Börger (1990) (except for $notX$ and $call(X)$) we could do well with rules of form $If\ b\ then\ F$ where b and F are variable free expressions of the given signature whereas for the description of dynamic code in Part II, Börger (1990) and of the built-in predicates considered in this paper it turned out as a considerable advantage to allow rule schemes instead of rules; the description with closed rules is possible but more complicated and less natural. For Occam (see Gurevich & Moss (1990)) it was natural to introduce quantification over small domains into the rules in relation with constructs like PAR.

(b) What are the relations between program building methods and model construction techniques? Here again experience shows that program constructors like PAR, SEQ etc. in Occam have some effect on the way the describing dynamic algebras are built; we realized a similar phenomenon when trying to extend the basic Prolog algebras for selection core and control to algebras which can handle dynamic Prolog code. Another important question is: what does modularity for programs mean in model theoretic terms and which model theoretic techniques can support modularity for construction and proof of properties and transformations of programs?

(c) What can be said about the relation between properties of programs of given program classes and model theoretic closure or non closure properties of classes of dynamic algebras? For ex. what does it mean in model theoretic terms that a process is perpetual or non deterministic? What are natural uniformity conditions for dynamic algebras which formalize the intuitive understanding of resource-boundedness in real computing systems?

(d) Which relations can be established between program transformations and transformations of algebras? For ex. what is the model theoretic meaning of optimization techniques? Is there any? Or more specifically: is there a model theoretic expression of stratification of Prolog programs?

There are more questions of this kind which we can expect from more experience with use of dynamic algebras for mathematical phenomena concerning real programming languages and running systems. The main and I think not at all trivial problem is to find methods (concepts and theorems) in the model theory of dynamic algebras which are useful for the solution of real problems in computer science.

Added in proof. (March 1991): Yuri Gurevich recently has discovered that "In a sense, the EA (read: dynamic algebra, called there evolving algebra) approach provides a foundation for the use of dynamic logic. Models of first-order dynamic logic can be viewed naturally as evolving algebras of a kind." (Gurevich 1991)

7. References to Related Work

In Part I (Börger 1990) we have given a detailed comparative discussion of other operational or denotational approaches to the semantics for the full language of Prolog. None of the papers mentioned there treat also the built-in predicates described in this paper. The easiness and naturalness with which it was possible to extend in this paper the formalization given in Part I and Part II to a formalization of basically purely procedural predicates (though embedded into the logical context characterized by unification) shows the strength of the dynamic algebra approach to a rigorous but natural semantics for the whole body of real programming languages.

Added in proof. (March 1991): Since the time of writing this paper the author together with Dean Rosenzweig has developed a complete formal description of the Warren Abstract Machine (WAM) by refining the Prolog algebras from Part II (Börger 1990) into WAM algebras (together with a correctness proof for the latter wrt the former). In doing this also the term representation had to be described abstractly whereby many of the abstract universes and functions used in this paper get a "concrete" meaning (model). See Börger & Rosenzweig 1991, Part I and II.

References

Arbab, B. & Berry, D. M. 1987, *Operational and denotational semantics of Prolog,* J. Logic Programming **4**, 309–329..

Blakley, R. 1991, *Ph.D. thesis,* University of Michigan (in preparation).

Börger, E. 1990, *A logical operational semantics of full Prolog. Part I. Selection core and control,* CSL '89. 3rd Workshop on Computer Science Logic, (Eds. E. Börger, H. Kleine Büning, M. Richter), Springer LNCS 440, pp. 36–64.

Börger, E. 1990, *A logical operational semantics of full Prolog. Part II. Built-in predicates for database manipulations,* MFCS '90. Proc. of the International Symposium on Mathematical Foundations of Computer Science, (Ed. B. Rovan), Springer LNCS 452, pp. 1–14.

Börger, E. & Rosenzweig, D. 1991, *From prolog algebras towards WAM—A mathematical study of implementation,* CSL '90. 4th Workshop on Computer Science Logic, (Eds. E. Börger, H. Kleine Büning, M. Richter, W. Schönfeld), Springer LNCS (to appear).

Börger, E. & Rosenzweig, D. 1991, *WAM algebras—A mathematical study of implementation, Part II,* (submitted).

Deransart, P. & Ferrand, G. 1987, *An operational formal definition of Prolog,* INRIA RR763 (revised version of the abstract in Proc. 4th Symposium on Logic Programming, San Francisco (1987),, 162–172).

Debray, S. K. & Mishra, P. 1988, *Denotational and operational semantics for Prolog,* J. Logic Programming **5**, 61–91.

DFPS 1989: Deransart, P., Folkjaer, P., Pique, J.-F., Scowen, R. S., *Prolog. Draft for Working Draft 2.0,* ISO/IEC YTC1 SC22 WG17 No. 40, VI + 96.

DFPS 1990: Deransart, P., Folkjaer, P., Pique, J.-F., Scowen, R. S., *Prolog. Draft for Working Draft 3.0,* ISO/IEC YTC1 SC22 WG17 No. 53, IV + 77.

Gurevich, Y. 1988, *Logic and the challenge of computer science,* Trends in Theoretical Computer Science, (E. Börger, ed.), Computer Science Press, 1988, pp. 1–57.

Gurevich, Y. 1988, *Algorithms in the world of bounded resources*, The Universal Turing Machine—a Half-Century Story, (Ed. R. Herken), Oxford University Press, pp. 407–416.

Gurevich, Y. 1991, *Evolving algebras. A tutorial introduction*, EATCS Bulletin **43** (February 1991).

Gurevich, Y. & Morris, J. M. 1988, *Algebraic operational semantics and Modula-2*, CSL '87. 1st Workshop on Computer Science Logic, (Eds. E. Börger, H. Kleine Büning, M. Richter), Springer LNCS 329, pp. 81–101.

Gurevich, Y. & Moss, L. S. 1990, *Algebraic operational semantics and Occam*, (Eds. E. Börger, H. Kleine Büning, M. Richter), Springer LNCS.

Jones, N. D. & Mycroft, A. 1984, *Stepwise development of operational and denotational semantics for Prolog*, Proc. Int. Symp. on Logic Programming 2/84, Atlantic City, IEEE, pp. 289–298.

Kappel, A. 1990, *Implementation of dynamic algebras with an application to Prolog*, Diploma Thesis, University of Dortmund, Fed. Rep. of Germany.

Lloyd, J.1989, *Current theoretical issues in logic programming*, Abstract. EATCS-Bulletin **39**, 211.

LPA 1988: Johns, N. & Spenser, C., *LPA Mac Prolog*^{TM} *2.5, Reference Manual*, Logic Programming Associates, London.

North, N. 1988, *A denotational definition of Prolog*, NPL Report DITC 106/88, National Physical Lab, Teddington, Middlesex.

Quintus 1987, *Quintus Prolog Reference Manual*, Version 10, February, 1987, Quintus Computer Systems, Mt. View, CA.

VM 1985, *VM/Programming in Logic*, Program Description and Operations Manual (July, 1985), IBM, 1st ed..

This work was written when the author was visiting IBM Germany, Heidelberg Scientific Center, Institute for Knowledge Based Systems, Tiergartenstr. 15, P.O. Box 10 30 68, D-6900 Heidelberg 1, Federal Republic of Germany, on sabbatical from: Dipartimento di Informatica, Università di Pisa, Cso Italia 40, I-56000 PISA, ITALIA

Research at MSRI supported in part by NSF Grant DMS-8505550.

COMPUTABILITY AND COMPLEXITY OF HIGHER TYPE FUNCTIONS

STEPHEN A. COOK

ABSTRACT. Several classical approaches to higher type computability are described and compared. Their suitability for providing a basis for higher type complexity theory is discussed. A class of polynomial time functionals is described and characterized. A new result is proved in Section 8, showing that an intuitively polynomial time functional is in fact not in the class described.

1. INTRODUCTION

Higher type functions occur very naturally in computer science. Early programming languages, such as FORTRAN, already allowed functions to be passed as input parameters to subroutines. Complex types are fundamental to more modern languages, such as ML.

We are interested in studying the computational complexity of such higher type functions. In order to develop a crisp mathematical theory, we restrict attention to domains built from the natural numbers. The functions of type (level) one take tuples of natural numbers to natural numbers, and their complexity has been well studied. Type two functionals take numbers and type one functions as input, and their complexity has been studied using the model: Turing machine with oracle. For type three and above, there has been little if anything in the complexity literature, until very recently.

There is, however, a substantial literature on the computability theory for higher type functions. Unlike the situation for type one functions, where Church's thesis supposes that all reasonable approaches to defining *computable function* agree, there exist several mutually inconsistent approaches to defining the computable higher type numerical functions. These approaches differ both in choosing different function domains to serve as inputs, and in the way in which a function is presented as an input to an algorithm. Before attempting to develop a complexity theory, it is worth summarizing some of these approaches. The reader is referred to the excellent paper by Gandy and Hyland [11] for more details, as well as to the other references cited here.

We start by defining the set of types we will consider. Not all treatments refer to use exactly this set, but the differences are not important.

Definition.

- 0 is a type
- $(\sigma \to \tau)$ is a type if σ and τ are types.

Each type σ has associated with it a set (domain) A_σ of objects of type σ. The exact definition of A_σ depends on which approach we are following, but in general $A_0 = \mathbb{N}$, and $A_{(\sigma\to\tau)}$ is a set of (possibly partial) functions from A_σ to A_τ.

Note that by repeatedly decomposing the right side of $\to$, every type τ can be written uniquely in the normal form

$$\tau = (\tau_1 \to (\tau_2 \to \cdots \to (\tau_k \to 0)\cdots)),$$

which we write as

$$\tau = \tau_1 \to \tau_2 \to \cdots \to \tau_k \to 0,$$

with the convention of association to the right when parentheses are omitted. Thus it is natural to think of a function F of type τ above as a functional taking arguments $X_1, \dots, X_k$, with X_i of type τ_i, and returning a natural number value. Accordingly, we write $F(X_1, \dots, X_k) = y$, with $y \in \mathbb{N}$. We use the word "functional" when referring to F in this form.

The *level* of a type is defined inductively by

- $level(0) = 0$
- $level(\tau_1 \to \tau_2 \to \cdots \to \tau_k \to 0) = 1 + \max_{1\le i\le k} level(\tau_i)$.

2. Kleene's Partial Recursive Functionals

Kleene [17], [19] introduced the "partial recursive functionals" in 1959. In this approach, the domain A_σ is taken to be HT_σ, the set of "hereditarily total" functionals of type σ. Thus $HT_0 = \mathbb{N}$, and in general $HT_{(\sigma\to\tau)}$ consists of all total functions from HT_σ to HT_τ. We set $HT = \bigcup_\sigma HT_\sigma$. (Actually Kleene considered only certain "pure" types τ, but this is not important. Gandy and Hyland [11] present Kleene's approach using our type structure.) Kleene gave an inductive definition of *partial recursive functional* on these domains, using nine schemes $S1, \dots, S9$. He defined a *general recursive functional* to be a total partial recursive functional. The class of partial recursive functionals has been well studied both by Kleene and others (for example [10], [11], [15]). For type level one it yields the ordinary partial recursive functions, and for type level two it yields the functionals computable by Turing machines with input functions presented by oracles, in the usual way.

When the schemas $S1$–$S9$ are used to compute a functional $G(X_1, \ldots, X_k)$, the way in which the arguments are presented is by oracles. That is, the X_i's are queried at certain computed functionals. When G has type three or more, then the X_i's may have type two or more. The schema used to access X_i's is $S8$, which states that if F is a partial recursive functional of a suitable type, so is G, where

2.1. $$G(X_1, \ldots, X_k) = X_1(\lambda Y.F(Y, X_1, \ldots, X_k)),$$

and it is understood that G is defined only for those values of $X_1, \ldots, X_k$ such that $F(Y, X_1, \ldots, X_k)$ is total in Y. This last condition must be put in, because the domain of X_1 contains only hereditarily total functions, but it leads to the strange property that the Kleene partial recursive functionals are not closed under substitution of Kleene recursive functionals, even when the latter are total ([15], [17]).

This suggests that Kleene's development would be more natural if the domains HT_σ were extended to include partial functions, and indeed Platek [25] (see Moldestad [23]) developed such a theory.

Nevertheless, the class of Kleene general recursive functionals is robust (it is closed under substitution), and has many characterizations, including machine characterizations (Kleene [18], Gandy [10]).

Furthermore, it may serve as a natural starting point for higher type complexity theory. This is because in general, type one complexity classes contain only total functions, so it seems plausible that their generalizations to higher types need contain only total functionals defined on total objects, and indeed these functionals should be Kleene recursive. For example, Kleene defines the primitive recursive functionals to be those generated by his schemas $S1$–$S8$ (omitting $S9$, the general recursion schema), and this class of hereditarily total functionals seems to be a perfectly natural generalization of the type one primitive recursive functions.

3. The Recursively Countable Functionals

In our second approach, we take the domain A_σ of objects of type σ to be C_σ, the so-called countable functionals of type σ ([16], [21]; see [11] and [24]). A countable functional in $C_{(\sigma \to \tau)}$ is total (defined on all objects of C_σ), but in general not all total functionals $F : C_\sigma \to C_\tau$ are countable, but only those which are continuous in a certain sense. More precisely, $C_0 = \mathbb{N}$, and C_τ is all total numerical functions of type τ when $level(\tau) = 1$. When $\tau = (0 \to 0) \to 0$, then C_τ consists of those total functions $F : C_{(0 \to 0)} \to 0$ such that for each $f : \mathbb{N} \to \mathbb{N}$ there is a finite restriction α of f such that for all total extensions g of α, $F(f) = F(g)$. In other words, $F(f)$ is determined by a finite amount of information about f. This last property holds for countable functionals of all types. Thus the definition is extended to other types τ in such a way that each functional F in C_τ is determined by a type one function α^F, called an "associate" of F. In fact, $F(f) + 1 = \alpha^F(x)$, when x codes a sufficiently long initial segment of some associate α^f of f. It follows that for every nonzero type τ, C_τ has the cardinality of the continuum. This is in contrast to the domains HT_σ of the previous section, where the cardinality of HT_σ increases with each increase in $level(\sigma)$.

We set $C = \bigcup_\sigma C_\sigma$.

A countable functional F is said to be *recursively countable* iff F has a recursive (type one) associate. It is interesting to compare this notion with that of Kleene's general recursive functional discussed in the previous section. In fact, if $level(\tau) \leq 2$, then $C_\tau \subseteq HT_\tau$, and a functional of type τ is recursively countable iff it is general recursive.

If $level(\tau) = 3$, then strictly speaking C_τ and HT_τ are disjoint, because a countable functional is defined only on countable inputs, but it is reasonable to ask whether the recursively countable type three functionals coincide with the restrictions to countable inputs of the Kleene general recursive functionals. The answer (discussed below) is that the latter form a proper subset of the former.

Another way to compare the two forms of computability is to interpret Kleene's schemas $S1$–$S9$ so that they define functionals on the domain C instead of HT. Let us call these functionals *Kleene computable functionals.*

All total Kleene computable functionals are countable. In fact, because of $S8$, it would not make sense to apply $S1$–$S9$ to the domain C if this were not so.

Proposition 3.1. *A countable functional of type level three or less is Kleene computable iff it is the restriction to countable arguments of a Kleene general recursive functional (defined on the domain HT).*

A proof of the "if" direction (in a more general context) appears in [24], page 24. The converse can be proved by similar methods; a crucial observation is that any application of schema $S8$ (see 2.1) in which G is type 3 must have Y of type 0.

It can be shown that every countable Kleene computable functional is recursively countable, but not conversely. In particular, the so-called type three "fan functional" is recursively countable but not Kleene computable. Here we define a simpler type three example:

Definition 3.2. The functional Φ of type $((0 \to 0) \to 0) \to 0$ is given by

$$\Phi(X) = \begin{cases} 1 \ \textit{if} \ \exists f \leq 1 \ \textit{so} \ X(f) = 0 \\ 0 \ \textit{otherwise} \end{cases}$$

where $f \leq 1$ means range $(f) \subseteq \{0, 1\}$.

Proposition 3.3. *Φ is recursively countable but not Kleene computable.*

That Φ is recursively countable can be seen intuitively because $\Phi(X)$ can be effectively determined from a finite amount of information about X, provided that X is countable. Since X is countable, each $f : \mathbb{N} \to \mathbb{N}$ has a finite restriction α such that $X(f)$ is determined by $X(\alpha)$ (assuming X is extended in the natural way to sufficiently large finite functions). The total 0–1 valued functions $f : \mathbb{N} \to \mathbb{N}$ form the branches of an infinite binary tree. Since every branch f can be pruned at a point sufficient to determine $X(f)$, it follows from König's Lemma that values attached to the leaves of a finite part of the tree are sufficient to determine X on all 0–1 valued f, and hence this finite amount of information determines $\Phi(X)$.

That Φ is not Kleene computable is plausible. If the input X to Φ can be accessed only by oracle calls $X(f)$ for f total (as is the case for schemas $S1$–$S9$) then there seems to be no way of determining $\Phi(X)$. A formal proof that Φ is not Kleene computable is easily adapted from page 418 of [11].

4. Partial Functionals as Inputs

The partial functionals discussed so far allow only total objects as arguments (although Kleene's partial recursive functionals need not be defined on all total objects of a given type). (An exception is Platek's hereditarily consistent functionals [25], [23], briefly mentioned earlier.) All general formalisms for computability define partial objects as well as total objects, so if we want the computable objects to be closed under substitution, we must allow partial objects as arguments.

Here we describe the domain $\dot{C}$ pf partial continuous functionals ([28], [8]; see [11] and [26]). For each type τ, $\dot{C}_\tau$ consists of the partial continuous functionals of type τ. These functionals are, roughly speaking, the continuous extensions of the countable functionals to partial objects. They are partially order by $\sqsubseteq$, where intuitively $\phi \sqsubseteq \psi$ iff ψ extends ϕ. The system $(\dot{C}_\tau, \sqsubseteq)$ is a *complete* partial order (cpo), in the sense that every increasing chain $\phi_1 \sqsubseteq \phi_2 \sqsubseteq \cdots \sqsubseteq \phi_n \sqsubseteq \cdots$ of functions in $\dot{C}_\tau$ has a least upper bound ϕ in $\dot{C}_\tau$. We denote ϕ by $\bigsqcup_i \phi_i$. Further, each $\dot{C}_\tau$ has a least element $\perp_\tau$, corresponding to the nowhere defined function. We denote $\perp_0$ by $\perp$, the "undefined number".

We can now define the domains $\dot{C}_\tau$ inductively as follows: $\dot{C}_0 = \mathbb{N} \cup \{\perp\}$, where distinct elements of $\mathbb{N}$ are incomparable with respect to $\sqsubseteq$, and $\perp \sqsubseteq n$ for all $n \in \mathbb{N}$. In general, $\dot{C}_{(\sigma \to \tau)}$ consists of all (total) continuous functions from $\dot{C}_\sigma$ to $\dot{C}_\tau$, where ϕ is *continuous* iff ϕ is monotone with respect to $\sqsubseteq$, and $\phi(\bigsqcup_i \psi_i) = \bigsqcup_i \phi(\psi_i)$, for all increasing chains $\psi_1 \sqsubseteq \psi_2 \sqsubseteq \cdots$ in $\dot{C}_\sigma$. If ϕ and ϕ' are in $\dot{C}_{(\sigma \to \tau)}$, then we write $\phi \sqsubseteq \phi'$ iff $\phi(\psi) \sqsubseteq \phi'(\psi)$ for all ψ in $\dot{C}_\sigma$. Under this definition, it is easy to check that $\dot{C}_{(\sigma \to \tau)}$ is a cpo.

We write $\dot{C}$ for $\bigcup_\sigma \dot{C}_\sigma$.

Notice that each ϕ in $\dot{C}_{(\sigma \to \tau)}$ is total on $\dot{C}_\sigma$, although we think of ϕ as partial, because it can take on the "undefined" value $\perp_\tau$. If ϕ is in $\dot{C}_\tau$, where $\tau = \tau_1 \to \tau_2 \to \cdots \to \tau_k \to 0$, then we think of ϕ as a "partial" functional from $\dot{C}_{\tau_1} \times \cdots \times \dot{C}_{\tau_k}$ to $\mathbb{N}$, by thinking of $\perp$ as "undefined". By the *domain* of ϕ we mean the set of all tuples $(\psi_1, \dots, \psi_k)$ such that $\phi(\psi_1, \dots, \psi_k) \neq \perp$.

The so-called *finite* elements of $\dot{C}$ are those which can be specified by a finite amount of information. All elements of $\dot{C}_0$ are finite. If the functional ϕ has type level one, then ϕ is finite iff its domain is finite. If ϕ has level two or more, then ϕ is finite iff ϕ can be specified by giving its (numerical) value $\phi(\alpha_1, \dots, \alpha_k)$ on finitely many tuples $(\alpha_1, \dots, \alpha_k)$ of finite elements α_i. In this case, the domain of ϕ is infinite in general, because $\phi(\alpha_1, \dots, \alpha_k) = \phi(\alpha'_1, \dots, \alpha'_k)$ whenever $\alpha_i \sqsubseteq \alpha'_i$, $1 \leq i \leq k$. However, the range of a finite functional ϕ is always a finite set of numbers.

If $\alpha \sqsubseteq \phi$ we say α *approximates* ϕ. It is not hard to show, by induction on the type level, that every element of $\dot{C}$ is the least upper bound of its finite approximations.

We say that a functional ϕ in $\dot{C}$ is *computable* iff its set of finite approximations is recursively enumerable. (The latter notion is well-defined, since finite elements can be assigned Gödel numbers in a natural way, using their finite descriptions.) Note that for type one functions, this definition of computable is consistent with the usual definition.

Plotkin [26] gave a simple functional programming language he calls $\mathcal{L}_{PA+\exists}$ and proved that its programs compute precisely the computable elements of $\dot{C}$. The language is based on LCF, Scott's logic of computable functions, extended by a "parallel" conditional, and a type two functional denoted by $\exists$ which approximates the existential quantifier. The latter has type $(0 \to 0) \to 0$ and can be defined by

4.1.
$$\exists(f) = \begin{cases} 0 \; if \; f(\bot) = 0 \\ 1 \; if \; f(n) = 1 \; for \; some \; n \in \mathbb{N} \\ \bot \; otherwise \end{cases}$$

The programs in the language are type 0 terms of the typed lambda calculus, built from variables, constants for simple numerical functions (including the parallel conditional), the constant $\exists$, and the constant Y_σ for each type σ. Here Y_σ has type $(\sigma \to \sigma) \to \sigma$, and takes an "operator" of type $(\sigma \to \sigma)$ to its least fixed point.

It is no surprise that the partial continuous functionals $\dot{C}$ and the countable functionals C are closely related. A crucial property in each case is that the value of a functional $\phi(\psi_1, \ldots, \psi_k)$ is determined by a finite amount of information about its arguments $\psi_1, \ldots, \psi_k$. In fact the countable functionals correspond to the "everywhere defined" partial continuous functionals; i.e. those which are defined on all "everywhere defined" inputs. To make the last notion precise, Ershov [9] defined for each $\tau \neq 0$ a map $t = t_\tau$ from $\dot{C}_\tau$ to C^*_τ, where C^*_τ contains not only the countable functionals in C_τ but also partial objects with the same domain. Here we follow the description of t given in [11] (except we generalize t from pure types to all types).

Let $C^*_0 = C_0 = \mathbb{N}$, and in general let $C^*_{(\sigma \to \tau)}$ be the set of partial functions from C_σ to C^*_τ. Thus $C_\tau \subseteq C^*_\tau$ for each τ. We define the map $t = t_\tau : \dot{C}_\tau \to C^*_\tau$ by induction on $level(\tau)$ $(\tau \neq 0)$. If τ has level one, and $\dot{\phi} \in \dot{C}_\tau$, then

4.2.
$$t(\dot{\phi})(m_1, \ldots, m_k) = \dot{\phi}(m_1, \ldots, m_k), \quad m_i \in \mathbb{N},$$

where it is understood that the left hand side is undefined if the right hand side is $\bot$. In general, if $\tau = \tau_1 \to \cdots \to \tau_k \to 0$ $(k \geq 1)$ and $\dot{\phi} \in \dot{C}_\tau$ then

4.3.
$$\begin{aligned} &t(\dot{\phi})(\psi_1, \ldots, \psi_k) \\ &\quad = \mu y[\dot{\phi}(\dot{\psi}_1, \ldots, \dot{\psi}_k) = y \; for \; all \; \dot{\psi}_i \in t^{-1}(\psi_i), 1 \leq i \leq k], \end{aligned}$$

$$for \; \psi_i \in C_{\tau_i}, 1 \leq i \leq k.$$

Then every countable functional is in the image of t. In fact if ϕ is countable then $t^{-1}(\phi)$ contains many elements, corresponding to the ways in which ϕ can be extended to partial objects (these correspond roughly to the different associates of ϕ). We have ([11], page 434)

Proposition 4.4. *If ϕ is a countable functional, then ϕ is recursively countable iff $\phi = t(\dot{\phi})$ for some computable $\dot{\phi} \in \dot{C}$.*

Definition 4.5. A partial continuous functional $\dot{\phi} \in \dot{C}$ is *everywhere defined* iff $t(\dot{\phi})$ is countable.

Since countable functionals are total, an everywhere defined functional $\dot{\phi}$ must be defined on all everywhere defined arguments (of appropriate type). Furthermore, by 4.3 its values must be consistent, in the sense that if $\dot{\psi}_1, \ldots, \dot{\psi}_k$ are everywhere defined, and $t(\dot{\psi}_i) = t(\dot{\psi}'_i)$, $1 \leq i \leq k$, then $\dot{\phi}(\dot{\psi}_1, \ldots, \dot{\psi}_k) = \dot{\phi}(\dot{\psi}'_1, \ldots, \dot{\psi}'_k)$.

5. The Hereditarily Recursive Operations

Rather than presenting input functions by oracles, the HRO approach assumes that the input ψ is computable and presented by a Gödel number describing an algorithm for computing ψ. This method of input provides more information than oracles, so that more functions are computable. On the other hand, the domains of these functions are smaller, since they are restricted to computable functions.

Buss [2] used this method of input to define a class of polynomial time functionals, for the purpose of realizing his intuitionistic theory IS_2^1. However, his development is not straightforward.

To illustrate the difficulties, consider the type two functional $\mathit{Apply}(f, x) = f(x)$. If we restrict the argument $f : \mathbb{N} \to \mathbb{N}$ to be polynomial time computable, and present f by a Gödel number of a polynomial time machine computing it, then it is natural to compute $\mathit{Apply}(f, x)$ by simulating that machine on input x. But then there will be no fixed polynomial bounding the run time of *Apply*.

In order to circumvent this problem, Buss introduced a more complicated type structure, in which types are recursively labelled with run time bounds. Since we prefer to remain with our simple type structure and still have the *Apply* function be polynomial time computable, we will not further consider the HRO approach.

6. Programming Languages and Complexity

In ordinary computability theory, *computable* means computable by an algorithm, and this should be the case for higher type computability theory too. If the algorithms in question are sequential, they should yield a notion of time complexity. For all the approaches we have described, the type two computable objects can be described by Turing machines (at least when the inputs are total). The inputs can be presented by oracles in all cases except for HRO, where they must be presented by Gödel number. For types greater than two, the situation is less clear. Both Kleene [18] and Gandy [10] presented machine characterizations of Kleene's partial recursive functionals, but Kleene's schema $S8$ forces the computations of these machines to be infinite objects, even when the function computed is defined.

It is perfectly possible to interpret Kleene's schemas $S1$–$S9$ on the domain $\dot{C}$ of partial continuous functionals, resulting in the class we shall call KPRCF (Kleene partial recursive continuous functionals). For this class, continuity allows the computation trees to be finite objects. Although it is not hard to see [1] that the total Kleene computable functionals (see section 3) all correspond (under the map t of section 4) to functionals in KPRCF, it is not clear that all everywhere defined functionals in KPRCF get mapped by t to Kleene computable functionals.

As explained in section 4, Plotkin's programming language $\mathcal{L}_{PA+\exists}$ provides a nice characterization of the computable elements of $\dot{C}$. Plotkin also defines a programming language $\mathcal{L}_{DA}$ [26] (D for "deterministic") which consists of $\mathcal{L}_{PA+\exists}$ without the parallel conditional and without the continuous existential operator $\exists$ (see 4.1). Bellantoni has shown [1] that the objects computable in this language are precisely those in KPRCF (see above). This gives a nice programming language characterization of Kleene's schemas $S1$–$S9$ with partial continuous semantics, but leaves open the following question: Are the Kleene computable functionals ($S1$–$S9$ interpreted over C) exactly the images under t of $\mathcal{L}_{DA}$ computable functionals? The answer is yes for type level at most two since (after applying the map t to the $\mathcal{L}_{DA}$ computable functionals) both computable classes can be characterized by Turing machines with oracles.

However, for type 3 we do not know the answer. In particular, if Φ is the type three functional given by 3.2, then Φ is not Kleene computable, but by 3.3, 4.4, and the completeness of $\mathcal{L}_{PA+\exists}$, we know that $\Phi = t(\dot{\Phi})$ for some $\dot{\Phi}$ computable by a $\mathcal{L}_{PA+\exists}$ program. Could it be that also some such $\dot{\Phi}$ is computable by $\mathcal{L}_{DA}$? (This would yield a negative answer to the question above.) Could it be that in fact *every* everywhere defined (see 4.5) functional computable by $\mathcal{L}_{PA+\exists}$ is already computable by $\mathcal{L}_{DA}$? Bellantoni [1] has recently shown that the answer to the last question is no. Specifically, he defined a $\mathcal{L}_{PA+\exists}$ computable functional $\dot{\Phi}$ in $t^{-1}(\Phi)$ such that $\exists$ (see 4.1) is $\mathcal{L}_{DA}$ computable from $\dot{\Phi}$. Since Plotkin showed $\exists$ is not $\mathcal{L}_{DA}$ computable, it follows that $\dot{\Phi}$ is not $\mathcal{L}_{DA}$ computable.

We note that although $\dot{\Phi}$ is everywhere defined, $\exists$ is not everywhere defined. In particular, referring to 4.1, if $f(n) = 0$ for $n \in \mathbb{N}$ and $f(\bot) = \bot$, then f is everywhere defined but $\exists(f) = \bot$.

From the point of view of complexity theory, functional programming languages such as $\mathcal{L}_{DA}$ and $\mathcal{L}_{PA+\exists}$ are not completely satisfactory, because they do not provide any intrinsic notion of time complexity. Plotkin does provide an "operational semantics" for his languages, by associating finite reduction sequences whenever a program returns a value for specific inputs. This notion of computation is nondeterministic, because the actual reduction sequence generated and its length depend heavily on the order in which the reduction rules are applied, even though the results are the same.

There may be no help for this difficulty for functionals which essentially require recursion, such as the type three iterator defined in 6.2 below. Nevertheless, time complexity is clearer for machines and languages with sequential facilities.

For example, Turing machines with oracles can be used for type two functionals. For measuring time, it is usual to charge just one step for an oracle call. However, we prefer to charge a number of steps equal to the length of the value returned by the oracle, on the grounds that this much work must be done by the tape head even if the oracle dictates the answer. For type three and above, Turing machines can still be used (for example [18]), but the definition becomes complicated and perhaps contrived.

A more natural way of defining time complexity for higher type functions may be to use a procedural language, in the style of Algol 60. This was done in [7], where "typed while-programs" were introduced. Their syntax is patterned after a very simple version of PASCAL, with the addition of higher type variables. Assignment statements can assign only to type 0 variables. Input variables of type two or more are accessed by giving them procedure names as arguments. Here is an outline of the syntax:

A *statement* is either an assignment statement or a while statement. An assignment statement has one of the four kinds

$$x \leftarrow 1$$

$$x \leftarrow y + z$$

$$x \leftarrow y \dot{-} z$$

6.1. $$x \leftarrow X(X_1, \dots, X_r).$$

Here lower case letters stand for type 0 variables, and upper case letters stand for variables of arbitrary (specified) type. Each variable on the right hand side of 6.1 is either an input variable, a subprogram name (see below), or a "parameter".

A while statement has the form

While $x \neq 0$ Do Begin I End

where I is a finite list of instructions.

In addition to statements, typed while-programs can have subprograms. (Typed while-programs themselves are a special case of subprograms.) Each subprogram consists of a declaration section and an instruction section. the declaration section consists of three declarations, which have the following forms:

- Input declaration

  `Input` $Y_1, \ldots, Y_k$

- Output declaration

  `Output` x

- Subprogram declarations

  $D_1, \ldots, D_\ell$

  where $\ell \geq 0$,

Each subprogram declaration D_i has the form

`Subprogram` $V : Q$

where Q is a program and V is a typed variable called the program name.

A subprogram name can appear as any of the variables on the right hand side of the assignment statement 6.1 of the main program. However recursion is disallowed: no subprogram can call itself; directly or indirectly.

Also "side effects" are disallowed, and in fact any variable assigned to must be local to that subprogram.

To give semantics for typed while-programs, one must first select a suitable collection of domains A_σ, as explained in section 1. The domains chosen in [7] were the hereditarily total functions HT (section 2), but it is perfectly possible to choose the countable functions C or the partial continuous functionals $\dot{C}$. When HT is chosen, the functionals computed by typed while-programs (i.e. *while-computable* functionals) can be compared to the Kleene partial recursive functionals. It turns out [14] that the former are the proper subset of the latter corresponding to Kleene's μ-recursive functionals ([17], sec 8). For types one and two, all partial recursive functionals are μ-recursive, but not so for type three. An example of a recursive functional not while-computable is the type three iterator IT, which takes an operator F of type $(0 \to 0) \to (0 \to))$ and number x and composes F with itself x times. That is

6.2. $$IT(F, x) = F^{(x)}(g_0)$$

where $g_0 = \lambda x.0$.

If we allowed programs to call themselves under suitable restrictions, it seems likely that while-computability would coincide with Kleene partial recursiveness. However, this would greatly complicate the definition of running time, which is a major reason for introducing our programming language.

We define the execution time of a program by charging $|x|$ steps for execution of each assignment statement with x on the left hand side (here $|x| = \lceil \log_2(x + 1) \rceil$ is the length of the binary notation for x). This is roughly consistent with the time taken by an oracle Turing machine to execute the algorithm, provided all variables are type level zero or one, and provided no subprogram name can appear as X in the assignment statement $x \leftarrow X(X_1, \ldots, X_k)$ (see 6.1). Allowing subprogram names to appear as X (in the role of an oracle call) without charging for their execution time is a bit of a cheat, but it provides a simple definition of time complexity. Also, for a robust complexity class such as polynomial time, this cheating does no harm, as we shall see in the next section.

7. Polynomial Time

For type two functionals, when the input functions are 0–1 valued (i.e. sets), a Turing machine with an oracle that costs one step per query provides a perfectly natural model. This notion was introduced in [6] in the form of polynomial time reducibility. This definition was generalized (in different ways) by Constable [4] and Mehlhorn [22] to capture the class of "polynomial operators", which allow more general functions $f : \mathbb{N} \to \mathbb{N}$ as arguments. More recently Buss [2] introduced a notion of polynomial time functional of all types using Turing machines that accept Gödel numbers as inputs, in the style of HRO (see section 5).

Our own interest in functionals originated in joint work with Urquhart [5], in which our aim was to find a simpler class of functionals realizing Buss's system IS_2^1, as well as to present a "feasible" version of Gödel's *Dialectica* interpretation [12] for IS_2^1. The resulting class of functionals was further studied in [7], where it was called "basic feasible functionals".

Before defining this class, we should point out that there is no single obvious definition for even the type two polynomial time functionals, when arbitrary type one functions are allowed as inputs. A Turing machine with oracle is a good model, and one should require the run time to be bounded by a polynomial, but the question is, a polynomial in what? A necessary condition is that the run time be bounded by a polynomial in the lengths of the input numbers and the lengths of the numbers returned by the oracle queries, but this is not sufficient (see section 8).

The approach taken in both [22] and [5] is motivated by Cobham's inductive characterization [3] of the type one polynomial time functions as the least class of functions containing certain simple initial functions, and closed under composition and limited recursion on notation. This is generalized to higher types in [5] by using terms in a formal system PV^{ω} to express functionals of all types. The terms are typed λ-expressions built from variables of all types and certain constants. The constants are numerals for numbers, function symbols for all type one polynomial time computable functions, and the single type two recursor $\mathcal{R}$. Here $\mathcal{R}$ represents limited recursion on binary notation. The arguments of $\mathcal{R}$ are as follows:

y of type 0 (the initial value),

g of type $0 \to 0 \to 0$ (the next step function),

h of type $0 \to 0$ (the bounding function)

x of type 0 (the recursion variable).

$$\mathcal{R}(y, g, h, x) = \begin{cases} \mathit{if}\ x = 0\ \mathit{then}\ y \\ \mathit{else\ if}\ |t| \leq |h(x)|\ \mathit{then}\ t \\ \mathit{else}\ h(x) \end{cases}$$

$$\text{where } t \equiv g(x, \mathcal{R}(y, g, h, \lfloor \tfrac{1}{2}x \rfloor))$$

(Recall the notation $|x|$ for the length of the binary notation for x.) Note that by Cobham's theorem, if g and h are fixed polynomial time computable functions, then $\lambda xy.\mathcal{R}(y, g, h, x)$ represents a polynomial time computable function. However, from our point of view the recursor $\mathcal{R}$ represents a type two polynomial time functional in (y, g, h, x).

These PV^{ω} terms have a natural denotational semantics, once the domains A_{σ} are chosen. In [7], the hereditarily total functionals HT_{τ} are chosen for the domains (section 2), but the domains C_{τ} or $\dot{C}_{\tau}$ (sections 3, 4) could have been chosen as well. In the following, we assume a semantics based on HT.

Definition 7.1. A functional of HT is *basic feasible* iff it is represented by a closed term of PV^{ω}.

As a simple example, we see that the type two functional *Apply* is basic feasible, since $\mathit{Apply} = \lambda fx.f(x)$.

It is easy to see that all basic feasible functionals are total and computable by typed while-programs (section 6) and hence are general recursive in the sense of Kleene. Obviously they are closed under substitution.

Now let $\phi(X_1, \ldots, X_k)$ be any type two basic feasible functional, where the arguments $X_1, \ldots, X_k$ have type one or zero. Then we can apply the PV^ω term representing ϕ to the variables $X_1, \ldots, X_k$ and place the result in β-normal form to obtain a type zero PV^ω term $t = \phi(X_1, \ldots, X_k)$. Thus the term t has no variables of type level one or higher except possibly the free variables $X_1, \ldots, X_k$. From this it is not hard to see that a multitape Turing machine M using oracles for the type one inputs X_i can compute ϕ in time polynomial in the lengths of the number inputs and the lengths of the oracle values returned. Therefore, if the input functions are 0–1 valued (i.e. sets) then M runs in time bounded by a polynomial in the lengths of its numerical inputs.

Conversely, one can show using the techniques of [7] that any polynomial time Turing machine with oracles for sets can be represented by a PV^ω term.

As an example, since the type two functional

7.2. $$Max(f, x) = \max_{y \leq x} f(y)$$

clearly cannot be computed by an oracle Turing machine in polynomial time, it is not basic feasible. On the other hand, the functional

7.3. $$Smax(f, x) = \max_{y \subseteq x} f(y)$$

is basic feasible, where $y \subseteq x$ means that y is an initial segment of x (referring to binary notation). To construct a PV^ω term representing $Smax$ requires a little thought, since there is no obvious way to bound $Smax$ when attempting to use the recursor $\mathcal{R}$ to compute it directly. The trick is to first compute

$$ArgSmax(f, y) = \mu y \subseteq x. f(y) = Smax(f, x)$$

using $\mathcal{R}$, and then observe that $Smax(f, x) = f(ArgSmax(f, x))$.

We note that PV^ω is a functional programming language rather like Plotkin's $\mathcal{L}_{DA}$ (section 6). One can give an operational semantics similar to Plotkin's by using the proof rules of PV^ω given in [5], section 6.

In order to address the question of whether PV^ω captures the intuitive polynomial time functionals, we should characterize the basic feasible functionals with some programming language which has a convincing notion of time complexity. This was carried out in [7] using typed while-programs. At the end of the previous section, the execution time of such a program was given by adding up the execution times of all the assignment statements executed, where the latter time is the length of the (numerical) value assigned. In this way, the time complexity of a program which computes a functional ϕ is defined to be a functional T_ϕ with the same type as ϕ.

What we would like is a theorem which states that ϕ is basic feasible iff there is some program computing ϕ such that T_ϕ is "polynomially bounded". Unfortunately, to define the last notion we seem to need the basic feasible functionals again. Thus theorem 7.4 below is a robustness result, in the style of Ritchie [27] and Cobham [3].

Let us say that (typed while-) program $\mathcal{P}$ which computes a functional $\phi(X_1, \ldots, X_k)$ with run-time T_ϕ is *feasibly length-bounded* iff there is a basic feasible functional Θ such that for all $\psi_1, \ldots, \psi_k$,

$$T_\phi(\psi_1, \ldots, \psi_k) \leq |\Theta(\psi_1, \ldots, \psi_k)|.$$

We say inductively that $\mathcal{P}$ is *feasible* iff it is feasibly length-bounded and all its subprograms are feasible.

Theorem 7.4. *A functional ϕ is basic feasible iff ϕ is computed by some feasible typed while-program.*

The proof appears in [7]. A similar result, for type two functionals, was proved earlier in [22].

A second characterization of the basic feasible functionals, which is not incestuous like 7.4, is also proved in [7]. For this the notion of a *bounded typed loop program* (BTLP) is introduced. A BTLP is like a typed while-program, except we need two additional assignment statements

$$\begin{aligned} x &\leftarrow y\#z \\ x &\leftarrow Smax(Y, z) \end{aligned}$$

and we replace while-statements with the bounded loop statement

$$Loop\ x, y\ Begin\ I\ End$$

where I is a list of assignment and bounded loop statements.

Here $y\#z = 2^{|y|\cdot|z|}$, $Smax$ is defined in 7.3, and the bounded loop statement means that I is interated exactly $|x|$ times, with the constraint that no variable is assigned a value with length greater than $|y|$ in any assignment statement inside the loop. We then have

Theorem 7.5. *A functional ϕ is basic feasible iff it is computed by some BTLP.*

8. A Polynomial Time Counter-Example

Let us return to the question of whether the basic feasible functionals provide the "right" definition of the polynomial time functionals. One argument supporting this view, at least for type two functionals, is that Mehlhorn's polynomial operators [22] are essentially the same as the type two basic feasible functionals. This is so even though the motivation behind the two definitions is quite different (defining polynomial time reducibility among functions, as opposed to realizing feasible logical theories), and even though the definitions and results of [5] and [7] were developed without knowledge of [22].

Another argument in support of "rightness" is the characterizations 7.4 and 7.5.

Nevertheless, we have serious doubts that all intuitively polynomial time computable functionals are basic feasible. For one thing, theorem 7.4 falls short of what we are looking for, as explained before 7.4. Here we state some necessary conditions that a class P of functionals must satisfy in order to be the intuitively correct class of polynomial time functionals:

8.1. *P must include the basic feasible functionals.*

8.2. *P must be closed under λ-abstraction and application.*

8.3. *Every type two functional in P must be computable by an oracle Turing machine in time polynomial in the lengths of both the numerical inputs and the values returned by the oracle queries.*

Note that the basic feasible functionals do satisfy 8.1–8.3. However, we now give an example of a natural type two functional L which is not basic feasible, but L satisfies 8.3, and when L is added to the basic feasible functionals and the result is closed under 8.2, then all the resulting type one functions are still polynomial time computable.

The functional L arises from the fact that $\mathbb{N} \times \mathbb{N}$ is well ordered by the lexicographical ordering. In order to make L polynomial time, we must define the ordering on the lengths of the numbers. That is, we define the relation $\preceq$ on $\mathbb{N} \times \mathbb{N}$ by $(a, b) \preceq (a', b')$ iff $|a| < |a'|$ or ($|a| = |a'|$ and $|b| \leq |b'|$). Then $\preceq$ defines a well ordering on the length equivalence classes of $\mathbb{N} \times \mathbb{N}$. By using a standard pairing function, we can assume $\preceq$ is defined on $\mathbb{N}$ instead of $\mathbb{N} \times \mathbb{N}$.

Definition 8.4. Let L be the functional of type $(0 \to 0) \to 0$ defined by $L(g) =$ least i such that for some $j < i \quad g(j) \preceq g(i)$.

Theorem 8.5. *L is not basic feasible.*

This follows from theorem 8.6 below.

Theorem 8.6. *Let G be a basic feasible functional of type two, taking as arguments $X^{0\to 0}$, $y_1^0, \dots, y_k^0$ $(k \geq 0)$ and returning a type 0 value. Then there exists $m \in \mathbb{N}$ and a polynomial p such that for all $g^{0\to 0}$ and all $c_1, \dots, c_k \in \mathbb{N}$ there are numbers $d_0, \dots, d_m,\ K_0, \dots, K_m$ such that*

8.7. $$K_0 = \max\{|c_1|, \dots, |c_k|\} \quad (K_0 = 0 \textit{ if } k = 0)$$

8.8. $$|d_i| \leq K_i,\ 0 \leq i \leq m$$

8.9. $$K_i = p(\max\{K_{i-1}, |g(d_{i-1})|\}),\ 1 \leq i \leq m$$

8.10. $$d_m = G(g, c_1, \dots, c_k)$$

Proof of Theorem 8.5 from Theorem 8.6: We apply 8.6 for the case $G = L$ (so $k = 0$). Let g be the function whose successive values $g(0), g(1), g(2), \dots$ are (for arbitrary $a, b_0, \dots, b_a$)

8.11. $$\begin{array}{l} <2^a, 2^{b_0}>, <2^a, 2^{b_0-1}>, \dots, <2^a, 2^1>, \\ <2^{a-1}, 2^{b_1}>, <2^{a-1}, 2^{b_1-1}>, \dots, <2^{a-1}, 2^1>, \\ <2^{a-2}, 2^{b_2}>, <2^{a-2}, 2^{b_2-1}>, \dots, <2^{a-2}, 2^1>, \\ \vdots \\ <2^0, 2^{b_a}>, <2^0, 2^{b_a-1}>, \dots, <2^0, 2^1>, \\ <2^1, 2^1>, \dots . \end{array}$$

Then the first $b_0 + \cdots + b_a$ values of g are strictly decreasing with respect to $\preceq$, so in fact

8.12. $$L(g) = b_0 + b_1 + \cdots + b_a.$$

Assume L is basic feasible and let p and m be as given by theorem 8.6 for L. We may assume p is monotone increasing. Let $a = m$, $b_0 = 2^{p(0)}$, and for $i = 1, \dots, a$ let $b_i = 2^{p(|<2^a, 2^{B_i}>|)}$, where $B_i = \max\{b_0, \dots, b_{i-1}\}$. Then by induction on i we see by 8.7–8.9, that for $i = 0, \dots, m$,

$$\begin{cases} d_i < b_i \leq b_0 + b_1 + \cdots + b_i & \text{and} \\ g(d_i) \leq <2^a, 2^{B_{i+1}}> . \end{cases}$$

Thus in particular by 8.10

$$L(g) = d_m < b_m \leq b_0 + b_1 + \cdots + b_a$$

which contradicts 8.12 above. □

Proof of Theorem 8.6: Let t be any type 0 PV^ω term in normal form, all of whose free variables are among $X^{0\to 0}$, $y_1, \ldots, y_k$, for some $k \geq 0$. Then t defines a functional G by the equation

8.13. $$G(X, y_1, \ldots, y_k) = t,$$

and in fact the functionals G to which theorem 8.6 applies are precisely those definable by such terms t. We will prove the theorem by induction on the length of t.

The base case is when t has length 1. Then t is either a constant symbol or a variable y_i, and in either case the theorem is obvious.

If t has length greater than 1, then t has one of the following forms:

(i) $X(t')$ where t' satisfies the induction hypothesis with parameters m' and p'. Let $m = m' + 1$ and $p = p' + \lambda x.x$

(ii) $f(t_1, \ldots, t_\ell)$ where f is a polynomial time function symbol and $t_1, \ldots, t_\ell$ are terms which satisfy the induction hypothesis with, say, parameters m_i, p_i, $i = 1, \ldots, \ell$. Let $m = m_1 + \cdots + m_\ell + 1$ and $p = p_1 + \cdots + p_\ell + q$ where q is a polynomial length bound for f (i.e. $|f(x_1, \ldots, x_\ell)| \leq q(|x_1|, \ldots, |x_\ell|)$).

(iii) $\mathcal{R}(t_1, U_1, U_2, t_2)$, where t_1 and t_2 satisfy the induction hypothesis with parameters, say (m_1, p_1) and (m_2, p_2), respectively, and U_2 is a term of type $0 \to 0$ in normal form. Let z be a new type 0 variable. Then the normal form of the term $U_2(z)$ is shorter than t, so the induction hypothesis applies to $U_2(z)$ with parameters, say, m_3, p_3. Now let G, G_1, G_2, G_3 be the functionals defined by the four terms in question; that is

$$\begin{aligned} G(X, \vec{y}) &= t = \mathcal{R}(t_1, U_1, U_2, t_2) \\ G_1(X, \vec{y}) &= t_1 \\ G_2(X, \vec{y}) &= t_2 \\ G_3(X, \vec{y}, z) &= U_2(z). \end{aligned}$$

Then by the semantics of the recursor, we have

$$|G(X, \vec{y})| \leq \max\{|G_1(X, \vec{y})|, |G_3(X, \vec{y}, G_2(X, \vec{y}))|\}.$$

Thus t satisfies the theorem with parameters $m = m_1 + m_2 + m_3 + 1$ and $p = p_1 + p_2 + p_3$.

□

Theorem 8.14. *Let F be a function symbol for the functional L of theorem 1. Then any type 1 function represented by a PV^ω term in which F may occur is a polynomial time computable function.*

Proof: By placing the term in normal form, it is easy to see that it suffices to show that if a function f is defined by

8.15. $$f(\vec{x}) = F(\lambda y.g(\vec{x}, y))$$

where g is polynomial time computable, then f is polynomial time computable. To show this, it suffices to show that for some $k \in \mathbb{N}$

8.16. $$f(\vec{x}) = O(n^k), \text{ where } n = \max_i |x_i|$$

(here $\vec{x} = (x_1, \ldots, x_k)$), because then $f(\vec{x})$ can be computed by "brute force" in polynomial time.

To prove 8.16, we may assume (since g is polynomial time computable) that there is a polynomial p such that for all $\vec{x}, y$

8.17. $$|g(\vec{x}, y)| \leq p(n),$$

where $n = \max\{|x_1|, \ldots, |x_k|, |y|\}$. Now let

8.18. $$n = \max\{|x_1|, \ldots, |x_k|\}.$$

We wish to bound how far the sequence

$$g(\vec{x}, 0), g(\vec{x}, 1), \ldots$$

can be strictly decreasing with respect to $\preceq$. While it is strictly decreasing, it must "look" like a subsequence of 8.11.

More formally, let π_1 and π_2 be the left and right projection functions with respect to the pairing $< \cdot, \cdot >$. That is, $\pi_1(< r, s >) = r$ and $\pi_2(< r, s >) = s$. Let

8.19. $$a = |\pi_1(g(\vec{x}, 0))| \leq p(n)$$

8.20. $$b_0 = |\pi_2(g(\vec{x}, 0))| \leq p(n).$$

Now define the elements $d_0, \ldots, d_a$ and $b_1, \ldots, b_a$ in $\mathbb{N} \cup \{\infty\}$ by

8.21. $$d_i = \text{least } d \text{ so } |\pi_1(g(\vec{x}, d)| \leq a - i$$

8.22. $$b_i = |\pi_2(g(\vec{x}, d_i))|.$$

Thus $d_0 = 0$, and in general if $f(\vec{x}) > d_{i-1} + b_{i-1}$ then for sufficiently large n for $i = 0, \ldots, a$

8.23. $$d_i \leq b_0 + \cdots + b_{i-1} \leq ip(n)$$

8.24. $$b_i \leq p(n).$$

These inequalities follow by induction on i, provided n is large enough that $|(p(n))^2| \leq n$, so that 8.23 implies $|d_i| \leq n$ and hence 8.24 follows from 8.23, 8.17, 8.18, and 8.22.

Assuming $f(\vec{x}) > d_{i-1} + b_{i-1}$ for $i = 1, \ldots, a$ (so 8.23, 8.24 hold) we have by definition of F

$$f(\vec{x}) \leq b_0 + \cdots + b_a + 1 \leq (p(n))^2 + 1$$

so 8.16 is proved. If $f(\vec{x}) \leq d_{i-1} + b_{i-1}$ then 8.16 follows by considering the first such i. □

9. Further Work

As explained at the beginning of section 6, there seems to be a gap in the computability theory of the 1970's, perhaps caused by lack of communication between recursion theorists and computer scientists. The question concerns characterizing the power of Plotkin's programming language $\mathcal{L}_{DA}$ on everywhere defined functionals; and perhaps relating this to Kleene's general recursive functionals.

The work of Ko and Friedman [20] and Hoover [13] on polynomial time analysis involves functionals on the reals similar in spirit to our basic feasible functionals. It would be nice to study the relationship more carefully.

Kapron [14] is studying classes of functionals satisfying 8.1–8.3, but more work needs to be done to either find the "right" definition of polynomial time functional, or give a convincing argument that there isn't any such thing.

Of course we have hardly scratched the surface of higher type complexity theory. No doubt all of the familiar type one complexity classes have higher type analogs.

Acknowledgment

I am grateful to my PhD students Steve Bellantoni and Bruce Kapron, who have contributed substantially to the ideas and points of view expressed here.

REFERENCES

1. S. Bellantoni, *Comparing two notions of higher type computability*, Manuscript, University of Toronto (in preparation).
2. Samuel R. Buss, *The polynomial hierarchy and intuitionistic bounded arithmetic*, *Structure in Complexity Theory*, Springer-Verlag Lecture Notes in Computer Science No. 223, pages 77–103, 1986.
3. Alan Cobham, *The intrinsic computational difficulty of functions*, Proc. of the 1964 International Congress for Logic, Methodology, and the Philosophy of Sciences (Y. Bar-Hillel, ed.), North Holland, Amsterdam, pp. 24–30.
4. R. Constable, *Type two computational complexity*, Proc. 5th STOC, pp. 108–121.
5. S. Cook and A. Urquhart, *Functional Interpretations of Feasibly Constructive Arithmetic*, Technical Report 210/88, University of Toronto, 1988. Extended Abstract in *Proceedings 21st ACM Symposium on Theory of Computing*, May 1989, pp. 107–112.
6. S.A. Cook, *The complexity of theorem-proving procedures*, Proc. 3rd STOC, pp. 151–158.
7. S.A. Cook and B.M. Kapron, *Characterizations of the basic feasible functionals of finite type*, Proc. MSI Workshop on Feasible Mathematics (S. Buss and P. Scott, eds.), Also Tech Rep 228/90, Dept. of Computer Science, University of Toronto, January, 1990. Abstract in Proc. 30th FOCS (1989) pp. 154–159.
8. Yu.L. Ershov, *Computable functionals of finite type*, Algebra and Logic **11** (1972), 203–242, (367–437 in Russian).
9. Yu.L. Ershov, *Maximal and everywhere defined functionals*, Algebra and Logic **13** (1974), 210–225, (374–397 in Russian).
10. R.O. Gandy, *Computable functionals of finite type I*, Sets, Models, and Recursion Theory (J. Crossley, ed.), North Holland, Amsterdam, 1967, (Proc. of the Summer School in Mathematical Logic and Logic Colloquim, Leicester, England, 1965), pp. 202–242.
11. R.O. Gandy and J.M.E. Hyland, *Computable and recursively countable functions of higher type*, Logic Colloquium 76, North Holland, 1977, pp. 407–438.
12. Kurt Gödel, *Über eine bisher noch nicht benützte erweiterung des finiten standpunktes*, *Dialectica* 12, 280-287. English translation: *J. of Philosophical Logic* 9 (1980), 133–142. Revised and expanded English translation to appear in Volume II of Gödel's *Collected Works*.
13. H.J. Hoover, *Feasible real functions and arithmetic circuits*, SIAM J. Comp. **19** (1990), 182–204.
14. B. Kapron, *Feasible computation in higher types*, Ph.D. thesis, University of Toronto, 1990 (in preparation).
15. A. Kechris and Y.N. Moschovakis, *Recursion in higher types*, Handbook of Mathematical Logic (J. Barwise, ed.), North Holland, 1977, pp. 681–737.
16. S.C. Kleene, *Countable functionals*, Constructivity in Mathematics (Amsterdam), North Holland, Amsterdam, 1959.
17. S.C. Kleene, *Recursive functionals and quantifiers of finite types I*, Trans. Amer. Math. Soc. **91** (1959), 1–52.
18. S.C. Kleene, *Turing-machine computable functionals of finite types I*, Proc. of the 1960 Congress for Logic, Methodology, and the Philosophy of Science (P. Suppes, ed.), pp. 38–45.
19. S.C. Kleene, *Recursive functionals and quantifiers of finite types II*, Trans. Amer. Math. Soc. **108** (1963), 106–142.
20. K. Ko and H. Friedman, *Computational complexity of real functions*, Theoretical Computer Science **20** (1982), 323–352.
21. G. Kreisel, *Interpretation of analysis by means of functionals of finite type*, Constructivity in Mathematics (Amsterdam), North Holland, Amsterdam, 1959.
22. K. Mehlhorn, *Polynomial and abstract subrecursive classes*, JCSS **12** (1976), 147–148.

23. J. Moldestad, *Computations in higher types*, Springer-Verlag Lecture Notes in Mathematics, No. 574, 1977.
24. D. Normann, *Recursion on the countable functions*, Springer-Verlag Lecture Notes in Mathematics, No. 811.
25. R.A. Platek, *Foundations of recursion theory*, Ph.D. thesis, Stanford University, 1966.
26. G.D. Plotkin, *LCF considered as a programming language*, Theoretical Computer Science **5** (1977), 223–255.
27. R.W. Ritchie, *Classes of predictably computable functions*, Trans. Amer. Math. Soc. **106** (1963), 139–173.
28. D. Scott, *Outline of a mathematical theory of computation*, Proc. 4th Annual Princeton conference on Information Science and systems, pp. 169–176.

University of Toronto
April 20, 1990

CONSTRUCTIVELY EQUIVALENT PROPOSITIONS AND ISOMORPHISMS OF OBJECTS, OR TERMS AS NATURAL TRANSFORMATIONS

ROBERTO DI COSMO AND GIUSEPPE LONGO

1. INTRODUCTION

In these notes, we sketch a recent application of the typed and type-free λ-calculus to Proof Theory and Category Theory. The proof is fairly incomplete and we refer the reader interested in the lengthy technical details to Bruce & DiCosmo & Longo [1990]. Our main purpose here is to hint a logical framework for the result below, in a rather preliminary and problematic form. The occasion is provided by the kind invitation to deliver a lecture at a meeting with such a stimulating title.

Indeed, the result in §3 (or, more precisely, our proof of it) may be viewed as part of the influence of Computer Science in Logic. First, because, in Bruce & Longo [1985] we were originally motivated by the work on recursive definitions of data types. Second, because λ-calculus today is a lively subject in Computer Science: its impact into mathematics and its foundation has been very little and very few researchers would work today at it or at the equivalent theory of combinators, if these theories hadn't turn out to be powerful tools for the understanding and the design of functional language. Their role in mathematics probably ended once they contributed to formulate the Church Thesis and helped by this to understand the general relevance of Gödel's incompleteness theorem (i.e. by suggesting, with Turing machines, that all "finitistically computable" functions are expressible in Peano Arithmetic, a fact of which Gödel was not aware in 1931). The point is that poverty of the reductionist view upon which they are based has little to say about the mathematical language and deduction, but seems to be able to focus the main theoretical concerns of the programmers in a functional style. Also the striking relevance of normalization theorems in λ-calculus for Proof Theory, e.g. the one of II order λ-calculus, is only (and largely) quoted, today, in discussions on polymorphisms in programming.

Indeed, we view at this as at a positive fact as it enlarged the scope and motivations of λ-calculus and its peculiar mathematics; moreover, many, like us, started to "play with" the challenging syntax of λ-calculus and combinatory logic without sharing at all the formalist or reductionist motivations of the founding fathers (for more discussion, see Longo [1990]).

A further motivation to our work has been recently provided by Rittri [1990a, 1990b] and it directly derives from an interesting application in Computer Science. In functional programming, it may be wise to store programs in a library according to their types. When retrieving them, the search should be done "up to isomorphisms of types", as the same program may have been coded under different, but isomorphic types; for example, a search program for lists of a given length, may be typed by $\mathrm{INT} \times \mathrm{LISTS} \to \mathrm{LISTS}$ or, equivalently, by $\mathrm{LISTS} \to (\mathrm{INT} \to \mathrm{LISTS})$. Indeed, we will provide a decidable theory of isomorphic functional types.

In a sense, the application we are going to discuss goes in the other direction with respect to the prevailing perspective, as, so far, most results connecting Proof Theory and Type Theory to Category Theory applied categorical tools to the understanding of deductive systems and their calculus of proofs, e.g. λ-calculus. One of the exceptions to this is the work by Mints and his school (see references).

The questions we raise here are the following: when two propositions can be considered as truly equivalent? Or, when is it that types may be viewed at as isomorphic under all circumstances?

In proof-theoretic terms this may be understood as in definition 0.1 below, on the grounds of the "types-as-propositions" analogy between the typed λ-calculus and intuitionistic propositional logic (**IPL**).

Recall first that IPL has conjunction and implication ($\times$ and $\to$) as connectives and that a (well-typed) term $M : A$, may be called an (effective) **proof** when A is understood as a proposition. A variable $x : A$, will be called a generic proof of A. An intuitionistic sequent has the following syntactic structure:

$$x_1 : A_1, \dots, x_n : A_n \vdash M : B$$

where $x_1 : A_1, \dots, x_n : A_n$ is a finite (possibly empty) list of distinct generic proofs, and $M : B$ is a proof of B whose free variables are among $x_1, \dots, x_n$. Every formula in the left-hand-side (l.h.s.) has an associated distinct variable; thus no confusion can arise between formulae, even if they have the same name.

The intuitive interpretation of the sequent $x_1 : A_1, \ldots, x_n : A_n \vdash M : B$ is that of a process which builds a proof M of B, as soon as it has proofs for $A_1, \ldots, A_n$, that is, a function f of type $A_1 \times \ldots \times A_n \to B$.

1.1 Definition. Propositions S and R are **constructively equivalent** if the composition of $x : S \vdash M : R$ and $y : R \vdash N : S$ (and of $y : R \vdash N : S$ and $x : S \vdash M : R$) reduce, by cut- elimination, to the axiom $x : S \vdash x : S$ (and $y : R \vdash y : R$, respectively).

Consider now the following equational theory of types. It is given by axiom schemata plus the obvious inference rules that turn "=" into a congruence relation.

1.2 Definition. $\mathfrak{Th}$ is axiomatized as follows, where T is a constant symbol:

(1) $A \times T = A$
(2) $A \times B = B \times A$
(3) $A \times (B \times C) = (A \times B) \times C$
(4) $(A \times B) \to C = A \to (B \to C)$
(5) $A \to (B \times C) = (A \to B) \times (A \to C)$
(6) $A \to T = T$
(7) $T \to A = A$.

1.3 Theorem (Proof-theoretic version). *S and R are constructively equivalent in* IPL *iff* $\mathfrak{Th} \vdash S = R$.

The proof sketched in §3 will be based on the following idea: assume that the equivalence is effectively proved by $x : S \vdash M : R$ and $y : R \vdash N : S$. Then $\vdash \lambda x.M : S \to R$ and $\vdash \lambda y.N : R \to S$ are such that $\lambda\beta\eta\pi^t \vdash \lambda x.M \circ \lambda y.N = \lambda x.x$ and $\lambda\beta\eta\pi^t \vdash \lambda y.N \circ \lambda x.M = \lambda x.x$ in a suitable extension $\lambda\beta\eta\pi^t$ of the typed λ-calculus, essentially because β-reductions correspond to cut-eliminations (we omit types for simplicity, whenever convenient). Then, we prove that $R = S$ is derivable in $\mathfrak{Th}$, by investigating the syntactic structure of the terms $\lambda x.M : S \to R$ and $\lambda y.N : R \to S$. The converse implication is easy.

We next discuss the categorical side of the same issue.

2. (Natural) Isomorphisms of Objects

There is an equivalent way to state Theorem 1.3. At first thinking, in category theoretic terms, Theorem 1.3 can be roughly stated as follows, by looking at functional types as objects of cartesian closed categories (CCC's),

2.1 Theorem (categorical version). *"$\mathfrak{Th}$ gives exactly the valid isomorphisms in all* CCC*'s"*.

Also in this case the motivation is clear: by the theorem, a simple and decidable equational theory would detect valid isomorphisms in all CCC's. (Indeed, one may easily prove that $\mathfrak{Th}$ is decidable; it may be interesting to check the complexity of $\mathfrak{Th}$.) Note that a proof of 2.1 has already been given by Soloviev [1983] and is stated in Martin [1972]. They use, though, a rather different technique, as their crucial argument is based on an interpretation of $\times$ and $\to$ as product and exponent over the natural numbers. In view of the connections between CCC's and IPL, in Bruce & DiCosmo & Longo [1990] we give a more constructive proof, based on λ-calculus, by arguing as after 1.3 (see §3). Our proof led us to theorem 1.3 and the remarks in this section.

It should be clear that, when $\times$ and $\to$ are interpreted as cartesian product and exponent, the provable equations of $\mathfrak{Th}$ hold as isomorphisms in any CCC. Note also that there are categorical models which realize $\mathfrak{Th}$, but are not CCC's. Take, for example, a cartesian category and a bifunctor "$\to$" that is constant in the second argument: this clearly yields a model of $\mathfrak{Th}$.

The point we want to make here, though, is that our *proof* of 2.1 suggests more than what is stated. Because of lack of time and the strict deadlines, we can only sketch some preliminary ideas.

Observe first that λ-terms provide "uniform" denotations for morphisms: roughly, they denote morphisms in lots of categories and, in a given category, between different objects. It is probably always so when dealing with a language or at a theoretical level: linguistic entities give uniform representations of structures and objects. But, in this case, what uniformity is given by λ-terms? Given a CCC, do λ-terms define natural transformations between their types viewed as functors, i.e. when the atomic types may also "range" over the objects of the intended CCC?

There is an immediate difficulty to this approach: the categorical meaning of λ-terms uses the map

$$\text{eval} : B^A \times A \to A$$

which is not natural in A and B, simply because the map on objects

$(A, B) \mapsto B^A \times A$ does not extend to a (bi)functor. Indeed, it should be at the same time contravariant and covariant in A.

In order to discuss this more precisely, we recall here a few notions.

The (pure) λ-calculus may be extended by adding fresh types and constants as well as consistent sets of equations. Consider now the extension of $\lambda\beta\eta\pi^t$, the typed λ-calculus with surjective pairing, by adding:

(1) a special atomic type T (the terminal object);

(2) an axiom schema

$$^*A : A \to T$$

which gives a constant of that type;

(3) a rule

$$\frac{M : A \to T}{M = {}^*A}$$

that gives the unicity of $^*A$.

Call $\lambda\beta\eta\pi^{*t}$ this extended calculus and Tp^* its collection of types. As we write types also as superscripts, we may use for them small greek letters as well. Then the usual categorical interpretation of $\lambda\beta\eta\pi^{*t}$ goes as follows:

2.2 Definition. Let $\mathbb{C}$ be a CCC, with terminal object t and projections fst, snd. Suppose one has a map I associating every atomic type A with an object of $\mathbb{C}$, with $I(T) = t$. Set then

- **Types**: $[\sigma] = I(\sigma)$ if σ is atomic
$[\sigma \to \tau] = [\tau]^{[\sigma]}$
$[\sigma \times \tau] = [\sigma] \times [\tau]$

- **Terms**: let M^σ be a term of $\lambda\beta\eta\pi^{*t}$, with

$$FV(M^\sigma) \subseteq \Delta = \{x^{\sigma}{}_1, \ldots, x^{\sigma}{}_n\},$$

and assume that $A, A_1, \ldots, A_n$ interpret $\sigma, \sigma_1, \ldots, \sigma_n$ (we omit typing, when unambiguous, and write $p_i \in \mathbb{C}[t \times A_1 \times \ldots \times A_n, A_i]$ for the projection $p_i : t \times A_1 \times \ldots \times A_n \to A_i$). Then $[M^\sigma]_\Delta$ is the morphism in $\mathbb{C}[t \times A_1 \times \ldots \times A_n, A]$ defined by

$[\sigma^*]$ is the unique morphism in the intended type
$[x^{\sigma_i}]_\Delta = p_i$
$[MN]_\Delta = \text{eval} \circ \langle [M]_\Delta, [N]_\Delta \rangle$
$[\lambda x^\tau . M]_\Delta = \Lambda([M]_{\Delta \cup \{x^\tau\}})$
$[\langle M, N \rangle]_\Delta = \langle [M]_\Delta, [N]_\Delta \rangle$
$[fst(M)]_\Delta = fst \circ [M]_\Delta$
$[snd(M)]_\Delta = snd \circ [M]_\Delta.$

Our aim now is to set a basis for an understanding of terms as natural transformations. In the following two subsections, this will be done by, first, focusing only on types "which yield functors", then by generalizing the notion of functor.

Functor Schemata

Note that, even in the case of the simply typed λ-calculus, which we discuss here, some parametricity or implicit polymorphism is lost by the usual description and in the interpretation in 2.2. For example, the fact that two types are provably isomorphic by λ-terms is "typically ambiguous", in the sense that the isomorphism holds for any coherent substitution instance of base types. More generally, in ML this sort of polymorphism is tidely taken care of by the notion of *type schemata.* In the case of ML, though, type schemata are inhabited by type free terms: e.g. the identity $\lambda x.x$ has type schema $X \to X$. By this, one needs an underlying type-free structure in order to deal with the meaning of ML style programming. Following the "ML idea" (indeed, Curry's type-assigment), a propositional typed λ-calculus may be described by having also type variables, as atomic types. So, $\lambda x : X.x$ would have "type schema" $X \to X$. In categorical terms, $\lambda x : X.x$ should be considered a name for the (polymorphic) identity; that is, we should interpret $\lambda x : X.x$ as a natural transformation from the identy functor to itself:

$$\lambda x : X.x : X \dot{\to} X.$$

This is a clear meaning for what we wanted above: $\lambda x : X.x$ uniformely names the identity, in all models, and its "implicit polymorphism" is categorically understood in terms of natural transformations. The idea then would be to interpret types as functors and terms as natural transformations. Unfortunately, this interpretation does not work for all types and λ-terms, by the remark above on the "innaturality" of eval, a crucial morphism in functional languages and CCC's! (But note that also the map $A \mapsto (A \to A)$ does not extend to a functor, for example). In particular, there is no way to restate Theorem 2.1 as "$\mathfrak{Th}$ *gives exactly the natural isomorphisms which are valid in all* CCC's".

However, something can be done, by restricting types to functor schemata, as we do here, or by generalizing functors to weak functors, as we will do next. As usual, we identify types and propositions of IPL. Consider, though, a "parametric" version of IPL, where propositional or type variables are allowed (but no quantification over them: this would lead us to higher order propositional calculi, such as Girard's system F). In a sense, in this typed calculus, atomic types include type variables. Its sematics, at first thought, may be given as in 2.2, by allowing type variables to range over objects, i.e., by interpreting type schemata as functions from objects to objects, with the aim to provide a more categorical understanding of its implicit polymorphism. We hint how this can be partly done when considering isomorphims, our current topic. Quite generally, though, propositions (2.5 and) 2.10 suggest a functorial semantics for (certain) type schemata.

2.3 Definition. Let X be a type variable. We say that an occurence of X is **positive** in a proposition A if $\text{sign}_A(X) = +1$; it is **negative** if $\text{sign}_A(X) = -1$, where

$$\begin{aligned}
&\text{sign}_C(X) = +1 \text{ if } X \text{ does not occur in } C \text{ or } C \equiv X;\\
&\text{sign}_{A\times B}(X) = \underline{\text{if}} \text{ the occurence of } X \text{ is in } A \ \underline{\text{then}}\ \text{sign}_A(X)\\
&\qquad\qquad \underline{\text{else}}\ \text{sign}_B(X);\\
&\text{sign}_{A\to B}(X) = \underline{\text{if}} \text{ the occurence of } X \text{ is in } A \ \underline{\text{then}}\ -\text{sign}_A(X)\\
&\qquad\qquad \underline{\text{else}}\ \text{sign}_B(X).
\end{aligned}$$

2.4 Definition. A proposition of IPL is a **functor schema** if all occurences of the same variable have the same sign.

2.5 Proposition. *Given a CCC $\mathbb{C}$, any functor schema F, with n type variables, can be extended to a functor $F : C^{(n)} \to C$, which is covariant in the (n) type variables which occur positively, contravariant in the others.*

Proof (Hint). Clearly, each functor schema specializes into a function over the objects of $\mathbb{C}$ when interpreting formal $\times$ and $\to$ as product and exponent in $\mathbb{C}$. Then proceed by induction on the syntactic structure of F, by observing that, if $H \times G$ or $G \to H$ (i.e. H^G) is a functor schema, so are H and G. As for the typical case, consider H^G. The behaviour of (H^G) on objects is given in the usual way for exponentiation w.r.t. H and G. As for morphisms, set

$$(H^G)(f_1, \ldots, f_n) = \Lambda(H(f_1, \ldots, f_n) \circ \text{eval} \circ \text{id} \times G(f_1, \ldots, f_n)).$$

□

2.6 Lemma. *Let F and G, F' and G' be such that F^G and $F'^{G'}$ are functor schemata. Assume also that there exist natural transformations*

$$\phi : F \dot{\to} F' \qquad \phi' : F' \dot{\to} F \qquad \gamma : G \dot{\to} G' \qquad \gamma' : G' \dot{\to} G$$

Then there exist two natural transformations $\delta : (F^G) \dot{\to} (F'^{G'})$ and $\delta' : (F'^{G'}) \dot{\to} (F^G)$.

Proof. Set $\delta = \Lambda(\phi \circ \mathrm{eval} \circ \mathrm{id} \times \gamma')$ and $\delta' = \Lambda(\gamma \circ \mathrm{eval} \circ \mathrm{id} \times \phi')$. □

We say that functor schemata F and G **yield naturally isomorphic functors** in a CCC, if F is naturally isomorphic to G.

2.7 Corollary. *Let F and G be functor schemata. Then $\mathfrak{Th} \vdash F = G$ iff F and G yield naturally isomorphic functors in all* CCC*'s.*

Proof. ($\Leftarrow$) By 1.1. ($\Rightarrow$) The propositions in each side of the axioms of $\mathfrak{Th}$ are functor schemata and, when specified in a CCC, they yield naturally isomorphic functors. The inductive step, then, only requires to look at the inference rules. Transitivity of "=" gives no problem, as natural transformations compose. Neither does substitutivity, when the outermost connective is $\times$. Otherwise, i.e. when the outermost connective is $\to$, observe that, if F and F' and G and G' yield naturally isosmorphic functors, respectively, then one obtains a natural isomorphism $\delta : (F^G) \dot{\to} (F'^{G'})$, by the lemma. □

Question. Is it the case that any term $M : F \to G$ gives a natural transformation in all models, when F and G are functor schemata? This strong fact would also imply Corollary 2.7. Note, however, the essential role of theory $\mathfrak{Th}$ in the inductive proof of our corollary, by Lemma 2.6.

Terms as "Natural Transformations"

In this subsection, we suggest a way to weaken the interpretation of propositions and reinforce Theorem 2.1, as stated. In this context, terms do turn out to be "natural transformations", but w.r.t. to a suitably generalized notion of functor. We only sketch the basic idea of an approach to be still worked out.

2.8 Definition. Given categories $\mathbb{C}$ and $\mathbb{D}$, a **weak-functor** F from $\mathbb{C}$ to $\mathbb{D}$ is a function on objects and a partial function on morphisms such that, if $i : A \cong B$ and $j : B \cong C$ are isomorphisms in $\mathbb{C}$, one has:

- $F(i)$ is defined
- $F(i \circ j) = F(i) \circ F(j)$
- $F(\mathrm{id}_A) = \mathrm{id}_F(A)$.

Clearly, weak functors take isomorphic objects to isomorphic objects. Indeed, one may understand weak functors in the following way. Given a category $\mathbb{C}$, call Ciso the subcategory whose morphisms are just the isomorphisms in $\mathbb{C}$. Then a weak functor is a functor from Ciso to $\mathbb{C}$.

2.9 Lemma. (i) *Every functor is a weak functor.*
(ii) *Weak functors compose.*

Proof. Obvious. □

The idea now is that the crucial map $A \mapsto (A \to A)$ extends to a weak functor, in every CCC. As isomorphisms go two ways, the "variance" of a weak functor is irrelevant.

2.10 Theorem. *Let* $\mathbb{C}$ *be a* CCC. *Then each proposition of* IPL, *with* n *type variables, can be extended to a weak functor from* $\mathbb{C}^{(n)}$ *to* $\mathbb{C}$.

Proof (Hint). We only sketch the critical case, i.e. the "$\to$" case when a type variable occurs with different signs. The rest is easy, in view of the lemma, by induction. Assume that $i : A \cong B$ is an isomorphism. Define then the value of the $\to$ on i as $\Lambda(i \circ \mathrm{eval} \circ \mathrm{id} \times i^{-1}) : A^A \to B^B$. Compositionality w.r.t. isomorphisms is immediate and the identity is taken to the identity. □

The definition of natural transformation can be extended to weak functors. One only has to keep in mind that the usual square diagram is required to commute only when vertical arrows are isomorphisms. In short, a **natural transformation** τ **between weak functors** $F, G : \mathbb{C} \to \mathbb{D}$ is a collection of maps $\{\tau_A \mid A \in \mathbb{C}\}$ such that, when $i : A \cong B$, $G(i) \circ \tau_A = \tau_B \circ F(i)$. Clearly, natural transformations between weak functors compose. Note that, even under these restricted circumstances, given two weak functors, the (large) set of natural transformations between them is not uninformative: for example, it is remote from truth to claim that any collection of morphisms of the intended type would yield a natural tranformation between them.

Remark. Given a category $\mathbb{C}$ and a CCC $\mathbb{C}'$, the weak functors from $\mathbb{C}$ to $\mathbb{C}$' form a category, call it WFunct$(\mathbb{C}, \mathbb{C}')$, with natural transformations as maps. Indeed, and that is the point, WFunct$(\mathbb{C}, \mathbb{C}')$ is a CCC. Just define products and exponents objectwise; e.g., $G^F(A) = G(A)^{F(A)}$ and use the argument in 2.10 to extend the exponent to morphisms. Similarly, for the maps $\mathrm{eval}_{F,G}$ and $\Lambda(\tau)$, when $t : F \times H \to G$, set $(\mathrm{eval}_{F,G})_{A,B} = \mathrm{eval}_{F(A),G(B)}$ and $\Lambda(\tau)_A = \Lambda(\tau_A)$ (see the lemma below).

Note also that covariant (or contravariant) functors with natural transformations form a subcategory of WFunct$(\mathbb{C}, \mathbb{C}')$. However, this is not full, if we leave the definition of weak functors as it is. It should be possible, though, to restrict them slightly in order to turn them into "a conservative extension" of functors, i.e., in such a way that Funct$(\mathbb{C}, \mathbb{C}')$ is full in WFunct$(\mathbb{C}, \mathbb{C}')$.

2.11 Lemma. *Let $\mathbb{C}$ be a CCC and F, G and H be weak functors. Then* eval $: G^H \times H \to G$ *is a natural transformation between weak functors. Moreover, if $\tau : F \times H \to G$ is a natural transformation, then also $\Lambda(\tau) : F \to G^H$ is a natural transformation between weak functors.*

Proof. As for the naturality of eval, let $i : A \cong B$ be an isomorphism. Then one has

$$G(i) \circ \mathrm{eval} = \mathrm{eval} \circ \Lambda(G(i) \circ \mathrm{eval} \circ \mathrm{id} \times H(i^{-1})) \times H(i),$$

by an easy computation. Moreover, assume that $G(i) \circ \tau_A = \tau_B \circ F(i) \times H(i)$, i.e., that τ is natural. Then one has

$$\begin{aligned}
\Lambda(G(i) \circ \mathrm{eval} \circ \mathrm{id} \times H(i^{-1})) \circ \Lambda(\tau_A) = & \\
&= \Lambda(G(i) \circ \mathrm{eval} \circ \Lambda(\tau_A) \times H(i^{-1})) \\
&= \Lambda(G(i) \circ \tau_A \circ \mathrm{id} \times H(i^{-1})) \\
&= \Lambda(\tau_B \circ F(i) \times H(i) \circ \mathrm{id} \times H(i^{-1})) \\
&\quad \text{by the naturality of } \tau \\
&= \Lambda(\tau_B \circ F(i) \times \mathrm{id}) \\
&= \Lambda(\tau_B) \circ F(i).
\end{aligned}$$

□

We continue with some "handwaving", as we still didn't specify formally the semantics of parametric types and terms, in our sense, in full details. This is just hinted before Definition 2.3 and above.

2.12 Theorem. *Given a* CCC, *the interpretation of each term* $M : F \to G$ *is a natural transformation between weak functors.*

Proof. Notice that, by the lemma, the interpretation of terms is inductively obtained by composing natural transformations, possibly between weak functors (cf. Definition 2.2). □

Consider now arbitrary propositions F and G of IPL. Then one has:

2.13 Corollary. $\mathfrak{Th} \vdash F = G$ *iff* F *and* G *yield naturally isomorphic weak functors in all* CCC*'s.*

Proof. ($\Leftarrow$) By 2.1. ($\Rightarrow$) If $F = G$ is a an axiom of $\mathfrak{Th}$, it is easy to give an invertible term $M : F \to G$. The inference rules of $\mathfrak{Th}$ constructively give terms that actually prove the equality as an isomorphism of types (see Remark 3.4 below or Bruce & DiCosmo & Longo [1990] for more details). Then, by the theorem, we are done. □

Note that the first part of the remark before 2.11 provides an indirect, but equivalent way to interpret types and terms as we wish: just give the usual interpretation of the typed λ-calculus over the CCC of weak functors and natural transformations. Of course, one should interpret atomic constant by constant functors, type variables as projections functors etc...

Problem. It would be interesting to find a uniform "category-theoretic" condition, relating weak functors and their natural transformations over different pairs of CCC's, such that each of these triples—natural tranformation, source and target functors—are named by a λ-term and its (arrow) type. In other words, we claim that not only λ-terms express a uniformity that is categorically meaningful, as we tried to show, but that there must exist a structural property of "uniformity" or "polymorphisms beyond a specific category", that is exactly captured by λ-terms.

Intermezzo

While we were struggling with the innaturality of λ-terms w.r.t. ordinary functors, we noticed that also Bainbridge & al [1989] deals with the problem. (That paper aims at the understanding of higher order calculi, while here we refer to their proposal only for the propositional fragment, with type variables, but no quatification.). Unfortunately, the elegant approach in Bainbridge & al [1989] does not help in stating our result in any further generality than we could do here, but its motivation is given exactly by an investigation of the same "uniformity" we wanted to capture. The idea is to understand terms as dinatural transformations; the backup is that, in general, dinatural transformations do not compose. Thus, they cannot provide a general semantics to λ-terms. However, in an extremely relevant class of categories, dinatural transformations, between suitable functors, happen to compose and the interpretation of λ-terms is sound. These models are the PER's, i.e. the categories of partial equivalence relations on a (possibly partial) combinatory algebra (see Asperti & Longo [1990] for a recent unified treatment).

Remark. Scott constructive domains—CD—may be fully and faithfully embedded into PER over Klenee's $(\omega, \cdot)$ by a functor which preserves products and exponents, in view of (generalized versions of) Myhill–Shepherdson theorem (see Longo [1988], for a simple proof and references); then, as for the propositional calculus with type variables, the interpretation of λ-terms as dinatural transformations should also work over CD.

As usual, the PER models proposed in Bainbridge & al [1989] are built over a type-free structure. This takes us back to the implicit polymorphism of ML, as the existence of an underlying type-free model is at the core of the semantics of ML style type-assignement. Indeed, the meaning of a term M in Bainbridge & al [1989] is precisely given as "the equivalence class, in its type as a p.e.r., of the type-free term $\mathrm{er}(M)$ obtained from M by erasing all type information" (see also Cardelli & Longo [1990]). Thus, polymorphims is given meaning by the interplay "type-free/typed structures" needed for the semantics of type-assignement. And this seems essential, so far.

The PER models have other nice features. First, by a theorem in Asperti & Longo [1986], the valid equations between simply typed λ-terms are exactly the provable ones. It should be easy to extend this result to the propositional calculus we deal with here. Thus, as for *definable* isomorphisms of types, this model should be complete w.r.t. the equational theory $\mathfrak{Th}$. Second, in Bainbridge & al [1989], λ-terms are interpreted as dinatural transformations by an essential use of the lattice structure of PER, where $A \leq B$ iff "n A-equivalent to m" implies "n B-equivalent to m". This is the meaning of subtyping proposed in Bruce & Longo [1988] and subtyping is the other, orthogonal, form of polymorphism, since a term may have more then one type as "element" of larger and larger types. Thus and once more,

PER models seem to provide the semantic basis for a unified understanding of various aspects of polymorphism, if we could fully relate the actual use of the lattice structure of PER in the two approaches. (For example, by stating the right theorems by which that specific partial order, at the same time, realizes contravariance of the arrow functor, i.e. antimonotonicity w.r.t. the partial order, and allows to mix up its contravariance and covariance in the two arguments in the meaning of terms as dinatural transformations. This would require a categorical generalization of subtyping as meant in Bruce & Longo [1988], which we still miss).

3. Isomorphisms in the Term Model

We finally hint the basic steps of the proof in Bruce & DiCosmo & Longo [1990].

3.1 Remark. The crucial point, again, is that the typed λ-calculus is at the same time:

(a) the "theory" of CCC's;
(b) the calculus of proofs of the intuitionistic calculus of sequents.

Thus, both 1.3 and 2.1 are shown by observing that the isomorphic types in the (closed) term model of typed λ-calculus are provably equal in $\mathfrak{Th}$.

A term model is closed, when terms in it contain no free variables. More precisely:

3.2 Definition. Given (an extension of) the typed λ-calculus, λ, the (closed) term model is the type structure

$$|\lambda| = \{|M : A| \mid M \text{ is a (closed) term of type } A\}$$

where $|M : A| = \{N \mid \lambda \vdash M = N\}$.

Clearly, the type structure is non trivial, if a collection of ground or atomic types is given. Indeed, the closed term model of $\lambda\beta\eta\pi^{*t}$ (and its extensions) forms a CCC, as the reader may check as an exercise. Then the provable equations of $\mathfrak{Th}$ are realized in $|\lambda\beta\eta\pi^{*t}|$, as isomorphisms. We give an explicit name to these isomorphisms, as λ-terms provide the basic working tools.

3.3 Definition. Let $A, B \in Tp^*$. Then A and B are provably isomorphic ($A \cong_p B$) iff there exist closed λ-terms $M : A \to B$ and $N : B \to A$ such that $\lambda\beta\eta\pi^{*t} \vdash M \circ N = I_B$ and $\lambda\beta\eta\pi^{*t} \vdash N \circ M = I_A$, where I_A and I_B are the identities of type A and B. We then say that M and N are invertible terms in $\lambda\beta\eta\pi^{*t}$.

3.4 Remark. By general categorical facts, we then have the easy implication of the equivalence we want to show; namely, $\mathfrak{Th} \vdash A = B \Rightarrow A \cong_p B$. It may be worth for the reader to work out the details and construct the λ-terms which actually give the isomorphisms. Indeed, they include the "abstract" verification of cartesian closure; for example, "currying" is realized by $\lambda z.\lambda x.\lambda y.z\langle x, y\rangle$ with inverse $\lambda z.\lambda x.z(p_1\ x)(p_2\ x)$, that prove $(A \times B) \to C \cong_p A \to (B \to C)$; the term $\lambda z.\langle \lambda x.(p_1\ (zx)), \lambda x.(p_2\ (zx))\rangle$ and its inverse $\lambda z.\lambda x.\langle (p_1 z)x, (p_2\ z)x\rangle$ prove $A \to (B \times C) \cong_p (A \to B) \times (A \to C)$. The others are easily derived.

The proof of the other implication, i.e., $A \cong_p B \Rightarrow \mathfrak{Th} \vdash A = B$, roughly goes as follows. As a first step, types are reduced to "type normal forms", in a "type rewrite" system. This will eliminate terminal types and bring products at the outermost level. Then one needs to show that isomorphisms between type normal forms are only possible between types with an equal number of factors and, then, that they yield componentwise isomorphisms. This takes us to the pure typed λ-calculus (i.e., no products nor T's). Then a characterization of the invertible terms of the pure type-free calculus is easily applied in the typed case, as the invertible type-free terms happen to be typable. The syntactic structure of the invertible terms gives the result. Note then, that this implies both 1.3 and 2.1, by the discussion above (and Remark 3.1).

The axioms of $\mathfrak{Th}$ suggest the following rewrite system $\mathcal{R}$ for types (essentially $\mathfrak{Th}$ "from left to right", with no commutativity):

3.5 Definition. (Type rewriting $\mathcal{R}$). Let "$>$" be the transitive and substitutive type-reduction relation given by:

(1) $A \times T > A$
(1′) $T \times A > A$
(3) $A \times (B \times C) > (A \times B) \times C$
(4) $(A \times B) \to C > A \to (B \to C)$
(5) $A \to (B \times C) > (A \to B) \times (A \to C)$
(6) $A \to T > T$
(7) $T \to A > A$.

The system $\mathcal{R}$ yields an obvious notion of normal form for types (type normal form), i.e., when no type reduction is applicable. Note that 1, 1′, 6 and 7 "eliminate the T's", while 4 and 5 "bring ouside $\times$". It is then easy to observe that each type normal form is identical to T or has the structure $S_1 \times \ldots \times S_n$ where each S_i does not contain T nor "$\times$". We write $\mathbf{nf}(\mathbb{S})$ for the normal form of S (there is exactly one, see 3.6) and say that a normal form is non trivial if it is not T.

3.6 Proposition. *$\mathcal{R}$ is Church–Rosser and each type has a unique type normal form in $\mathcal{R}$.*

Proof. Easy exercise. □

By the implication discussed in Remark 3.4, since $\mathcal{R} \vdash S > R$ implies $\mathfrak{Th} \vdash S = R$, it is clear that any reduction $\mathcal{R} \vdash S > R$ is wittnessed by an invertible term of type $S \to R$.

3.7 Corollary. *Given types S and R, one has:*

(1) *$\mathfrak{Th} \vdash S = nf(S)$ and, thus,*
(2) *$\mathfrak{Th} \vdash S = R \Leftrightarrow \mathfrak{Th} \vdash nf(S) = nf(R)$.*

In conclusion, when $\mathfrak{Th} \vdash S = R$, either we have $nf(S) \equiv T \equiv nf(R)$, or $\mathfrak{Th} \vdash nf(S) \equiv (S_1 \times \ldots \times S_n) = (R_1 \times \ldots \times R_m) \equiv nf(R)$. A crucial lemma below shows that, in this case, one also has $n = m$.

The assertion in the corollary can be reformulated for invertible terms in a very convenient way:

3.8 Proposition (commuting diagram). *Given types A and B, assume that $F : A \to nf(A)$ and $G : B \to nf(B)$ prove the reductions to type n.f.. Then a term $M : A \to B$ is invertible iff there exist an invertible term $M' : nf(A) \to nf(B)$, such that $M = G^{-1} \circ M' \circ F$.*

Proof. $\Leftarrow$) Set $M^{-1} \equiv (G^{-1} \circ M' \circ F)^{-1} \equiv F^{-1} \circ M'^{-1} \circ G$, then M is invertible. $\Rightarrow$) Just set $M' = G \circ M \circ F^{-1}$. Then $M'^{-1} \equiv F \circ M^{-1} \circ G^{-1}$ and M' is invertible. □

A diagram easily represents the situation in the proposition:

$$\begin{array}{ccc} A & \xrightarrow{M} & B \\ {\scriptstyle F}\big\downarrow & & \big\downarrow{\scriptstyle G} \\ A_1 & & B_1 \\ \times & & \times \\ \vdots & \xrightarrow{M'} & \vdots \\ \times & & \times \\ A_n & & B_m \end{array}$$

We now state a few lemmas that should guide the reader through the basic ideas of this application of λ-calculus to Category Theory. Most technical proofs, indeed λ-calculus proofs, are omitted (and the reader should consult the reference).

Recall first that, when $\mathfrak{Th} \vdash S = R$, one has

$$nf(S) \equiv T \equiv nf(R),$$
$$\text{or } \mathfrak{Th} \vdash nf(S) \equiv (S_1 \times \ldots \times S_n) = (R_1 \times \ldots \times R_m) \equiv nf(R).$$

Notice that, in the latter case, there cannot be any occurrence of T in either type. Indeed, a non trivial type normal form cannot be provably equated to T, as it can be easily pointed out by taking a non trivial model. Thus it may suffice to look at equations such as $(S_1 \times \ldots \times S_n) = (R_1 \times \ldots \times R_m)$ with no occurrences of T and, hence, to invertible terms with no occurrences of the type constant T in their types. We can show that these terms do not contain any occurrence of $^*A$ either, for no type A, via the following lemma.

3.9 Lemma. *Let M be a term of $\lambda\beta\eta\pi^{*t}$ in n.f..*

(1) *(Terms of a product type) If $M : A \times B$, then either $M \equiv \langle M_1, M_2\rangle$, or there is $x : C$ such that $x \in FV(M)$ and $A \times B$ is a type subexpression of C.*

(2) *(Every term, whose type contains no T, has no occurrence of $^*A$ constants) Assume that in M there is an occurrence of $^*A$, for some type A. Then there is some occurrence of the type constant T in the type of M or in the type of some free variable of M.*

Proof. By induction on the structure of M. □

Note now that (the equational theory of) $\lambda\beta\eta\pi^{*t}$ is a conservative extension of (the equational theory of) $\lambda\beta\eta\pi^{*t}$. Similarly for $\lambda\beta\eta\pi^{*t}$ w.r.t. $\lambda\beta\eta^{t}$. Thus, invertibility in the extended theory, given by terms of a purer one, holds in the latter.

3.10 Proposition (Isomorphisms between type normal forms are given by terms in $\lambda\beta\eta\pi^{t}$). *Assume that S and R are non trivial type normal forms. If the closed terms M and N prove $S \cong_p R$ in $\lambda\beta\eta\pi^{*t}$, then their normal forms contain no occurrences of the constants $^*A$. (Thus, M and N are actually in $\lambda\beta\eta\pi^{t}$).*

Proof. By the previous lemma, as the terms are closed and no T occurs in their type. □

So we have factored out the class of constants $^*A$, and we restricted the attention to $\lambda\beta\eta\pi^{t}$. By the next step, we reduce the problem to the pure calculus, i.e., we eliminate pairing as well, in a sense.

3.11 Proposition (Isomorphic type normal forms have equal length). *Let $S \equiv S_1 \times \ldots \times S_m$ and $R \equiv R_1 \times \ldots \times R_n$ be type normal forms. Then $S \cong_p R$ iff $n = m$ and there exist $M_1, \ldots, M_n$; $N_1, \ldots, N_n$ such that*

$$x_1 : S_1, \ldots x_n : S_n \vdash \langle M_1, \ldots, M_n \rangle : (R_1 \times \ldots \times R_n)$$
$$y_1 : R_1, \ldots y_n : R_n \vdash \langle N_1, \ldots, N_n \rangle : (S_1 \times \ldots \times S_n)$$

with

$$M_i[\underline{x} := \underline{N}] =_{\beta\eta} y_i, \text{ for } 1 \leq i \leq n$$
$$N_j[\underline{y} := \underline{M}] =_{\beta\eta} x_j, \text{ for } 1 \leq j \leq n$$

and there exist permutations σ, π over n such that

$$M_i = \lambda\underline{u}_i \cdot x_{\sigma i}\underline{P}_i \text{ and } N_j = \lambda\underline{v}_j \cdot y_{\pi j}\underline{Q}_j$$

($\underline{M}$ is a vector of terms; substitution of vectors of equal length is meant componentwise).

Proof. (Not obvious, see reference). □

By induction, one may easily observe that terms of $\lambda\beta\eta\pi^{t}$ whose type is arrow-only belong to $\lambda\beta\eta^{t}$. Thus, one may look componentwise at terms that prove an isomorphism. The next point is to show that each component, indeed a term of $\lambda\beta\eta^{t}$, yields an isomorphism. This will be done by using a characterization of invertible terms in the pure calculus. The same result will be applied once more in order to obtain the result we aim at.

The characterization below has been given in the type-free calculus, as an answer to an old question of Church on the group of type-free terms. We follow the type-free notation, also for notational convenience.

3.12 Definition. Let M be a type-free term. Then M is a *finite hereditary permutation* (f.h.p.) iff either

(i) $\lambda\beta\eta \vdash_u M = \lambda x.x$, or

(ii) $\lambda\beta\eta \vdash_u M = \lambda z.\lambda \underline{x}.z\underline{N}_\sigma$, where if $|\underline{x}| = n$ then σ is a permutation over n and $z\underline{N}_\sigma = zN_{\sigma_1}N_{\sigma_2}\ldots N_{\sigma_n}$, such that, for $1 \le i \le n$, $\lambda x_i.N_i$ is a finite hereditary permutation.

For example, $\lambda z.\lambda x_1.\lambda x_2.zx_2x_1$ and $\lambda z.\lambda x_1.\lambda x_2.zx_2(\lambda x_3.\lambda x_4.x_1x_4x_3)$ are f.h.p.'s. The structure of f.h.p.'s is tidily desplayed by Böhm-trees. The *Böhm-tree* of a term M is (informally) given by:

$BT(M) = W$ if M has no head normal form

$$BT(M) = \begin{array}{c} \lambda x_1\ldots x_n.y \\ /\ldots\backslash \\ BT(M_1) \quad BT(M_p) \end{array} \qquad \text{if } M =_\beta \lambda x_1\ldots x_n.yM_1\ldots M_p$$

(see references).

It is easy to observe that a $BT(M)$ is finite and Ω-free iff M has a normal form. Then one may look at f.h.p.'s as Böhm-trees, as follows:

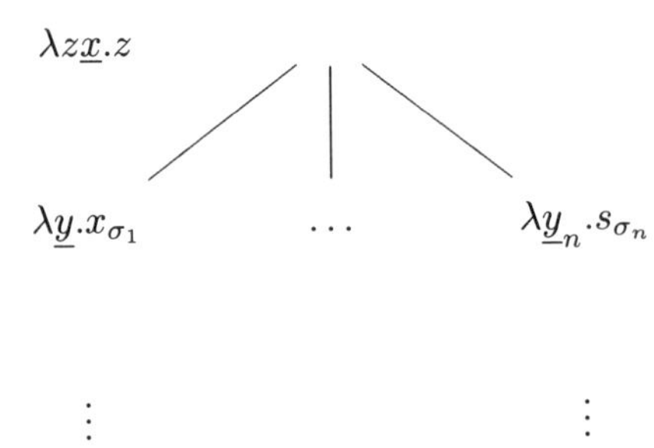

and so on, up to a finite depth (note that $\underline{y}_i$ may be an empty string of variables). Clearly, the f.h.p.'s are closed terms and possess a normal form. In particular, exactly the abstracted variables at level $n+1$ appear at level $n+2$, modulo some permutation of the order (note the special case of z at level 0). The importance of f.h.p.'s arises from the following classic theorem of λ-calculus. (Clearly, the notion of invertible term given in 3.3 easily translates to type-free λ-calculi).

3.13 Theorem. *Let M be an untyped term possessing normal form. Then M is $\lambda\beta\eta$-invertible iff M is a f.h.p.*

Recall now that all typed terms possess a (unique) normal form (see references). Let then M be a typed λ-term and write $e(M)$ for the *erasure* of M, i.e. for M with all type labels erased.

Remark. Observe that the erasures of all axioms and rules of the typed lambda calculus are themselves axioms and rules of the type-free lambda calculus. Then, if M and N are terms of $\lambda\beta\eta^t$ and $\lambda\beta\eta^t \vdash M = N$, one has $\lambda\beta\eta \vdash e(M) = e(N)$. Thus, in particular, if $M : \sigma \to \tau$ and $N : \tau \to \sigma$ are invertible terms in $\lambda\beta\eta^t$, $e(M)$ and $e(N)$ are f.h.p.'s.

Exercise. Show that the f.h.p.'s are typable terms (Hint: Just follow the inductive definition and give z, for instance, type $A1 \to (A2 \cdots \to B)$, where the Ai's are the types of the N_{σ_i}.) Then, by the a small abuse of language, we may talk also of typed f.h.p.'s. Observe that these are exactly the typed invertible terms in Definition 3.3.

The first application of 3.13 we need is the following.

3.14 Proposition. *Let $M_1, \ldots, M_n$, $N_1, \ldots, N_n$ and permutation σ be as in Lemma 2.11. Then, for all i, $\lambda x_{\sigma_i}.M_i : S_{\sigma_i} \to R_i$ and $\lambda y_i.N_{\sigma_i} : R_i \to S_{\sigma_i}$ are invertible terms.*

Proof. For a suitable typing of the variables it is possible to build the following terms of $\lambda\beta\eta^t$ (we erase types for convenience):

$$M = \lambda z x_1 \ldots x_n . z M_1 \ldots M_n$$
$$N = \lambda z y_1 \ldots y_n . z N_1 \ldots N_n .$$

It is an easy computation to check, by the definition of the M_i's and of the N_i's, that M and N are invertible. Moreover, they are (by construction) in normal form, thus, by Theorem 3.13, M and N are f.h.p.'s. This is enough to show that every M_i has only one occurrence of the x_i's (namely x_{σ_i}); similarly for the N_i's.

Thus we obtain

$$M_i[\underline{x} := \underline{N}] \equiv M_i[x_\sigma(i) := N_\sigma(i)] =_{\beta\eta} y_i, \text{ for } 1 \le i \le n$$
$$N_i[\underline{y} := \underline{M}] \equiv N_i[y_\pi(i) := M_\pi(i)] =_{\beta\eta} x_i, \text{ for } 1 \le i \le n$$

and, hence, for each i, $\lambda x_\sigma(i).M_i : S_\sigma(i) \to R_i$ and $\lambda y_i.N_\sigma(i) : R_i \to S_\sigma(i)$ are invertible. □

As a result of the work done so far, we can then focus on invertible terms whose types contain only "$\to$" i.e., investigate componentwise the isomorphisms of type normal forms. Of course, these isomorphisms will be given just by a fragment of theory $\mathfrak{Th}$.

Call $\mathcal{S}$ the subtheory of $\mathfrak{Th}$ given by just one proper axiom (plus the usual axioms and rules for "="), namely

$$\text{(swap)} \qquad A \to (B \to C) = (B \to (A \to C)).$$

$\mathcal{S}$ is a subtheory of $\mathfrak{Th}$ by axioms 2 and 4 of $\mathfrak{Th}$.

3.15 Proposition. *Let A, B be type expressions with no occurences of T nor $\times$. Then*

$$A \cong_p B \Rightarrow S \vdash A = B.$$

Proof. Suppose $A \cong_p B$ via M and N. As usual, we may assume without loss of generality that M and N are in normal form. By Lemma 3.9 and the observation after 3.11, M and N actually live in $\lambda\beta\eta^t$ and, by Theorem 3.13, they are f.h.p.'s. We prove $S \vdash A = B$ by induction on the depth of the Böhm-tree of M.

Depth 1: $M \equiv \lambda z : C.z$. Thus $M : C \to C$, and $S \vdash C = C$ by reflexivity.
Depth $n+1$: $M \equiv \lambda z : E.\lambda \underline{x} : \underline{D}.z\underline{N}_\sigma$. Recall $z\underline{N}_\sigma = zN_{\sigma_1} \ldots N_{\sigma_n}$ where if the ith abstraction in $\lambda \underline{x} : \underline{D}$ is $\lambda x_i : D_i$ then the erasure of $\lambda x_i : D_i.N_{\sigma_i}$ is a f.h.p.. Thus $\lambda x_i : D_i.N_{\sigma_i}$ give (half of) a provable isomorphism from D_i to some F_i. Hence the type of N_{σ_i} is F_i. In order to type check, we must have $E = (F_{\sigma_1} \to \cdots \to F_{\sigma_n} \to B)$ for some B. Thus the type of M is $(F_{\sigma_1} \to \cdots \to F_{\sigma_n} \to B) \to (D_1 \to \cdots \to D_n \to B)$. By induction, since the height of the Böhm tree of (of the erasure of) each $\lambda x_i : D_i.N_{\sigma_i}$ is less than the height of the Böhm tree of M, one has $\mathcal{S} \vdash D_i = F_i$ for $1 \leq i \leq n$. By repeated use of the rules for "=", we get

$$\mathcal{S} \vdash (F_{\sigma_1} \to \cdots \to F_{\sigma_n} \to B) = (D_{\sigma_1} \to \cdots \to D_{\sigma_n} \to B).$$

Hence it suffices to show

$$\mathcal{S} \vdash (D_{\sigma_1} \to \cdots \to D_{\sigma_n} \to B) = (D_1 \to \cdots \to D_n \to B).$$

This is quite simple to show by repeated use of axiom (swap) above in conjunction with the rules. □

Clearly, also the converse of Proposition 3.15 holds, since the "$\Leftarrow$" part in 3.15 is provable by a fragment of the proof hinted in 3.4. Thus one has:

$$\mathcal{S} \vdash A = B \Leftrightarrow A \cong_p B \text{ by terms in } \lambda\beta\eta^t.$$

The result we aim at is just the extension of this fact to $\mathfrak{Th}$ and $\lambda\beta\eta\pi^{*t}$.

3.16 Main Theorem. $S \cong_p R \Leftrightarrow \mathfrak{Th} \vdash S = R$.

Proof. In view of 3.4, we only need to prove $S \cong_p R \Rightarrow \mathfrak{Th} \vdash S = R$. As we know, this is equivalent to proving $nf(S) \cong nf(R) \Rightarrow \mathfrak{Th} \vdash nf(S) = nf(R)$.

Now, by Proposition 3.11, for $nf(S) \equiv (S_1 \times \ldots \times S_n)$ and $(R_1 \times \ldots \times R_m) \equiv nf(R)$, we have

$$\begin{aligned} nf(S) \cong nf(R) \Rightarrow n = m &\text{ and there exist } M_1, \ldots, M_n, N_1, \ldots, N_n \\ &\text{and a permutation } \sigma \text{ such that } \lambda x_{\sigma i}.M_i : S_{\sigma i} \to R_i \\ &\text{and } \lambda y_i.N_{\sigma i} : R_i \to S_{\sigma i}. \end{aligned}$$

By 3.14, these terms are invertible, for each i. Thus, by .15, $\mathcal{S} \vdash R_i = S_{\sigma i}$ and, hence, by the rules, $\mathfrak{Th} \vdash S = R$. □

This concludes the proof of the main theorem.

Remark. By going back at the theme in the Intermezzo, one could hope that the PER model constructed over the term model of the type-free λ-calculus may allow to extend Theorem 2.1 to dinatural isomorphisms, in the sense of 2.13. Namely, that $\mathfrak{Th}$ gives exactly the dinatural isomorphisms of types, as this specific PER model allows the interpretation of terms as dinatural transformations and its morphism, after all, are (given by type-free) terms. Unfortunately, the type structure PER, even over a type-free term model, is much richer than the term model of *typed* λ-calculus, which we used in our proof.

References

Asperti A., Longo G. [1986], *Categories of partial morphisms and the relation between type-structures*, Mathematical Problems in Computation Theory, Warsaw, December 1985; Banach Center Publications, **21, 1987**.

Asperti A., Longo G. [1991], *Categories, Types and Structures: An Introduction to Category Theory for the Working Computer Scientist*, M.I.T. Press, to appear.

Bainbridge E., Freyd P., Scedrov A., P.J. Scott [1990], *Functorial polymorphism*, Theoretical Computer Science **70**, 35–64.

Bruce K., Di Cosmo R., Longo G. [1990], *Provable isomorphisms of types*, LIENS, technical report 90-14. (to appear: MSCS).

Bruce K., Longo G. [1985], *Provable isomorphisms and domain equations in models of typed languages*, (Preliminary version) 1985 A.C.M. Symposium on Theory of Computing (STOC 85), Providence (R.I.).

Bruce K., Longo G. [1990], *Modest models for inheritance and explicit polymorphism*, Info& Comp. **87,1-2.**

Cardelli L., Longo G. [1990], *A semantic basis for Quest*, DIGITAL, System Research Center, Report 55 (preliminary version: Conference on Lisp and Funct. Progr., Nice, June 1990) (to appear: Journal of Functional Programming).

Longo G. [1988], *From type-structures to type theories lecture notes*, Spring semester 1987/8, Computer Science Dept., C.M.U.

Longo G. [1990], *The new role of mathematical logic: a tool for computer science*, Information Sciences, special issue on "Past, Present and Future of C.S.", to appear.

Martin C.F. [1972], *Axiomatic bases for equational theories of natural numbers*, (abstract) Notices Am. Math. Soc. **19,7**, 778.

Mints G.E. [1977], *Closed categories and the theory of proofs*, Translated from Zapiski Nauchnykh Seminarov Leniongradslogo Otdeleniya Matematicheskogo Instituta im V.A. Steklova AN SSSR, **68**, 83–114.

Mints G.E. [198?], *A simple proof of the coherence theorem for cartesian closed categories*, Bibliopolis (to appear in translation from russian).

Rittri M. [1990a], *Using types as search keys in function libraries*, Journal of Functional Programming, Vol. 1 (1990).

Rittri M. [1990b], *Retrieving library identifiers by equational matching of types*, X Int. Conf. on Automated Deduction, Kaiserslautern, Germany (Proceedings in LNCS, Spriger-Verlag).

Soloviev S.V. [1983], *The category of finte sets and Cartesian closed categories*, J. Soviet Mathematics 22 **3**, 1387–1400.

Dipartimento di Informatica, Universitá di Pisa and
LIENS, Ecole Normale Supérieure, 45 R. D'Ulm, 75005 PARIS

Work partly supported by CNR (Stanford Grant #89.00002.26). Research at MSRI supported in part by NSF Grant DMS-8505550.

LOGICS FOR TERMINATION AND CORRECTNESS OF FUNCTIONAL PROGRAMS

SOLOMON FEFERMAN

1. INTRODUCTION

In the literature of logic and theoretical computer science, types are considered both semantically as some kinds of mathematical objects and syntactically as certain kinds of formal terms. They are dealt with in one way or the other, or both, in all programming languages, though most attention has been given to their role in functional programming languages. One difference in treatment lies in the extent to which types are explicitly represented in programs themselves and, further, can be passed as values. Familiar examples, more or less at the extremes, are provided by LISP—an untyped language—and ML—a polymorphic typed language. The aim of this paper (and the more detailed ones to follow) is to provide a logical foundation for the use of type systems in functional programming languages and to set up logics for the termination and correctness of programs relative to such systems. The foundation provided here includes LISP and ML as special cases. I believe this work should be adaptable to other kinds of programming languages, e.g. those of imperative style.

The approach taken here is an extension of that in [F 1990] [1], which itself grew out of a general approach to varieties of explicit (constructive and semi-constructive) mathematics initiated in [F 1975] by means of certain *theories of variable type* (VT). The paper [F 1990] should be consulted initially for background and motivation and, further on, for certain specifics; however, much of this paper can be read independently. As this is in the nature of a preliminary report, the intent is to explain the leading ideas and scope of the work, with various details to be spelled out in subsequent publications.

2. STRENGTH OF FORMAL SYSTEMS

In my work on explicit mathematics I have been at pains to show that relatively weak formal systems, as measured in proof-theoretical terms, suf-

[1] References to my papers in the bibliography are by the letter 'F'; otherwise authors' names are indicated in full.

fice to account for mathematical practice. In particular, the last section of [F 1990] was devoted to showing that systems no stronger than Peano Arithmetic (PA) account for the use of polymorphic typed programming languages in practice. This is opposite to the direction of much current work on polymorphism which requires, *prima-facie,* extremely strong principles proof-theoretically, at least that of analysis (= 2nd order arithmetic, PA_2). But the latter in turn flies in a direction opposite to that of computational practice, where the primary concern is to keep the complexity of algorithms low. In this respect it is more natural to measure strength of formal systems used, in terms of the complexity of their provably computable functions.

3. Theories of Variable Type: Main Results

What is advanced here is not a single formal system but a group of systems or *logics,* among which one makes choices for different purposes. These come in two levels of logical complexity, the higher of which uses the full language of *quantificational 2nd order logic* (QL) and the lower of which is *quantifier-free* or *propositional* (PL) in form. Additionally, we provide a choice among *user types* (U-Types) indicated by grades $i(= 0, 1, 2, \dots)$ as subscripts, thus yielding QL_i, PL_i, resp. U-types are distinguished from those which describe the general or global behaviour of algorithms (across the whole range of user types) and are thus called G-types.[2] The latter include *polymorphic types* of various kinds; a further refinement to be described depends on whether these are represented explicitly (Λ) or not (λ) in the programs themselves, thus leading to logics $QL_i(\Lambda), QL_i(\lambda), PL_i(\Lambda), PL_i(\lambda)$.

The base designation QL, PL for the logics used should be considered as temporary. A more informative designation would be something like Q-VT and QF-VT, but happier choices should be possible. The logics $QL_1(\Lambda)$ and $QL_1(\lambda)$ are improved versions of the theories EOC_Λ and EOC_λ, resp., of [F 1990]; all the other logics introduced here are new.

If T is a formal logic for which we can talk about the provably computable functions on the type N of natural numbers, we write $\mathrm{Comp}_N(T)$ for this class of functions. Then if T_1, T_2 are two such logics, we write $T_1 \equiv T_2$ if $\mathrm{Comp}_N(T_1) = \mathrm{Comp}_N(T_2)$. In all our examples this relation coincides with the relation of proof-theoretical equivalence (cf. [F 1988] p.368). The main results are that:

(1) (i) $QL_0(\Lambda) \equiv QL_0(\lambda) \equiv PL_0(\Lambda) \equiv PL_0(\lambda) \equiv PRA$

---

[2]This is related to the separation of a typing system for XML (Explicit ML) into two universes U_1 and U_2 in [Mitchell and Harper 1988], with U like U_1 and G like U_2 in certain respects.

(ii) $QL_1(\Lambda) \equiv QL_1(\lambda) \equiv PL_1(\Lambda) \equiv PL_1(\lambda) \equiv PA$,
(iii) $QL_2(\Lambda) \equiv QL_2(\lambda)(\equiv (?)PL_2(\Lambda) \equiv PL_2(\lambda)) \equiv RA_{<\epsilon_0}$,

where PRA = Primitive Recursive (Skolem) Arithmetic, PA = Peano Arithmetic and $RA_{<\epsilon_0}$ is Ramified Analysis in all levels $< \epsilon_0$. Characterizations of the provably computable functions of PRA and PA are well known as follows:

(2) (i) $Comp_N(PRA)$ = the class of primitive recursive functions
(ii) $Comp_N(PA)$ = the functions of type $N \to N$ generated by the primitive recursive functionals of [Gödel 1958].

The characterization of $Comp_N(RA_{<\epsilon_0})$ is more recondite and will not be given here (cf. [F 1977]). I expect that the work here can be extended to natural logics below QL_0, PL_0 to obtain match-ups of the provably computable functions with levels of the polynomial complexity hierarchy.

For simplicity, the main pattern of our work will be explained for the results (1)(ii), followed by an outline of how it is modified for (1)(i) and a bare indication in the case of (1)(iii) (part of which is still conjectural).

4. Syntax of $QL_1(\lambda)$

We begin with the quantificational system for $i = 1$ in which the behaviour of individual terms relative to types can only be inferred ("implicit typing") rather than carried explicitly as in $QL_1(\Lambda)$, below.

In all the logics considered, there is an official infinite list of *individual variables* indicated by the meta-variables x, y, z with or without subscripts and there is an official infinite list of *type variables* indicated by X, Y, Z with or without subscripts. In general s, t, t_1, t_2 range over *individual terms* and S, T, T_1, T_2 over *general type terms* (G-Types), and ϕ, ψ over (well-formed) *formulas,* in each logic, but what these syntactic classes are will vary from logic to logic.

In $QL_1(\lambda)$ the **individual terms** are just those of one of the following forms:

$x, y, z, \ldots$
$0, c_n (n > 0)$
$st, \lambda x.t$
$p(t_1, t_2), p_i(t)$ $(i = 1, 2)$
$sc(t), pd(t)$
$eq(s, t)$, $cond(s, t_1, t_2)$
$rec(t)$.

Then the G-**Type terms** are just those of one of the following forms:

$X, Y, Z, \ldots$
N
$\{x \mid \phi\}$

and the **formulas** are just those of one of the following forms:

$t \downarrow, s = t, s \in T, s \dot{\in} t$
$\neg\phi, \phi \wedge \psi, \phi \vee \psi, \phi \Rightarrow \psi$[3]
$\forall x \phi, \exists x \phi, \forall X \phi, \exists X \phi.$

Note that the type terms and formulas are given by a simultaneous inductive definition.

4.1. Substitution and other syntactic notation. We write $t[s/x]$ for the result of substituting s for x at all its free occurrences in t, renaming bound variables if necessary to avoid collisions. Often we write $t[x]$ for t which may contain x at some free occurrences, and then write $t[s]$ for $t[s/x]$. The notation is extended in the very same way to simultaneous substitution, i.e. $t[x_1, \ldots, x_n]$ for t and then $t[s_1, \ldots, s_n]$ for $t[s_1/x, \ldots, s_n/x_n]$ (simultaneous substitution, renaming bound variables if necessary). Similar notation applies to type terms and formulas, e.g. $T[X], T[S], \phi[x], \phi[s]$, etc.

We write $x(X)$ for a sequence of like variables $x_1, \ldots, x_m (X_1, \ldots, X_m)$. $\ulcorner \cdot \urcorner$ associates with each syntactic object its Gödel number. We shall also use the following abbreviations:

(i) $(s \simeq t) := [s \downarrow \vee t \downarrow \Rightarrow s = t]$
(ii) $(s, t) := p(s, t)$
(iii) $t' := sc(t)$
(iv) $S \subseteq T := \forall x[x \in S \Rightarrow x \in T]$
(v) $S = T := \forall x[x \in S \Leftrightarrow x \in T]$
(vi) $Rep(t, T) := \forall x[x \dot{\in} t \Leftrightarrow x \in T]$
(vii) $Rep(T) := \exists z \, Rep(z, T)$
(viii) $Rep(t) := \exists X \, Rep(t, X)$
(ix) $Q\text{-}Typ(T) := \exists X[X = T]$

We read $Rep(t, T)$ as '*t represents T*', $Rep(T)$ as '*T is representable*' and $Rep(t)$ as '*t is a (type) representation*'. In (ix), the 'Q' is used to indicate that T is in the range of the type quantifier.

[3]We write $(\phi \Rightarrow \psi)$ instead of $(\phi \to \psi)$ to avoid confusion with arrow types; the same implication symbol may be used informally. (Similarly for $\Leftrightarrow$.)

5. Informal Interpretation

The individual variables are conceived of as ranging over a universe V of computationally amenable objects, closed under pairing. Some of these objects may themselves represent programs, others may represent types (which can then be passed as values). 0 is a specific element of V which is supposed to satisfy $x' \neq 0$, where $x' = sc(x)$; we write 1 for $0'$. pd is supposed to be the predecessor operation, $pd(x') = x$. The c_n's will be used to represent certain types in a way to be explained later. The term st is used to denote the value of a program (given by) s at input t, if the computation at that input terminates, i.e. if $(st)\downarrow$; $\lambda x.t$ or $\lambda x.t[x]$ represents the program whose value at each input x is given by $t[x]$ when defined. $p(t_1, t_2)$ is the pair of t_1 and t_2, also simply written (t_1, t_2), and when $t = (t_1, t_2)$, $p_i(t)$ is t_i (otherwise, p_i is assumed to take some conventional value such as 0). When s, t, t_1, t_2 are all defined, $eq(s,t) = 1$ or 0 according as $s = t$ or not, and $cond(s, t_1, t_2) = t_1$ or t_2 according as $s = 1$ or not. Here we thus identify 1, 0 with the Boolean values *true, false,* resp. Finally, $rec(t)$ is supposed to denote a program for defining a function given by a recursion equation $fx \simeq tfx$.

The type terms denote subclasses of V. However, $X, Y, Z, \ldots$ are *not* to be thought of as ranging over arbitrary subclasses of V, but only those denoted by the U-type terms, as defined below. The intended interpretation of N is the least class containing 0 and closed under successor. $\{x \mid \phi[x]\}$ is the class of x in V satisfying ϕ, for any formula ϕ (and any given assignment to the free variables of ϕ other than x).

The formula $(t\downarrow)$, associated with each individual term t, is considered to hold at an assignment in V to the free variables of t if the value of t is defined there. Thus $(x\downarrow)$ holds for all x in V. The relation $s = t$ is taken to hold (at an assignment) just in case both s, t are defined and have identical values. $s \in T$ holds if s is defined and (its value) belongs to the class denoted by T. The relation $s \dot{\in} t$ holds if s, t are defined and the value of t is an object in V representing a type T, e.g. by means of a Gödel number, and the value of s belongs to T. The meaning of the propositional operations $\neg, \wedge, \vee, \Rightarrow$ and quantifiers $\forall x, \exists x$, where 'x' ranges over V, is as usual in model-theoretic semantics for classical logic. (But the logics dealt with may also be considered in intuitionistic logic.) The meaning of $\forall X(\ldots)$ and $\exists X(\ldots)$ is only determined when we tell what the U-types are, since as noted above, the type variables range only over the extensions of these. For the logic $QL_1(\lambda)$ this is fixed next.

6. User-Type Terms

The *U-Type terms* of $QL_1(\lambda)$ are defined by a simultaneous inductive definition with the class of *pure elementary formulas.* The basic point is that the internal membership relation $\dot{\in}$ does not appear in pure formulas and that type quantification is not applied in elementary formulas. We use $A, B, \ldots$ to range over pure elementary formulas, thus given with U-terms as follows:

U-Type terms

$X, Y, Z, \ldots$
N
$\{x \mid A\}$

Pure elementary formulas
$t\downarrow, s = t, s \in T$ (for T a U-Type term)
$\neg A, A \wedge B, A \vee B, A \Rightarrow B$
$\forall x A, \exists x A.$

As we have already noted, the informal interpretation is that type variables range over the (extensions of) U-Types, not the more general G-Types. Moreover, only U-Type terms will be assumed to be represented by objects of the universe and thus can be passed (via their representations) as inputs to programs.

7. Logic of Partial Terms

The logic of partial terms, LPT, which governs the use of the formulas $t\downarrow$, is derived from [Beeson 1985], ch.V.1; cf. also [F 1990] p.107. The following are axioms of LPT:

(1) (i) $x\downarrow$
(ii) $c\downarrow$ for each individual constant symbol
(iii) $F(s_1, \ldots, s_n)\downarrow \Leftrightarrow (s_1\downarrow) \wedge \cdots \wedge (s_n\downarrow)$ for $F = p, p_1, p_2, sc, pd, eq$, *cond* and *rec*, and n as appropriate in each case.
(iv) $(st)\downarrow \Rightarrow (s\downarrow) \wedge (t\downarrow)$
(v) $(\lambda x.t)\downarrow$
(vi) $(s = t) \Rightarrow (s\downarrow) \wedge (t\downarrow)$, $(s \dot{\in} t) \Rightarrow (s\downarrow) \wedge (t\downarrow)$
(vii) $(s \in T) \Rightarrow (s\downarrow)$

(2) The propositional part of our logic may be classical or intuitionistic. Our models verify classical logic, while realizability interpretations for intuitionistic logic give useful closure conditions for that case. The main results on strengths of the logics $QL_i(\lambda)$, $PL_i(\lambda)$ (etc.) described in §3 above hold whichever basic logic we take.

(3) The axioms and rules for quantification over individuals are:

(i) $\forall x\phi[x] \wedge (t\downarrow) \Rightarrow \phi[t]$, (ii) $\dfrac{\psi \Rightarrow \phi[x]}{\psi \Rightarrow \forall x\phi[x]}$, for x not free in ψ,

and their duals for $\exists x$.

(4) The axioms and rules for quantification over types are:

(i) $\forall X\phi[X] \Rightarrow \phi[Y]$ (ii) $\dfrac{\psi \Rightarrow \phi[X]}{\psi \Rightarrow \forall X\phi[X]}$ for X not free in ψ,

and their duals for $\exists X$.

Remarks.

1°. The axiom (1)(iii) for *cond*:

$$cond(s,t_1,t_2)\downarrow \Leftrightarrow (s\downarrow) \wedge (t_1\downarrow) \wedge (t_2\downarrow)$$

can be replaced by

$$cond(s,t_1,t_2)\downarrow \Leftrightarrow (s=1 \wedge (t_1\downarrow)) \vee (s=0 \wedge (t_2\downarrow)) .$$

2°. The axiom (1)(iv) corresponds to "call-by-value". There is a question whether it can be weakened to

$$(st)\downarrow \Rightarrow (s\downarrow)$$

for the various applications below, in order to allow for "call-by-name".

3°. *N.B.* that we *do* not in general have $\forall X\phi[X] \Rightarrow \phi[T]$ for G-Type terms T. However, we will be able to derive from the axioms of $QL_1(\lambda)$

$$(*) \qquad \forall X\phi[X] \Rightarrow \phi[T]$$

for each U-Type T, (and its dual for $\exists$). This comes about through the following observations.

(i) Type terms have only extensional occurrences in arbitrary formulas ϕ, i.e. only to the right of $\in$. Thus we may prove by induction on ϕ,

$$T_1 = T_2 \Rightarrow (\phi[T_2] \Leftrightarrow \phi[T_2]) .$$

(Recall that $T_1 = T_2 := \forall x[x \in T_1 \Leftrightarrow x \in T_2]$.)

(ii) We may thus conclude

$$\forall X\phi[X] \wedge Q\text{-}Typ(T) \Rightarrow \phi[T] .$$

For suppose $\forall X\phi[X]$ and $Q\text{-}Typ(T)$, then by definition of the latter (§4), $Y = T$ for some Y. Since from $\forall X\phi[X]$ we may conclude $\phi[Y]$, we have $\phi[T]$ by (i).

(iii) The axioms of $QL_1(\lambda)$ (namely the Representation Axiom VII) will allow us to conclude

$$Q\text{-}Typ(T)$$

for each U-Type term T.

Combining (ii) and (iii) will then give us $(*)$ above.
4°. The system $QL_1(\lambda)$ and its relatives described in this paper thus violate Quine's dictum "To be is to be the value of a variable". Whatever the philosophical significance of that, it has no formal necessity.

8. Axioms of the Logic $QL_1(\lambda)$

I. Abstraction-reduction

$(\lambda x.t[x])y \simeq t[y]$

II. Pairing, projections, successor, predecessor

(i) $p_1(x,y) = x \wedge p_2(x,y) = y$
(ii) $x' \neq 0 \wedge pd(x') = x$

III. Equality and conditional definition

(i) $eq(x,y) = 0 \vee eq(x,y) = 1$
(ii) $x = y \Leftrightarrow eq(x,y) = 1$
(iii) $cond(1,y,z) = y \wedge cond(0,y,z) = z$

IV. Recursion

(i) $y = rec(z) \Rightarrow yx \simeq zyx$

V. Natural numbers

(i) $0 \in N \wedge (x \in N \Rightarrow x' \in N)$
(ii) $0 \in X \wedge \forall x(x \in X \Rightarrow x' \in X) \Rightarrow N \subseteq X$

VI. General type construction

$y \in \{x \mid \phi[x]\} \Leftrightarrow \phi[y]$, ϕ arbitrary

VII. Representation

(i) $Rep(T) \Leftrightarrow Q\text{-}Typ(T)$
(ii) $Rep(N)$
(iii) $Rep(z,Z) \Rightarrow c_{\ulcorner A \urcorner}(y,z)\downarrow \wedge Rep(c_{\ulcorner A \urcorner}(y,z), \{x \mid A[x,y,Z]\})$, for each pure elementary formula A with free variables among $x, y = y_1,\ldots,y_n$ and $Z = Z_1,\ldots,Z_n$, where $z = z_1,\ldots,z_n$.

Remarks.

1°. By §7(3) we can derive from Axiom I:

$$(s\downarrow) \Rightarrow (\lambda x.t[x])s \simeq t[s]\ .$$

2°. In place of II(ii), we could assume $(x,y) \neq 0$ and define $sc(x) := (x,0)$ and $pd(x) := p_1(x)$ to satisfy the conditions there. (This was done in [F 1990], for example.)

3°. Some metatheorems in intuitionistic logic require that Axiom III(i) for eq be restricted to $x, y \in N$. Our main results are insensitive to such a modification.

4°. An operation rec can be defined as a λ-term so that Axiom IV is derivable from I, the definition being given as a variant of the Curry Y-combinator; this is demonstrated in [F 1990] p.118. However, many programming languages in practice include a form of rec directly.

5°. Assumptions of the form VI are usually called *Comprehension Axioms* CA for classes. Thus we shall also refer to the schema VI as GCA and to its restriction to pure elementary formulas as ECA.

6°. To understand the Representation Axiom VII, recall the abbreviations from §4:

$$Rep(t,T) := \forall x(x\dot{\in}t \Leftrightarrow x \in T),\ Rep(T) := \exists x Rep(x,T)$$
$$\text{and } Q\text{-}Typ(T) := \exists X(X = T).$$

Thus VII(i) expresses that the range of the type variables are exactly the internally representable types. Now Axiom VII(ii), (iii) tell us that the representable types, and hence the terms satisfying the Q-Typ predicate, are closed under ECA. It follows that every U-Type term T satisfies Q-Typ. Hence we have, as promised in Remark 4° of §7, the principle

$$(*) \qquad \forall X\phi[X] \Rightarrow \phi[T]$$

holds for every such term.

7°. Returning now to Axiom V for the natural numbers, we infer immediately the principle of induction for U-Types T,

$$0 \in T \wedge \forall x(x \in T \Rightarrow x' \in T) \Rightarrow N \subseteq T\ .$$

Then by ECA applied to $T = \{x \mid A[x]\}$ we have

$$A[0] \wedge \forall x(A[x] \Rightarrow A[x']) \Rightarrow \forall x(x \in N \Rightarrow A[x])$$

for each pure elementary formula A.

8°. The relation Rep plays here the same role as the "naming" relation Nam in [Jäger 1988], the only difference being that in the latter, Nam is taken as a basic relation, whereas here it is defined in terms of $\dot{\in}$. One can, of course, define $\dot{\in}$ in terms of the naming relation by $(x\dot{\in}z) := \exists Z[Nam(z,Z) \wedge x \in Z]$. Thus the two approaches are essentially equivalent. However, certain proof-theoretical arguments are simpler under the present approach, namely those having to do with the eliminability of Axiom VI (GCA).

9°. In [F 1975] up to [F 1990] I made use, alternatively, of an "Ontological Axiom" which literally identified types (or classes) with certain elements of the universe of individuals in the form:

Ont $$\forall X \exists x(X = x)\ .$$

(Obviously this requires a mixed 2-sorted logic.) Under the assumption of *Ont*, types must be treated as intensional objects. In the present sort of axiomatization, following [Jäger 1988], types may be considered extensionally; it is only their internal representation by individuals which is intensional. The representation (or naming) axioms are perhaps more natural conceptually than the ontological axiom, but the latter is sometimes more convenient, as we shall see below.

9. Semantics of $QL_1(\lambda)$

(a) **The applicative part.** Note that the axioms for $QL_1(\lambda)$ divide into those which concern individual term constructions only, namely I-IV, and those which concern type constructions, namely V-VII. By an *applicative pre-model for* $QL_1(\lambda)$ we mean a structure of the form

(1) $\mathcal{A} = \langle V, \cdot, P, P_1, P_2, Sc, Pd, Eq, Cond, Rec, 0\rangle$

and a term evaluation map

(2) $\|t\|_\alpha$ for each $\alpha : Free\,Var(t) \to V$,

for which $\cdot$ is a partial operation on $V \times V$ to V (the *application* operation of $\mathcal{A}$) and $\|\text{-}\|$ satisfies the obvious conditions for evaluating the individual terms of $\mathcal{L}(QL_1(\lambda))$, e.g. $\|st\|_\alpha = \|s\|_\alpha \cdot \|t\|_\alpha$, $\|\lambda x.t\|_\alpha \cdot a \simeq \|t\|_{\alpha*(x\mapsto a)}$, $\|p(s,t)\|_\alpha = P(\|s\|_\alpha, \|t\|_\alpha)$, etc. Note that $\|t\|_\alpha$ need not always be defined, since application is partial.

We call $\mathcal{A}$ *an applicative model* (under an evaluation map (2)) if it satisfies each of the Axioms I-IV of $QL_1(\lambda)$. A great variety of such models may be constructed by the methods of [F 1979], p.199–204, starting with models of the partial untyped combinatory calculus of the form

(3) $\mathcal{C} = \langle V, \cdot, k, s, f_P, f_{P_1}, \ldots, 0\rangle$

where

(4) $kab = a,\ sab\downarrow,\ sabc \simeq ac(bc)$

and for each of the operations $F = P, P_1$, etc. we have an associated element f_F of V with $f_F a_1 \cdots a_n = F(a_1, \ldots, a_n)$ $(a, b, c, a_1, \ldots, a_n \in V$, and application associated to the left). Evaluation of λ-terms is then defined using the usual translation of the λ-calculus into the combinatory calculus.

(b) **Expansion to standard models.** Any applicative model $\mathcal{A} = \langle V, \cdot, \dots \rangle$ can be expanded to a model of the form

(5) $\mathcal{M} = \langle V, \cdot, \dots, N, \mathcal{D}, \in, Rep, \langle c_n \rangle_{n>0} \rangle$

of all of $QL_1(\lambda)$ which is standard for N as follows. First take N to be the least subclass of V containing 0 and closed under the successor operation Sc of $\mathcal{A}$. Then take $\mathcal{D}$ to be the collection of all subclasses of $\langle V, \cdot, \dots, N \rangle$ which are first-order definable from parameters in V; the type variables $X, Y, Z, \dots$ are now interpreted as ranging over $\mathcal{D}$. Note that if $A[x, y, Z]$ is any pure elementary formula and α is an assignment to y in V and β is any assignment to Z in $\mathcal{D}$, then $\{a \mid A[x, y, Z]$ is satisfied at $\alpha * (x \mapsto a), \beta\}$ is again in $\mathcal{D}$.

Next, choose $c_n \in V$ so that $c_n \cdot a = (n, a)$; this gives sense to $c_{\ulcorner A \urcorner}(y, z)$ as $(\ulcorner A \urcorner, (y, z))$. We generate Rep as the least relation satisfying

(6) $Rep(z, Z) \Rightarrow Rep(c_{\ulcorner A \urcorner}(y, z), \{x \mid A[x, y, Z]\})$,
for each pure elementary A, at any assignment α, β as above.

By construction, no $d = c_{\ulcorner A \urcorner}(y, z)$ represents two different classes. We may then define $(x \dot{\in} z) := \exists Z[Rep(z, Z) \wedge x \in Z]$. We now have arrived at an expansion of $\mathcal{A}$ to a structure $\mathcal{M}$ of the form (5) which is checked to satisfy all of the axioms of $QL_1(\lambda)$ except GCA(VI). But now, for that, notice that we have obtained at the same time a semantics for arbitrary formulas ϕ of the language $QL_1(\lambda)$[4], and we can simply interpret $\{x \mid \phi[x, y, Z]\}$ at any given assignment α, β (as above) to be the class of all $a \in V$ satisfying ϕ under the assignment $\alpha * (x \mapsto a), \beta$. By definition this, then, verifies Axiom VI.

(c) **Expansion of non-standard models.** Even when N does not receive a standard interpretation, we can still carry out the final part of the construction just described. Namely, suppose we have a structure

(6) $\mathcal{M}_0 = \langle V, \cdot, \dots, N, \mathcal{D}, \in \rangle$
satisfying Axioms I-V when the type variables are taken to range over $\mathcal{D}$, and such that $\mathcal{D}$ is closed under ECA. Then, whether or not N is standard, we can still define $\dot{\in}$, the c_n's and the $\{x \mid \phi\}$'s exactly as above, so as to satisfy Axioms VI and VII in addition.

(d) **Models in arithmetic.** We put the observation (c) to immediate use. Every model

[4]To be more precise, this is a semantics for arbitrary ϕ which do not contain any general type terms of the form $T = \{x \mid \psi\}$. The construction described is actually carried out by induction on the complexity of ϕ containing such subterms, according to their degree of nesting.

(7) $\mathcal{N} = \langle V, ', +, \cdot, \ldots, 0\rangle$

of PA (Peano Arithmetic with symbols for all primitive recursive functions) can be turned into a model of $QL_1(\lambda)$ of the form (5) as follows. We first choose as the application operation for V,

(8) $a \cdot b \simeq \{a\}(b)$,

where $\{a\}(b)$ is defined in o.r.t. (ordinary recursion theory) for an enumeration of the partial recursive functions (e.g. by means of Kleene normal form). Note that what we are working with is a translation of o.r.t. into PA, which then makes sense in any model $\mathcal{N}$ of PA. Standard facts from o.r.t. allow us to convert (8) into an applicative model $\mathcal{A} = \langle V, \cdot, P, P_1, P_2, Sc, Pd, Eq, Cond, Rec, 0\rangle$, where $P, P_1, P_2, Sc, Pd, Eq, Cond, Rec$ are all primitive recursive (Rec may be found from the Recursion Theorem) and Sc is the successor operation $'$ of $\mathcal{N}$, with Pd as its associated predecessor operation. Thus $\mathcal{A}$ consists entirely of operations definable in $\mathcal{N}$.

Next take $N = V$ and let $\mathcal{D}$ consist of all subclasses of V which are first-order definable in $\mathcal{N}$ from parameters in V. It is immediate that $\mathcal{M}_0 = \langle V, \cdot, \ldots, N, \mathcal{D}, \in\rangle$ is a model of Axioms I-V. Moreover, $\mathcal{D}$ is closed under ECA in $\mathcal{L}(QL_1(\lambda))$ since that reduces to closure under ECA in $\mathcal{L}(PA)$. Hence by (c), we can expand this to a model of all $QL_1(\lambda)$. If we start with the standard model $\mathcal{N}$ of PA, i.e. where $V = \omega = \{0, 1, 2, \ldots\}$, then the resulting model $\mathcal{M}$ is trivially N-standard, since $N = V$; this is called *the recursion-theoretic model of* $QL_1(\lambda)$.

10. PA as a Subtheory of $QL_1(\lambda)$

Using the *cond* and *rec* operators in $QL_1(\lambda)$, given any g, h we can define

$$fx \simeq \begin{cases} g & \text{if } x = 0 \\ h(pd(x), f(pd(x))) & \text{if } x \neq 0\,, \end{cases} \tag{1}$$

so that

$$f0 = g, \qquad fx' \simeq h(x, fx) \quad \text{if } x \neq 0\,. \tag{2}$$

It follows that we can define $+, \cdot$, and all further primitive recursive functions to satisfy their recursive defining axioms on N: $x{+}0 = x$, $x{+}y' = (x{+}y)'$, etc. Then every arithmetical formula A is translated by a pure elementary formula $A^{(N)}$ in the language of $QL_1(\lambda)$. This translation is such that $(\forall x A)^{(N)} = (\forall x \in N)A^{(N)}$ and $(\exists x A)^{(N)} = (\exists x \in N)A^{(N)}$.

Theorem 1. *For closed A in the language of arithmetic,*

$$PA \vdash A \Leftrightarrow QL_1(\lambda) \vdash A^{(N)}\,.$$

Proof. (i) By the above translation, if A is provable in PA then $QL_1(\lambda)$ proves its translation $A^{(N)}$.
(ii) A quick model-theoretic argument serves to establish the converse. Let $\mathcal{N} = \langle V,', +, \cdot, \dots, 0\rangle$ be any model of PA. Then by §9(d) we can use this to define a model of $QL_1(\lambda)$ in which $N = V$. Thus if $\mathcal{N} = \langle V, \dots\rangle$ satisfies a closed formula of the form $\neg A$ then the corresponding model $\mathcal{M} = \langle V, \cdot, \dots\rangle$ of $QL_1(\lambda)$ also satisfies $\neg A$, which is in this case equivalent to $A^{(N)}$. Hence if $QL_1(\lambda) \vdash A^{(N)}$ then A is satisfied in all models of PA, so $PA \vdash A$. $\square$

Remark 1. One can also give a finitary proof-theoretic argument for the conservativity of $QL_1(\lambda)$ over PA in the sense of Theorem 1, but this would be somewhat longer.

Theorem 2. $QL_1(\lambda) \equiv PA$.

For, by definition, a function $F : \omega \to \omega$ belongs to $Comp_N(QL_1(\lambda))$ just in case

$$(3) \qquad QL_1(\lambda) \vdash \forall x \in N\ \exists y \in N\ R(x, y)\ ,$$

where in the recursion-theoretic model $\mathcal{M} = \langle \omega, \cdot, \dots\rangle$ for $QL_1(\lambda)$ (in which $N = \omega$), we have

$$(4) \qquad F(x) = (\mu y)R(x, y) \qquad \text{for each } x \in \omega\ .$$

By Theorem 1, (3) holds just in case

$$(5) \qquad PA \vdash \forall x\ \exists y R(x, y)\ ,$$

and the associated provably computable function is the same F.

The class $Comp(PA)$ of *provably computable* (or *recursive*) *functions of natural numbers in* PA has been characterized in a number of interesting ways. We shall use below one such characterization due to [Gödel 1958].

11. User-Type Constructions in $QL_1(\lambda)$

The U-types (and the G-types more generally) are closed under the following constructions, defined as usual. For more details on these cf. also [F 1990] pp. 110–111.

(a) Non-parametric constructions.

(i) $\phi, V,\ B(= \{0, 1\})$
(ii) $S \cap T, S \cup T, V\text{-}S$
(iii) $S \times T$
(iv) $S \to T\ (= \{z \mid \forall x(x \in S \Rightarrow zx \in T)\})$
(v) $S \underset{p}{\to} T\ (= \{z \mid \forall x(x \in S \wedge zx \downarrow \Rightarrow zx \in T)\}$

(b) Parametric constructions. For $T[x](= T[x, \ldots])$,

(vi) $\bigcap\limits_{x \in S} T[x], \quad \bigcup\limits_{x \in S} T[x]$

(vii) $\prod\limits_{x \in S} T[x] \ (= \{z \mid \forall x(x \in S \Rightarrow zx \in T[x])$

(ix) $\sum\limits_{x \in S} T[x] \ (= \{z \mid \exists x, y(x \in S \wedge y \in T[x] \wedge z = (x, y))\})$

(c) Finite sequences. For $n \in N$ let $[0, n) = \{i \mid i \in N \wedge 0 \leq i < n\}$.

$$S^{<\omega} := \{z \mid \exists n, x(n \in N \wedge x : [0, n) \to S \wedge z = (n, x)\} .$$

In other words we identify a sequence $\langle x_0, \ldots, x_{n-1} \rangle$ of length n in S with a pair (n, x) where $x : [0, n) \to S$ and $x_i = xi$ for $0 \leq i < n$. $S^{<\omega}$ is a U-type for each U-type S.

(d) Recursively generated types. These are also called finitarily inductively generated types (cf. [F 1989]). A typical construction, denoted $I(S, T)$, produces the least X satisfying the closure conditions

(1) $S \subseteq X$ and $\dfrac{y_1, y_2 \in X}{x \in X} \quad (x, y_1, y_2) \in T.$

This suggests a definition "from above", as:

(2) $\{z \mid \forall X[S \subseteq X \wedge \forall x, y_1, y_2(y_1 \in X \wedge y_2 \in X \wedge (x, y_1, y_2) \in T \Rightarrow x \in X) \Rightarrow z \in X]\}$

However, this has the form $\{z \mid \phi[z]\}$ where ϕ is not elementary, since it contains the quantifier '$\forall X$'. The following definition "from below" is pure elementary and hence the U-types are closed under this construction:

(3)
$$\begin{aligned} I(S, T) := \{z \mid \exists n \exists x[n \in N \wedge x : [0, n) \to V \wedge n > 0 \wedge \\ \wedge \forall i(i < n \Rightarrow xi \in S \vee \exists j, k(j < i \wedge k < i \wedge (xi, xj, xk) \in T)) \\ \wedge z = x(n-1)\} \end{aligned}$$

Then provably in $QL_1(\lambda)$ we have: if X satisfies the closure conditions (1) then $I(S, T) \subseteq X$, and $I(S, T)$ itself satisfies (1) in place of X. See [F 1989] and [F 1990] §5.5 for many syntactic type constructions which are special cases of $I(S, T)$. In particular, from the latter we may define the type S^* of S-strings (given by $S^* = \bigcup\limits_{n \in N} S^n$, where $S^0 = \{0\}, S^{n+1} = S^n \times S$), in the form $S^* = I(S^0, T)$.

12. Implicit Polymorphism in $QL_1(\lambda)$

We begin with an example. Let $i = \lambda x.x$. Then $\forall X[i \in (X \to X)]$ since $\forall X \forall x[x \in X \Rightarrow ix \in X]$. We can assign a G-type to the global behaviour of i, namely $\bigcap_X(X \to X)$. More generally, we define

(1) $\bigcap_X T[X] := \{x \mid \forall X(x \in T[X])\}$

By the general CA, we have

(2) $x \in \bigcap_X T[X] \Leftrightarrow \forall X(x \in T[X])$.

Examples of implicit polymorphism using this construction are offered by:

(3) (i) $(\lambda x.x) \in \bigcap_X(X \to X)$
(ii) $\lambda x.\lambda y.x(xy) \in \bigcap_X[(X \to X) \to (X \to X)]$
(iii) $\lambda x.\lambda y.(y, x) \in \bigcap_X \bigcap_Y[X \to (Y \to X \times Y)]$
(iv) $\lambda z.(P_2(z), P_1(z)) \in \bigcap_Z[Z \to P_2(Z) \times p_1(Z)]$,
where $P_i(Z) = \{x \mid \exists z(z \in Z \wedge x = P_i(z))\}$.

The construction of $I(S,T)$ from above in §11(2) is not of the form $\bigcap_X T[X]$. For that one wants the more general construction

(4) $\bigcap_{X \mid \phi[X]} T[X] := \{x \mid \forall X[\phi[X] \Rightarrow x \in T[X]\}$.

In particular, $I(S,T)$ can be put in the form $\bigcap_{X \mid \phi[X]} X$ for suitable ϕ.

We shall delay describing the extension of $QL_1(\lambda)$ to the system $QL_1(\Lambda)$ for explicit polymorphism until after dealing with the propositional system $PL_1(\lambda)$, to which we turn next.

13. The Quantifier-Free Logic $PL_1(\lambda)$: Syntax

We actually consider two forms of this logic: the first form, denoted $PL_1(\lambda)$, makes use only of certain U-types, while the second form, denoted $PL_1^+(\lambda)$ and described in a later section, adds certain parametric types as well as some G-types for implicit polymorphism. The syntax of $PL_1(\lambda)$ is specified as follows:

Individual terms $(s, t, \dots)$ are the same as for $QL_1(\lambda)$ (§4)
Type terms $(S, T, \dots)$ are of one of the following forms:

$$N, S \times T, S \to T$$

Atomic formulas are of the form $t : T$ and $t_1 \simeq t_2$
Sequents are of the form

$$\Gamma \vdash t : T \text{ and } \Gamma \vdash t_1 \simeq t_2$$

where $\Gamma = s_1 : S_1, \dots, s_n : S_n \qquad (n \geq 0)$.

The logic provides certain sequents as axioms, and rules of inference for inferring sequents from one or more sequents. One may consider an axiom as a rule with no hypotheses. Then the rules are divided into two kinds, called *type inference* and *equality inference*, according as to whether the sequent in the conclusion of the rule is of the form $\Gamma \vdash t : T$ or $\Gamma \vdash t_1 \simeq t_2$, resp.

We here regard the atomic formula $t : T$ as being a syntactic variant of $t \in T$, and identify the two.

Discussion

The syntactic form of the logic $PL_1(\lambda)$ constitutes an extension of usual type (and equality) inference systems of the sort considered in the literature on polymorphic type systems in the following respects.

(i) Usual inference systems derive sequents of the form $\Gamma \vdash t : T$ and $\Gamma \vdash t_1 = t_2 : T$ where $\Gamma = x_1 : S_1, \ldots, x_n : S_n$ provides a *context* consisting of *variable declarations.* Moreover, the constant N (if it appears), is not given a special role. Many such systems are *decidable.*

(ii) $PL_1(\lambda)$ could be massaged to the form where Γ is taken as in (i), by adding a type construction $s^{-1}S$ with the intended meaning $s^{-1}S = \{x \mid sx : S\}$.

(iii) N is given a distinctive role in $PL_1(\lambda)$ by its rules which correspond to the closure conditions for 0 and $'$, and the principle of induction on N. This, together with the treatment of *rec* (discussed next) leads to *undecidability of* $PL_1(\lambda)$. The work on decidability under (i) is still useful in $PL_1(\lambda)$ when considering derivations or parts of derivations not involving induction or recursion.

(iv) The basic reason for allowing possibly undefined terms and equalities of the form $t_1 \simeq t_2$ between such terms is in the use of the recursion operator, which may lead to non-terminating computations of $rec(t)s$. Of course, λ-terms in general may be undefined. Since $t : T$ is identified with $t \in T$, this implies $(t\downarrow)$ in our logic. Thus if $\Gamma \vdash t_1 : T$ then we can be sure that $(\Gamma \vdash t_1 \simeq t_2)$ implies $\Gamma \vdash t_1 = t_2 : T$.

14. Inference Rules of $PL_1(\lambda)$

(a) Type inference rules. These come in *Introduction* and *Elimination forms,* abbreviated I and E, respectively.

$$(\to I) \qquad \frac{\Gamma, x : S \vdash t[x] : T}{\Gamma \vdash \lambda x.t[x] : (S \to T)} \qquad \text{`}x\text{' fresh}$$

$$(\to E) \qquad \frac{\Gamma \vdash t : (S \to T) \quad \Gamma \vdash s : S}{\Gamma \vdash ts : T}$$

$(\times I)$
$$\frac{\Gamma \vdash t_1 : T_1 \quad \Gamma \vdash t_2 : T_2}{\Gamma \vdash (t_1, t_2) : (T_1 \times T_2)}$$

$(\times E)$
$$\frac{\Gamma \vdash t : (T_1 \times T_2)}{\Gamma \vdash p_i(t) : T_i} \quad (i = 1, 2)$$

$(N\text{-}I)$
$$\vdash 0 : N \qquad x : N \vdash x' : N$$

$(N\text{-}E)$
$$\frac{\Gamma \vdash t[0] : T \qquad \Gamma, t[x] : T \vdash t[x'] : T}{\Gamma, x : N \vdash t[x] : T} \quad \text{`}x\text{' fresh}$$

In addition to these, we have *structural rules*

$(Struct_1)$
$$s : S \vdash s : S$$

$(Struct_2)$
$$\frac{\Gamma \vdash t : T}{\Gamma' \vdash t : T} \qquad \text{for } \Gamma \subseteq \Gamma' \text{ (as sets)},$$

as well as *substitution*

$(Subst)$
$$\frac{\Gamma[x] \vdash t[x] : T}{\Gamma[s], s : S \vdash t[s] : T}$$

and

(Cut)
$$\frac{\Gamma \vdash s : S \quad \Gamma, s : S \vdash t : T}{\Gamma \vdash t : T}$$

Finally we have

$(\simeq)$
$$\frac{\Gamma \vdash t_1 \simeq t_2 \quad \Gamma \vdash t_2 : T}{\Gamma \vdash t_1 : T}$$

Remarks.

(i) In the substitution rule, the additional hypothesis $s : S$ ensures $s \downarrow$.

(ii) We can derive from $(N\text{-}E), (Subst)$ and (Cut),

$$\frac{\Gamma \vdash t[0] : T \quad \Gamma, t[x] : T \vdash t[x'] : T \quad \Gamma \vdash s : N}{\Gamma \vdash t[s] : T}.$$

(b) Equality inference rules. Here there are quite a few more rules to consider and we shall not exhibit them all, but merely explain what they are supposed to look like. Besides structural, substitution, and cut rules of the sort just considered, these fall into the following groups:

(i) **General $\simeq$ rules.** These correspond to the reflexive, symmetric and transitive laws for $\simeq$.

(ii) **Compounds.** Pairwise equalities are closed under application, abstraction, and under each operator $F = p, p_1, p_2, sc, pd, eq, cond, rec$

(iii) **Abstraction-reduction** $\vdash (\lambda x.t[x])y \simeq t[y]$

(iv) **Equality rules for the operators** $(p, \ldots, rec)$.
For example $\vdash p_i(x_1, x_2) \simeq x_i$, $\vdash cond(1, x_1, x_2) \simeq x_1$, $\vdash cond(0, x_1, x_2) \simeq x_2$, $x : N \vdash eq(x', 0) \simeq 0$, $\vdash rec(z)x \simeq z(rec(z))x$, etc.

15. The Computational Strength of $PL_1(\lambda)$: Upper Bounds

We have a translation of $PL_1(\lambda)$ into $QL_1(\lambda)$, as follows. First of all, each individual term of $PL_1(\lambda)$ is the same as a term of $QL_1(\lambda)$. Secondly, using the definitions of $S \times T$ and $(S \to T)$ in $QL_1(\lambda)$, each type term of $PL_1(\lambda)$ is identified with a term of the form N or $\{x \mid \phi\}$. Next, atomic formulas of the form $t : T$ are translated into (and may be taken as identified with) the corresponding formulas of the form $t \in T$, while formulas of the form $s \simeq t$ are translated into (and may be identified with) the same formula as defined in $QL_1(\lambda)$, namely as $(s \downarrow \wedge\, t \downarrow \Rightarrow s = t)$. The final step is to translate sequents

(1) $\qquad s_1 : S_1, \ldots, s_n : S_n \vdash t : T$ and $s_1 : S_1, \ldots, s_n : S_n \vdash t_1 \simeq t_2$

into formulas of $QL_1(\lambda)$, of the form,

(2) $(s_1 \in S_1 \wedge \cdots \wedge s_n \in S_n \Rightarrow t \in T)$ and $(s_1 \in S_1 \wedge \cdots \wedge s_n \in S_n \Rightarrow t_1 \simeq t_2)$

respectively. This translation of sequents into formulas may again be considered as an identification. Then we can assert the following:

Theorem 3. *$PL_1(\lambda)$ is contained in $QL_1(\lambda)$.*

This is proved by a straightforward induction on the axioms and rules of $PL_1(\lambda)$.

It follows from Theorem 3 that $Comp_N(PL_1(\lambda)) \subseteq Comp_N(QL_1(\lambda))$, where $Comp_N(PL_1(\lambda))$ is the class of functions $F : \omega \to \omega$ defined in the recursion-theoretic model by a closed term t with $\vdash t : (N \to N)$ or $x : N \vdash tx : N$ in $PL_1(\lambda)$.

16. The Computational Strength of $PL_1(\lambda)$: Lower Bounds

Theorem 4. *$Comp(PA) \subseteq Comp_N(PL_1(\lambda))$.*

Proof. For this we make use of the characterization of $Comp(PA)$ obtained from [Gödel 1958], when combined with [Tait 1965]. The introductory note by Troelstra to the former in [Gödel 1990] (on pp.217–241) gives the basic facts needed for our purpose. Very briefly, they may be described as follows: First of all, Gödel defines a class of *primitive recursive functionals of finite type*, that we denote here by $PR^{(\omega)}$. The type hierarchy for this is generated from N (or type 0) by closing under $\times$ and $\to$, i.e. is given exactly by the type terms of $PL_1(\lambda)$. Each specific element F of $PR^{(\omega)}$ is assigned a definite type

$S = Typ(F)$, and we write $F \in PR_S^{(\omega)}$ or $F : S$ in that case. The functionals in $PR^{(\omega)}$ are generated by explicit definition from 0 and sc, using also pairing and projections in each combination $S \times T$ of types, together with primitive recursion Rec_S for each type S, determined by:

$$
\begin{aligned}
&(1) \qquad Rec_S ab0 = a \\
&\qquad\quad Rec_S abx' = b(Rec_S(abx))x \ , \\
&\qquad\quad \text{where } a : S, b : S \to N \to S, x : N \\
&\qquad\quad \text{and } Rec_S : S \to (S \to N \to S) \to N \to S
\end{aligned}
$$

(arrows associated to the right).

Next, a quantifier-free formal system T is set up by Gödel, whose informal interpretation is $PR^{(\omega)}$. There is a closed term F for each member of $PR^{(\omega)}$, and variables $a^S, b^S, \ldots, x^S, y^S, z^S$ ranging over each type S. The logic is that of many-sorted intuitionistic quantifier-free predicate calculus with equality (in each type); the axioms are the usual ones for 0 and sc, the defining equations for each F in $PR^{(\omega)}$, and one has a rule of induction on N.

Gödel then associates with each sentence A of the language of Heyting Arithmetic (HA) a quantifier-free formula $R_A[f, x]$ where $f = f_1^{S_1}, \ldots, f_n^{S_n}$, $x = x_1^{T_1}, \ldots, x_m^{T_m}$, and an interpretation ("Dialectica")

$$(2) \qquad A^D = \exists f \, \forall x \, R_A[f, x]$$

in such a way that

$$
\begin{aligned}
&(3) \qquad \text{if } HA \vdash A \text{ then } T \vdash R_A[F, x] \text{ for} \\
&\qquad\qquad \text{some } F = F_1, \ldots, F_n \text{ with } F_i \in PR^{(\omega)}(S_i) \ .
\end{aligned}
$$

(The sequence F is found finitarily from a proof of A.)

In particular, for A of the form $\forall x \exists y R(x, y)$ with R primitive recursive (in the usual sense), we have

$$(4) \qquad (\forall x \exists y R(x, y))^D = \exists f^{N \to N} \forall x^N R(x, fx) \ .$$

Hence, by (3), if HA proves $\forall x \exists y R(x, y)$ we have an F of type $(N \to N)$ such that $T \vdash R(x^N, Fx^N)$.

By Gödel's 1933 "double-negation" translation of PA into HA,

$$(5) \qquad PA \vdash \forall x \exists y R(x, y) \text{ implies } HA \vdash \forall x \neg\neg \exists y R(x, y) \ .$$

As it happens, $(\forall x \neg\neg \exists y R(x, y))^D$ is equivalent to $(\forall x \exists y R(x, y))^D$. Hence, by (4), if $PA \vdash \forall x \exists y R(x, y)$ then the function $F(x) = \mu y R(x, y)$ is in $PR^{(\omega)}(N \to N)$. Put in other terms,

$$(6) \qquad Comp(PA) \subseteq PR^{(\omega)}(N \to N) \ .$$

A proof that

$$PR^{(\omega)}(N \to N) \subseteq Comp(PA) \tag{7}$$

can be given in several ways. The first proof is due to [Tait 1965], where it is shown how to normalize terms representing the members of PR^{ω}, using transfinite induction for ordinals $< \epsilon_0$. It follows that each element of $PR^{(\omega)}(N \to N)$ can be defined by a recursion on some ordinal $\alpha < \epsilon_0$. Gentzen showed that induction up to each $\alpha < \epsilon_0$ is provable in PA, so PA is provably closed under such recursions. A second proof of (7) is simpler and more conceptual; it formalizes in PA the model of $PR^{(\omega)}$ in HRO, the so-called hereditarily recursive operations of finite type (cf. Troelstra, *op.cit.*, pp.233–234). This associates with each (term) $F \in PR^{(\omega)}(S)$ a proof in PA that $F \in HRO_S$. In particular, since $HRO_{N \to N} = \{e \mid e \text{ is total recursive on } N\}$, this shows each $F \in PR^{(\omega)}(N \to N)$ to be provably recursive in PA. In this way the proof of (7) is completed. Actually, for the rest of the proof of our theorem, only the part (6) of the characterization is needed. For, by this, in order to prove

$$Comp(PA) \subseteq Comp_N(PL_1(\lambda)) \tag{8}$$

it is sufficient to show that

$$PR^{(\omega)}(N \to N) \subseteq Comp_N(PL_1(\lambda)) \ . \tag{9}$$

This is established by associating with each closed term F of PR^{ω} a term t_F of $PL_1(\lambda)$ such that

(10) (i) t_F defines the same (number, function or) functional as F when restricted to the type of arguments of F,
and (ii) $PL_1(\lambda)$ proves the sequent $\vdash t_F : Typ(F)$.

The construction of t_F follows that of F quite directly: (i) when the latter is given by a *typed* lambda term as an explicit definition from 0 and sc, we simply take t_F to be the corresponding *untyped* lambda term with all type symbols suppressed; (ii) when $F = Rec_S$, we first take r to be a term defined essentially by the same means as in §10, so as to satisfy

$$\begin{aligned} &\vdash r[a,b,0] \simeq a \\ &\vdash r[a,b,x'] \simeq b(r[a,b,x])x \end{aligned} \tag{11}$$

for arbitrary a, b, x. Then we prove by the induction rule (N-E) in $PL_1(\lambda)$

$$a : S, b : S \to N \to S, x : N \vdash r[a,b,x] : S \tag{12}$$

Thus t_F may be defined in this case as $\lambda a \lambda b \lambda x.r$, in order to satisfy, provably

$$(13) \qquad \vdash t_F : (S \to (S \to N \to S) \to N \to S) \ .$$

This completes the (outline of the) proof of Theorem 4. □

We now have as a corollary from Theorems 3 and 4:

Theorem 5. $PL_1(\lambda) \equiv QL_1(\lambda) \equiv PA$.

A main part of the result promised in (1)(ii) of §3 is thus established.

17. Extensions of $PL_1(\lambda)$ Having the Same Strength

Since $PL_1(\lambda)$ is contained in $QL_1(\lambda)$, any extension of the former by rules for U-types, or even G-types, constructed in the latter will not increase computational strength. Every such system is determined by the closure conditions for generating the type terms, and by the type inference rules for such. No new forms of equality rules are needed, since the individual terms of $PL_1(\lambda)$ are the same as those of $QL_1(\lambda)$, and hence are the same in any intermediate system.

The following description of type terms for an intermediate system $PL_1^+(\lambda)$ is typical.

The type terms of $PL_1^+(\lambda)$ are those of the forms

$X, Y, Z, \ldots$
$N, V, \{0,1\}$
$\{x : S \mid t_1 = t_2 : T\}$ $\qquad$ x not free in S, T
$I(S,T)$
$\sum_{x:S} T \ , \ \prod_{x:S} T$ $\qquad$ (x not free in S)
$\bigcap_X T$.

Only the last takes us out of U-types to G-types. We could consider a weaker system, sufficient for application of polymorphism in practice, in which one forms only intersections $\bigcap_{X_1} \cdots \bigcap_{X_n} T$, where T is a U-Type generated by the preceding rules.[5] Note that $S \times T$ may be identified with $\sum_{x:S} T$ and $S \to T$ with $\prod_{x:S} T$, for x not free in both S, T.

We leave it to the reader to provide appropriate type-inference rules for this system.[6] Two points need special attention, though, when one is using G-Types:

---

[5] In effect, this is what is done in ML's polymorphism; cf. [Mitchell and Harper 1988].
[6] And to think up other potentially useful intermediate systems.

(i) The (N-E) rule

$$\frac{\Gamma \vdash t[0] : T \quad \Gamma, t[x] : T \vdash t[x'] : T}{\Gamma, x : N \vdash t[x] : T}$$

must still be restricted to U-Types T (as the corresponding principle of induction in $QL_1(\lambda)$ applies only to U-Types).

(ii) The $\bigcap$ rules are to be formulated in the following form:

($\bigcap$-I)
$$\frac{\Gamma \vdash t : T[X]}{\Gamma \vdash t : \bigcap_X T[X]} \quad \text{`}X\text{' fresh}$$

($\bigcap$-E)
$$\frac{\Gamma \vdash t : \bigcap_X T[X]}{\Gamma \vdash t : T[S]} \quad \text{for each } U\text{-type } S$$

In other words, in ($\bigcap$-E) we can only instantiate by U-types S.

The other rules do not require such restrictions.

18. Explicit Polymorphism: the Systems $QL_1(\Lambda)$ and $PL_1(\Lambda)$

Explicit polymorphism is a notion which applies to systems with explicitly typed lambda terms, i.e. in which each abstract is of the form $(\lambda x : S)t$, and where one has type variables, so that S (and t) may depend on type parameters. The simplest example is given by $i_X = (\lambda x : X)x$, which represents the identity operation of type $(X \to X)$. Here an individual term may be constructed which "packages" the global behaviour of i_X, by abstracting with respect to the type variable X. We use capital lambda for such abstraction, as $\Lambda X.(\lambda x : X)x$; this term is assigned a new G-type, $\prod_X(X \to X)$. To specialize the term to any specific type S, we need another application operation $t \cdot S$; in general this obeys $(\Lambda X.t[X]) \cdot S \simeq t[S]$. For example, $(\Lambda X.(\lambda x : X)x) \cdot S = (\lambda x : S)x$ is the identity operation of type $(S \to S)$. Principles for such a type system were formulated in a logic of QL-style in [F 1990], within a theory denoted IOC_Λ, of strength (at least) that of second-order arithmetic PA_2 ("analysis"). In the same paper, a subtheory EOC_Λ of IOC was considered which has the same strength as PA, obtained essentially by restricting all types to be of the form $\{x \mid \phi\}$ for ϕ pure elementary. In EOC_Λ we can form $(\lambda x : S)t$ and $t \cdot S$ only for elementary types S, and though we can form terms there of the form $\Lambda X.t[X]$ we do not have their types of the forms $\Pi X.T[X]$. Nevertheless, it was shown at the end of [F 1990] that we could still translate into EOC_Λ a system of inferences $\Gamma \vdash t : T$ where t need not be restricted to the terms of EOC_Λ, but could be taken to be "semi-predicative" in a certain sense. Moreover, it was conjectured there that all applications of polymorphism met in practice are covered by these inferences.

The systems $QL_1(\Lambda)$ and $PL_1(\Lambda)$ introduced here are intended to clarify the last part of that work, by the explicit use of the distinction between U-Type terms and G-Type terms. The following constitutes a simultaneous inductive definition of the individual, U-Type and G-Type terms, and of the formulas and pure elementary formulas:

The individual terms of $QL_1(\Lambda)$ $(s, t, \dots)$ are just those of the form

$x, y, z, \dots$
$0, c_n (n > 0)$
$st, (\lambda x : S)t$ (S a G-Type term)
$s \cdot T$, $\Lambda X.t$ (T a U-Type term)
$p(t_1, t_2), p_i(t)(i = 1, 2), sc(t), pd(t), eq(s, t), cond(s, t), rec(t).$

The G-Type terms of $QL_1(\Lambda)$ $(S, T, \dots)$ are just those of the form
$X, Y, Z, \dots$
N
$\{x \mid \phi\}$, ϕ a formula

The U-Type terms are just those of the preceding form in which ϕ is a pure elementary formula.

The formulas $(\phi, \psi, \dots)$ are just those of the form
$t\downarrow,\ s = t,\ s \in T,\ s\dot{\in}t$
$\neg\phi,\ \phi \wedge \psi,\ \phi \vee \psi,\ \phi \Rightarrow \psi,$
$\forall x\phi, \exists x\phi, \forall X\phi, \exists X\phi$

The pure elementary formulas $(A, B, \dots)$ are just those of the form
$t\downarrow,\ s = t,\ s \in T$ (for T a U-Type term)
$\neg A, A \vee B, A \vee B, A \Rightarrow B$
$\forall xA,\ \exists xA$

The basic logic of $QL_1(\Lambda)$ is just like that of $QL_1(\lambda)$ (§7 above), and the *Axioms for* $QL_1(\Lambda)$ look just like those for $QL_1(\lambda)$ (§8) except that the *Abstraction-reduction axiom* I is modified to:

I′ (i) $y \in S \Rightarrow ((\lambda x : S)t[x])y \simeq t[y]$
(ii) $(\Lambda X.t[X]) \cdot Y \simeq t[Y]$.

We can identify the language of $QL_1(\lambda)$ with a part of that for $QL_1(\Lambda)$ by identifying $\lambda x.t$ with $(\lambda x : V)t$, where $V = \{x \mid x = x\}$. Then it is trivial that $QL_1(\lambda)$ is contained (as a theory) in $QL_1(\Lambda)$.[7] Clearly all the U-Type

[7]Also the system EOC_Λ of [F 1990] is contained in $QL_1(\Lambda)$.

and G-Type constructions of the former carry over to the latter. But now we have a new G-Type construction:

$$\Pi_X T[X] = \{z \mid \forall X(z \cdot X \in T[X])\} . \tag{1}$$

By Axiom VI (GCA) of $QL_1(\Lambda)$,

$$z \in \Pi_X T[X] \Leftrightarrow \forall X(z \cdot X \in T[X]) . \tag{2}$$

Thus, e.g. we have, as expected, $\Lambda X(\lambda x : X)x \in \prod_X(X \to X)$, and so on. But our restriction on the logic of type quantification only permits the inference

$$z \in \prod_X T[X] \Rightarrow (z \cdot S) \in T[S] \tag{3}$$

for S a U-Type term, giving e.g. $(\lambda x : S)x \in (S \to S)$.

Theorem 6. *$QL_1(\Lambda)$ is a conservative extension of $QL_1(\lambda)$.*

The proof of this theorem, which is not given here, is by the same method as that of [F 1990] (pp.124–126) for the translation of IOC_Λ into IOC_λ and of EOC_Λ into EOC_λ, namely by an extension of the Girard-Troelstra "erasing" translation $(-)^*$ which takes $((\lambda x : S)t)^* := \lambda x.t^*$, $(\Lambda X.t)^* := \lambda u, t^*$ (with 'u' fresh) and $(t \cdot S)^* := t^*0$ (following a correction of [F1990] by M. Felleisen).

Now the choice of the system $PL_1(\Lambda)$ is somewhat free, given only that it should include as individual terms all those of $QL_1(\Lambda)$, and as type terms at least those of the form

$X, Y, Z, \ldots,$
$N,$
$S \times T,\ S \to T$
$\Pi_X T$

The U-Types of $PL_1(\Lambda)$ are just those which do not contain the Π operation symbol. Again we leave formulation of an inference system for $PL_1(\Lambda)$ to the reader. As in §17 for $PL_1^+(\lambda)$, the only points requiring special attention are

(i) The $(N\text{-}E)$ rule

$$\frac{\Gamma \vdash t[0] : T \qquad \Gamma, t[x] : T \vdash t[x'] : T}{\Gamma, x : N \vdash t[x] : T}$$

is restricted to U-Types T,
and

(ii) the Π rules are formulated as follows:

$$(\Pi\text{-}I) \qquad \frac{\Gamma \vdash t[X] : T[X]}{\Gamma \vdash \Lambda X.t[X] : \Pi_X T[X]}, \quad \text{`}X\text{' fresh}$$

$$(\Pi\text{-}E) \qquad \frac{\Gamma \vdash t : \Pi_X T[X]}{\Gamma \vdash t \cdot S : T[S]}, \quad \text{for } S \text{ a } U\text{-Type term.}$$

Theorem 7. *$PL_1(\Lambda)$ is a conservative extension of $PL_1(\lambda)$.*

This is proved by the same translation as for Theorem 6. We may thus conclude from Theorems 5, 6, and 7, the result promised in (§3)(1)(ii):

Theorem 8. $QL_1(\lambda) \equiv PL_1(\lambda) \equiv QL_1(\Lambda) \equiv PL_1(\Lambda) \equiv PA$.

Obviously, $PL_1(\Lambda)$ can be enriched to systems $PL_1^+(\Lambda)$ contained in $QL_1(\Lambda)$ as $PL_1(\lambda)$ was enriched to $PL_1^+(\lambda)$.

Remark. So far, no use has been made of the representation relation $Rep(z, Z)$ and the Representation Axiom VII beyond the general consequences for instantiation of type quantification. It would be natural to look to these for an alternative interpretation of a system for explicit polymorphism (like $QL_1(\Lambda)$) in a system with implicit polymorphism (like $QL_1(\lambda)$). This is possible but not as simple as the $(-)^*$ translation.

19. Logics of Strength PRA

We begin by describing a modification $QL_0(\lambda)$ of $QL_1(\lambda)$ which is of the same computational strength as PRA. The simplest way to obtain such is by adding a new predicate symbol $x \leq y$ to the language, with the following further axioms for the natural numbers:

V (iii) $x \leq y \Rightarrow x \in N \wedge y \in N$
 (iv) $x \leq 0 \Leftrightarrow x = 0$
 (v) $y \in N \Rightarrow (x \leq y' \Leftrightarrow x \leq y \vee x = y')$.

Then we may introduce *bounded (numerical) quantification* as usual by

(1) $(\forall x \leq y)\phi := \forall x[x \leq y \Rightarrow \phi]$ and $(\exists x \leq y)\phi := \exists x(x \leq y \wedge \phi)$.

Note that we can define $(\forall x < y)\phi$ as $(\forall x \leq y)[eq(x, y) = 1 \vee \phi]$ and $(\exists x < y)\phi$ as $(\exists x \leq y)[eq(x, y) = 0 \wedge \phi]$.

The individual terms of $QL_0(\lambda)$ are taken to be the same as in $QL_1(\lambda)$. The G-Type terms and general formulas are also generated in the same way, except that we now have $(s \leq t)$ as an additional basic atomic formula.

Next, the U-Type terms of $QL_0(\lambda)$ and the $(\exists, Bd)^+$-formulas (or, to be more precise, the *Existential-Bounded-Positive formulas*) are generated simultaneously as follows:

The *U-Type terms* are those of one of the forms

$X, Y, Z, \ldots$
N
$\{x \mid A\}$ where A is an $(\exists, Bd)^+$ formula

The *$(\exists, Bd)^+$-formulas* $(A, B, \ldots)$ are those of one of the forms

$s\downarrow,\ s = t,\ s \in T$ for T a U-Type term
$A \wedge B,\ A \vee B$
$(\exists x)A,\ (\forall x \leq y)A.$

The logic of partial terms in $QL_0(\lambda)$ looks exactly like that of $QL_1(\lambda)$ in §7 above (adding, of course, $s \leq t \Rightarrow s\downarrow \wedge\, t\downarrow$).

The *Axioms of* $QL_0(\lambda)$ are now taken to be the same as those of $QL_1(\lambda)$ in §8 above, with the understanding that Axiom V for N is expanded by the defining conditions for $\leq$ given in V(iii)-(v) above, and that the formulas A appearing in the Representation Axiom VII are the $(\exists, Bd)^+$ formulas as just defined. It follows that we have type instantiation $\forall X \phi[X] \Rightarrow \phi[T]$ for each U-type term in the new sense.

The next step is to establish the computational strength of $QL_0(\lambda)$. This is done by use of the fragment $(\Sigma^0_1 - IA)$ of PA based on the *Σ^0_1-Induction Axiom schema.* We take this system formulated in a form with a symbol for each primitive recursive function and associated defining axiom, in addition to the usual axioms for 0 and $'$, and with the induction scheme

$$B[0] \wedge \forall x(B[x] \Rightarrow B[x']) \Rightarrow \forall x B[x] \tag{2}$$

taken only for Σ^0_1-formulas B, i.e. those of the form $\exists y R(x, y, \ldots)$ with R primitive recursive.

Theorem 9. $QL_0(\lambda) \equiv (\Sigma^0_1\text{-}IA)$.

Proof. $(\Sigma^0_1 - IA)$ is translated directly into $QL_0(\lambda)$ by the same translation as for PA in $QL_1(\lambda)$. Let, as before, for each sentence A of the language of arithmetic, $A^{(N)}$ be its image under this translation. Thus we have

$$(\Sigma^0_1\text{-}IA) \vdash A \Rightarrow QL_0(\lambda) \vdash A^{(N)}\ . \tag{3}$$

Now, for the converse, we apply a semantical argument as for Theorem 1 of §10. Let $\mathcal{N} = \langle V, ', +, \cdot, \ldots, 0\rangle$ be any model of $(\Sigma^0_1 - IA)$. We define a recursion-theoretic applicative model $\langle V, \cdot, \ldots\rangle$ of Axioms I-IV of $QL_0(\lambda)$

from this in the way described in §9(d) above. Then we interpret N as V in this model. Finally, let $\mathcal{D}$ be the collection of Σ_1^0-definable subsets of V definable in $\mathcal{N}$ from parameters in V. We take the range of the type variables in $QL_0(\lambda)$ to be $\mathcal{N}$. It may be shown that every $(\exists, Bd)^+$-formula without type variables defines a member of $\mathcal{D}$, and every U-Type term without type variables denotes a member of $\mathcal{D}$; moreover, we have closure under substitution by the latter for type variables in $(\exists, Bd)^+$ formulas in general. Thus if we take $\mathcal{M} = \langle V, \cdot, \ldots, N, \mathcal{D}, \in \rangle$, we have obtained a model of the logic of $QL_0(\lambda)$ and Axioms I-V. The proof is completed by defining $\dot{\in}$ and the denotations of G-Type terms in general so as to satisfy Axioms VI and VII as before. It follows that if $QL_0(\lambda) \vdash A^{(N)}$ then $\mathcal{M} \models A$ for any such $\mathcal{M}$; hence $\mathcal{N} \models A$ for all models $\mathcal{N}$ of $(\Sigma_1^0 - IA)$ and therefore $(\Sigma_1^0 - IA) \models A$. □

Corollary 10. $QL_0(\lambda) \equiv PRA$.

This is by the well known result of [Parsons 1970] that $(\Sigma_1^0\text{-}IA) \equiv PRA$; cf. also [Sieg 1985].

A sublogic $PL_0(\lambda)$ of $PL_1(\lambda)$ whose provably computable functions are just the primitive recursive functions is easily obtained. We simply *omit* closure under $(S \to T)$ in the type formation rules of §13, and the corresponding rules $(\to I)$ and $(\to E)$ in §14. Thus the only types that may be constructed in this minimal system are of the form $N \times \cdots \times N$. The forms of the retained type inference and equality rules are exactly as before, but restricted to the new system of types. It suffices to apply the induction rule $(N\text{-}E)$ to the type $T = N$. Then we establish by straightforward induction, closure of the operators on N^k into N under primitive recursion, as defined using the *rec* operator. Since $PL_0(\lambda)$ is obviously contained in $QL_0(\lambda)$ we may supplement Theorem 10 by:

Theorem 11. $PL_0(\lambda) \equiv QL_0(\lambda)$.

Now there are many useful logics between $PL_0(\lambda)$ and $QL_0(\lambda)$, given by suitable type constructions from the latter. For example, referring to the closure conditions of §11, the U-Types of $QL_0(\lambda)$ are closed under

(i) $V, \{0, 1\}$
(ii) $\sum_{x \in S} T[x]$
(iii) $I(S, T)$,

and the G-types are closed under $\bigcap_X T$.

We can define a system $PL_0^+(\lambda)$ which adds these and associated inference rules to the type generating rules of $PL_0(\lambda)$. Then from $PL_0(\lambda) \subseteq PL_0^+(\lambda) \subseteq$

$QL_0(\lambda)$, we have $PL^+(\lambda) \equiv PRA$ too. The general construction $I(S,T)$ is most useful in practice.

Implicit polymorphism is provided by the same type construction $\bigcap_X T[X]$ in $QL_0(\lambda)$ as in $QL_1(\lambda)$. To obtain systems $QL_0(\Lambda)$ and $PL_0(\Lambda)$ of strength PRA for explicit polymorphism, we simply modify the systems $QL_1(\Lambda)$ and $PL_1(\lambda)$ of §18 in the same way as $QL_1(\lambda)$ and $PL_1(\lambda)$ were modified to $QL_0(\lambda)$ and $PL_0(\lambda)$ resp. (In the case of $PL_0(\Lambda)$ this simply comes down to omitting the construction $S \to T$ and associated rules.) Then the reduction of each Λ system to the corresponding λ system goes through by the same $*$-translation as before. In this way one may arrive at the result promised in §3(1)(i):

Theorem 12. $QL_0(\lambda) \equiv PL_0(\lambda) \equiv QL_0(\Lambda) \equiv PL_0(\Lambda) \equiv PRA$.

20. THE LOGIC $QL_2(\lambda)$ AND ITS RELATIVES

The material of this section has been dealt with only in part and requires further work; it is thus indicated more sketchily. In $QL_1(\lambda)$ we can form joins (disjoint unions) $\sum_{x\in S} T[x]$ and products $\prod_{x\in S} T[x]$ only when $T[x]$ is a single type term containing the variable ("parameter") x. The motivation for $QL_2(\lambda)$ is that we should be able to form $\sum_{x\in S} T_x$ and $\prod_{x\in S} T_x$ whenever we have a sequence of types $\langle T_x\rangle_{x\in S}$ given by an operation f from S to the universe of types with $fx = T_x$ for each $x \in S$. This makes sense directly in systems with the ontological axiom *Ont* (cf. §8, Remark 9^0). In the set-up of the $QL(\lambda)$'s, it must be reformulated using the *Rep* relation. One adds a constant symbol j (for "join") to the language and then adds the following closure condition to the Representation Axiom VII:

$$VII(iv)\ Rep(a) \wedge \forall x(x\dot{\in}a \Rightarrow Rep(fx)) \Rightarrow j(a,f)\downarrow \wedge Rep(j(a,f))$$
$$\wedge\, \forall z[z\dot{\in}j(a,f) \Leftrightarrow \exists x,y(z=(x,y) \wedge x\dot{\in}a \wedge y\dot{\in}fx)]\ .$$

Thus, under the hypothesis, if $Rep(a,S)$ and $Rep(fx,T_x)$ for each $x \in S$ then $Rep(j(a,f), \sum_{x\in S} T_x)$. Once we have joins in this sense we have products, since $\prod_{x\in S} T_x = \{z \mid \forall x(x \in S \Rightarrow zx \in T_x)\} = \{z \mid \forall x(x \in S \Rightarrow (x,zx)\dot{\in}j(a,f))\}$.

Now if no further modification than adding VII(iv) is made to the system $QL_1(\lambda)$, the resulting theory is denoted $QL_2(\lambda)\restriction$; the sign '$\restriction$' indicates that the induction axiom on N is still restricted to Q-Types. It turns out that $QL_2(\lambda)\restriction$ is no stronger than $QL_1(\lambda)$. For, in order to prove the existence of Q-Types which are not U-Types of $QL_1(\lambda)$ we need to be able to apply induction on N to stronger properties. Informally speaking, the U-Types of $QL_1(\lambda)$ are found in the finite type hierarchy over N. We can pass to transfinite types using joins (or products) by forming, for example, $\sum_{k\in N} N_k$ (or $\prod_{k\in N} N_k$) where $N_0 = N, N_{k+1} = N_k \to N$. To obtain a representation of $\sum_{k\in N} N_k$, let $z_1 \dot{\to} z_2$ be the operation $c_{\ulcorner A \urcorner}(z_1, z_2)$ with $Rep(c_{\ulcorner A \urcorner}(z_1, z_2), Z_1 \to Z_2)$ whenever $Rep(z_i, Z_i)(i = 1, 2)$; then define $f0 = a,\ fk' = (fk \dot{\to} a)$ primitive recursively, where $Rep(a, N)$. Now to prove $Rep(j(a, f), \sum_{k\in N} N_k)$ from Axiom VII(iv) we need to prove $(\forall k \in N) Rep(fk)$ by induction on N, i.e. $(\forall k \in N)\exists X\, Rep(fk, X)$. This is a non-elementary property and so the induction in question cannot be carried out without admitting such properties to the induction axiom. The full system $QL_2(\lambda)$ is obtained from $QL_2(\lambda)\restriction$ by changing the induction axiom V(ii) to:

$V'(ii)$ $\qquad 0 \in T \wedge \forall x[x \in T \Rightarrow x' \in T] \Rightarrow N \subseteq T$, for any G-Type T.

The system $QL_2(\lambda)$ is closely related to a theory $EM_0 + J$ introduced in [F 1979], and shown there to be proof-theoretically equivalent to the subsystem $\Sigma^1_1 - AC$ of analysis. The essential difference of $EM_0 + J$ from $QL_2(\lambda)$ is that the former does not contain G-Type terms and that the Ontological Axiom Ont is used instead of the Replacement Axiom as here; thus a join axiom J is formulated directly as indicated above. Similarly $QL_2(\lambda)\restriction$ is related to the subsystem $EM_0\restriction +J$ where induction is restricted to classes (= types), and the latter system is proof-theoretically equivalent to $(\Sigma^1_1 - AC)\restriction$. Now the computational strength of the systems $\Sigma^1_1 - AC$ and $(\Sigma^1_1 - AC)\restriction$ is known from results of H. Friedman, re-established by proof-theoretical methods in [Feferman and Sieg 1981]. These lead to the following:

Theorem 13. *(i)* $QL_2(\lambda) \equiv RA_{<\epsilon_0}$ *(Ramified Analysis in all levels* $<\epsilon_0$*), and (ii)* $QL_2(\lambda)\restriction \equiv PA$.

The idea of the proof is simply to reduce $QL_2(\lambda)$ to $EM_0 + J$ by eliminating the GCA (Axiom VI). This is accomplished by a translation first of $QL_2(\lambda)$ into $QL_2(\lambda) - (VI)$ which replaces, step by step, each occurrence of a formula of the form $t \in T$ where $T = \{x \mid \phi[x]\}$ by $\phi[t]$.[8] Then it is easy to interpret $QL_2(\lambda) -$VI into $EM_0 + J$ by taking the range of the type variables to be the objects satisfying Cl and taking $x \dot{\in} y$ to be the relation $Cl(y) \wedge x \in y$. The same translations serve to reduce $QL_2(\lambda)\restriction$ to $EM_0\restriction +J$. The theorem then follows by the known proof-theoretical results for $EM_0 + J, EM_0\restriction +J$ mentioned above.

The characterization of $Comp(RA_{<\epsilon_0})$ and thence of $Comp_N(QL_2(\lambda))$ may be given in terms of the provably recursive ordinals of ramified analysis,

[8] If we had started with the representation relation Rep instead of $\dot{\in}$ as basic this reduction would have been more complicated; cf. §8, Remarks 8° and 9°.

as obtained by Schütte and Tait; cf., e.g. [F 1977] pp.957–959. This will not be described here.

Using the basic idea of $QL_2(\lambda)$, one may set up a corresponding quantifier-free logic $PL_2(\lambda)$ by introducing a type symbol U whose intended interpretation is the class (or type) of all U-types. In addition, one has a formal operation $\sum_S t$ (or $\sum_{x:S} tx$) satisfying inference rules like:

$$(1) \qquad \frac{\Gamma \vdash S : U \quad \Gamma, x : S \vdash tx : U}{\Gamma \vdash \sum_S t : U}$$

$$(2)(\textstyle\sum\text{-}I) \qquad \frac{\Gamma \vdash S : U \quad \Gamma, x : S \vdash tx : U \quad \Gamma \vdash s_1 : S \quad \Gamma \vdash s_2 : ts_1}{\Gamma \vdash (s_1, s_2) : \sum_S t}$$

and a corresponding rule ($\sum$-E) for $\sum$-elimination. One introduces similarly a formal operation $\prod_S t$ with associated rules. Types in general (the G-Types in $PL_2(\lambda)$) are also taken to be closed under $\sum$ and $\prod$. Finally, the induction rule (N-E) in $PL_2(\lambda)$ is now allowed to apply to *all* types of the system.

A precise formulation of such a system $PL_2(\lambda)$ can be extracted from that of [Mitchell and Harper 1988], whose logic XML employs two universes U_1 and U_2, with U_1 like our U and U_2 like our "types in general", except that the closure conditions on U_1 are strengthened to (1), and rules (N-I) and *full* (N-E) are to be added (but individual terms are not explicitly typed).

The system $PL_2(\lambda)$ thus described is interpretable in $QL_2(\lambda)$ and hence $Comp_N(PL_2(\lambda)) \subseteq Comp_N(QL_2(\lambda))$. I have not gone through a proof of the reverse inclusion, but am convinced this should be so. It is for this reason that the statement

$$(3) \qquad Comp_N(QL_2(\lambda)) \equiv Comp_N(PL_2(\lambda))$$

must be treated as a conjecture (though highly plausible).

There is no problem to expand $QL_2(\lambda)$ to a system $QL_2(\Lambda)$ admitting explicit polymorphism and of the same computational strength, as shown by an extension of the Girard-Troelstra "erasing" translation. The step from $PL_2(\lambda)$ to $PL_2(\Lambda)$ of the same strength is similar; then $PL_2(\Lambda)$ has an appearance closer to the logic XML of [Mitchell and Harper 1988].[9]

[9] John Mitchell has brought to my attention the paper [Harper, Mitchell and Moggi 1990] which provides a refinement of the earlier Mitchell and Harper work, by building in the distinction between inferences at compile time and at run time. It is a question whether logics of QL style can be designed to incorporate such distinctions.

21. Algebraic Varieties and Abstract Data Types

An *algebraic variety* in a rather general sense is a collection $\mathcal{K}_\phi$ of many-sorted structures $\mathcal{A} = (X_1, \ldots, X_m, z_1, \ldots, z_n)$ of the same similarity type[10] satisfying some property $\phi[X, z]$, that is

$$\mathcal{K}_\phi := \{(X, z) \mid \phi[X, z]\} \ . \tag{1}$$

These may be treated as G-Types in the $QL_i(\lambda)$ systems using the representation relation as follows, when ϕ is a formula of such a system. Let

$$\mathcal{K}_\phi^{(U)} := \{(x, z) \mid \exists X(Rep(x, X) \wedge \phi[X, z])\} \ . \tag{2}$$

Then $\mathcal{K}_\phi^{(U)}$ is a G-Type which represents $\mathcal{K}_\phi$. In the simplest case, $\mathcal{K}_\phi$ consists of single-sorted structures $\mathcal{A} = (X, z)$. In this case let

$$T_\phi[x] := \{z \mid \exists X(Rep(x, X) \wedge \phi[X, z]\} \ . \tag{3}$$

Then we may rewrite $\mathcal{K}_\phi^{(U)}$ in the form

$$\mathcal{K}_\phi^{(U)} = \sum_{x:U} T_\phi[x] \tag{4}$$

where

$$U := \{x \mid Rep(x)\} \ . \tag{5}$$

By abuse of notation we may thus regard $\mathcal{K}_\phi$ as being given in the form

$$\mathcal{K}_\phi = \sum_X T[X] \ . \tag{6}$$

Hence the use of the $\mathcal{K}_\phi^{(U)}$'s allows us to generalize the approach to *abstract data types* of [Mitchell and Plotkin 1984] (cf. also [Cardelli and Wegner 1985]).

In the $QL_i(\Lambda)$ systems we can define operations F on $\mathcal{K}_\phi$ by Λ-abstraction. To fix matters, take $i = 1$. An example is given in [F 1990] pp.121–122, to describe polymorphic sorting as an operation which applies to the collection $\mathcal{K}_{\phi_{Lin}}$ where $\phi_{Lin}[X, z]$ holds if (X, z) has $z : X^2 \to \{0, 1\}$, a characteristic function of a linear ordering on X. Here $T_\phi[X] = \{z \mid \phi[X, z]\}$ in what was denoted $Lin[X]$ loc.cit. Then if $y \in X^n$ and sort$[z, y]$ is any uniform way of sorting y w.r. to z, the polymorphic sorting operation may be given as

$$s = \Lambda X(\lambda z : Lin[X])(\lambda n : N)(\lambda y : X^n)\text{sort}[z, y] \tag{7}$$

[10]Given by specifying for each z_i that $z_i : X_{k_1} \times \cdots \times X_{k_{p_i}} \to X_{j_i}$ (z_i possibly partial, and $z_i \in X_{j_i}$ when $p_i = 0$), or $z_i : X_{k_1} \times \cdots \times X_{k_{p_i}} \to \{0, 1\}$ $(p_i > 0)$.

This may be regarded as of the form $(\Lambda(X, z) : \mathcal{K}_{\phi_{Lin}})t[X, z]$ for an operation on $\mathcal{K}_{\phi_{Lin}}$, and we can also make sense of it in the form $(\lambda(x, z) : \mathcal{K}^{(U)}_{\phi_{Lin}})t[x, z]$ using the representation (2) above. In any case, we can establish in $QL_1(\Lambda)$ (or even $QL_0(\Lambda)$),

$$s \in \prod_X (Lin[X] \to \prod_{n \in N} (X^n \to X^n)) \tag{8}$$

Other examples from computer science come readily to mind (cf. also [Mitchell and Plotkin 1984] and [Cardelli and Wegner 1985]. For example, the class of structures which correspond to stacks X over a type Y is given by the class $\mathcal{K}_{\phi_{Stack}}$ of structures (X, Y, z_1, z_2, z_3) where $z_1 : X \to Y$ is partial and corresponds to "pop" on X, $z_2 : X \times Y \to X$ is "push", and $z_3 : X \to \{0, 1\}$ is the predicate "is nil", satisfying the usual conditions. We can also form the operation $\mathcal{K}_{\phi_{Stack}}(Y) = \{(X, z) \mid (X, Y, z) \in \mathcal{K}_{\phi_{Stack}}\}$ of stack structures over Y. Similarly we can form representing G-Types $\mathcal{K}^{(U)}_{\phi_{Stack}}(Y)$.

Elaboration of these ideas, with further applications, is intended for a subsequent publication.

22. Open Questions

(1) Can the main results for $i = 0, 1$ be improved to:

$$QL_i(\lambda) \vdash (s_1 \in S_1 \wedge \cdots \wedge s_n \in S_n \to t \in T) \Leftrightarrow$$
$$PL_i(\lambda) \text{ proves } (s_1 : S_1, \ldots, s_{n_1} S_n \vdash t : T)$$

for U-Types $S_1, \ldots, S_n, T$? (Similarly for equality inferences.) Note this is not established by the present work even for $n = 0$ and $T = N \to N$. The main theorems here only say that the provably computable functions on N, considered extensionally, are the same for $PL_i(\lambda)$ as for $QL_i(\lambda)$.

(2) Is there a suitable pair of logics $QL^{(-)}, PL^{(-)}$ below $QL_0(\lambda)$, with

$Comp_N(QL^{(-)}) = Comp_N(PL^{(-)}) =$ polynomial-time computable functions?

References

[M. Beeson 1985] *Foundations of Constructive Mathematics*, Springer-Verlag, New York.

[L. Cardelli and P. Wegner 1985] *On understanding types, data abstraction and polymorphism,* ACM Computing Surveys **17** (1985), 471–522.

[S. Feferman 1975] *A language and axioms for explicit mathematics*, in *Algebra and Logic,* Lecture Notes in Mathematics **450**, 87–139.

[________1977] *Theories of finite type related to mathematical practice*, in *Handbook of Mathematical Logic*, North-Holland, Amsterdam, 913–971.

[________1979] *Constructive theories of functions and classes*, in *Logic Colloquium '78*, North-Holland, Amsterdam, 159–224.

[________1989] *Finitary inductively presented logics*, in *Logic Colloquium '88*, North-Holland, Amsterdam, 191–220.

[________1988] *Hilbert's program relativized*, J. Symbolic Logic **53**, 364–384.

[________1990] *Polymorphic typed lambda-calculi in a type-free axiomatic framework*, in *Logic and Computation*, Contemporary Mathematics **104**, A.M.S., Providence, 101–136.

[S. Feferman and W. Sieg 1981] *Proof theoretic equivalences between classical and constructive theories for analysis*, in *Iterated inductive definitions and subsystems of analysis*, Lecture Notes in Mathematics **897**, 78–142.

[K. Gödel 1958] *Über eine bisher noch nicht benützte Erweiterung des finiten Standpunktes*, Dialectica **12**, 280–287. (Reproduced, with English translation in [Gödel 1990], 240–251).

[________1990] *Collected Works, Vol. II* Oxford University Press, New York.

[R. Harper, J.C. Mitchell and E. Moggi 1990] *Higher-order modules and the phase distinction*, ACM Principles of Programming Languages (to appear).

[G. Jäger 1988] *Induction in the elementary theory of types and names*, (preprint).

[J.C. Mitchell and R. Harper 1988] *The essence of ML*, Proc. 15th Annual ACM Symp. on Principles of Programming Languages, 28–46.

[J.C. Mitchell and G.D. Plotkin 1984] *Abstract types have existential type*, Proc. 12th Annual ACM Symposium on Principles of Programming Languages, 37–51.

[C. Parsons 1970] *On a number-theoretic choice schema and its relation to induction*, in *Intuitionism and Proof Theory*, North-Holland, Amsterdam, 459–473.

[W. Sieg 1985] *Fragments of Arithmetic*, J. of Pure and Applied Logic **28**, 33–72.

[W.W. Tait 1965] *Infinitely long terms of transfinite type*, in *Formal systems and recursive functions* North-Holland, Amsterdam, 176–185.

[A.S. Troelstra 1990] Introductory note to [Gödel 1958] in [Gödel 1990], 217–241.

Department of Mathematics, Stanford University, Stanford, CA 94305–2125, USA

TRANSPARENT GRAMMARS

KIT FINE

1. INTRODUCTION

'Cat' is a word which occurs in 'cattle', but it does not occur as a word; '1+2' is a term which occurs in '1+2.3', but it does not occur as a term. All such occurrences of expressions might be said to be accidental, since they are accidents of how the syntax of the language happens to be realized.

The notion of accidental occurrence is significant in various areas of thought. In logic, it greatly aids the formulation and proof of meta-logical results if it can be assumed that the underlying language contains no accidental occurrences. For example, a subformula can then simply be defined as a formula which occurs within a given formula rather than as an expression which is thrown up by a parsing of that formula. In philosophy, the issue of whether one can quantify into modal contexts has been seen to turn on such questions as to whether the occurrence of '9' in 'necessarily, $9 > 7$' is accidental or not; and the absence of accidental occurrence has been regarded as a condition on any "ideal language". In computer science and in linguistics, the presence of accidental occurrences has an obvious relevance to parsing, since they lead to the danger that a parser might mistake an apparent constituent of the expression to be parsed for a genuine constituent.

Let us say that a language or grammar is transparent if it permits no accidental occurrences. It is the main purpose of the present paper to investigate the conditions under which a context-free grammar is transparent. It is shown how any accidental occurrence reduces to a certain "primitive" case; and it is shown how such primitive occurrences might be detected. On the basis of these results on reduction and detection, an effective test for transparency is then given.

The concept of transparency represents a strengthening of the more familiar concept of nonambiguity. Any transparent grammar, at least of a well-behaved sort, is unambiguous, though not every unambiguous grammar is transparent. Moreover, what is required for many purposes is not merely an unambiguous but a transparent grammar. It is therefore significant in this regard that there is an effective test for the stronger property even though there is no effective test for the weaker one.

The plan of the paper is as follows. The first two sections introduce the relevant notions from the theory of context-free grammars. The third section explains the connection between nonambiguity and transparency. The fourth section establishes the reduction of accidental occurrence to the primitive case. The next three sections deal with the question of detecting the primitive accidental occurrences: a more fully articulated or canonical version of the given grammar is introduced; it is shown how accidental occurrences in the given grammar correspond to certain kinds of expression in the canonical grammar; and a precedence analysis is given of those expressions in the canonical grammar which correspond to the primitive accidental occurrences in the given grammar. An effective test for transparency is then provided in the final section.

The treatment of transparency in this paper has been very brief. Many of the results can be extended; and I have given a much fuller account in Fine [1].

2. Grammars

I adopt standard notation and terminology from the theory of strings and context-free grammars. The reader may consult Harrison [2] for further details.

Some of my terminology is not so standard. I use *precedes* and *succeeds* to mean immediately precedes and immediately succeeds. The string β is said to *begin* the string α if α is of the form $\beta\gamma$ for some string γ; and β is said to *end* α if α is of the form $\gamma\beta$ for some string γ. β is said to *properly* begin (or end) α if it begins (ends) α but is not identical to α.

I talk about occurrences in an intuitively evident, though not always rigorous, way. I take the null string to have an occurrence at any position in a string. Thus in the string ab, there are three occurrences of the null string: one at the beginning; one between a and b; and one at the end. Any occurrence of the null string is said to be a *null* occurrence.

Quite often, $\underline{\beta}$ will be used to symbolize an occurrence of a string β. A string α containing an occurrence $\underline{\beta}$ will be represented by $\alpha(\underline{\beta})$ (or sometimes by $\alpha(\beta)$). The result of replacing the occurrence $\underline{\beta}$ by γ will then be symbolized by $\alpha(\beta)$.

Two occurrences are said to be *disjoint* if they have no non-null occurrence in common. The *union of* two occurrences in a string is the shortest occurrence to contain them both. Suppose that two occurrences in a string have an occurrence in common (this includes the case in which they are adjacent and hence have a null occurrence in common). The *intersection of* the two occurrences is then taken to be the longest occurrence contained in them both.

Each string of symbols from the alphabet of a context-free grammar G is uniquely of the form $u_0A_1u_1 \ldots u_{n-1}A_nu_n, n \geq 0$, where $A_1, \ldots, A_n$ are variables and $u_0, u_1, \ldots, u_n$ are (possibly null) strings of terminals. Hence each production P is uniquely of the form $A \to u_0A_1u_1 \ldots u_{n-1}A_nu_n$. We call n the *degree* of the production P and, for $0 \leq i \leq n$, call u_i the i-th *operator of* P and denote it by o_i^P. Some, even all, of the operators for a given production may be the null string. The sequence $\langle u_0, u_1, \ldots, u_n\rangle$ of operators is called the *operator stencil of* P and the sequence $\langle A_1, \ldots, A_n\rangle$ of variables is called the *variable imprint of* P.

Recall that a *null* production is one of the form $A \to \wedge$ and that a *chain* production is one of the form $A \to B$. A grammar G is *null-free* if it contains no null productions; G is *invertible* if no two of its productions have the same variable on the left; and G is *chain-free* if it contains no chain productions. It may be shown by a straightforward induction that each of the terms of a null-free grammar G is non-null.

Any string which can be derived (in zero or more steps) from a variable of the grammar is a *term* or *phrase*. A phrase may contain variables and, in particular, each variable is itself a phrase. If a phrase contains no variables, it is said to be *ground*. A phrase t is said to be an A-phrase or a phrase of *category* A if it is derivable from the variable A. The A-phrases can be defined directly by means of the following induction:

(i) each variable A is an A-phrase;

(ii) if $\alpha B\gamma$ is an A-phrase and $B \to \beta$ is a production of the grammar, then $\alpha\beta\gamma$ is an A-phrase.

A *subphrase* s of a phrase t is any substring of t which is itself a phrase. A phrase is *elementary* if it appears on the right-hand side of some production.

We define what it is for $\mathbf{t}$ to be a *parse* or *syntactic analysis of* the phrase t. (A parse, as defined here, corresponds to the usual notion of a derivation tree.)

(i) if A is a variable, then $\langle A \rangle$ is an A-parse of A;

(ii) if $P = A \to u_0 A_1 u_1 \ldots u_{n-1} A_n u_n$ is a production of the grammar and $\mathbf{t}_1, \ldots, \mathbf{t}_n$ are respectively A_1-, ..., A_n-parses of $t_1, \ldots, t_n$, then $\langle P, \mathbf{t}_1, \ldots, \mathbf{t}_n \rangle$ is an A-parse of $u_0 t_1 u_1 \ldots u_{n-1} t_n u_n$.

It should be evident that each A-phrase has an A-parse. Parses, as so defined, correspond in the obvious way to derivation-trees.

We shall generally use $\mathbf{t}$ to symbolize a parse of the phrase t; and we sometimes talk loosely of $\mathbf{t}$ itself as a phrase. We usually write a parse $\langle P, \langle A_1 \rangle, \ldots, \langle A_n \rangle \rangle$, where $A_1, \ldots, A_n$ are variables, in the simplified form $\langle P \rangle$.

Suppose that the parse $\mathbf{t}$ is of the form $\langle P, \mathbf{t}_1, \ldots, \mathbf{t}_n \rangle$ (as under clause (ii)). P is then called the *principal* production of $\mathbf{t}$; and the parses $\mathbf{t}_1, \ldots, \mathbf{t}_n$ (and the corresponding parse occurrences) are called the *immediate components* of $\mathbf{t}$. The *subparses* of a parse are defined by induction in the obvious way: each parse is a subparse of itself; and each subparse of an immediate component is a subparse.

There is a clear sense in which a parse $\mathbf{t}$ can contain *occurrences* of subparses. But I shall no more bother to make talk of occurrence precise in this case than in the case of subphrases or substrings.

The *complexity* of a parse $\mathbf{t}$ is given by the number of times the second clause above has been applied in its generation:

(i) $\text{complexity}(\langle A \rangle) = 0$;

(ii) $\text{complexity}(\langle P, \mathbf{t}_1, \ldots, \mathbf{t}_n \rangle) = 1 + \text{complexity}(\mathbf{t}_1) + \cdots + \text{complexity}(\mathbf{t}_n)$.

A parse $\mathbf{t}$ is said to be *simple* if it is of complexity 0, i.e. if it is of the form $\langle A \rangle$ for A a variable; otherwise it is said to be *complex.* A parse is said to be *elementary* if it is of complexity 1, i.e. of the form $\langle P, \langle A_1 \rangle, \ldots, \langle A_n \rangle \rangle$.

3. Accidental Occurrence

If $\mathbf{t}$ is a parse of t, then to each occurrence $\underline{\mathbf{s}}$ of a subparse $\mathbf{s}$ in $\mathbf{t}$ there corresponds an occurrence $\underline{s}$ of a subphrase s in t. The occurrences which so correspond under the given parse $\mathbf{t}$ are said to be the *syntactic* or *constituent* occurrences in t *under* the parse $\mathbf{t}$; and those which correspond under some parse of t are said to be the *syntactic* or *constituent* occurrences, simpliciter, in t. Occurrences which correspond to occurrences of A-parses are said to be A-constituent occurrences. Of course, every constituent occurrence is of a subphrase; but not every occurrence of a subphrase need be constituent occurrence.

A somewhat more rigorous definition of constituency is as follows:

(i) the occurrence $\underline{t}$ in t is a constituent occurrence of t under the parse $\mathbf{t}$;

(ii) if $\mathbf{t}$ is the parse $\langle P, \mathbf{t}_1, \ldots, \mathbf{t}_n \rangle$ and t is the phrase $u_0 t_1 u_1 \ldots u_{n-1} t_n u_n$, then the constituent occurrences in $t_1, \ldots, t_n$ under the respective parses $\mathbf{t}_1, \ldots, \mathbf{t}_n$ are constituent occurrences in t under the parse $\mathbf{t}$.

(Strictly, it is not the constituent occurrences in $t_1, \ldots, t_n$ which are constituent occurrences in t, but rather the occurrences in t which correspond to those in $t_1, \ldots, t_n$. In general, we will feel free to identify an occurrence within a substring with the corresponding occurrence within the given string). If we think in terms of the more usual concept of a derivation-tree, then the constituent occurrences will be determined by the subtrees whose root is labelled by a variable.

We state some results concerning the connection between typographic and syntactic occurrence. Suboccurrence and disjointness are reflected typographically:

Lemma 1. *Suppose that* $\underline{\mathbf{r}}$ *and* $\underline{\mathbf{s}}$ *are subparse occurrences in* $\mathbf{t}$. *Then:*

(i) $\underline{r}$ *is a suboccurrence of* $\underline{s}$ *if* $\underline{\mathbf{r}}$ *is a suboccurrence of* $\underline{\mathbf{s}}$*;*

(ii) $\underline{r}$ *is disjoint from* $\underline{s}$ *if neither* $\underline{\mathbf{r}}$ *nor* $\underline{\mathbf{s}}$ *is a suboccurrence of the other.*

Proof. By a straightforward induction on the complexity of $\mathbf{t}$. □

It follows from this result that constituent occurrences cannot properly overlap: if $\underline{r}$ and $\underline{s}$ are two constituent occurrences of $\mathbf{t}$, then either one is a suboccurrence of the other or they are disjoint.

A fundamental feature of constituent occurrences is their syntactic replaceablity:

Theorem 2 (Replacement). *Suppose that* $\mathbf{t} = \mathbf{t}(\underline{\mathbf{s}})$ *is a parse for* $t(\underline{s})$ *and that* $\mathbf{s}$ *and* $\mathbf{s}'$ *are both A-parses. Then* $\mathbf{t}(\mathbf{s}')$ *is a parse for* $t(s')$.

Proof. By induction on the complexity of $\mathbf{t}$. □

From this result it follows that $t(r)$ is a phrase whenever $\underline{s}$ is an A-constituent occurrence of $t = t(\underline{s})$ and r is an A-phrase.

Suppose that s is a phrase and that $\underline{s}$ is an occurrence of s in the phrase with parse $\mathbf{t}$; $\underline{s}$ is then said to be an *accidental* or *merely typographic* occurrence under the parse $\mathbf{t}$ if it is not a constitutent occurrence under $\mathbf{t}$. Thus the constituent occurrences of phrases occur *as* phrases, while the accidental occurrences do not. They have the status of a phrase when considered on their own, but not when considered in the context of the given phrase.

It should be emphasized that the notion of accidentality, as it has been defined, is relative to a parse $\mathbf{t}$ of the phrase t. Thus it is perfectly possible that a given occurrence in a phrase might be accidental under one of its parses and not under another.

A parsed phrase is *transparent* if it contains no accidental occurrences of subphrases. A phrase is *transparent* if it contains no accidental occurrence under any of its parses. And a grammar is *transparent* if each of its phrases is transparent. Thus a transparent grammar is one that admits no accidental occurrences of phrases. A parsed phrase, phrase or grammar is said to be *opaque* if it is not transparent.

We shall need two results on the preservation of accidental occurrences under embedding and under the replacement of constituents.

Lemma 3. *Suppose that* $\underline{\mathbf{s}}$ *is a subparse occurrence in* $\mathbf{t}$ *and that* $\underline{r}$ *is an accidental occurrence of the non-null phrase* r *in* $\mathbf{s}$. *Then* $\underline{r}$ *(or rather the corresponding occurrence in* t*) is an accidental occurrence of* r *in* $\mathbf{t}$.

Proof. Suppose, for reductio, that $\underline{r}$ is a constituent occurrence in $\mathbf{t}$. Then $\underline{r}$ corresponds to some subparse occurrence $\underline{\mathbf{r}}$ in $\mathbf{t}$. Now $\underline{\mathbf{r}}$ is not a subparse occurrence of $\underline{\mathbf{s}}$, since otherwise $\underline{r}$ is a constituent occurrence in $\mathbf{s}$ after all. Nor is $\underline{\mathbf{s}}$ a subparse occurrence of $\underline{\mathbf{r}}$; since otherwise, by lemma 1(i), $\underline{s}$ is a suboccurrence in $\underline{r}$ and so $\underline{s}$ and $\underline{r}$ are the same. Given that neither $\underline{\mathbf{r}}$ nor $\underline{\mathbf{s}}$ has an occurrence within the other, it follows by lemma 1(ii) that $\underline{r}$ and $\underline{s}$ are disjoint. But this is impossible given that $\underline{r}$ is a non-null suboccurrence in $\underline{s}$. □

Lemma 4. *Suppose that* $\underline{\mathbf{r}}$ *is an A-parse occurrence in* $\mathbf{t} = \mathbf{t}(\underline{\mathbf{r}})$, *that* $\underline{s} = \underline{s}(\underline{r})$ *is an accidental occurrence in* $\mathbf{t}$ *containing the occurrence* $\underline{r}$ *as an A-constituent, and that* $\mathbf{r}'$ *is an A-parse. Then* $\underline{s}(r')$ *is an accidental occurrence in* $\mathbf{t}(\mathbf{r}')$.

Proof. By Replacement, $\underline{s}(r')$ is a phrase. Now suppose $\underline{s}(r')$ corresponded to a subparse occurrence in $\mathbf{t}(\mathbf{r}')$. Then again by Replacement, $\underline{s} = \underline{s}(\underline{r})$ would correspond to a subparse occurrence in $\mathbf{t} = \mathbf{t}(\mathbf{r})$. □

4. Opacity and ambiguity

A phrase of a grammar is said to be *unambiguous* if it does not possess two distinct parses. The grammar itself is said to be *unambiguous* if each of its phrases is unambiguous. Note that the present definition of ambiguity is stronger than the standard one in that each phrase, and not merely each ground S-phrase, is required to be unambiguous.

Our main concern in this section is to show that, under certain very weak conditions, transparency implies nonambiguity. We need a critical lemma, which in its turn will rest on two simple facts.

Lemma 5. *Any transparent grammar is null-free.*

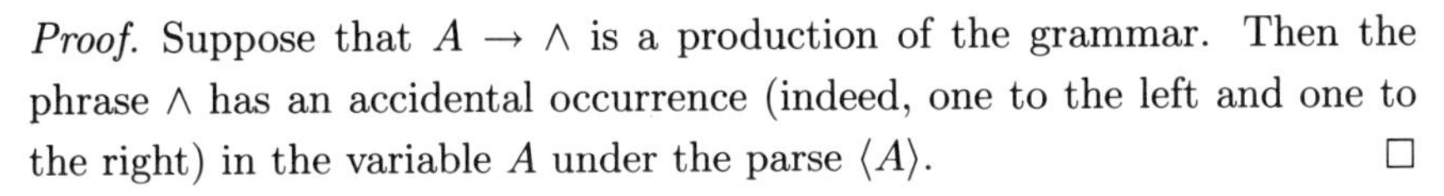

Proof. Suppose that $A \to \wedge$ is a production of the grammar. Then the phrase $\wedge$ has an accidental occurrence (indeed, one to the left and one to the right) in the variable A under the parse $\langle A \rangle$. □

Lemma 6. *Let G be chain- and null-free grammar. Then each variable is unambiguous.*

Proof. Suppose that an arbitrary phrase t has a parse of complexity > 0. Then it may be shown by an easy induction that either t contains a terminal symbol or is of length > 1. So any parse of a variable A must be of complexity 0; and so the only parse of A is $\langle A \rangle$. □

We can now show that there is no difference in parses without a difference in constituents.

Lemma 7 (Discrepancy)**.** *Let G be an invertible, chain-free and null-free grammar. Suppose that* $\mathbf{r}$ *and* $\mathbf{s}$ *are two distinct parses of the same phrase t. Then their constituent occurrences are not the same.*

Proof. By induction on the complexity k of $\mathbf{r}$. If $k = 0$, then t is a variable; and so by lemma 6, it is unambiguous and cannot possess two distinct parses.

Suppose that $k > 0$. Let $P = A \to u_0 A_1 u_1 \ldots u_{m-1} A_m u_m$ be the principal production of $\mathbf{r}$ and let $\mathbf{r}_1, \ldots, \mathbf{r}_m$ be its immediate components. Thus $r = t$ is of the form $u_0 r_1 u_1 \ldots u_{m-1} r_m u_m$. We can assume, given the basis of the induction applied to $\mathbf{s}$, that the complexity of $\mathbf{s}$ is also > 0; and so let $Q = B \to v_0 B_1 v_1 \ldots v_{n-1} B_n v_n$ be the principal production of $\mathbf{s}$ and let $\mathbf{s}_1, \ldots, \mathbf{s}_n$ be its immediate components. Thus $s = t$ is of the form $v_0 s_1 v_1 \ldots v_{n-1} s_n v_n$.

Suppose, for reductio, that the constituent occurrences of **r** and **s** are the same. We may then prove by a subinduction on $j \geq 0$ that the j-th term of the sequence $u_0, r_1, u_1, \ldots, u_{m-1}, r_m, u_m$ exists iff the j-th term of the sequence $v_0, s_1, v_1, \ldots, v_{n-1}, s_n, v_n$ exists and that, given their existence, they are the same. The case $j = 0$ is trivial (since we start counting the members of the sequences from 1). So suppose that $j > 0$. There are two cases depending upon whether j is odd or even.

Suppose first that j is odd and that the j-th term u^p of the first sequence exists (the case in which the j-th term v^p of the second sequence exists is analogous). By the hypothesis of the subinduction, the $(j-1)$-th term of the second sequence exists (should $j > 1$). But it corresponds to a variable; and so since the last term of the sequence must correspond to a string of terminals, the j-th term of the second sequence must exist.

Suppose that $u_p \neq v_p$. Then given the subinductive hypothesis, either u_p properly begins v_p or v_p properly begins u_p. Suppose u_p properly begins v_p (the other case is similar). Then the phrase r_{p+1} exists and either has an accidental occurrence in the parsed phrase **s** or is the whole phrase t. But in the latter case, given that no phrase is null, the principal production P of **r** must have been of the form $A \rightarrow A_{p+1}$, contrary to the supposition that the grammar is chain-free.

Suppose now that j is even and that the j-th term r_p of the first sequence exists. Since the grammar is null-free, r_p is not the null string. So since, by the subinduction hypothesis, the concatenation of the first $(j-1)$ phrases of the two sequences is the same, the j-th phrase s_p of the second sequence must exist.

Suppose that $r_p \neq s_p$. Then either r_p properly begins s_p or s_p properly begins r_p. Suppose that r_p properly begins s_p (the other case is similar). Then either s_p has an accidental occurrence in **r** or else s_p is the whole phrase t. But in the latter case, given that no phrase is null, the principal production Q of **s** must have been of the form $B \rightarrow B_j$, contrary to the grammar being chain-free.

We conclude that the two sequences $u_0, r_1, u_1, \ldots, u_{m-1}, r_m, u_m$ and $v_0, s_1, v_1, \ldots, v_{n-1}, s_n, v_n$ are the same. Suppose now that, for some i, $\mathbf{r}_i$ and $\mathbf{s}_i$ are distinct parses of the same phrase $r_i = s_i$. In this case, it follows by the main induction that a constituent occurrence $\underline{t}'$ of one of $\mathbf{r}_i$ and $\mathbf{s}_i$, say of $\mathbf{r}_i$, is an accidental occurrence of the other. But clearly $\underline{t}'$ is then a constituent occurrence of **r**; and it follows by lemma 3 that $\underline{t}'$ is an accidental occurrence of $\mathbf{s}$.

We may therefore suppose that, for each i, the parses $\mathbf{r}_i$ and $\mathbf{s}_i$ are the same. But then if the parses $\mathbf{r}$ and $\mathbf{s}$ are to be distinct, their principal productions must be distinct. But the right-hand sides of the productions must be the same, since otherwise the component parses $\mathbf{r}_i$ and $\mathbf{s}_i$ could not be the same. They therefore differ in their left-hand sides. But this is contrary to the supposition that the grammar is invertible. □

From the discrepancy lemma follows the main result of the section.

Theorem 8 (For invertible and chain-free grammars). *If the grammar G is transparent then it is unambiguous.*

Proof. Given that G is transparent, it follows from lemma 5 that it is null-free. Suppose now that the phrase r had two distinct parses. By lemma 7, their constituent occurrences could not be the same; and so a constituent occurrence of one would be an accidental occurrence of the other. □

The converse of this theorem does not hold. A simple counter-example arises from the practice of dropping outermost brackets. Thus suppose the production rules are: $S \rightarrow S' \vee S', S' \rightarrow (S' \vee S'), S' \rightarrow p$. The grammar is then unambiguous. But is is not transparent, since $p \vee p$ has an accidental occurrence in $(p \vee p)$.

5. Reduction

We establish a reduction of the question of transparency for a given grammar. It is shown, for any suitable grammar, that if one phrase occurs accidentally then so does a phrase of an especially simple sort.

Say that a phrase of a grammar G is *primitive* if it properly contains no elementary phrases. Then it may be shown that accidental occurrence can always be located in the primitive phrases of a grammar.

Theorem 9 (Reduction). *Suppose that the grammar G is chain-free and invertible. Then if some phrase of G occurs accidentally, some primitive phrase of G occurs accidentally.*

Proof. Let us first deal with the case in which G is not null- free. It is then clear that the null string $\wedge$ is a primitive phrase and that the parsed phrase $\langle A \rangle$, where A is any variable of the grammar, contains an accidental occurrence of $\wedge$.

So from henceforth let us assume that G is null-free. We may then prove by a double induction on $l =$ length$(s) > 0$ and the minimal complexity k of some parse $\mathbf{s}$ of s that: if s has an accidental occurrence in some phrase $\mathbf{t}$, then some primitive phrase has an accidental occurrence. (In the induction, we allow ourselves to decrease l while possibly increasing k or to decrease k without increasing l). For $l = 0$, the result is trivial, since no phrase of a null-free grammar is of length 0 (i.e., is the null string).

So suppose $l > 0$. If s is primitive, we are done. So suppose that s properly contains an elementary phrase r. Thus we can write s in the form $s(\underline{r})$, where length$(r) <$ length$(s) = l$. Let $\mathbf{s}$ be a parse of minimal complexity for s. If $\underline{r}$ is an accidental occurrence in either $\mathbf{s}$ or $\mathbf{t}$, we are done by the I.H. So we may suppose that r has a constituent occurrence in both $\mathbf{s}$ and $\mathbf{t}$.

Suppose $\underline{r}$ is an A-constituent in $\mathbf{s}$ and a B-constituent in $\mathbf{t}$. If A and B are distinct, then r has two parses, within $\mathbf{s}$ and $\mathbf{t}$ respectively; and so, by the Discrepancy Lemma, one of them contains an accidental occurrence $\underline{u}$. But length$(u) \leq$ length(r) and so, by the I.H., we are done. So suppose A and B are the same. By lemma 4, $s(A)$ has an accidental occurrence in $\mathbf{t}(\mathbf{s}(A))$. Given that the grammar G is chain-free, r is not a variable. So complexity$(\mathbf{s}(A)) <$ complexity$(\mathbf{s}(\mathbf{r}))$, and the I.H. applies. $\square$

Each primitive phrase (in any grammar) is either a variable or is elementary. For if it is not a variable then it must contain an elementary phrase; and given that it is primitive, it must improperly contain the elementary phrase, i.e. be identical to it. Moreover, no variable is capable of an accidental occurrence. So we see from the theorem that accidental occurrence can always be located in elementary phrases which are not variables and which do not themselves properly contain any elementary phrase.

The result is quite useful on its own in establishing that particular grammars are transparent. Suppose, for example, that we wished to show that the standard bracketed notation for sentential logic is transparent. Then it would follow from the theorem that if the notation were not transparent there would have to be an accidental occurrence of a primitive phrase, such as $-S$ or $(S\&S)$. But it is readily shown that there can be no such accidental occurrence.

From the Reduction Theorem we immediately obtain:

Corollary 10 (For G invertible and chain-free). *G is transparent iff it contains no accidental occurrences of primitive phrases.*

There is a way of lifting the restriction on the grammar in this and in some of the subsequent results; for each grammar G may effectively be transformed into a grammar G' which is invertible, chain-free and equi-transparent with G. The transformation requires a sensitive use of the standard techniques for putting a grammar into normal form and will not be given here.

6. Canonical Grammars

We now embark on the task of identifying accidental occurrences of primitive phrases. Rather than do this directly, we will find it useful to "track" the symbol occurrences in the given grammar by means of a more fully articulated grammar. In order to keep the two grammars apart, we shall refer to them as the *base* and the *canonical* grammars respectively. In this section we define the canonical grammars and characterize the sense in which they correspond to the base grammar.

A grammar G is said to be *canonical* if it satisifies the following two conditions: (i) each operator is a single symbol; (ii) any two operators o_i^P and o_j^Q are distinct when either P is distinct from Q or i is distinct from j. For example, if c, d, e, f, g and h are distinct symbols, then the grammar with rules $S \to aSbSc, S \to dSe, S \to f$ is canonical.

To each grammar G may be associated, in a natural manner, a canonical grammar G^c. With each production $P = A \to u_0 A_1 u_1 \ldots u_{n-1} A_n u_n$ of G, associate a production $P^c = A \to c_0 A_1 c_1 \ldots c_{n-1} A_n c_n$, where each c_i is a new symbol, not appearing in the grammar G or elsewhere in the associated productions. The *canonical grammar G^c for G* then consists of the associated productions P^c. (Its initial variable, if this is of any concern, can be taken to be the same as the initial variable of G).

It is evident that G^c is a canonical grammar and that it is unique up to the choice of the symbols c_i. If, for example, the grammar G is defined by the rules $S \to (S\&S)$, $S \to -S$ and $S \to p$, then the canonical grammar above is a canonical grammar for G. If G^c is the canonical grammar associated with G, then we may denote the i-th operator of P^c by c_i^P. (Thus in associating a canonical grammar G^c with a base grammar G, we must also set up an appropriate association between the operators of the two grammars).

There is a natural homomorphism h^c (with respect to concatenation) from the strings of a canonical grammar G^c into the strings of the base grammar G. The map h^c is determined by the conditions that $h^c(A) = A$ for each variable A of G and $h^c(c_i^P) = o_i^P$. Thus the i-th operator of P^c is mapped onto the i-th operator of P. In the case of our examples above, h^c is given by: $h^c(a) = ($, $h^c(b) = \&$, $h^c(c) =)$, $h^c(d) = -$, $h^c(e) = \wedge$ and $h^c(f) = p$.

It should be clear that h^c induces a natural one-one correspondence between the parses of G^c and the parses of G. Since the grammar G^c is unambiguous, there is a one-one correspondence between the phrases of G^c and their parses and, consequently, there is a one-one correspondence between the phrases of G^c and the parses of G. I shall talk about all such relationships of correspondence in an inexact but, what I hope, is an intuitively evident way; I shall also use $\mathbf{t}^c$ to represent the phrase from G^c which corresponds to the parse $\mathbf{t}$ of G.

We shall find it useful to deal with grammars in which the beginning and end of phrases is explicitly marked. Given a grammar G, the corresponding *framed* grammar G^f is obtained by adding to G two new terminal symbols [and] (called *markers*) and by adding, for each variable A of G, the new variable A' and the production $A' \to [A]$. There is again a natural homorphism h^f from the strings of the framed grammar G^f onto the strings of the given grammar G. This is obtained by mapping each symbol of G into itself, each new variable A' into A, and each of the marker symbols [and] into the null string. Under this map, each phrase t of G^f will map onto a phrase of G. Indeed, the A-phrases of G^f, for A a variable of G, are simply the A-phrases of G, and the A'-phrases of G^f are the strings of the form $[t]$ where t is an A-phrase of G.

The canonical grammar G^c may itself be framed. We thereby obtain the grammar $(G^c)^f$ and a natural map from its phrases into those of G. If $\mathbf{t}$ is a parse of the phrase t in G, then we take $\mathbf{t}^f$ to be the phrase $[\mathbf{t}^c]$.

7. Irregular Correspondence

We are interested in finding an intrinsic characterization of those strings of the canonical grammar which correspond to accidental occurrences in the base grammar.

Let $\mathbf{t}$ be a parse of a phrase t from the base grammar and let $\mathbf{t}^c$ be the corresponding phrase from the canonical grammar. Then each occurrence $\underline{u}^c$ of a substring of $\mathbf{t}^c$ corresponds to the occurrence $\underline{u}$ of a substring of t. However, the converse is not true; some occurrences of substrings of t do not correspond to any occurrences of substrings of $\mathbf{t}^c$. This is because an operator of G^c might be mapped into a string of two or more symbols. So if an occurrence of a substring of t "cut through" the image of such an operator, then no substring of $\mathbf{t}^c$ would correspond to it. Moreover, several different occurrences of a substring of $\mathbf{t}^c$ might correspond to the same occurrence of t. This is because an operator of G^c might be *degenerate*, i.e. be mapped onto the null string.

In order to take account of these possibilities, we will need to distinguish two somewhat more refined concepts of correspondence. Say that the string α of $(G^c)^f$ *frames* the string α' of $(G^c)^f$ if α is of the form $X\alpha'Y$, where X is either [or a nondegenerate symbol of G^c and Y is either] or a nondegenerate symbol of G^c. We call the pair $\langle X, Y\rangle$ the *frame* of α, and α' the *framed* string. Similarly, say that the occurrence $\underline{\alpha}$ *frames* the occurrence $\underline{\alpha}'$ if $\underline{\alpha}$ is of the form $\underline{X\alpha'Y}$, with X and Y as above. We can then call the pair $\langle \underline{X}, \underline{Y}\rangle$ the *frame of* $\underline{\alpha}$, and $\underline{\alpha}'$ the *framed* string occurrence.

We say that a string α of the framed canonical grammar $(G^c)^f$ *loosely corresponds* to a string β of the base grammar G if α frames a string α' which corresponds to β and if, in addition, no subphrase of α' (possibly identical to α' corresponds to β. We say that a string α *inexactly corresponds to* β if it corresponds to a string properly containing β and if any proper substring of α corresponds to a proper substring of β.

These notions may be extended to occurrences. Let $\mathbf{t}^f$ be a phrase of the framed canonical grammar which corresponds to the phrase t of the base grammar. We say that a substring occurrence $\underline{\alpha}$ in $\mathbf{t}^f$ *loosely corresponds to* a substring occurrence $\underline{\beta}$ in t if it frames an occurrence $\underline{\alpha}'$ which corresponds to $\underline{\beta}$ and which contains no subphrase occurrence which corresponds to $\underline{\beta}$. We say that a substring occurrence $\underline{\alpha}$ in $\mathbf{t}^f$ *inexactly corresponds to* a substring occurrence $\underline{\beta}$ in t if it corresponds to a string occurrence properly containing $\underline{\beta}$ and if any proper substring occurrence in $\underline{\alpha}$ corresponds to a proper substring occurrence in $\underline{\beta}$. Under loose correspondence, we take the longest occurrence corresponding to the given occurrence and then frame it. Under inexact correspondence, we take the *shortest* occurrence corresponding to a cover for the given occurrence.

Although an occurrence in $\mathbf{t}^f$ which loosely corresponds to the occurrence $\underline{\beta}$ in t may contain one or both of the markers [and], any occurrence in $\mathbf{t}^f$ which inexactly corresponds to $\underline{\beta}$ must be an occurrence in $\mathbf{t}^c$. Also, any occurrence in $\mathbf{t}^f$ which inexactly corresponds to $\underline{\beta}$ must be non-null, for otherwise it would correspond to a null occurrence. Moreover, no occurrence $\underline{\alpha}$ in $\mathbf{t}^f$ can either loosely or inexactly correspond to a variable occurrence $\underline{A}$ in t; for the occurrence of A would have to derive from a corresponding occurrence of A within $\underline{\alpha}$.

The two non-standard notions of correspondence can be combined. We say that a string α of $(G^c)^f$ *irregularly corresponds to* a string β of G if α either loosely or inexactly corresponds to β; and we say that an occurrence $\underline{\alpha}$ in the phrase $\mathbf{t}^f$ *irregularly corresponds to* an occurrence $\underline{\beta}$ in t if $\underline{\alpha}$ either loosely or inexactly corresponds to $\underline{\beta}$. Thus the irregularity either shows up in the inexactitude of the correspondence or in the absence of a corresponding phrase.

The accidental occurrences within the base grammar can now be "tracked" within the canonical grammar.

Lemma 11. *An occurrence $\underline{s}$ of the phrase s in $\mathbf{t}$ is accidental iff it irregularly corresponds to an occurrence $\underline{\alpha}$ in $\mathbf{t}^f$.*

Proof. Suppose that the occurrence $\underline{s}$ is accidental in $\mathbf{t}$. There are two cases:

(1) Some occurrence of a substring in $\mathbf{t}^c$ corresponds to the occurrence $\underline{s}$ in t. There is therefore a longest occurrence $\underline{\alpha}'$ of a substring of $\mathbf{t}^c$ which contains the given occurrence of $\mathbf{t}^c$ and which corresponds to the occurrence $\underline{s}$. Now let $\underline{\alpha}$ be $\underline{X\alpha'Y}$, where $\underline{X}$ is the symbol occurrence immediately to the left of $\underline{\alpha}'$ in $\mathbf{t}^f$ and $\underline{Y}$ is the symbol occurrence immediately to the right of $\underline{\alpha}'$ in $\mathbf{t}^f$. Then $\underline{\alpha}$ loosely corresponds to $\underline{s}$; for if either X or Y were degenerate symbols of G^c, the string occurrence $\underline{\alpha}'$ could be made longer. Also $\underline{\alpha}'$ contains no subphrase occurrence which corresponds to $\underline{s}$, since otherwise $\underline{s}$ would be a constituent occurrence in $\mathbf{t}$.

(2) No occurrence of a substring of $\mathbf{t}^c$ corresponds to the occurrence $\underline{s}$ in t. Let $\underline{\beta}$ be the shortest substring occurrence in t which both contains $\underline{s}$ and corresponds to a substring occurrence of $\mathbf{t}^c$. Let $\underline{\alpha}$ be the shortest substring occurrence of $\mathbf{t}^c$ to correspond to $\underline{\beta}$. It is then clear that $\underline{\alpha}$ inexactly corresponds to $\underline{s}$.

For the other direction, first suppose that the occurrence $\underline{s}$ loosely corresponds to the occurrence $\underline{\alpha}$ in $\mathbf{t}^c$. Thus $\underline{\alpha}$ is of the form $\underline{X\alpha'Y}$. where

$\langle \underline{X}, \underline{Y} \rangle$ is a frame for $\underline{\alpha}'$. If $\underline{s}$ were a constituent occurrence, it would correspond to the occurrence $\underline{r}$ of a subphrase r of $\mathbf{t}^c$. Since $\underline{\alpha}'$ and $\underline{r}$ both correspond to $\underline{s}$ and since $\underline{\alpha}'$ is framed, $\underline{r}$ must be then be a suboccurrence of $\underline{\alpha}'$, contrary to the definition of loose correspondence.

Next suppose that the occurrence $\underline{s}$ inexactly corresponds to the substring occurrence $\underline{\alpha}$ in $\mathbf{t}^c$. If $\underline{s}$ were a constituent occurrence, then it would correspond to the occurrence $\underline{r}$ of a subphrase of $\mathbf{t}^c$. If $\underline{\alpha}$ and $\underline{r}$ had no intersection, then $\underline{s}$ would be null; and so, since $\underline{\alpha}$ is non-null, it would contain a proper substring which corresponded to $\underline{s}$. Therefore $\underline{\alpha}$ and $\underline{r}$ have an intersection. Let it be $\underline{\gamma}$. Then it is clear that $\underline{\gamma}$ corresponds to $\underline{s}$ and is a proper suboccurrence of $\underline{\alpha}$, contrary to the definition of inexact correspondence. □

Lemma 11 has a formulation for phrases as opposed to occurrences. Say that α is a *phrase-part* if it is a substring of a phrase. Then:

Lemma 12. *A phrase s of G has an accidental occurrence iff it irregularly corresponds to a phrase-part of $(G^c)^f$.*

Proof. Suppose s has an accidental occurrence $\underline{s}$ in a phrase $\mathbf{t}$. By lemma 11, $\underline{s}$ irregularly corresponds to a phrase-part $\underline{\alpha}$ of $\mathbf{t}^f$. But then s irregularly corresponds to the phrase-part α. Now suppose s irregularly corresponds to a phrase-part α. Let $\mathbf{t}^f$ be a phrase containing an occurrence $\underline{\alpha}$ of α; and let $\underline{s}$ be the corresponding occurrence of s in $\mathbf{t}$. Then by lemma 11, $\underline{s}$ is accidental in $\mathbf{t}$. □

From lemma 12 we then obtain a criterion for transparency in terms of iregular correspondence:

Theorem 13 (Irregular Correspondence). *The grammar G is transparent iff no phrase-part of $(G^c)^f$ irregularly corresponds to a phrase of G.*

This criterion may be simplified with the aid of the reduction to primitive occurrence:

Theorem 14. *Suppose G is chain-free and invertible. Then G is transparent iff no phrase-part of $(G^c)^f$ irregularly corresponds to a primitive phrase of G.*

Proof. By definition, G is transparent iff no phrase has an accidental occurrence. But by theorem 9, this holds iff no primitive phrase has an accidental occurrence; and by lemma 12, this holds iff no phrase-part of $(G^c)^f$ irregularly corresponds to a primitive phrase of G. □

Being primitive has an intrinsic significance within the canonical grammar. Say that the string β *embeds* α if, for some non-null strings γ_1 and γ_2, α is of the form $\gamma_1\beta\gamma_2$; and say that a string (not necessarily a phrase) of the framed canonical grammar is *quasi-primitive* if it embeds no elementary phrase. We then have:

Lemma 15. *Suppose that the string α of $(G^c)^f$ irregularly corresponds to the primitive phrase s of G. Then α is quasi-primitive.*

Proof. Suppose that α is not quasi-primitive. Then it is of the form $\beta u\gamma$, for u an elementary phrase of G^c and β and γ non-null strings. So s contains an elementary phrase v corresponding to the occurrence of u in α. Suppose α loosely corresponds to s. Then α is of the form $X\beta'u\gamma'Y$, where $\langle X, Y\rangle$ is a frame for $\alpha' = \beta'u\gamma'$. But β' and γ' cannot both correspond to the null string, since then the subphrase u of α' would also correspond to s. So s properly contains u. On the other hand, suppose α inexactly corresponds to s. Then since u is embedded in α, it must correspond to a proper substring of s. □

The criterion for transparency can now be further simplified:

Corollary 16. *Suppose G is chain-free and invertible. Then G is transparent iff no quasi-primitive phrase-part of $(G^c)^f$ irregularly corresponds to a phrase of G.*

Proof. The left-to-right direction follows from theorem 13. The right-to-left direction follows from theorem 14 and lemma 15. □

8. The Rules of Precedence

In this section, we analyse symbol order within a canonical grammar and then use the analysis to characterize phrase-parts and quasi-primitivity. We shall show that there is an intrinsic criterion for when a quasi-primitive string is a phrase-part. This criterion yields a very simple test for phrase-parthood, one which requires only that we scan the string from left to right.

With each operator u_i in a production $P = A \rightarrow u_0A_1u_1 \ldots u_{n-1}A_nu_n$ may be associated three types: a governing, a successor, and a predecessor type. The *governing type of u_i in P* is the variable A. The *successor type of u_i in P* is A_{i+1} if $i < n$ (and otherwise does not exist). The *predecessor type of u_i in P* is A_{i-1} if $i > 0$ (and otherwise does not exist). An operator with no predeccesor type and with a successor type is said to be a *prefix* (in the given production); an operator with no successor type and with a predecessor type is said to be a *suffix*; an operator with both a predeccesor and a successor type is said to be an *infix*; and an operator with neither a predecessor nor a successor type is said to be a *constant*. A prefix (suffix, infix, constant) of governing type A is said to be an A-prefix (A-suffix, A-infix, A-constant). An *A-atom* is taken to be either the variable A itself or an A-constant, i.e. a string u of terminals which occurs in a production of the form $A \to u$. In a canonical grammar, the production is determinable from the operator and so reference to it may be dropped.

With each operator might be associated a *comprehensive type*, which is an ordered triple $\langle X, Y, Z\rangle$ consisting of the predecessor type, the successor type and the governing type (where a blank '–' is left whenever a type is undefined). Thus the comprehensive type of a prefix is of the form $\langle -, B, C\rangle$, that of a suffix is of the form $\langle A, -, C\rangle$, and that of an infix is of the form $\langle A, B, C\rangle$.

We require the following elementary results on the occurrence of operators in the terms of a canonical grammar:

Lemma 17 (For G canonical). *Any A-phrase, if it is not an A-atom, begins with an A-prefix and ends with an A-suffix.*

Lemma 18 (For G canonical). *Suppose that c_i occurs in a phrase t. Then it occurs as part of a syntactic occurrence of a subphrase of the form $c_0 t_1 c_1 \ldots c_{n-1} t_n c_n$.*

The proofs are straightforward and are left to the reader.

We may now define the rules of precedence for a framed canonical grammar:

(i) an operator of successor type A can only be succeeded by an A-atom or an A-prefix;

(ii) an A-atom or A-suffix can only be succeeded by an operator of predecessor type A or by];

(iii) (a) an atom which is preceded by the operator c_i, for $i < n$, can only be succeeded by the related operator c_{i+1}; (b) an atom which is preceded by [can only be succeeded by];

(iv) [can only be succeeded by a prefix or atom;] cannot be succeeded by anything.

We say that a string is conforming if it conforms to the rules (i)–(iv).

The rules (i)–(iv) are of a local character. To check that a string conforms to the them we need only know that it does not contain certain three-element substrings. It is therefore evident that any substring of a conforming string is also conforming. Also notice that in the rules we take no cognizance of the governing type of infixes.

Lemma 19 (For framed canonical grammars). *Any phrase conforms to the rules of precedence.*

Proof. It suffices to prove the result for phrases of the form $[t]$. This is done by induction on the length $l \geq 1$ of t. If $l = 1$, then t is an atom; and so it is evident that $[t]$ conforms to the rules.

Now suppose $l > 1$. Let the principal production in the parse $\mathbf{t}$ of t at $\mathbf{t}$ be $A \rightarrow c_0 A_1 c_1 \ldots c_{n-1} A_n c_n$, $n > 0$, and let the immediate components be $\mathbf{t}_1, \ldots, \mathbf{t}_n$. Thus t itself is of the form $c_0 t_1 c_1 \ldots c_{n-1} t_n c_n$. We wish to show that each occurrence of a symbol X in $[t]$ conforms to the rules (i), (ii), (iii) and (iv). There are five cases:

(1) X is one of the operators c_i, $i < n$. Then X has successor type A_{i+1} and is succeeded in t by the phrase t_{i+1}. By lemma 17, t_{i+1} is either an A_{i+1}-atom or begins with an A_{i+1}-prefix. So in either case, conformity to rule (i) is assured (and conformity to the other rules is trivial).

(2) X is the suffix c_n. Then X is succeeded by] and hence conforms to rule (ii).

(3) X occurs in one of the phrases t_i, but not at the end. Conformity then follows by I.H.

(4) X occurs at the end of one of phrases t_i. Since t_i is an A_i-phrase, it follows by lemma 17 that X is either an A_i-atom or an A_i-suffix. Suppose that X is an A_i-suffix. Then since X is succeeded by the operator c_i, it conforms to rule (ii). Now suppose that X is an A_i-atom. Then it must be the phrase t_i itself; since otherwise t_i ends with a suffix. But then X is preceded by c_{i-1} and succeeded by c_i, in conformity to rule (iii).

(5) X is either [or]. [is succeeded by c_0, which is a prefix;] is succeeded by nothing. Thus conformity to rule (iv) is assured. □

A weak converse of the above lemma obtains:

Lemma 20 (For G framed canonical). *Any quasi-primitive string α which conforms to the rules of precedence is a phrase-part.*

Proof. By induction on the length $l \geq 0$ of α. For $l = 0$, the result is trivial; and for $l = 1$, α is either an operator or an atom or one of the markers [or]; and it is clear in all of these cases that α is a phrase-part.

So suppose $l = k + 1$, for $k > 0$. Let β be the string of the first k symbols of α. By I.H., β is a substring of some phrase t. There are four cases.

(1) β ends with an operator c_i of sucessor type A. By lemma 18, the end occurrence of c_i in β occurs in t in a constituent s of the form $c_0 s_1 c_1 \ldots c_i s_i c_{i+1} \ldots c_{n-1} s_n c_n$; and so s_i is a constituent A-phrase. It follows by the rules that the last symbol of α is an A-atom X or an A-prefix d. If the last symbol of α is an A-atom X, replace s_i in s with X; and if the last symbol is a prefix d_0, replace s_i with a phrase which begins with d_0. The result s' of the replacement then contains α as a substring. Moreover, since s_i is an A-constituent of s, it follows by the replacement theorem that s' is also a phrase.

(2) β is an A-atom X or ends with an A-suffix c_n. In either case, β occurs within an A-phrase s of t. For in the former case, X occurs within the A-phrase X itself. While in the latter case, the end occurrence of c_n in β must, by lemma 18, occur in t in an A-constituent s of the form $c_0 s_1 c_1 \ldots c_{n-1} s_n c_n$. But if s properly ends β, then α is not quasi-primitive; and so β must end s and hence occur within s.

By rule (ii), the last symbol of α must be an operator d_j of predecessor type A or]. In the former case, it follows by replacement that α is a substring of the phrase $d_0 B_1 d_1 \ldots d_{j-1} s d_j \ldots d_{m-1} B_m d_m$, for an appropriate choice of variables $B_1, \ldots, B_{j-1}, B_{j+1}, \ldots, B_m$. In the latter case, α is a substring of $[s]$.

(3) β properly ends with an A-atom X. It follows from the rules that X must be preceded by [or by an operator c_i of sucessor type A. In the former case, it follows from rule (iii) that α is of the form $[X]$. But then α is itself a phrase. In the latter case, it follows by rule (iii) that α must end with the operator c_{i+1}. By lemma 18, the penultimate symbol c_i of β occurs in a constituent phrase $s = c_0 s_1 c_1 \ldots c_i s_i c_{i+1} \ldots c_{n-1} s_n c_n$ of t, where s_i is a constituent A-phrase of t. Let t' be the result of replacing the given occurrence of s_i in t with the atom X. Then α is a substring of t'; and by Replacement, t' is a phrase.

(4) β ends with [or]. Given that β properly begins α, it follows by

rule (iv) that β cannot end with]. So β ends with [. It follows from the rules that nothing can precede [in a conforming string; and so [is the sole symbol of β. But then by rule (iv) the last symbol X of α is either an atom or a prefix. Let t be a phrase which begins with X. Then α is a substring of the phrase $[t]$. □

Putting the two results together we obtain a necessary and sufficient condition for a quasi-primitive string to be a phrase-part:

Theorem 21 (Precedence). *Let α be a quasi-primitive string of the framed canonical grammar. Then α is a phrase-part iff it conforms to the rules of precedence.*

Call α an *operator (prefix, infix, suffix) block* if it is a string of non-atomic operators (respectively, prefixes, infixes, suffixes). Then a special case of the above theorem is when α is an operator block.

Corollary 22. *The operator block α is a phrase-part only if it is of the form SIP, where S, I and P are respectively suffix, infix and prefix blocks, and where I contains at most one infix and contains at least one infix if S contains at least one suffix.*

Proof. Suppose α is an operator block. Then it is quasi-primitive, since any elementary phrase must contain an atom. Therefore by the theorem, α is conforming. But the rules of precedence have the following consequences for an operator block: a suffix can only be succeeded by a suffix or by an infix; an infix can only be succeeded by a prefix; and a prefix can only be succeeded by a prefix. The condition then follows. □

Fortunately, our precedence analysis applies to exactly those phrase-parts required to test for transparency. We therefore obtain from theorem 21 and corollary 16:

Theorem 23 (Transparency). *(For G chain-free and invertible). G is transparent iff no quasi-primitive conforming string of the framed canonical grammar $(G^c)^f$ irregularly corresponds to a phrase of G.*

It may also be shown, given that a string is conforming, that it will be quasi-primitive iff it does not embed a string which begins with a c_0 and ends with the related c_n, $n > 0$. We are thereby able to effectively construct from the framed canonical grammar a finite state automaton which will recognize exactly the set of conforming quasi-primitive strings.

9. Deciding Transparency

It is possible to extract an effective test for transparency from theorem 23. It is straightforward to effectively perform the following tasks: enumerate the phrases of $(G^c)^f$ of an arbitrary given length; determine whether they are quasi-primitive and conform to the rules of precedence; determine whether they irregularly correspond to a phrase of the grammar G. Therefore by our previous results, it only remains to set an effective upper bound on the length of the quasi-primitive strings, i.e. given a primitive phrase t of the base grammar we need to effectively determine from it a number n which is such that if any conforming string of the base grammar irregularly corresponds to t then a string of length $< n$ irregularly corresponds to t.

Now the markers [and] can occur at most once in any quasi- primitive conforming string; and there can be at most two occurrences of constants, for if there were more then any middle occurrence of a constant would prevent the string from being quasi-primitive. By corollary 22, the number of occurrences of infixes is no greater than the number of maximally long occurrences of operator blocks. But the number of maximally long occurrences of operator blocks is at most one more than the number of occurrences of atoms, since any two such blocks must be separated by an atom. It therefore follows that we need only set an upper bound on the length of the degenerate prefix and the degenerate suffix blocks.

But it is readily seen that k is such an upper bound, where k is the number of degenerate prefixes (respectively, suffixes). For suppose that $c_0 \dots d_0$ is a block of more than k degenerate prefixes. Then two of the occurrences in the block are of the same prefix e_0, so the block is of the form $c_0 \dots e_0 \dots e_0 \dots d_0$ (where the first e_0 might be identical to the initial c_0 and the second e_0 might be identical to the terminal d_0). Now replace this block in the conforming string by $c_0 \dots e_0 \dots d_0$. It is then evident that the resulting string is still conforming and that it irregularly corresponds to the same phrases of the base grammar.

We therefore obtain:

Theorem 24. *It is decidable whether an invertible and chain-free context-free grammar is transparent.*

By using the reduction of an arbitrary grammar to one that is invertible and chain-free, it is possible to establish an unrestricted result:

Theorem 25 (Decidability). *It is decidable whether an arbitrary context-free grammar is transparent.*

The above procedures for deciding transparency may be made more efficient in various ways. Indeed, it may be shown that the question of opacity for invertible and chain-free grammars is NP. However, it is most unlikely that any such result holds for the general case.

In addition to the problem of deciding transparency, there is the problem of deciding whether an arbitary phrase t of a given grammar has an accidental occurrence. It is natural to try to solve this problem by appeal to lemma 12: we look for a phrase-part α of the framed canonical grammar which irregularly corresponds to t. But there still remains the question of finding an effective upper bound on the length of the phrase-parts α which need be considered. The number of non-degenerate symbol occurrences in α must, of course, be bound by the length of s itself. But there is no obvious way of setting a bound on the number of non-degenerate symbols which can occur in α.

There are some fairly sophisticated techniques which may be used to find an effective upper bound. This gives:

Theorem 26. *It is decidable whether a given phrase of a context- free grammar has an accidental occurrence.*

10. Extensions and Refinements

Let me briefly indicate some of the ways in which the concepts and results of this paper can be extended.

(1) It is usual to restrict the "words" of a grammar to those strings that (a) are derived from the initial variable S and (b) are ground. It is possible to extend our results to the cases in which either one or both of these additional restrictions are imposed. Thus to accord with restriction (a), we may define a grammar G to be S-*transparent* if each S-parsed phrase of G is transparent, i.e. contains no accidental occurrence of a subphrase; and with each grammar G we may then effectively associate a grammar G^S which is S-transparent iff G is transparent.

(2) Our results have been proved for context-free grammars. But in fact they hold for a much wider class of grammars. First of all, we may

note that nothing (except the proofs of decidability) turn upon the set of productions being finite. But, more significantly, we may restrict the set of parses sanctioned by the grammar.

Let a *constrained grammar* G be an ordered pair $\langle H, \mathbf{P} \rangle$, where H is a context-free grammar and $\mathbf{P}$ is a set of parses from H. Intuitively, we think of the constraints as permitting, by unspecified means, only the parses in $\mathbf{P}$. A parse of H is said said to be *permissible* in the constrained grammar if it belongs to $\mathbf{P}$; and a phrase of H is *permissible* just in case it has a permissible parse. A constrained grammar is then taken to be transparent if no permissible phrase occurs accidentally in a permissibly parsed phrase.

By analysing our proofs, we are able to see that they only require rather weak closure conditions on the set $\mathbf{P}$ of permissible parses.

(3) There is a natural distinction between two kinds of accidental occurrence. On the one hand there is the kind which it typified by 'cat' in 'cattle'; and, on the other hand, there is the kind that is typified by '$1+2$' in '$1+2.3$'. The first is orthographic in nature and rests entirely on how the individual words are "spelt". The second is partly syntactic in nature and also also rests on how the phrases are parsed.

It is possible to define an *orthographic* accident as one that is not always preserved under an orthographic revision of the grammar, i.e. under a system of respelling; and a separate test for orthographic accident may then be developed.

(4) Under certain procedures for parsing, only some accidental occurrences may matter, i.e. be a potential danger. Suppose, for example, that we parse bottom-up and from left to right. Then certain accidental occurrences may "disappear" under reduction, while others may not. Accordingly, we may define and determine when an occurrence s is accidental in t under left-to-right parsing; and similarly for other types of parsing.

References

[1] Fine, K., *Transparency I and II*, submitted to Language and Control.
[2] Harrison, M. A., *Introduction to Formal Language Theory*, Addison-Wesley, Massachusetts.

Department of Philosophy, U.C.L.A.

DESIGNING UNIFICATION PROCEDURES USING TRANSFORMATIONS: A SURVEY

JEAN H. GALLIER AND WAYNE SNYDER

1. INTRODUCTION

Unification is a very general computational paradigm that plays an important role in many different areas of symbolic computation. For example, unification plays a central role in

- Automated Deduction (First-order logic with or without equality, higher-order logic);
- Logic Programming (Prolog, λ-Prolog);
- Constraint-based Programming;
- Type Inferencing (ML, ML^+, etc.);
- Knowledge-Base Systems, Feature structures; and
- Computational Linguistics (Unification grammars).

In this survey, we shall focus on unification problems arising in methods for automated theorem proving. This covers at least the kind of unification arising in resolution and E-resolution (Robinson [103], Plotkin [99]), matings (Andrews [4], Bibel [12], [13], [14]), equational matings (Gallier, Plaisted Narendran, Raatz, Snyder [38], [37]), and ET-proofs (Miller, Pfenning [94], [82], [84]). Clearly, many other important parts of unification theory are left out, and we apologize for this. In particular, we will not cover the classification theory of the Siekmann school (for example, [106], [119], [16], [17]), the many unification procedures for special theories (AC, etc., see Siekmann [106]), the combination of unification procedures (for example, Yelick [122], Schmidt-Schauss [111], Boudet, Jouannaud, and Schmidt-Schauss [15]) order-sorted unification (for example, Meseguer, Goguen, and Smolka [81], and Isakowitz [60]), semi-unification (see [68] for references), unification applied to type-inferencing (for example, Milner [88], Kfoury, Tiuryn, and Urzyczyn, [67],[69], and Remy [100], [101]), unification in computational linguistics (for example, Shieber [108]), and unification in feature structures (for example,

Aït-Kaci [1], [2]). Fortunately for the uninitiated reader, there are other very good survey papers covering significant parts of the above topics: the survey by Siekmann [106], the survey by Knight [70], and the survey by Kirchner and Jouannaud [61]. The most elementary and having the broadest coverage is probably Knight's survey. Our account has a narrower focus, but it also sketches some of the proof techniques, which are usually missing from the other surveys. The topics that we will cover are:

- Standard Unification and E-Unification (E a finite set of first-order equations);
- Rigid E-Unification (E a finite set of first-order equations), a decidable form of E-unification recently introduced in the framework of Andrews and Bibel's method of matings; and
- Higher-Order Unification (in the simply-typed λ-calculus).

All these unification problems will be tackled using the *method of transformations* on term systems, already anticipated in Herbrand's thesis [48] (1930), and revived very effectively by Martelli and Montanari [79] for standard unification. In a nutshell, the method is as follows:

A unification problem is gradually transformed into one whose solution is (almost) obvious.

This approach is an instance of a very old method in mathematics, but a fairly recent trend in computer science, namely, the specification of procedures and algorithms in terms of inference rules. There are a number of significant advantages to this method. (1) A clean separation of logic and control is achieved. (2) The correctness of the procedure obtained this way is often easier to establish, and irrelevant implementation issues are avoided. (3) The actual design of algorithms from the procedure specified by rules can be viewed as an optimization process.

Another benefit of this approach to the design of algorithms is that one often gains a deeper understanding of the problem being solved, and one understands more easily the differences between algorithms solving a same problem. The effectiveness of this method for tackling unification problems was first shown by Martelli and Montanari [79] (although, as we said earlier, it was anticipated by Herbrand [48]). Similarly, Bachmair, Dershowitz, Hsiang, and Plaisted [8],[9],[10] showed how to describe and study Knuth-Bendix completion procedures [71] in terms of proof rules. Presented in terms of proof rules, completion procedures are more transparent, and their correctness proofs are significantly simplified. Many other examples of the effectiveness of the method of proof rules can be easily found (for example, in type inference problems, and Gentzen-style automated deduction, see Gallier [30] for the latter). Jouannaud and Kirchner [61] also emphasize the proof rules method, and provide many more examples of its use. We venture to say that the success of the method of proof rules for specifying procedures and algorithms is one of the most striking illustrations of the fact that the "invasion" of computer science by logic is healthy. We believe that this trend will continue with great success, if adequately controlled to remain at a fairly computational level.

Although in this paper the perspective is to discuss unification in terms of transformations (proof rules), we should not forget about the history of unification theory, and the major turning points.[1] Undoubtedly, the invention of the resolution method and of the first unification algorithm by Alan Robinson in the early sixties [103] (1965) marks the beginning of a new era. In the post Robinson era, we encounter Plotkin's seminal paper on building-in equational theories [99] (1972), in which E-unification is introduced, and then Gérard Huet's thesis [54] (1976) (with Huet [50] as a precursor). Huet's thesis (1976) makes a major contribution to the theory of higher-order unification, in that it shows that a restricted form of unification, preunification, is sufficient for most theorem-proving applications. The first practical higher-order (pre)unification algorithm is defined and proved correct. Huet also gives a quasi-linear unification algorithm for standard unification, and an algorithm for unifying infinite rational trees. In the more recent past, in our perspective, we would like to mention Martelli and Montanari's paper showing the effectiveness of the method of transformations [79] (1982), and Claude Kirchner's thesis [64] in which the method of transformations is systematically applied to E-unification. We also would like to mention that most of the results on E-unification and higher-order unification discussed in this paper originate from Wayne Snyder's thesis [112] (1988). A more comprehensive presentation of these results will appear in Snyder [116].

Unification theory is a very active field of research, and it is virtually impossible to keep track of all the papers that have appeared on this subject. As evidence that unification is a very active field of research, two special issues of the *Journal of Symbolic Computation* are devoted to unification

[1] The following list is by no means exclusive, and only reflects the perspective on unification adopted in this paper.

theory (Part I in Vol. 7(3 & 4), and Part II in Vol. 8(1 & 2), both published in 1989). It is our hope that this paper will inspire other researchers to work in this area.

The paper is organized as follows. Section 2 provides a review of background material relevant to unification. Section 3 is devoted to standard unification and E-unification. The general E-unification problem is solved in terms of a set of transformations due to Gallier and Snyder [35]. Section 4 is devoted to rigid E-unification, a restricted version of E-unification arising in the framework of Andrews and Bibel's theorem proving method of matings. We give an overview of results of Gallier, Narendran, Plaisted, and Snyder [37], showing among other things that rigid E-unification is decidable and NP-complete. Section 5 is devoted to a presentation of higher-order unification in terms of transformations. This section is a summary of results from Snyder and Gallier [113]. Directions for further research are discussed in section 6.

2. Algebraic Background

We begin with a brief review of algebraic background material. The purpose of this section is to establish the notation and the terminology used throughout this paper. As much as possible, we follow Huet [56] and Gallier [30].

Definition 2.1. Let $\longrightarrow \subseteq A \times A$ be a binary relation on a set A. The *converse* (or *inverse*) of the relation $\longrightarrow$ is the relation denoted as $\longrightarrow^{-1}$ or $\longleftarrow$, defined such that $u \longleftarrow v$ iff $v \longrightarrow u$. The symmetric closure of $\longrightarrow$, denoted by $\longleftrightarrow$, is the relation $\longrightarrow \cup \longleftarrow$. The transitive closure, reflexive and transitive closure, and the reflexive, symmetric, and transitive closure of $\longrightarrow$ are denoted respectively by $\xrightarrow{+}$, $\xrightarrow{*}$, and $\xleftrightarrow{*}$.

Definition 2.2. A relation $\longrightarrow$ on a set A is *Noetherian* or *well founded* iff there are no infinite sequences $\langle a_0, \ldots , a_n, a_{n+1}, \ldots\rangle$ of elements in A such that $a_n \longrightarrow a_{n+1}$ for all $n \geq 0$.

Definition 2.3. A *preorder* $\preceq$ on a set A is a binary relation $\preceq \subseteq A \times A$ that is reflexive and transitive. A *partial order* $\preceq$ on a set A is a preorder that is also antisymmetric. The converse of a preorder (or partial order) $\preceq$ is denoted as $\succeq$. A *strict ordering* (or *strict order*) $\prec$ on a set A is a transitive and irreflexive relation. Given a preorder (or partial order) $\preceq$ on a set A, the strict ordering $\prec$ associated with $\preceq$ is defined such that $s \prec t$ iff $s \preceq t$

and $t \not\preceq s$. Conversely, given a strict ordering $\prec$, the partial ordering $\preceq$ associated with $\prec$ is defined such that $s \preceq t$ iff $s \prec t$ or $s = t$. The converse of a strict ordering $\prec$ is denoted as $\succ$. Given a preorder (or partial order) $\preceq$, we say that $\preceq$ is well founded iff $\succ$ is well founded.

Definition 2.4. Let $\longrightarrow \subseteq A \times A$ be a binary relation on a set A. We say that $\longrightarrow$ is *locally confluent* iff for all $a, a_1, a_2 \in A$, if $a \longrightarrow a_1$ and $a \longrightarrow a_2$, then there is some $a_3 \in A$ such that $a_1 \xrightarrow{*} a_3$ and $a_2 \xrightarrow{*} a_3$. We say that $\longrightarrow$ is *confluent* iff for all $a, a_1, a_2 \in A$, if $a \xrightarrow{*} a_1$ and $a \xrightarrow{*} a_2$, then there is some $a_3 \in A$ such that $a_1 \xrightarrow{*} a_3$ and $a_2 \xrightarrow{*} a_3$. We say that $\longrightarrow$ is *Church-Rosser* iff for all $a_1, a_2 \in A$, if $a_1 \xleftrightarrow{*} a_2$, then there is some $a_3 \in A$ such that $a_1 \xrightarrow{*} a_3$ and $a_2 \xrightarrow{*} a_3$. We say that $a \in A$ is *irreducible* iff there is no $b \in A$ such that $a \longrightarrow b$. It is well known (Huet [56]) that a Noetherian relation is confluent iff it is locally confluent and that a relation is confluent iff it is Church-Rosser. A relation $\longrightarrow$ is *canonical* iff it is Noetherian and confluent. Given a canonical relation $\longrightarrow$, it is well known that every $a \in A$ reduces to a unique irreducible element $a\!\downarrow \in A$ called the *normal form of* a, and that $a \xleftrightarrow{*} b$ iff $a\!\downarrow = b\!\downarrow$ (Huet [56]).

Definition 2.5. Terms are built up inductively from a *ranked alphabet* (or *signature*) Σ of constant and function symbols, and a countably infinite set $\mathcal{X}$ of *variables*. For simplicity of exposition, we assume that Σ is a one-sorted ranked alphabet, i.e., that there is a *rank function* $r\colon \Sigma \to N$ assigning a *rank* (or *arity*) $r(f)$ to every symbol $f \in \Sigma$ ($\mathbf{N}$ denotes the set of natural numbers). We let $\Sigma_n = \{f \in \Sigma \mid r(f) = n\}$. Symbols in Σ_0 (of rank zero) are called *constants*.

Definition 2.6. We let $T_\Sigma(\mathcal{X})$ denote the set of terms built up inductively from Σ and $\mathcal{X}$. Thus, $T_\Sigma(\mathcal{X})$ is the smallest set with the following properties:

- $x \in T_\Sigma(\mathcal{X})$, for every $x \in \mathcal{X}$;
- $c \in T_\Sigma(\mathcal{X})$, for every $c \in \Sigma_0$;
- $f(t_1, \ldots, t_n) \in T_\Sigma(\mathcal{X})$, for every $f \in \Sigma_n$ and all $t_1, \ldots, t_n \in T_\Sigma(\mathcal{X})$.

Given a term $t \in T_\Sigma(\mathcal{X})$, we let $Var(t)$ be the set of variables occurring in t. A term t is a *ground term* iff $Var(t) = \emptyset$.

It is well known that $T_\Sigma(\mathcal{X})$ is the term algebra freely generated by $\mathcal{X}$, and this allows us to define substitutions.

Definition 2.7. A *substitution* is a function $\varphi\colon \mathcal{X} \to T_\Sigma(\mathcal{X})$ such that $\varphi(x) \neq x$ for only finitely many $x \in \mathcal{X}$. The set $D(\varphi) = \{x \in \mathcal{X} \mid \varphi(x) \neq x\}$ is the *domain* of φ, and the set $I(\varphi) = \bigcup_{x \in D(\varphi)} Var(\varphi(x))$ is the *set of variables introduced* by φ.

A substitution $\varphi\colon \mathcal{X} \to T_\Sigma(\mathcal{X})$ with domain $D(\varphi) = \{x_1, \ldots, x_n\}$ and such that $\varphi(x_i) = t_i$ for $i = 1, \ldots, n$, is denoted as $[t_1/x_1, \ldots, t_n/x_n]$. Since $T_\Sigma(\mathcal{X})$ is freely generated by $\mathcal{X}$, every substitution $\varphi\colon \mathcal{X} \to T_\Sigma(\mathcal{X})$ has a unique homomorphic extension $\widehat{\varphi}\colon T_\Sigma(\mathcal{X}) \to T_\Sigma(\mathcal{X})$. For every term $t \in T_\Sigma(\mathcal{X})$, we denote $\widehat{\varphi}(t)$ as $t[\varphi]$ or even as $\varphi(t)$ (with an intentional identification of φ and $\widehat{\varphi}$).

Definition 2.8. Given two substitutions φ and ψ, their *composition* denoted $\varphi\,;\psi$ is the substitution defined such that $\varphi\,;\psi(x) = \widehat{\psi}(\varphi(x))$ for all $x \in \mathcal{X}$. Thus, note that $\varphi\,;\psi = \varphi \circ \widehat{\psi}$, but not $\varphi \circ \psi$, where $\circ$ denotes the composition of functions (written in diagram order). A substitution φ is *idempotent* iff $\varphi\,;\varphi = \varphi$. It is easily seen that a substitution φ is idempotent iff $I(\varphi) \cap D(\varphi) = \emptyset$. A substitution φ is a *renaming* iff $\varphi(x)$ is a variable for every $x \in D(\varphi)$, and φ is injective over its domain. Given a set V of variables and a substitution φ, the *restriction of φ to V* is the substitution denoted $\varphi|_V$ defined such that, $\varphi|_V(x) = \varphi(x)$ for all $x \in V$, and $\varphi|_V(x) = x$ for all $x \notin V$.

There will be occasions where it is necessary to replace a subterm of a given term with another term. We can make this operation precise by defining the concept of a tree address originally due to Gorn.

Definition 2.9. Given a term $t \in T_\Sigma(\mathcal{X})$, the set $Tadd(t)$ of *tree addresses* in t is a set of strings of positive natural numbers defined as follows (where ϵ denotes the null string):

- $Tadd(x) = \{\epsilon\}$, for every $x \in \mathcal{X}$;
- $Tadd(c) = \{\epsilon\}$, for every $c \in \Sigma_0$;
- $Tadd(f(t_1, \ldots, t_n)) = \{\epsilon\} \cup \{iw \mid w \in Tadd(t_i),\ 1 \leq i \leq n\}$.

Definition 2.10. Given any $\beta \in Tadd(t)$, the *subtree* rooted at β in t is denoted as t/β. Given $t_1, t_2 \in T_\Sigma(\mathcal{X})$ and $\beta \in Tadd(t_1)$, the tree $t_1[\beta \leftarrow t_2]$ obtained by replacing the subtree rooted at β in t_1 with t_2 can be easily defined.

Definition 2.11. Let $\longrightarrow$ be a binary relation $\longrightarrow \subseteq T_\Sigma(\mathcal{X}) \times T_\Sigma(\mathcal{X})$.

(i) The relation $\longrightarrow$ is *monotonic* (or *stable under the algebra structure*) iff for every two terms s, t and every function symbol $f \in \Sigma$, if $s \longrightarrow t$ then $f(\ldots, s, \ldots) \longrightarrow f(\ldots, t, \ldots)$.

(ii) The relation $\longrightarrow$ is *stable* (under substitution) if $s \longrightarrow t$ implies $s[\sigma] \longrightarrow t[\sigma]$ for every substitution σ.

Definition 2.12. A strict ordering $\prec$ has the *subterm property* iff $s \prec f(\ldots, s, \ldots)$ for every term $f(\ldots, s, \ldots)$. A *simplification ordering* $\prec$ is a strict ordering that is monotonic and has the subterm property (since we are considering symbols having a fixed rank, the deletion property is superfluous, as noted in Dershowitz [22]). A *reduction ordering* $\prec$ is a strict ordering that is monotonic, stable (under substitution), and such that $\succ$ is well founded. With a slight abuse of language, we will also say that the converse $\succ$ of a strict ordering $\prec$ is a simplification ordering (or a reduction ordering). It is shown in Dershowitz [22] that there are simplification orderings that are total on ground terms.

Definition 2.13. A *set of rewrite rules* is a binary relation $R \subseteq T_\Sigma(\mathcal{X}) \times T_\Sigma(\mathcal{X})$ such that $Var(r) \subseteq Var(l)$ whenever $\langle l, r\rangle \in R$. A rewrite rule $\langle l, r\rangle \in R$ is usually denoted as $l \to r$. A rewrite rule $s \to t$ is a *variant* of a rewrite rule $u \to v \in R$ iff there is some renaming ρ with domain $Var(u) \cup Var(v)$ such that $s = u[\rho]$ and $t = v[\rho]$.

Let $R \subseteq T_\Sigma(\mathcal{X}) \times T_\Sigma(\mathcal{X})$ be a set of rewrite rules.

Definition 2.14. The relation $\longrightarrow_R$ over $T_\Sigma(\mathcal{X})$ is defined as the smallest stable and monotonic relation that contains R. This is the *rewrite relation* associated with R. This relation is defined explicitly as follows: Given any two terms $t_1, t_2 \in T_\Sigma(\mathcal{X})$, then

$$t_1 \longrightarrow_R t_2$$

iff there is some variant $l \to r$ of some rule in R, some tree address β in t_1, and some substitution σ, such that

$$t_1/\beta = l[\sigma], \quad \text{and} \quad t_2 = t_1[\beta \leftarrow r[\sigma]].$$

The concept of an equation is similar to that of a rewrite rule, but equations can be used oriented forward or backward, and the restriction $Var(r) \subseteq Var(l)$ is dropped.

Definition 2.15. A *set of equations* is a binary relation $E \subseteq T_\Sigma(\mathcal{X}) \times T_\Sigma(\mathcal{X})$. An equation $\langle l, r\rangle \in E$ is usually denoted as $l \doteq r$, to emphasize the difference with a rewrite rule. An equation $s \doteq t$ is a *variant* of an equation $u \doteq v \in E$ iff there is some renaming ρ with domain $Var(u) \cup Var(v)$ such that $s = u[\rho]$ and $t = v[\rho]$.

Definition 2.16. The relation $\longleftrightarrow_E$ over $T_\Sigma(\mathcal{X})$ is defined as the smallest symmetric relation containing E that is stable, and monotonic. This relation is defined explicitly as follows: Given any two terms $t_1, t_2 \in T_\Sigma(\mathcal{X})$, then

$$t_1 \longleftrightarrow_E t_2$$

iff there is some variant $l \doteq r$ of some equation in $E \cup E^{-1}$, some tree address β in t_1, and some substitution σ, such that

$$t_1/\beta = l[\sigma], \quad \text{and} \quad t_2 = t_1[\beta \leftarrow r[\sigma]].$$

Note that an equation can be used oriented forward or backward, since E^{-1} consists of all $r \doteq l$ such that $l \doteq r \in E$.

Definition 2.17. The reflexive and transitive closure of $\longrightarrow_R$ is denoted as $\stackrel{*}{\longrightarrow}_R$, and the reflexive and transitive closure of $\longleftrightarrow_E$ as $\stackrel{*}{\longleftrightarrow}_E$. Sometimes, $\stackrel{*}{\longleftrightarrow}_E$ is denoted as $=_E$. It is easily seen that $\stackrel{*}{\longleftrightarrow}_E$ is an equivalence relation. In fact, $\stackrel{*}{\longleftrightarrow}_E$ is the smallest congruence containing E that is stable under substitution, and $\stackrel{*}{\longleftrightarrow}_E = (\longrightarrow_E \cup \longrightarrow_E^{-1})^*$. It can be shown that $E \models u \doteq v$ iff $u \stackrel{*}{\longleftrightarrow}_E v$ (a form of Birkhoff's completeness theorem). A set R of rewrite rules is called Noetherian, confluent, Church-Rosser, or canonical, iff the relation $\longrightarrow_R$ has the corresponding property.

3. Standard Unification and E-Unification

Standard unification was brought to the fore as a seminal component of automated deduction systems by Robinson in 1964, and has been studied by numerous researchers since that time. In this section we shall treat this problem as a special case of a more general form of first-order unification, namely, E-unification. The study of E-unification was triggered mostly by the inherent inefficiency of theorem proving methods in the presence of equality. Robinson [104] and then Plotkin [99] suggested that theorem provers be stratified into a (non-equational) refutation mechanism, and an E-unification mechanism, which performs equational reasoning during unification steps. More recently, E-unification has also been proposed as the theoretical basis of the incorporation of functional and equational languages into the basic paradigm of logic programming [31], [41]. We now define the E-unification problem and present an abstract view of E-unification as a set of non-deterministic rules for transforming a unification problem into an explicit representation of its solution, if such exists. This elegant approach is due to Martelli and Montanari [79], but, as mentioned above, was in fact implicit in Herbrand (1930).[2]

It is natural to define a unification problem as a set $\{\langle u_1, v_1\rangle, \dots, \langle u_n, v_n\rangle\}$ of ordered pairs of terms. This is fine, but it turns out that it is more convenient to allow repetitions of pairs $\langle u_i, v_i\rangle$, and to allow $\langle u_i, v_i\rangle$ to be unordered. Thus, we adopt the following definition of a unification problem.

Definition 3.1. A *term pair* or just a *pair* is a multiset of two terms, denoted by $\langle u, v\rangle$. A *term system* (or *system*) is a (finite) multiset of term pairs. In denoting term systems, we will often drop the curly brackets and simply write $\langle u_1, v_1\rangle, \dots, \langle u_n, v_n\rangle$.

The reason why multisets are more convenient than sets is that they lead to a simpler statement of the transformations. This shows up in two ways. Firstly, since multiset union is not idempotent (contrary to set union), when we write $S \cup \{\langle u, v\rangle\}$, we mean the multiset consisting of all pairs in S distinct from $\langle u, v\rangle$, and of the $m+1$ pairs $\langle u, v\rangle$, where m is the number of occurrences of $\langle u, v\rangle$ in S. Thus, in a transformation $S \cup \{\langle u, v\rangle\} \Longrightarrow S \cup R$, where S and R are multisets of (unordered pairs) and $\langle u, v\rangle$ does not belong to R, there are fewer occurrences of the pair $\langle u, v\rangle$ on the right-hand side of the transformations than there are on the left-hand side. If S, R were interpreted as sets, and $\cup$ as set union, we would have to stipulate that $\langle u, v\rangle$ does not belong to S, or explicitly remove it from S on the right-hand side. Secondly, if pairs $\langle u, v\rangle$ are considered ordered, then a special transformation switching pairs $\langle u, x\rangle$ to $\langle x, u\rangle$ is needed when x is a variable but u is not. If we treat $\langle u, v\rangle$ as unordered, such a transformation is unnecessary. We can now state the *E-unification problem*:

Definition 3.2. Given a set E of equations and a term system $S = \langle u_1, v_1\rangle, \dots, \langle u_n, v_n\rangle$, find some (all) substitution(s) σ s.t.

$$E \models u_i[\sigma] \doteq v_i[\sigma], \quad i = 1, \dots, n.$$

[2]It is remarkable that in his thesis Herbrand gave all the steps of a (nondeterministic) unification algorithm based on transformations on systems of equations. These transformations are given at the end of the section on property A, p. 148 of [48].

Equivalently, find some (all) substitution(s) σ s.t.

$$u_i[\sigma] \stackrel{*}{\longleftrightarrow}_E v_i[\sigma], \quad i = 1, \ldots, n,$$

where $\stackrel{*}{\longleftrightarrow}_E$ is equality modulo E.

We let $U_E(S)$ denote the set of all E-unifiers of S. When $E = \emptyset$, what we have is just *standard unification*, the problem of finding some (all) substitution(s) σ such that $u_i[\sigma] = v_i[\sigma]$ (where $=$ denotes the identity of terms). Historically, the standard unification problem was studied long before E-unification, even though we defined it as a special case of E-unification. For a very good historical account and technical details on standard unification, the reader is referred to Knight's survey article [70], Jouannaud and Kirchner's survey article [61], and Lassez, Maher, and Marriot [75].

Example 3.3. Let $E = \{x + y \doteq y + x,\ x + (y + z) \doteq (x + y) + z\}$, and

$$S = \langle \big(f(a) + g(f(u))\big) + v,\ f(u) + \big(f(b) + g(f(a))\big)\rangle.$$

An E-unifier of S is $\theta = [a/u,\ f(b)/v]$, since

$$\theta(S) = \langle \big(f(a) + g(f(a))\big) + f(b),\ f(a) + \big(f(b) + g(f(a))\big)\rangle.$$

Example 3.4. Let $E = \{g(f(x, y)) \doteq g(y)\}$, and $S = \langle g(z),\ g(0)\rangle$.

Some E-unifiers are listed below:

$[0/z]$, $[f(x_1, 0)/z]$, $[f(x_2, f(x_1, 0))/z]$, $[f(x_n, f(x_{n-1}, \ldots\ f(x_1, 0)\ \ldots))/z]$, ...

Observe that in example 3.4, there is an infinite number of E-unifiers. This leads us to the following question.

Question: How do we compare E-unifiers?

Answer: Define a preorder $\leq_E$ on substitutions.

Definition 3.5. Let V be a set of variables. We write $\sigma =_E \theta[V]$ (*σ is equal to θ over V*) iff

$$\sigma(x) \stackrel{*}{\longleftrightarrow}_E \theta(x) \quad \text{for all } x \in V.$$

We write $\sigma \leq_E \theta[V]$ (*σ is more general than θ over V*) iff there exists a substitution η such that $\theta =_E \sigma\,;\eta[V]$.

The intuitive idea behind these definitions is that σ is more general than θ (over V) when each $\theta(x)$ can be obtained from $\sigma(x)$ by instantiating some of the variables occurring in $\sigma(x)$, and considering equality of terms modulo E-congruence. The reason for relativizing the definition of $=_E$ and $\leq_E$ to a set V of variables is technical. For one thing, some of the results are incorrect if V is left out. Also, in theorem proving applications, it is often desirable to compare substitutions with respect to a "protected set" of variables V. A crucial concept in E-unification theory is that of a complete set of E-unifiers, originally due to Plotkin [99].

Definition 3.6. Given a set E of equations, a term system S, and a finite set V of "protected" variables, a set U of substitutions is a *complete set of E-unifiers for S away from V* iff:

(i) For every $\sigma \in U$, $D(\sigma) \subseteq Var(S)$ and $I(\sigma) \cap (V \cup D(\sigma)) = \emptyset$;

(ii) Every $\sigma \in U$ is an E-unifier of S, i.e. $U \subseteq U_E(S)$;

(iii) For every E-unifier θ of S, there is some $\sigma \in U$ such that $\sigma \leq_E \theta[Var(S)]$.

Note that condition (i) implies that every $\sigma \in U$ is idempotent. When U has a single element, say σ, we say that σ is a *most general unifier* for S away from V (*an mgu away from* V). When V is not significant, we just call σ an *mgu*. A number of questions now arise naturally.

Some Questions:

- 1. Is $\leq_E$ well-founded?
- 2. Given E and S, can we decide whether S is E-unifiable?
- 3. Given E and S, if S is E-unifiable, is there a finite complete set of E-unifiers?
- 4. Given E and S, if S is E-unifiable, is there a complete set of E-unifiers minimal in some sense?
- 5. Given E and S, if S is E-unifiable, can we enumerate a complete set of E-unifiers which is reasonably nonredundant?

Unfortunately, general E-unification is a rather nasty problem, as the following answers will reveal.

Some Answers:

- When $E = \emptyset$, the answer to questions 1-5 is YES. In fact, when S is unifiable, it has an mgu.

That $\leq$ is well-founded in the case of standard unification is shown in Huet [54]. The decidability of standard unification is implicit in Herbrand's thesis [48] (1930), and it is also settled by Robinson [103] (1965), who gives the first algorithm to find mgu's, settling questions (3-5).

Robinson's original version of the unification algorithm ([103]) can run in exponential time. Since Robinson's seminal discovery, several polynomial-time algorithms for standard unification have been given, including one by Robinson himself. Among them, we single out a quasi-linear algorithm due to Huet [54] (1976), a linear-time algorithm due to Paterson and Wegman [92] (1978), and quasi-linear and linear algorithms due to Martelli and Montanari [79] (1982). For an excellent account of standard unification, the reader is referred to Knight's survey article [70], and to Jouannaud and Kirchner's survey article [61]. Martelli and Montanari's important contribution ([79]), perhaps even more than their algorithm itself,[3] is to have demonstrated with perfect clarity that the method of transformations is remarkably well suited for tackling unification problems. In some sense, Martelli and Montanari revived Herbrand's approach, which led to new important work by Kirchner and others.

- When $E \neq \emptyset$, the situation is grim, since the answer to questions 1-4 is NO. On the other hand, the answer to question 5 is essentially YES.

The undecidability of E-unification is a trivial consequence of the undecidability of the word problem for semigroups (see Machtey and Young [78]). However, this undecidabilty has really nothing to do with unification, but with the fact that the equational theory in question is undecidable. Nevertheless, there are decidable equational theories E for which E-unification is undecidable. The most famous is perhaps Hilbert's tenth problem (decide whether a diophantine equation has an integer solution). The undecidability of E-unification has been shown for simpler sets E by Siekmann and Szabo [107], [119]. For example, E-unification is undecidable for associativity and distributivity, and for associativity, commutativity, and distributivity. There are even simpler problems, for example the problem of DA-unification, see Tiden and Arnborg [120] (i.e., unification for a theory stating associativity of

[3]Their algorithm is not so different from Paterson and Wegman's algorithm.

a symbol +, and left and right distributivity of a symbol $*$ with respect to +, with at least one free constant). Schmidt-Schauss [110] has also shown that there are permutative theories whose unification problem is undecidable.

The following example due from Fages and Huet [25] shows that $\leq_E$ is not always well-founded and also gives a negative answer to questions 3-4.

Example 3.7. Let $E = \{g(f(x,y)) \doteq g(y),\ f(0,x) \doteq x\}$, and $S = \langle g(z),\ g(0)\rangle$.

The preorder $\leq_E$ is **not** well founded:

$$\theta_{n+1} = [f(x_{n+1}, f(x_n, \dots\ f(x_1, 0)\ \dots))/z] \leq_E \theta_n$$
$$= [f(x_n, f(x_{n-1}, \dots\ f(x_1, 0)\ \dots))/z],$$

since $\theta_{n+1}\,;[0/x_{n+1}] =_E \theta_n$, and it can be shown that $\theta_{n+1} <_E \theta_n$ (i.e. $\theta_n \not\leq_E \theta_{n+1}$). Every complete set for S is infinite, and S has no minimal complete set of E-unifiers.

We will now show that question 5 has a positive answer. A key point is the similarity between solving a unification problem and solving a system of linear equations:

$$u_1 = v_1$$
$$\vdots$$
$$u_n = v_n$$

and

$$x_1 = a_{1\,1}x_1 + \cdots + a_{1\,n}x_n$$
$$\vdots$$
$$x_m = a_{m\,1}x_1 + \cdots + a_{m\,n}x_n.$$

However, there are some major differences.

- In E-unification, we cannot assume that the algebraic structure is a field.
- The u_i may not be variables.
- We may have $u_i = v_j$ for $i \neq j$.

Nevertheless, the basic idea of *variable elimination* (as in Gaussian elimination) applies. If u_i is a variable that does not occur in v_i, we can *substitute* v_i for u_i in the rest of the system, and preserve the set of solutions:

$$u_1[v_i/u_i] = v_1[v_i/u_i]$$

$$\vdots$$

$$u_i = v_i$$

$$\vdots$$

$$u_n[v_i/u_i] = v_n[v_i/u_i].$$

This leads to the idea of the *method of transformations* on term systems:

Attempt to transform S into a system S' which is obviously solved.

One of the critical issues is to decide what we mean by a solved system. Quite obviously, a solved system is one that should represent a unifying substitution. Since we are dealing with multisets of unordered pairs, we have to be a little careful in formalizing this idea.

Definition 3.8. A term pair $\langle u, v \rangle$ is in *solved form* in a system S iff either u or v is a variable, say x, and this variable x does not occur anywhere else in S; in particular, if $x = u$ then $x \notin Var(v)$ (and similarly if $x = v$ then $x \notin Var(u)$). The variable x is called a *solved variable*. A system is in solved form if all its pairs are in solved form; a variable is *unsolved* if it occurs in S but is not solved.

Note that a solved form system is always a *set* of solved pairs. Also, note that in a solved pair $\langle u, v \rangle$, it is possible that both u and v are variables, and in this case, it may be that only one of the two is solved in S, or that both are solved in S. Thus, ignoring the order in term pairs, a system is *solved* iff it is of the form

$$S = \langle x_1, v_1 \rangle, \ldots, \langle x_n, v_n \rangle,$$

where $x_1, \ldots, x_n$ are **distinct variables**, and $x_i \notin Var(v_j)$ for all i, j, $1 \leq i, j \leq n$. A system in solved form defines essentially a unique substitution as shown in the definition below.

Definition 3.9. Given a system S in solved form, we define the substitution σ_S as follows: if $S = \langle x_1, v_1 \rangle, \ldots, \langle x_n, v_n \rangle$, then $\sigma_S = [v_1/x_1, \ldots, v_n/x_n]$.

Actually, the above definition is ambiguous, and this is the one place where we might regret our definition of a term system where pairs $\langle u, v \rangle$ are unordered. Let us explain where the difficulty lies. There is no problem with a solved pair $\langle u, v \rangle$ in which **only one** of u, v, say u, is a solved variable, because then the substitution component must be $[v/u]$. But when **both** u and v are solved variables, we can use $[u/v]$ or $[v/u]$ interchangeably as a substitution component. Thus, σ_S is not uniquely defined. However, note for any two σ'_S and σ''_S obtained from S, there is a renaming permutation ρ such that $\sigma'_S = \sigma''_S\,; \rho$ (where ρ is determined by the pairs $\langle u, v \rangle$ where both u and v are solved in S). Thus, σ_S is uniquely defined, modulo some inessential renaming permutation. The important fact is that σ_S is an idempotent *mgu* of S, as we now show.

Lemma 3.10. *Let* $S = \langle x_1, t_1 \rangle, \ldots, \langle x_n, t_n \rangle$ *be in solved form, where the* $x_1, \ldots, x_n$ *are solved variables. If* $\sigma = [t_1/x_1, \ldots, t_n/x_n]$, *then* σ *is an idempotent mgu of* S. *Furthermore, for any unifier* θ *of* S, *we have* $\theta = \sigma; \theta$.

Proof. We simply observe that for any θ, $\theta(x_i) = \theta(t_i) = \theta(\sigma(x_i))$ for $1 \leq i \leq n$, and $\theta(x) = \theta(\sigma(x))$ otherwise. Clearly σ is an *mgu*, and since $D(\sigma) \cap I(\sigma) = \emptyset$ by the definition of solved forms, it is idempotent. $\square$

The next question is to find sets of transformations for solving E-unification problems. The following properties of such a set $\mathcal{T}$ of transformations are desirable:

1. (Soundness) Whenever $S \stackrel{*}{\Longrightarrow}_{\mathcal{T}} S'$, then $U_E(S') \subseteq U_E(S)$.
2. (Completeness) For any E-unifier θ of S, there is some solved S' s.t. $S \stackrel{*}{\Longrightarrow}_{\mathcal{T}} S'$, and $\sigma_{S'} \leq_E \theta[Var(S)]$.

When $\mathcal{T}$ satisfies (1) and (2), we say that $\mathcal{T}$ *is a complete set of transformations*. We also want the transformations to be as *deterministic as possible*, to reduce the search space.

To the best of our knowledge, Gallier and Snyder [35] were the first to give a set of transformations with some minimality properties and complete for arbitrary sets E. This set is given in the next definition.

Definition 3.11. (The Set of Transformations $\mathcal{T}$) Let S be any term system (possibly empty), and u, v two terms. The set $\mathcal{T}$ consists of the following transformations:

$$\{\langle u, u \rangle\} \cup S \Longrightarrow S \qquad (triv)$$

$$\{\langle f(u_1,\ldots,u_k), f(v_1,\ldots,v_k)\rangle\} \cup S \Longrightarrow \{\langle u_1,v_1\rangle,\ldots,\langle u_k,v_k\rangle\} \cup S \quad (dec)$$

$$\{\langle x,v\rangle\} \cup S \Longrightarrow \{\langle x,v\rangle\} \cup S[v/x], \quad (vel)$$

where x is a variable s.t. $\langle x,v\rangle$ is **not** solved in $\{\langle x,v\rangle\} \cup S$, and $x \notin Var(v)$;

$$\{\langle u,v\rangle\} \cup S \Longrightarrow \{\langle u/\beta, l\rangle, \langle u[\beta \leftarrow r], v\rangle\} \cup S, \quad (lazy)$$

where u/β is some subterm of u, and $l \doteq r$ is some variant of an equation in $E \cup E^{-1}$.

In this last rule (*lazy paramodulation*), the subterm u/β **is not** a variable. This is an important restriction which eliminates many redundant E-unifiers. It should be noted that in transformation (*vel*), one should not relax the condition "$\langle x,v\rangle$ is **not** solved" to "x is not solved". If this were allowed, one could eliminate the variable x even when v is also a solved variable, and this could have the effect that v could then become unsolved again, leading to an infinite (cyclic) sequence of transformations.

The set of transformations $\mathcal{ST}$ consisting only of (*triv*), (*dec*), and (*vel*), is just a variant of the Herbrand-Martelli-Montanari transformations for standard unification. Before proceeding with the general case, let us now examine the special case where $E = \emptyset$, and show that the set $\mathcal{ST}$ is a complete set of transformations for standard unification. The basic idea of the transformations is to transform the original problem into a solved form which represents its own solution.

Example 3.12.

$$\begin{aligned}
&\langle f(x, g(a,y)),\ f(x, g(y,x))\rangle \\
&\Longrightarrow_{dec}\ \langle x,x\rangle,\ \langle g(a,y), g(y,x)\rangle \\
&\Longrightarrow_{triv}\ \langle g(a,y), g(y,x)\rangle \\
&\Longrightarrow_{dec}\ \langle a,y\rangle,\ \langle y,x\rangle \\
&\Longrightarrow_{vel}\ \langle a,y\rangle,\ \langle a,x\rangle\,.
\end{aligned}$$

The reader can immediately verify that the substitution $[a/y, a/x]$ is a unifier of the original system (in fact, it is an *mgu*). The sense in which these transformations preserve the logically invariant properties of a unification problem is shown in the next lemma.

Lemma 3.13. *Let the set of all standard unifiers of a system S be denoted by $U(S)$. If $S \Longrightarrow S'$ using any transformation from $\mathcal{ST}$, then $U(S) = U(S')$.*

Proof. The only difficulty concerns (vel). Suppose

$$\{\langle x, v\rangle\} \cup S \Longrightarrow_{vel} \{\langle x, v\rangle\} \cup \sigma(S)$$

with $\sigma = [v/x]$. For any substitution θ, if $\theta(x) = \theta(v)$, then $\theta = \sigma;\theta$, since $\sigma;\theta$ differs from θ only at x, but $\theta(x) = \theta(v) = \sigma;\theta(x)$. Thus,

$$\begin{aligned} & \theta \in U(\{\langle x, v\rangle\} \cup S) \\ \text{iff} \;\; & \theta(x) = \theta(v) \;\text{ and }\; \theta \in U(S) \\ \text{iff} \;\; & \theta(x) = \theta(v) \;\text{ and }\; \sigma;\theta \in U(S) \\ \text{iff} \;\; & \theta(x) = \theta(v) \;\text{ and }\; \theta \in U(\sigma(S)) \\ \text{iff} \;\; & \theta \in U(\{\langle x, v\rangle\} \cup \sigma(S)). \end{aligned}$$

□

The point here is that the most important feature of a unification problem—its set of solutions—is preserved under these transformations, and hence we are justified in our method of attempting to transform such problems into a trivial (solved) form in which the existence of an *mgu* is evident.

We may now show the soundness and completeness of these transformations following [79].

Theorem 3.14. *(Soundness) If $S \stackrel{*}{\Longrightarrow} S'$ with S' in solved form, then $\sigma_{S'} \in U(S)$.*

Proof. Using the previous lemma and a trivial induction on the length of transformation sequences, we see that $U(S) = U(S')$, and so clearly $\sigma_{S'} \in U(S)$. □

Theorem 3.15. *(Completeness) Suppose that $\theta \in U(S)$. Then any sequence of transformations*

$$S = S_0 \Longrightarrow S_1 \Longrightarrow S_2 \Longrightarrow \ldots$$

must eventually terminate in a solved form S' such that $\sigma_{S'} \leq \theta$.

Proof. We first show that every transformation sequence terminates. For any system S, let us define a complexity measure $\mu(S) = < n, m >$, where n is the number of *unsolved* variables in the system, and m is the sum of the sizes of all the terms in the system. Then the lexicographic ordering on $< n, m >$ is well-founded, and each transformation produces a new system with a measure strictly smaller under this ordering: $(triv)$ and (dec) must decrease m and can not increase n, and (vel) must decrease n.

Therefore the relation $\Longrightarrow$ is well-founded, and every transformation sequence must end in some system to which no transformation applies. Suppose a given sequence ends in a system S'. Now $\theta \in U(S)$ implies by lemma 3.13 that $\theta \in U(S')$, and so S' can contain no pairs of the form $\langle f(t_1, \ldots, t_n), g(t'_1, \ldots, t'_m)\rangle$ or of the form $\langle x, t\rangle$ with $x \in Var(t)$. But since no transformation applies, all pairs in S' must be in solved form. Finally, since $\theta \in U(S')$, by lemma 3.10 we must have $\sigma_{S'} \leq \theta$. □

Putting these two theorems together, we have that the set $\mathcal{ST}$ can always find an *mgu* for a unifiable system of terms; as remarked in [79], this abstract formulation can be used to model many different unification algorithms, by simply specifying data structures and a control strategy.

In fact, we have proved something stronger than necessary in Theorem 3.15: it has been shown that all transformation sequences terminate and that *any* sequence of transformations issuing from a unifiable system must eventually result in a solved form. This is possible because the problem is decidable. Strictly speaking, it would have been sufficient for completeness to show that if S is unifiable then there exists *some* sequence of transformations which results in a solved form, since then a complete search strategy, such as breadth-first search, could find the solved form. This form of completeness, which might be termed *non-deterministic completeness*, will be used in finding results on E-unification, where the general problem is undecidable.

Now in the general case, where $E \neq \emptyset$, the rule $(lazy)$ is added to the set $\mathcal{ST}$ to account for the presence of equations in the unification problem. The use of the transformation $(lazy)$ is illustrated by the following example.

Example 3.16. Let $E = \{f(g(a)) \doteq a\}$, and $S = \{\langle g(f(x)), x\rangle\}$. We have the following sequence of transformations:

$$\begin{aligned}\langle g(f(x)), x\rangle &\Longrightarrow_{lazy} \langle f(x), f(g(a))\rangle, \langle g(a), x\rangle\\ &\Longrightarrow_{dec} \langle x, g(a)\rangle, \langle g(a), x\rangle\\ &\Longrightarrow_{vel} \langle x, g(a)\rangle, \langle g(a), g(a)\rangle\\ &\Longrightarrow_{triv} \langle x, g(a)\rangle.\end{aligned}$$

Thus, $[g(a)/x]$ is an E-unifier for $S = \{\langle g(f(x)), x\rangle\}$.

Since E-unification is significantly more complex than standard unification, it is remarkable that the addition of the transformation $(lazy)$ alone yields a complete set for general E-unification. The completeness of the set of transformations $\mathcal{T}$ is the major theorem in Gallier and Snyder [35].

Theorem 3.17. *The set $\mathcal{T}$ is a complete set of transformations for arbitrary E. Furthermore,*

$$\{\sigma_{S'}|_{Var(S)} \mid S \stackrel{*}{\Longrightarrow}_{\mathcal{T}} S', \;\; S' \textit{ solved}\}$$

is a complete set of E-unifiers for S.

It is also remarkable that the following additional restriction improving the determinism of the transformations can be imposed without losing the completeness of the set $\mathcal{T}$: if l is not a variable, then we must have $root(l) = root(u/\beta)$, and decomposition is immediately applied to $\langle u/\beta, l\rangle$. Thus, if l is not a variable, letting $l = f(l_1, \ldots, l_k)$ and $u/\beta = f(t_1, \ldots, t_k)$, Lazy Paramodulation can be specialized to:

$$\{\langle u[\beta \leftarrow f(t_1, \ldots, t_k)], v\rangle\} \cup S \Longrightarrow \{\langle t_1, l_1\rangle, \ldots, \langle t_k, l_k\rangle, \langle u[\beta \leftarrow r], v\rangle\} \cup S.$$

This restriction helps in *weeding out* redundant E-unifiers. Further restrictions can be added safely when E is a *canonical set* of rewrite rules (see definition 2.4). In this case, we can restrict the use of the lazy paramodulation rule so that it is applied only when $\beta = \epsilon$ (at the root), or when one of u, v is a variable (but not both). Example 3.16 shows that when u is a variable, it may be necessary to apply the rule (*lazy*) inside v (i.e. $\beta \neq \epsilon$).

In fact, when E is canonical, a new rule, *narrowing*, can be defined which, in conjunction with $\mathcal{ST}$, forms a complete set of transformations. We shall return to this important special case at the end of this section, after discussing the general case.

We now would like to give an overview of the proof of theorem 3.17. Two different proofs are given in Gallier and Snyder [35]. The first one uses a streamlined version of unfailing completion, and the second one an extension of narrowing. We only discuss the first proof. Assume that θ is an E-unifier of a pair $\langle u, v\rangle$. One of the major difficulties in the proof is to show that lazy paramodulation is complete with the restriction that u/β is a *nonvariable* term. In particular, it must be shown that given a pair $\langle x, v\rangle$ where x is a variable, (*lazy*) only needs to be applied to v at a nonvariable position. The key point is that if the equations in E were orientable and formed a canonical system R, then we could work with normalized substitutions, that is, substitutions such that $\theta(x)$ is irreducible for every $x \in D(\theta)$. If R is canonical, for every pair $\langle x, v\rangle$ where x is a variable, there is a proof of the form $\theta(v) \xrightarrow{*}_R w \xleftarrow{*}_R \theta(x)$ for some irreducible w, and if θ is normalized, then the proof is in fact of the form $\theta(v) \xrightarrow{*}_R \theta(x)$, where every rule $\rho(l) \rightarrow \rho(r)$ used in this sequence applies at some *nonvariable* address β in v. Hence, for any rule in this sequence applied at a topmost level, $\theta(v/\beta)$ and $\rho(l)$ must be E-congruent. This is the motivation for the lazy paramodulation rule

$$\{\langle u, v\rangle\} \cup S \Longrightarrow \{\langle u/\beta, l\rangle, \langle u[\beta \leftarrow r], v\rangle\} \cup S,$$

where β is a nonvariable occurrence in u.

However, not every set of equations is equivalent to a canonical system of rewrite rules, and even if it is orientable with respect to some reduction ordering (thus forming a Noetherian set of rules), it may not be confluent. Three crucial observations allow us to overcome these difficulties:

(1) There is no loss of generality in considering only ground substitutions;
(2) There are simplification orderings $\succ$ that are total on ground terms;
(3) Ground confluence (or equivalently, being ground Church-Rosser) is all that is needed.

These ingredients make possible the existence of *unfailing completion* procedures (Bachmair, Dershowitz, Hsiang, and Plaisted [8],[9],[10]). The main trick is that one can use *orientable ground instances of equations*, that is, ground equations of the form $\rho(l) \doteq \rho(r)$ with $\rho(l) \succ \rho(r)$, where $l \doteq r$ is a variant of an equation in $E \cup E^{-1}$. Even if $l \doteq r$ is not orientable, $\rho(l) \doteq \rho(r)$ always is if $\succ$ is total on ground terms. The last ingredient is that given a set E of equations and a reduction ordering $\succ$ total on ground terms, we can show that E can be extended to a set E^ω equivalent to E such that the set $R(E^\omega)$ of orientable instances of E^ω is ground Church-Rosser. Furthermore, E^ω is obtained from E by computing critical pairs (in a hereditary fashion), treating the equations in E as two-way rules.[4]

Our "plan of attack" for the completeness proof of the set of transformations $\mathcal{T}$ is the following.

(1) Show the existence of the ground Church-Rosser completion E^ω of E (theorem 3.24).

[4]Although a consequence of the existence of fair unfailing completion procedures proved by Bachmair, Dershowitz, Hsiang, and Plaisted [8],[9],[10], this result can be proved more directly and with less machinery.

(2) Assuming that E is ground Church-Rosser, show that the $\mathcal{T}$-transformations are complete.

(3) For an arbitrary E, show that the $\mathcal{T}$-transformations are complete using theorem 3.24 and a lemma which shows that the computation of critical pairs can be simulated by Lazy Paramodulation (lemma 3.26).

In (2), we also need the fact that given any E-unifier θ, there is another normalized E-unifier σ such that $\sigma =_E \theta$. It is actually more general (and more flexible) but no more complicated to deal with pairs (E, R) where E is a set of equations and R a set of rewrite rules contained in some given reduction ordering $\succ$. The set E represents the nonorientable part (w.r.t. $\succ$) of the system. Thus, as in Bachmair, Dershowitz, Hsiang, and Plaisted [8],[9],[10], we present our results for such systems.

Definition 3.18. Given a set R of rewrite rules, a proof of the form $u \xrightarrow{*}_R w \xleftarrow{*}_R v$ is called a *rewrite proof*. A proof of the form $u \longleftarrow_R w \longrightarrow_R v$ is called a *peak*. Clearly, an equational proof $u \xleftrightarrow{*}_{R\cup E} v$ is a rewrite proof iff it is a proof without peaks.

We also need the concepts of orientable instance, ground Church-Rosser, and critical pair.

Definition 3.19. Let E be a set of equations and $\succ$ a reduction ordering. Given a variant $l \doteq r$ of an equation in $E \cup E^{-1}$, an equation $\sigma(l) \doteq \sigma(r)$ is an *orientable instance* (w.r.t. $\succ$) of $l \doteq r$ iff $\sigma(l) \succ \sigma(r)$ for some substitution σ.[5] Given a reduction ordering $\succ$, the set of all orientable instances of equations in $E \cup E^{-1}$ is denoted by $R(E)$. Note that if $u \longrightarrow_{R(E)} v$, then $u \longrightarrow_{[\alpha,\sigma(l)\doteq\sigma(r)]} v$ for some variant of an equation $l \doteq r$ in $E \cup E^{-1}$ such that $\sigma(l) \succ \sigma(r)$, and since $\succ$ is a reduction ordering, $u \succ v$.

Definition 3.20. Let E be a set of equations, R a rewrite system, and $\succ$ a reduction ordering. The pair (E, R) is *ground Church-Rosser relative to* $\succ$ iff (a) $R \subseteq \succ$ and (b) for any two *ground* terms u, v, if $u \xleftrightarrow{*}_{E\cup R} v$, then there is a rewrite proof $u \xrightarrow{*}_{R(E)\cup R} w \xleftarrow{*}_{R(E)\cup R} v$ for some w. A reduction ordering $\succ$ is *total on E-equivalent ground terms* iff for any two distinct ground terms u, v, if $u \xleftrightarrow{*}_E v$, then either $u \succ v$ or $v \succ u$. A

[5]The interested reader might convince himself that because $\succ$ is stable and has the subterm property, for any two terms u and v, $u \succ v$ implies that $Var(v) \subseteq Var(u)$. This fact is sometimes glossed over. In the present case thus $Var(\sigma(r)) \subseteq Var(\sigma(l))$.

reduction ordering $\succ$ that is total on E-equivalent ground terms is called a *ground reduction ordering for E.*

It is important to note that for every set R of rewrite rules which is noetherian with respect to a given reduction ordering $\succ$, if R is Church-Rosser, then it is ground Church-Rosser relative to $\succ$, but in general the converse is not true. For example, consider the set of rewrite rules

$$\begin{aligned} R = \{\; & fx \to gx \\ & fx \to hx \\ & fa \to a \\ & ga \to a \\ & ha \to a\; \}, \end{aligned}$$

where $\Sigma = \{f, g, h, a\}$. It is easy to show that R is noetherian with respect to the recursive path ordering generated by the precedence $f \succ g \succ h \succ a$, and, since every ground term reduces to a, it is ground Church-Rosser relative to $\succ$. But R is *not* Church-Rosser, since $hy \longleftarrow_R fy \longrightarrow_R gy$, and hy and gy are irreducible. In general, being Church-Rosser is a stronger condition than being ground Church-Rosser.

Definition 3.21. Let E be a set of equations, R a rewrite system, and $\succ$ a reduction ordering containing R. Let $l_1 \to r_1$ and $l_2 \to r_2$ be variants of rules in $E \cup E^{-1} \cup R$ with no variables in common (viewing an equation $l \doteq r \in E \cup E^{-1}$ as the rule $l \to r$). Suppose that for some address β in l_1, l_1/β is *not* a variable and l_1/β and l_2 are unifiable, and let σ be the mgu of l_1/β and l_2. If $\sigma(r_1) \not\succeq \sigma(l_1)$ and $\sigma(r_2) \not\succeq \sigma(l_2)$, the *superposition* of $l_1 \to r_1$ on $l_2 \to r_2$ at β determines a *critical pair* $\langle g, d\rangle$ *of* (E, R), with $g = \sigma(r_1)$ and $d = \sigma(l_1[\beta \leftarrow r_2])$. The term $\sigma(l_1)$ is called the *overlapped term*, and β the *critical pair position.*

The importance of critical pairs lies in the fact that they can be used to eliminate peaks in proofs.

Lemma 3.22. *(Critical pair lemma, Knuth and Bendix,* [71], *Huet* [56]*) Let E be a set of equations, R a rewrite system, and $\succ$ a reduction ordering containing R. For every peak $s \longleftarrow_{R(E)\cup R} u \longrightarrow_{R(E)\cup R} t$, either there exists some term v such that $s \xrightarrow{*}_{R(E)\cup R} v \xleftarrow{*}_{R(E)\cup R} t$, or there exists a critical pair $\langle g, d\rangle$ of $E \cup R$, an address α in u (s.t. u/α is not a variable) and a substitution η such that, $s = u[\alpha \leftarrow \eta(g)]$ and $t = u[\alpha \leftarrow \eta(d)]$.*

We shall now prove that given a pair (E, R) and a reduction ordering $\succ\!\!\succ$ containing R that is a ground reduction ordering for $E \cup R$, there is a pair (E^{ω}, R^{ω}) containing (E, R) that is equivalent to (E, R) and is ground Church-Rosser relative to $\succ\!\!\succ$. The pair (E^{ω}, R^{ω}) can be viewed as an abstract completion of (E, R) (not produced by any specific algorithm). The existence of (E^{ω}, R^{ω}) follows from the existence of fair unfailing completion procedures proved by Bachmair, Dershowitz, Hsiang, and Plaisted [8],[9],[10]. However, this proof requires more machinery than we need for our purposes. We can give a more direct and simpler proof (inspired by their proof) that isolates clearly the role played by critical pairs. (In this proof, one will not be distracted by features of completion procedures that have to do with efficiency, like simplification of equations or rules by other rules.) The following definition is needed.

Definition 3.23. Let E be a set of equations, R a rewrite system, and $\succ$ a reduction ordering containing R. Let $CR(E, R)$ denote the set of all critical pairs of (E, R) (w.r.t. $\succ$). The sets E^n and R^n are defined inductively as follows: $E^0 = E$, $R^0 = R$, and for every $n \geq 0$,

$$
\begin{aligned}
R^{n+1} &= R^n \cup \{g \to d \mid \langle g, d\rangle \in CR(E^n, R^n) \text{ and } g \succ d\} \\
&\qquad \cup \{d \to g \mid \langle g, d\rangle \in CR(E^n, R^n) \text{ and } d \succ g\}, \quad \text{and} \\
E^{n+1} &= E^n \cup \{g \doteq d \mid \langle g, d\rangle \in CR(E^n, R^n),\ g \not\succ d \text{ and } d \not\succ g\}.
\end{aligned}
$$

We also let

$$
E^{\omega} = \bigcup_{n \geq 0} E^n \quad \text{and} \quad R^{\omega} = \bigcup_{n \geq 0} R^n.
$$

Thus, R^{ω} consists of orientable critical pairs obtained from (E, R) (hereditarily), and E^{ω} consists of nonorientable critical pairs obtained from (E, R) (hereditarily). As the next theorem shows, (E^{ω}, R^{ω}) is a kind of abstract completion of (E, R).

Theorem 3.24. *Let E be a set of equations, R a rewrite system, and $\succ$ a reduction ordering containing R that can be extended to a ground reduction ordering $\succ\!\!\succ$ for $E \cup R$. Then, (E^{ω}, R^{ω}) is equivalent to (E, R) and is ground Church-Rosser relative to $\succ\!\!\succ$.*

A number of technical details must be taken care of, in particular, it is necessary to show that given a ground Church-Rosser pair (E, R), there is no loss of generality in working with normalized substitutions (substitutions that are reduced w.r.t $R(E) \cup R$), and we arrive at the following theorem.

Theorem 3.25. *Let E be a set of equations, R a rewrite system, and $\succ$ a reduction ordering containing R, and assume that (E, R) is ground Church-Rosser relative to $\succ$. Given any system S, if θ is an (E, R)-unifier of S, then there is a sequence of $\mathcal{T}$-transformations $S \stackrel{*}{\Longrightarrow} \widehat{S}$ (using variants of equations in $E \cup E^{-1} \cup R$) yielding a solved system $\widehat{S}$ such that if $\sigma_{\widehat{S}}$ is the substitution associated with $\widehat{S}$, then $\sigma_{\widehat{S}} \leq_{E \cup R} \theta[Var(S)]$. Furthermore, Lazy Paramodulation can be restricted so that it is applied only when either $\beta = \epsilon$ or one of u, v is a variable (but not both).*

In Gallier and Snyder [35], the proof of theorem 3.25 is obtained using the notion of equational proof tree. This notion was introduced to obtain the completeness of another set of transformations. It is possible to give a more direct proof. Such a proof was obtained by Snyder but was not included in [35]. It will appear in [116], and another proof is sketched in Dougherty and Johann [23]. Theorem 3.25 also provides a rigorous proof of the correctness of the transformations of Martelli, Moiso, and Rossi [80] in the case where $E = \emptyset$ and R is canonical. In fact, we have shown the more general case where R is ground Church-Rosser w.r.t. $\succ$. In order to prove the completeness of the $\mathcal{T}$-transformations in the general case, the following lemma showing that the computation of critical pairs can be simulated by Lazy Paramodulation is needed.

Lemma 3.26. *Let E be a set of equations, R a rewrite system, and $\succ$ a reduction ordering containing R. For every finite system S, every sequence of $\mathcal{T}$-transformations $S \stackrel{*}{\Longrightarrow} \widehat{S}$ using equations in $E^{\omega} \cup (E^{\omega})^{-1} \cup R^{\omega}$ can be converted to a sequence $S \stackrel{*}{\Longrightarrow} \widehat{S'}$ using equations only in $E \cup E^{-1} \cup R \cup R^{-1}$, such that $\widehat{S}$ and $\widehat{S'}$ are in solved form and $\sigma_{\widehat{S}}|_{Var(S)} = \sigma_{\widehat{S'}}|_{Var(S)}$.*

The lemma is established by proving by induction on k that every sequence of $\mathcal{T}$-transformations $S \stackrel{*}{\Longrightarrow} \widehat{S}$ using equations in $E^k \cup (E^k)^{-1} \cup R^k$ can be converted to a sequence of $\mathcal{T}$-transformations $S \stackrel{*}{\Longrightarrow} \widehat{S'}$ using equations only in $E \cup E^{-1} \cup R \cup R^{-1}$, such that $\widehat{S}$ and $\widehat{S'}$ are in solved form and $\sigma_{\widehat{S}}|_{Var(S)} = \sigma_{\widehat{S'}}|_{Var(S)}$.

It should be noted that in [35], the authors forgot to ensure that the property

if l is not a variable, then we must have $root(l) = root(u/\beta)$, and decomposition is immediately applied to $\langle u/\beta, l\rangle$, is preserved in the proof.[6] However, this can be easily fixed (see [23] and [116]). Finally, we can prove the completeness of the $\mathcal{T}$-transformations in the general case.

Theorem 3.27. *Let E be a set of equations, R a rewrite system, and $\succ$ a reduction ordering containing R total on ground terms. Given any finite system S, if θ is an (E, R)-unifier of S, then there is a sequence of $\mathcal{T}$-transformations $S \stackrel{*}{\Longrightarrow} \widehat{S}$ (using variants of equations in $E \cup E^{-1} \cup R \cup R^{-1}$) yielding a solved system $\widehat{S}$ such that if $\sigma_{\widehat{S}}$ is the substitution associated with $\widehat{S}$, then $\sigma_{\widehat{S}} \leq_{E \cup R} \theta[Var(S)]$.*

Proof. By theorem 3.24, $E^\omega \cup R^\omega$ is equivalent to (E, R) and is ground Church-Rosser relative to $\succ$. By theorem 3.25, there is a sequence of $\mathcal{T}$-transformations $S \stackrel{*}{\Longrightarrow} \widehat{S}$ using variants of equations in $E^\omega \cup (E^\omega)^{-1} \cup R^\omega$ yielding a solved system $\widehat{S}$ such that if $\sigma_{\widehat{S}}$ is the substitution associated with $\widehat{S}$, then $\sigma_{\widehat{S}} \leq_{E \cup R} \theta[Var(S)]$. Finally, we use lemma 3.26 to eliminate uses of critical pairs, obtaining a sequence where all equations are in $E \cup E^{-1} \cup R \cup R^{-1}$. □

Note that when (E, R) is ground Church-Rosser, equations in E are used as two-way rules in Lazy Paramodulation, but rules in R can be used oriented. This means that in a step

$$\langle u, v \rangle \Longrightarrow \langle u/\beta, l \rangle, \langle u[\beta \leftarrow r], v \rangle,$$

where β is a nonvariable occurrence in u, then $l \doteq r \in E \cup E^{-1}$ if $l \doteq r$ is not in R, but $r \to l$ is not tried if $l \to r$ is in R, and similarly for a step

$$\langle u, v \rangle \Longrightarrow \langle u, v[\beta \leftarrow r] \rangle, \langle l, v/\beta \rangle,$$

where β is a nonvariable occurrence in v. Furthermore, Lazy Paramodulation can be restricted so that it applies only when either $\beta = \epsilon$ or one of u, v is a variable (but not both). This is in contrast to the general case where even rules in R may have to be used as two-way rules due to the computation of critical pairs. Also, Lazy Paramodulation may have to be applied with $\beta \neq \epsilon$ even when both u and v are not variables. This case only seems necessary to compute critical pairs. So far, we have failed to produce an example where Lazy Paramodulation needs to be applied in its full generality (that is, when neither u nor v is a variable and $\beta \neq \epsilon$). We conjecture that $\mathcal{T}$ is still

[6] We thank Dan Dougherty and Patricia Johann for bringing this lacuna to our attention.

complete if Lazy Paramodulation is restricted so that it applies only when either $\beta = \epsilon$ or one of u, v is a variable (but not both).

Since the publication of Gallier and Snyder [35], modifications of the set of transformations $\mathcal{T}$ have been given, and simpler proofs of theorem 3.27 have been claimed. Hsiang and Jouannaud presented a set of transformations inspired by our set $\mathcal{T}$ at the Workshop in Val d'Ajol, in June 1988 [59]. They added some failure conditions and split some of the rules to avoid provisos, and sketched a simpler proof of completeness. This set of transformations is also presented in Kirchner and Jouannaud [61]. So far, no completeness proof has been found for this set. Interestingly, it appears that Hsiang and Jouannaud have retracted their earlier claim that their set of transformations is complete using the strategy of "eager variable elimination". The strategy of eager variable elimination is the stategy of giving priority to rule (vel) whenever it can be applied, even if other rules apply. This strategy has been erroneously claimed to be correct a number of times in the past (including by us), and to the best of our knowledge, its validity remains an open problem. We should mention here that Dougherty and Johann [23] have given a refinement of Gallier and Snyder's transformations which directly generalizes narrowing, and adds an important restriction to the use of lazy paramodulation; basically they have shown that we may require that when lazy paramodulation is used, decomposition be applied to the new pair $\langle u/\beta, l\rangle$ until only pairs of the form $\langle x, t\rangle$ remain. Essentially, this iterates the original requirement that the top function symbols of u/β and l match (when l is not a variable) down into the two terms.

We will conclude our (somewhat biased) survey on E-unification with a short discussion regarding other methods. Since the work of Plotkin [99], most of the energy of researchers in this field has been directed either toward (i) isolating and investigating the E-unification problem in specific theories such as commutativity, associativity, etc., and various combinations of such specific axioms, and (ii) investigating the E-unification problem in the presence of canonical rewrite systems. There has been some work as well on various extensions to the latter.

Narrowing was first presented in Slagle [109] and Lankford [74], but the E-unification algorithm based on this technique first appeared in Fay [28] and was refined by Hullot [57]. (A good presentation of the important results concerning the algorithm can be found in Kirchner and Kirchner [66].) Since then the basic method has been developed by various researchers including

Kirchner, Jouannaud, Kaplan, Hussmann, Nutt, Rety, Smolka [64], [62], [58], [80], [91], [102].

Narrowing is generally defined on single pairs of terms rather than systems, since pairs are not decomposed, as in the system $\mathcal{T}$. Given a canonical E, the narrowing rule is either

$$\langle s, t\rangle \Longrightarrow_{narr} \sigma(\langle s[\alpha \leftarrow r], t\rangle)$$

where $l \doteq r \in E$, α is a non-variable address in s, and $\sigma = mgu(s/\alpha, l)$; or

$$\langle s, t\rangle \Longrightarrow_{narr} \sigma(\langle s, t[\beta \leftarrow r]\rangle),$$

where $l \doteq r \in E$, β is a non-variable address in t, and $\sigma = mgu(t/\beta, l)$. Let us denote a narrowing transformation which uses an equation $l \doteq r$ and an mgu σ by $\Longrightarrow_{[l \doteq r, \sigma]}$. The *paramodulation* rule is identical to the narrowing rule, except that $l \doteq r \in E \cup E^{-1}$ and, in general, paramodulation is allowed at variable occurrences unless otherwise restricted.

The basic idea of narrowing is that when E is canonical, and $\theta(s) \overset{*}{\longleftrightarrow}_E \theta(t)$, with θ reduced (i.e., $\theta(x)$ is in normal form for every $x \in D(\theta)$), then there exists a rewrite proof $\theta(s) \overset{*}{\longrightarrow}_E \overset{*}{\longleftarrow}_E \theta(t)$ where no rewrite takes place at a part of a term introduced by θ. This sequence can be expressed in terms of rewrites on pairs as

$$\theta(\langle s, t\rangle) = \langle s'_0, t'_0\rangle \longrightarrow_{[l_1 \to r_1]} \langle s'_1, t'_1\rangle \ldots \longrightarrow_{[l_n \to r_n]} \langle s'_n, t'_n\rangle$$

for some $n \geq 0$, where $s'_n = t'_n$. This rewrite sequence can be "lifted," using unification, instead of the matching which occurred in the rewrite proof, to obtain an analogous narrowing sequence which results in a pair $\langle u, v\rangle$ where u and v are unifiable; the answer substitution returned is the composition of all the mgus produced by narrowing steps with the mgu of u and v (suitably restricted to the variables in the original pair). This answer is in fact an E-unifier more general than θ.

Example 3.28. Let $\Sigma = \{a, \cdot\}$ (where "$\cdot$" represents concatenation of strings), $E = \{x_1(y_1z_1) \to (x_1y_1)z_1\}$, and consider the terms ax and xa. Then $\theta = [aa/x]$ is an E-unifier of ax and xa, as shown by the following rewrite sequence:

$$\theta(\langle ax, xa\rangle) = \langle a(aa), (aa)a\rangle \longrightarrow_{[x_1(y_1z_1) \to (x_1y_1)z_1, \rho]} \langle (aa)a, (aa)a\rangle,$$

where $\rho = [a/x_1, a/y_1, a/z_1]$. Now consider the narrowing sequence

$$\langle ax, xa\rangle \Longrightarrow_{[x_1(y_1z_1) \to (x_1y_1)z_1, \sigma]} \langle (ay_1)z_1, (y_1z_1)a\rangle,$$

where $\sigma = [a/x_1, y_1z_1/x]$ and the final term has a *mgu* $\mu = [a/y_1, a/z_1]$. Note that $\theta = \sigma; \mu|_{\{x\}}$.

The technical results on this form of E-unification are as follows.

Theorem 3.29. *(Soundness of Narrowing) Let E be a canonical set of rewrite rules, and s and t be two terms. For any narrowing sequence*

$$\langle s, t\rangle \Longrightarrow_{[l_1 \to r_1, \sigma_1]} \langle s_1, t_1\rangle \Longrightarrow_{[l_2 \to r_2, \sigma_2]} \cdots \Longrightarrow_{[l_n \to r_n, \sigma_n]} \langle s_n, t_n\rangle,$$

such that each $l_i \to r_i$ is a variant of a rule in E and s_n and t_n are (standard) unifiable, the substitution

$$\theta = \sigma_1; \ldots ; \sigma_n; \mu$$

is an E-unifier of s and t, where $\mu = mgu(s_n, t_n)$.

Theorem 3.30. *(Completeness of Narrowing) Let E be a canonical set of rewrite rules, and s and t be two terms. For any substitution θ, if $\theta(s) \stackrel{*}{\longleftrightarrow}_E \theta(t)$ then there exists a narrowing sequence*

$$\langle s, t\rangle \Longrightarrow_{[l_1 \to r_1, \sigma_1]} \langle s_1, t_1\rangle \Longrightarrow_{[l_2 \to r_2, \sigma_2]} \cdots \Longrightarrow_{[l_n \to r_n, \sigma_n]} \langle s_n, t_n\rangle,$$

for some $n \geq 0$, and a mgu μ of s' and t' such that

$$\sigma_1; \ldots ; \sigma_n; \mu \leq_E \theta[Var(s, t)].$$

This result gives us a complete strategy for E-unification in the case that E is canonical: we simply search the *narrowing tree* of all possible narrowing sequences from the term $\langle s, t\rangle$ in some complete fashion (say breadth-first) and whenever a unifiable pair of terms is found, return the appropriate answer substitution.

The fundamental idea behind the narrowing procedure is the *lifting* of rewrite proofs to narrowing sequences. The interesting feature of canonical rewrite systems is of course that rewriting is non-deterministic, so that *any* strategy (e.g. top-down, bottom-up, inner-most left-most) for rewriting will reduce a term to its normal form. It turns out that by examining the result of lifting rewrite proofs found under various strategies, we can improve the narrowing procedure by reducing the search space without sacrificing completeness. There are two principal improved versions of narrowing which have been defined. The first, *basic narrowing*, due to [57], reduces the search space by forbidding narrowing at addresses in those parts of the term introduced by the various narrowing substitutions, and is a lifting of an *innermost* rewrite proof (i.e., where no proper subterm of a redex is reducible). The second, *normalized narrowing*, due to [28], reduces the search space by normalizing a term before applying a narrowing step, and is a lifting of (roughly) a certain kind of *top-down* rewrite proof. Both these restrictions reduce the size of the narrowing tree and are complete, but unfortunately, as discussed in [102], the naive combination of these two is not complete.

It is important to note that narrowing is complete in the case of a canonical set of rules, but in the general case it is not; in fact, if the more general rule of paramodulation is restricted to applications at non-variable addresses, it is not complete in the general case either. For example,[7] if $E = \{f(a,b) \doteq a, a \doteq b\}$ and $S = \langle f(x,x), x \rangle$, then no paramodulation step can be applied at a non-variable address, and thus no substitution more general than the E-unifier $[a/x]$ can ever be generated; thus in the general case, lazy paramodulation is necessary if we are to apply inference rules only to the system (and not among the equations themselves). In fact, this is a special case of a more general problem, that in paramodulation theorem provers the set–of–support strategy (which, roughly, forbids forward reasoning from a set of hypotheses to the goal, and only allows backward reasoning from the goal statement) is not complete; a solution to this which involves an extension of the lazy paramodulation calculus to the general refutational case for first-order clauses is presented in [117].

Narrowing and its refinements represent a very clean and elegant solution to an important subclass of E-unification problems, and we do not claim to have improved upon these results. Instead we view our research as an attempt to place these results in a more general context, by showing in a very abstract way how the same proof techniques used in narrowing may be applied to our more general problem. We should in particular note that Martelli, Moiso, and Rossi [80] have presented an E-unification procedure using a set of transformations much the same as our set $\mathcal{T}$, but they attempted to prove completeness only in the context of canonical systems. The work of Kirchner [64] attempts to extend the basic paradigm of E-unification in canonical theories by adapting the approach of Martelli and Montanari [79] to standard unification which uses the operations of merging and decomposition over multiequations to find mgu's in ordered form; by respecting the ordering of variable dependencies among the various terms, one may avoid explicit application of substitutions, and so Variable Elimination is not used.

[7]We owe this example to Dan Dougherty.

Kirchner expands this basic method by defining conditions under which decomposition may be done in the presence of equations, and by defining a new operation on multiequations, called mutation, which is dependent upon the theory under consideration. He extends the procedure for canonical theories by showing that if a theory permits the use of variable dependency orderings to avoid explicit substitution (such a theory is termed *strict*), and if a mutation operation can be deduced, then his procedure returns a complete set of E-unifiers. He then gives a general strategy for deriving the mutation operation via a critical pair computation, and hence a way of automating the creation of specialized E-unification procedures. As an example this strategy is applied to the class of *syntactic* theories, which basically allow complete sets of E-unifiers to be found by allowing at most one rewrite at the root between any two terms. Our approach to E-unification owes much to Kirchner's initial inspiration to adapt the method of transformations to E-unification.

4. Rigid E-Unification

Rigid E-unification is a restricted kind of E-unification that arises naturally in extending Andrews and Bibel's theorem proving method of matings [4], [6], [12], [13], [14], to first-order languages with equality. Gallier started working on this generalization early in 1987, and in the Fall of 1987, he realized that the kind of unification needed by this method was not general E-unification. For the lack of a better name, this form of restricted E-unification was termed rigid E-unification. The generalized method of matings was first presented in Gallier, Raatz, and Snyder [32], where it was conjectured that rigid E-unification is decidable. Several months later, Gallier, Narendran, Plaisted, and Snyder proved that rigid E-unification is NP-complete and that finite complete sets of rigid E-unifiers always exist. These results were announced (without complete proofs) at LICS'88 [34]. Full details and proofs appear in Gallier, Narendran, Plaisted, and Snyder [37].

We now explain why this result is significant for theorem proving in first-order languages with equality. At first glance, a generalization of the method of matings to first-order languages with equality where equality is built-in in the sense of Plotkin [99] (thus, it is not the naive method where explicit equality axioms are added which is rejected for well known inefficiency reasons) requires general E-unification. Hence, there are two factors contributing to the undecidability of the method of matings for first-order languages with equality: (1) the fact that one cannot predict how many disjuncts will occur in a Herbrand expansion (which also holds for first-order languages without equality); (2) the undecidability of the kind of unification required (E-unification). However, it was shown in [32], [38] that the completeness of the method of equational matings is preserved if unrestricted E-unification is replaced by rigid E-unification. Since we also proved that rigid E-unification is decidable, the second undecidability factor is eliminated.

The NP-completeness of rigid E-unification also shows clearly how the presence of equality influences the complexity of theorem proving methods. For languages without equality, one can use standard unification whose time complexity is polynomial, and in fact $O(n)$. For languages with equality, the type of unification required is NP-complete.

Before launching into rigid E-unification, let us recall how it arises naturally in generalizing the method of matings to first-order languages with equality. For details, the reader is referred to Gallier, Raatz, and Snyder [32], [36], and Gallier, Narendran, Raatz, and Snyder [38]. The crucial observation due to Andrews and Bibel is that a quantifier-free formula without equality is unsatisfiable iff certain sets of literals occurring in A (called *vertical paths*) are unsatisfiable. Matings come up as a convenient method for checking that vertical paths are unsatisfiable. Roughly speaking, a *mating* is a set of pairs of literals of opposite signs (*mated pairs*) such that all these (unsigned) pairs are globally unified by some substitution. The importance of matings stems from the fact that a quantifier-free formula A has a mating iff there is a substitution θ such that $\theta(A)$ is unsatisfiable. For languages without equality, this can be checked using standard unification.

In the case of languages with equality, one needs to extend matings to *equational matings*, which is nontrivial and requires proving a generalization of Andrews's version of the Skolem-Herbrand-Gödel theorem [4], [5]. An equational mating is now a set of sets of literals (*mated sets*), where a mated set consists of several positive equations and a single negated equation (rather than pairs of literals as in Andrews's case). Checking that a family of mated sets is unsatisfiable, i.e. an equational mating, reduces to the following problem.

Problem. Given a finite set $E = \{u_1 \doteq v_1, \ldots, u_n \doteq v_n\}$ of equations and a pair $\langle u, v \rangle$ of terms, is there a substitution θ such that, treating $\theta(E)$ as a set of ground equations, $\theta(u) \stackrel{*}{\longleftrightarrow}_{\theta(E)} \theta(v)$, that is, $\theta(u)$ and $\theta(v)$ are congruent modulo $\theta(E)$ by congruence closure (Kozen [72], Nelson and Oppen [90])?

The substitution θ is called a *rigid E-unifier of u and v*.

Example 4.1. Let $E = \{fa \doteq a,\ ggx \doteq fa\}$, and $\langle u, v\rangle = \langle gggx, x\rangle$. Then, the substitution $\theta = [ga/x]$ is a rigid E-unifier of u and v. Indeed, $\theta(E) = \{fa \doteq a,\ ggga \doteq fa\}$, and $\theta(gggx)$ and $\theta(x)$ are congruent modulo $\theta(E)$, since

$$\begin{aligned} \theta(gggx) = gggga \quad &\longrightarrow gfa \qquad &&\text{using } ggga \doteq fa \\ &\longrightarrow ga = \theta(x) \qquad &&\text{using } fa \doteq a. \end{aligned}$$

Note that θ is not the only rigid E-unifier of u and v. For example, $[gfa/x]$ or more generally $[gf^n a/x]$ is a rigid E-unifier of u and v. However, θ is more general than all of these rigid E-unifiers (in a sense to be made precise later). Remarkably, it can be shown that there is always a finite set of most general rigid E-unifiers called a complete set of rigid E-unifiers.

It is interesting to observe that the notion of rigid E-unification arises by *bounding* the resources, in this case, the number of available instances of equations in E. In order to understand more clearly the concept of rigid E-unification, let us recall what (unrestricted) E-unification is. We are given a set of equations $E = \{u_1 \doteq v_1, \ldots, u_n \doteq v_n\}$, and (for simplicity) a pair of terms $\langle u, v\rangle$. The problem is to decide whether is there a substitution θ s.t. $\theta(u) \stackrel{*}{\longleftrightarrow}_E \theta(v)$.

Note that there is *no bound* on the number of instances of equations in E that can be used in the proof that $\theta(u) \stackrel{*}{\longleftrightarrow}_E \theta(v)$. Going back to definition 2.16, we observe that $\theta(u) \stackrel{*}{\longleftrightarrow}_E \theta(v)$ iff is there a *multiset* of equations (from E)

$$\left\{\begin{pmatrix} u_1' \doteq v_1' \\ n_1 \end{pmatrix}, \ldots, \begin{pmatrix} u_m' \doteq v_m' \\ n_m \end{pmatrix}\right\}$$

and m sets of substitutions $\{\sigma_{j,1}, \ldots, \sigma_{j,n_j}\}$, s.t., letting

$$E' = \{\sigma_{j,k}(u_j' \doteq v_j') \mid 1 \le j \le m,\ 1 \le k \le n_j\},$$

we have $\theta(u) \stackrel{*}{\longleftrightarrow}_{E'} \theta(v)$, considering E' as **ground**. Basically, the restriction imposed by rigid E-unification is that $n_1 = \ldots = n_m = 1$, i.e., at most a single instance of each equation in E can be used. In fact, these instances $\theta(u_1 \doteq v_1), \ldots, \theta(u_n \doteq v_n)$ must arise from the substitution θ itself. Also, once these instances have been created, the remaining variables (if any) are considered rigid, that is, treated as constants, so that it is not possible to instantiate these instances. Thus, rigid E-unification and Girard's linear logic [39] share the same spirit. Since the resources are bounded, it is not too surprising that rigid E-unification is decidable, but it is not obvious at all that the problem is in NP. The special case of rigid E-unification where E is a set of ground equations has been investigated by Kozen who has shown that this problem is NP-complete (Kozen, [72],[73]). Thus, rigid E-unification is NP-hard. We also showed that it is in NP, hence NP-complete.

Our plan for the rest of this section is to define precisely what complete sets of rigid E-unifiers are, and to sketch the decision procedure. The definitions of a rigid E-unifier, the preorder $\leq_E$, and complete sets of rigid E-unifiers, will parallel those given for E-unification, but equations are considered as ground in equational proofs. It will be convenient to write $u \overset{*}{\cong}_E v$ to express that $u \overset{*}{\longleftrightarrow}_E v$, treating the equations in E as *ground equations*.

Definition 4.2. Let $E = \{(s_1 \doteq t_1), \ldots, (s_m \doteq t_m)\}$ be a finite set of equations, and let $Var(E) = \bigcup_{(s \doteq t) \in E} Var(s \doteq t)$ denote the set of variables occurring in E.[8] Given a substitution θ, we let $\theta(E) = \{\theta(s_i \doteq t_i) \mid s_i \doteq t_i \in E,\ \theta(s_i) \neq \theta(t_i)\}$. Given any two terms u and v,[9] a substitution θ is a *rigid unifier of* u *and* v *modulo* E (for short, a *rigid* E*-unifier of* u *and* v) iff

> $\theta(u) \overset{*}{\cong}_{\theta(E)} \theta(v)$, that is, $\theta(u)$ and $\theta(v)$ are congruent modulo the set $\theta(E)$ considered as a set of *ground* equations.

The following example should help the reader grasp the notion of rigid E-unifica-
tion. The problem is to show that if $x \cdot x = 1$ in a monoid, then the monoid is commutative.

---

[8]It is possible that equations have variables in common.

[9]It is possible that u and v have variables in common with the equations in E.

Example 4.3.

$$\begin{aligned} E = \{u_1 \cdot 1 &\doteq u_1 \\ w_1 \cdot w_1 &\doteq 1 \\ x_1 \cdot (y_1 \cdot z_1) &\doteq (x_1 \cdot y_1) \cdot z_1 \\ x_2 \cdot (y_2 \cdot z_2) &\doteq (x_2 \cdot y_2) \cdot z_2 \\ w_2 \cdot w_2 &\doteq 1 \\ 1 \cdot v_1 &\doteq v_1 \\ x_3 \cdot (y_3 \cdot z_3) &\doteq (x_3 \cdot y_3) \cdot z_3 \\ x_4 \cdot (y_4 \cdot z_4) &\doteq (x_4 \cdot y_4) \cdot z_4 \\ w_3 \cdot w_3 &\doteq 1\}. \end{aligned}$$

$$\langle u, v \rangle = \langle a \cdot b,\ b \cdot a \rangle.$$

The reader can verify that θ below is a rigid E-unifier:

$$\begin{aligned} \theta = [&a/u_1,\ a/x_1,\ a/x_2,\ a/y_2,\ a/w_2,\ a/x_4, \\ &b/z_2,\ b/v_1,\ b/x_3,\ b/z_3,\ b/y_4,\ b/z_4,\ b/w_3, \\ &a \cdot b/w_1,\ a \cdot b/y_1,\ a \cdot b/z_1,\ a \cdot b/y_3]. \end{aligned}$$

Definition 4.4. Let E be a (finite) set of equations, and W a (finite) set of variables. For any two substitutions σ and θ, $\sigma =_E \theta[W]$ iff $\sigma(x) \overset{*}{\cong}_E \theta(x)$ for every $x \in W$. The relation $\sqsubseteq_E$ is defined as follows. For any two substitutions σ and θ, $\sigma \sqsubseteq_E \theta[W]$ iff $\sigma =_{\theta(E)} \theta[W]$. The set W is omitted when $W = \mathcal{X}$ (where $\mathcal{X}$ is the set of variables), and similarly E is omitted when $E = \emptyset$.

Intuitively speaking, $\sigma \sqsubseteq_E \theta$ iff σ can be generated from θ using the equations in $\theta(E)$. Clearly, $\sqsubseteq_E$ is reflexive. However, it is not symmetric as shown by the following example.

Example 4.5. Let $E = \{fx \doteq x\}$, $\sigma = [fa/x]$ and $\theta = [a/x]$. Then $\theta(E) = \{fa \doteq a\}$ and $\sigma(x) = fa \overset{*}{\cong}_{\theta(E)} a = \theta(x)$, and so $\sigma \sqsubseteq_E \theta$. On the other hand $\sigma(E) = \{ffa \doteq fa\}$, but a and fa are not congruent from $\{ffa \doteq fa\}$. Thus $\theta \sqsubseteq_E \sigma$ *does not* hold.

It is not difficult to show that $\sqsubseteq_E$ is also transitive. We also need an extension of $\sqsubseteq_E$ defined as follows.

Definition 4.6. Let E be a (finite) set of equations, and W a (finite) set of variables. The relation $\leq_E$ is defined as follows: for any two substitutions σ and θ, $\sigma \leq_E \theta[W]$ iff $\sigma\,;\eta \sqsubseteq_E \theta[W]$ for some substitution η (that is, $\sigma\,;\eta =_{\theta(E)} \theta[W]$ for some η).

Intuitively speaking, $\sigma \leq_E \theta$ iff σ is more general than some substitution that can be generated from θ using $\theta(E)$. Clearly, $\leq_E$ is reflexive. The transitivity of $\leq_E$ is also shown easily. When $\sigma \leq_E \theta[W]$, we say that *σ is (rigid) more general than θ over W*. It can be shown that if σ is a rigid E-unifier of u and v and $\sigma \leq_E \theta$, then θ is a rigid E-unifier of u and v. The converse is false. Finally, the crucial concept of a complete set of rigid E-unifiers can be defined.

Definition 4.7. Given a (finite) set E of equations, for any two terms u and v, letting $V = Var(u) \cup Var(v) \cup Var(E)$, a set U of substitutions is a *complete set of rigid E-unifiers for u and v* iff: For every $\sigma \in U$,

(i) $D(\sigma) \subseteq V$ and $D(\sigma) \cap I(\sigma) = \emptyset$ (idempotence),

(ii) σ is a rigid E-unifier of u and v,

(iii) For every rigid E-unifier θ of u and v, there is some $\sigma \in U$, such that, $\sigma \leq_E \theta[V]$.

Suppose we want to find a rigid E-unifier θ of u and v. There is an algorithm using transformations for finding rigid E-unifiers. Roughly, the idea is to use a form of unfailing completion procedure (Knuth and Bendix [71], Huet [56], Bachmair [8], Bachmair, Dershowitz, and Plaisted [9], Bachmair, Dershowitz, and Hsiang [10]). In order to clarify the differences between our method and unfailing completion, especially for readers unfamiliar with this method, we briefly describe the use of unfailing completion as a refutation procedure. For more details, the reader is referred to Bachmair [8].[10]

Let E be a set of equations, and $\succ$ a reduction ordering total on ground terms. The central concept is that of E being *ground Church-Rosser w.r.t.* $\succ$.[11] The crucial observation is that every ground instance $\sigma(l) \doteq \sigma(r)$ of an equation $l \doteq r \in E$ is orientable w.r.t. $\succ$, since $\succ$ is total on ground terms. Let $E^{\succ}$ be the set of all instances $\sigma(l) \doteq \sigma(r)$ of equations $l \doteq r \in E \cup E^{-1}$

[10] The reader may find it helpful to review the section containing definitions 3.19 and 3.20.

[11] This concept is defined in definition 3.20, and the reader may want to review this definition.

with $\sigma(l) \succ \sigma(r)$ (the set of *orientable instances*). We say that E is *ground Church-Rosser w.r.t.* $\succ$ iff for every two ground terms u, v, if $u \xleftrightarrow{*}_E v$, then there is some ground term w such that $u \xrightarrow{*}_{E^\succ} w$ and $w \xleftarrow{*}_{E^\succ} v$. Such a proof is called a *rewrite proof*.

An unfailing completion procedure attempts to produce a set E^∞ equivalent to E and such that E^∞ is ground Church-Rosser w.r.t. $\succ$. In other words, every ground equation provable from E has a rewrite proof in E^∞. The main mechanism involved is the computation of critical pairs. Given two equations $l_1 \doteq r_1$ and $l_2 \doteq r_2$ where l_2 is unifiable with a subterm l_1/β of l_1 which is not a variable, the pair $\langle \sigma(l_1[\beta \leftarrow r_2]), \sigma(r_1) \rangle$ where σ is a mgu of l_1/β and l_2 is a *critical pair*.

If we wish to use an unfailing completion procedure as a refutation procedure, we add two new constants T and F and a new binary function symbol eq to our language. In order to prove that $E \models u \doteq v$ for a ground equation $u \doteq v$, we apply the unfailing completion procedure to the set $E \cup \{eq(u, v) \doteq F,\ eq(z, z) \doteq T\}$, where z is a new variable. It can be shown that $E \models u \doteq v$ iff the unfailing completion procedure generates the equation $F \doteq T$. Basically, given any proof of $F \doteq T$, the unfailing completion procedure extends E until a rewrite proof is obtained. It can be shown that unfailing completion is a complete refutation procedure, but of course, it is not a decision procedure. It should also be noted that when unfailing completion is used as a refutation procedure, E^∞ is actually never generated. It is generated "by need", until $F \doteq T$ turns up.

We now come back to our situation. Without loss of generality, it can be assumed that we have a rigid E-unifier θ of T and F such that $\theta(E)$ is ground. In this case, equations in $\theta(E)$ are orientable instances. The crucial new idea is that in trying to obtain a rewrite proof of $F \doteq T$, we still compute critical pairs, but we **never rename variables**. If l_2 is equal to l_1/β, then we get a critical pair essentially by simplification. Otherwise, some variable in l_1 or in l_2 gets bound to a term *not* containing this variable. Thus the total number of variables in E keeps decreasing. Therefore, after a polynomial number of steps (in fact, the number of variables in E) we must stop or fail. So we get membership in NP. Oversimplifying a bit, we can say that our method is a form of lazy unfailing completion with no renaming of variables.

However, there are some significant departures from traditional Knuth-Bendix completion procedures, and this is for two reasons. The first reason is that we must ensure termination of the method. The second is that we want to show that the problem is in NP, and this forces us to be much more concerned about efficiency.

Our method can be described in terms of a single transformation on triples of the form $\langle \mathcal{S}, \mathcal{E}, \mathcal{O} \rangle$, where $\mathcal{S}$ is a unifiable set of pairs, $\mathcal{E}$ is a set of equations, and $\mathcal{O}$ is something that will be needed for technical reasons and can be ignored for the present. Starting with an initial triple $\langle \mathcal{S}_0, \mathcal{E}_0, \mathcal{O}_0 \rangle$ initialized using E and u, v (except for $\mathcal{O}$ that must be guessed), if the number of variables in E is m, one considers sequences of transformations

$$\langle \mathcal{S}_0, \mathcal{E}_0, \mathcal{O}_0 \rangle \Rightarrow^+ \langle \mathcal{S}_k, \mathcal{E}_k, \mathcal{O}_k \rangle$$

consisting of at most $k \leq m$ steps. It will be shown that u and v have some rigid E-unifier iff there is some sequence of steps as above such that the special equation $F \doteq T$ is in $\mathcal{E}_k$ and $\mathcal{S}_k$ is unifiable. Then, the most general unifier of $\mathcal{S}_k$ is a rigid E-unifier of u and v.

Roughly speaking, $\mathcal{E}_{k+1}$ is obtained by overlapping equations in $\mathcal{E}_k$ (forming critical pairs), as in unfailing Knuth-Bendix completion procedures, except that no renaming of variables takes place. In order to show that the number of steps can be bounded by m, it is necessary to show that some measure decreases every time an overlap occurs, and there are two difficulties. First, the overlap of two equations may involve the identity substitution when some equation simplifies another one. In this case, the number of variables does not decrease, and no other obvious measure decreases. Second, it is more difficult to handle overlap at variable occurrences than it is in the traditional case, because we are not allowed to form new instances of equations.

The first difficulty can be handled by using a special procedure for reducing a set of (ground) equations. Such a procedure is presented in Gallier et al. [33] and runs in polynomial time (see also [114]). Actually, one also needs a total simplification ordering $\prec$ on ground terms, and a way of orienting equations containing variables, which is the purpose of the mysterious component $\mathcal{O}$. The second difficulty is overcome by noticing that one only needs to consider ground substitutions, that the ordering $\prec$ (on ground terms) can be extended to ground substitutions, and that given any rigid E-unifier θ of u and v, there is always a least rigid E-unifier σ (w.r.t $\prec$) that is equivalent to θ (in a sense to be made precise).

Other complications arise in proving that the method is in NP, in particular, we found it necessary to represent most general unifiers (mgu's) by their triangular form as in Martelli and Montanari [79]. We now provide some technical details,

Definition 4.8. Given an idempotent substitution σ (i.e., $D(\sigma) \cap I(\sigma) = \emptyset$) with domain $D(\sigma) = \{x_1, \ldots, x_k\}$, a *triangular form for* σ is a finite set T of pairs $\langle x, t\rangle$ where $x \in D(\sigma)$ and t is a term, such that this set T can be sorted (possibly in more than one way) into a sequence $\langle\langle x_1, t_1\rangle, \ldots, \langle x_k, t_k\rangle\rangle$ satisfying the following properties: for every i, $1 \leq i \leq k$,

(1) $\{x_1, \ldots, x_i\} \cap Var(t_i) = \emptyset$, and

(2) $\sigma = [t_1/x_1]; \ldots; [t_k/x_k]$.

The set of variables $\{x_1, \ldots, x_k\}$ is called the *domain* of T. Note that in particular $x_i \notin Var(t_i)$ for every i, $1 \leq i \leq k$, but variables in the set $\{x_{i+1}, \ldots, x_k\}$ may occur in $t_1, \ldots, t_i$. It is easily seen that σ is an (idempotent) mgu of the term system T.

Example 4.9. Consider the substitution $\sigma = [f(f(x_3, x_3), f(x_3, x_3))/x_1, f(x_3, x_3)/x_2]$. The system $T = \{\langle x_1, f(x_2, x_2)\rangle, \langle x_2, f(x_3, x_3)\rangle\}$ is a triangular form of σ since it can be ordered as $\langle\langle x_1, f(x_2, x_2)\rangle, \langle x_2, f(x_3, x_3)\rangle\rangle$ and $\sigma = [f(x_2, x_2)/x_1]; [f(x_3, x_3)/x_2]$.

The triangular form $T = \{\langle x_1, t_1\rangle, \ldots, \langle x_k, t_k\rangle\}$ of a substitution σ also defines a substitution, namely $\sigma_T = [t_1/x_1, \ldots, t_k/x_k]$. This substitution is usually different from σ and not idempotent as can be seen from example 4.9. However, this substitution plays a crucial role in our decision procedure because of the following property.

Lemma 4.10. *Given a triangular form* $T = \{\langle x_1, t_1\rangle, \ldots, \langle x_k, t_k\rangle\}$ *for a substitution* σ *and the associated substitution* $\sigma_T = [t_1/x_1, \ldots, t_k/x_k]$, *for every unifier* θ *of* T, $\theta = \sigma_T; \theta$.

An other important observation about σ_T is that even though it is usually not idempotent, at least one variable in $\{x_1, \ldots, x_k\}$ does not belong to $I(\sigma_T)$ (otherwise, condition (1) of the triangular form fails). We will assume that a procedure TU is available, which, given any unifiable term system S, returns a triangular form for an idempotent mgu of S, denoted by $TU(S)$. When S consists of a single pair $\langle u, v\rangle$, $TU(S)$ is also denoted by $TU(u, v)$.

One of the major components of the decision procedure for rigid E-unification is a procedure for creating a reduced set of rewrite rules equivalent to a given (finite) set of ground equations. Given a set R of rewrite rules, we say that R *is rigid reduced* iff

(1) No lefthand side of any rewrite rule $l \to r \in R$ is reducible by any rewrite rule in $R - \{l \to r\}$ treated as a ground rule;

(2) No righthand side of any rewrite rule $l \to r \in R$ is reducible by any rewrite rule in R treated as a ground rule.

A procedure for creating a rigid reduced set of rewrite rules equivalent to a given (finite) set of rewrite rules was first presented in Gallier et al. [33] and runs in polynomial time. However, due to the possibility that variables may occur in the equations, we have to make some changes to this procedure. Roughly speaking, given a "guess" $\mathcal{O}$ (a preorder which we call an *order assignment*) of the ordering among all subterms of the terms in a set of equations E, we can run the reduction procedure R on E and $\mathcal{O}$ to produce a reduced rewrite system $R(E, \mathcal{O})$ equivalent to E, and whose orientation is dictated by the preorder $\mathcal{O}$. The precise definition of an order assignment $\mathcal{O}$ is too involved to be reproduced here, but this is not essential anyway. All we need to know is that we have an algorithm R such that, given a set E of equations and an order-assignment $\mathcal{O}$, a rigid-reduced set of rewrite rules $R(E, \mathcal{O})$ is returned, the rules in $R(E, \mathcal{O})$ being oriented by $\mathcal{O}$. We are now ready to define a procedure for finding rigid E-unifiers.

This method uses the reduction procedure just discussed, and a single transformation on certain systems defined next. First, the following definition is needed.

Definition 4.11. Given a set E of equations and some equation $l \doteq r$, the set of equations obtained from E by deleting $l \doteq r$ and $r \doteq l$ from E is denoted by $(E - \{l \doteq r\})^\dagger$. Formally, we let $(E - \{l \doteq r\})^\dagger = \{u \doteq v \mid u \doteq v \in E,\ u \doteq v \neq l \doteq r, \text{ and } u \doteq v \neq r \doteq l\}$.

Definition 4.12. Let $\prec$ be a total simplification ordering on ground terms. We shall be considering finite sets of equations of the form $\mathcal{E} = \mathcal{E}_\Sigma \cup \{eq(u, v) \doteq F,\ eq(z, z) \doteq T\}$,[12] where $\mathcal{E}_\Sigma$ is a set of equations over $T_\Sigma(\mathcal{X})$, and $u, v \in T_\Sigma(\mathcal{X})$. We define a transformation on systems of the form $\langle \mathcal{S}, \mathcal{E}, \mathcal{O}\rangle$, where $\mathcal{S}$ is a term system, $\mathcal{E}$ a set of equations as above, and $\mathcal{O}$ an order assignment:

$$\langle \mathcal{S}_0,\ \mathcal{E}_0,\ \mathcal{O}_0\rangle \Rightarrow \langle \mathcal{S}_1,\ \mathcal{E}_1,\ \mathcal{O}_1\rangle,$$

[12] eq, T, F are some new symbols not occurring in E, u, v.

where $l_1 \doteq r_1$, $l_2 \doteq r_2 \in \mathcal{E}_0 \cup \mathcal{E}_0^{-1}$, either l_1/β is *not* a variable or the equation $l_2 \doteq r_2$ is degenerate,[13] $l_1/\beta \neq l_2$, $TU(l_1/\beta, l_2)$ represents an mgu of l_1/β and l_2 in triangular form,[14] $\sigma = [t_1/x_1, \dots, t_p/x_p]$ where $TU(l_1/\beta, l_2) = \{\langle x_1, t_1\rangle, \dots, \langle x_p, t_p\rangle\}$,

$$\mathcal{E}_1' = \sigma((\mathcal{E}_0 - \{l_1 \doteq r_1\})^\dagger \cup \{l_1[\beta \leftarrow r_2] \doteq r_1\}),$$

$\mathcal{O}_1$ is an order assignment on $\mathcal{E}_1'$ compatible with $\mathcal{O}_0$, $\mathcal{S}_1 = \mathcal{S}_0 \cup TU(l_1/\beta, l_2)$, and $\mathcal{E}_1 = R(\mathcal{E}_1', \mathcal{O}_1)$.

Observe that $\sigma(l_1[\beta \leftarrow r_2] \doteq r_1)$ looks like a critical pair of equations in $\mathcal{E}_0 \cup \mathcal{E}_0^{-1}$, but it is not.[15] This is because a critical pair is formed by applying the mgu of l_1/β and l_2 to $l_1[\beta \leftarrow r_2] \doteq r_1$, but $[t_1/x_1, \dots, t_p/x_p]$ is usually not a mgu of l_1/β and l_2. It is the composition $[t_1/x_1]; \dots; [t_p/x_p]$ that is a mgu of l_1/β and l_2. The reason for not applying the mgu is that by repeated applications of this step, exponential size terms could be formed, and it would not be clear that the decision procedure is in NP. We have chosen an approach of "lazy" (or delayed) unification. Also note that we use the rigid reduced system $R(\mathcal{E}_1', \mathcal{O}_1)$ rather than $\mathcal{E}_1'$, and so, a transformation step is defined only if R does not fail. The method for finding E-unifiers is then is the following.

Definition 4.13. *(Method)* Let $E_{u,v} = E \cup \{eq(u,v) \doteq F,\ eq(z,z) \doteq T\}$, $\mathcal{O}_0$ an order assignment on $E_{u,v}$, $\mathcal{S}_0 = \emptyset$, $\mathcal{E}_0 = R(E_{u,v}, \mathcal{O}_0)$, m the total number of variables in $\mathcal{E}_0$, and $V = Var(E) \cup Var(u,v)$. For any sequence

$$\langle \mathcal{S}_0, \mathcal{E}_0, \mathcal{O}_0\rangle \Rightarrow^+ \langle \mathcal{S}_k, \mathcal{E}_k, \mathcal{O}_k\rangle$$

consisting of at most m transformation steps, if $\mathcal{S}_k$ is unifiable and $k \leq m$ is the first integer in the sequence such that $F \doteq T \in \mathcal{E}_k$, return the substitution $\theta_{\mathcal{S}_k}|_V$, where $\theta_{\mathcal{S}_k}$ is the mgu of $\mathcal{S}_k$ (over $T_\Sigma(\mathcal{X})$).

Example 4.14. Let E be the set of equations $E = \{fa \doteq a,\ ggx \doteq fa\}$, and $\langle u, v\rangle = \langle gggx, x\rangle$. We have

$$E_{u,v} = \{fa \doteq a,\ ggx \doteq fa,\ eq(gggx, x) \doteq F,\ eq(z,z) \doteq T\}.$$

---

[13]An equation $x \doteq v$ is degenerate if x is a variable and $x \notin Var(v)$.

[14]Note that we are requiring that l_1/β and l_2 have a *nontrivial* unifier. The triangular form of mgus is important for the NP-completeness of this method.

[15]Critical pairs are defined in definition 3.21.

The congruence closure Π of $E_{u,v}$ has three nontrivial classes $\{a, fa, ggx\}$, $\{eq(gggx, x), F\}$, and $\{eq(z, z), T\}$. Let $\mathcal{O}_0$ be the order assignment on $E_{u,v}$ such that

$$\begin{aligned} &T \prec_{\mathcal{O}_0} eq(gggx, x),\\ &F \prec_{\mathcal{O}_0} eq(z, z),\\ &a \prec_{\mathcal{O}_0} fa \prec_{\mathcal{O}_0} ggx, \end{aligned}$$

the least elements of classes being ordered in the order of listing of the classes. We have $\mathcal{S}_0 = \emptyset$, and the reduced system $\mathcal{E}_0 = R(E_{u,v}, \mathcal{O}_0)$ is

$$\mathcal{E}_0 = \{fa \doteq a,\ ggx \doteq a,\ eq(ga, x) \doteq F,\ eq(z, z) \doteq T\}.$$

Note that there is an overlap between $eq(ga, x) \doteq F$ and $eq(z, z) \doteq T$ at address ϵ in $eq(ga, x)$, and we obtain the triangular system $\{\langle x, ga\rangle, \langle z, ga\rangle\}$ and the new equation $F \doteq T$. Thus, we have

$$\langle \mathcal{S}_0, \mathcal{E}_0, \mathcal{O}_0\rangle \Rightarrow \langle \mathcal{S}_1, \mathcal{E}_1, \mathcal{O}_1\rangle,$$

where $\mathcal{S}_1 = \{\langle x, ga\rangle, \langle z, ga\rangle\}$

$$\mathcal{E}_1' = \{fa \doteq a,\ ggga \doteq a,\ eq(ga, ga) \doteq F,\ F \doteq T\},$$

and $\mathcal{O}_1$ is the restriction of $\mathcal{O}_0$ to the subterms in $\mathcal{E}_1'$. After reducing $\mathcal{E}_1'$, we have

$$\mathcal{E}_1 = \{fa \doteq a,\ ggga \doteq a,\ eq(ga, ga) \doteq T,\ F \doteq T\}.$$

Since $F \doteq T \in \mathcal{E}_1$ and $\mathcal{S}_1$ is unifiable, the restriction $[ga/x]$ of the mgu $[ga/x, ga/z]$ of $\mathcal{S}_1$ to $Var(E) \cup Var(u, v) = \{x\}$ is a rigid E-unifier of $gggx$ and x.

The following major results are proved in Gallier, Narendran, Plaisted, and Snyder [37].

Theorem 4.15. *The procedure given by definition 4.13 is a decision procedure for rigid E-unification. Furthermore, it belongs to NP.*

The soundness and completeness of the method are subsumed by the following result.

Theorem 4.16. *Let E be a set of equations over $T_\Sigma(\mathcal{X})$, u, v two terms in $T_\Sigma(\mathcal{X})$, m the number of variables in $E \cup \{u, v\}$, and $V = Var(E) \cup Var(u, v)$. There is a finite complete set of rigid E-unifiers for u and v given by the set*

$$\{\theta_{\mathcal{S}_k}|_V \mid \langle \mathcal{S}_0, \mathcal{E}_0, \mathcal{O}_0\rangle \Rightarrow^+ \langle \mathcal{S}_k, \mathcal{E}_k, \mathcal{O}_k\rangle,\ k \le m\},$$

for any order assignment $\mathcal{O}_0$ *on* $E_{u,v}$, *with* $\mathcal{S}_0 = \emptyset$, $\mathcal{E}_0 = R(E_{u,v}, \mathcal{O}_0)$, *and where* $\mathcal{S}_k$ *is unifiable,* $F \doteq T \in \mathcal{E}_k$, $F \doteq T \notin \mathcal{E}_i$ *for all* i, $0 \leq i < k$, *and* $\theta_{\mathcal{S}_k}$ *is the mgu of* $\mathcal{S}_k$ *over* $T_\Sigma(\mathcal{X})$.

Thus, we note another major difference between general E-unification and rigid E-unification. In rigid E-unification, there is always a finite complete set of (rigid) E-unifiers.

The above results have been improved by Isakowitz [60] and by Choi and Gallier [18]. Isakowitz has shown that order assignments can be dispensed with if a different reduction procedure is used. Isakowitz also studied the extension of rigid E-unification to order-sorted logic, and proved results analogous to those presented here for some subclasses of equations. Choi and Gallier have obtained a more direct proof of the NP-completeness of rigid E-unification that also avoids order assignments. This new proof is more algebraic and uses some key ideas from Kozen [72].

5. Higher-Order Unification

Higher-order unification is a method for unifying terms in the Simple Theory of Types [19], that is, given two typed lambda-terms e_1 and e_2, finding a substitution σ for the free variables of the two terms such that $\sigma(e_1)$ and $\sigma(e_2)$ are equivalent under the conversion rules of the calculus. This problem is fundamental to automating higher-order reasoning, as convincingly shown for example in the automated proof of Cantor's Theorem (that there is no surjection from a set to its powerset) found by the TPS system [6], where the higher-order unification procedure finds a term which corresponds to the diagonal set $\{a \in A \,|\, a \notin f(a)\}$ used in the standard proof (for details, see [6]). Higher-order unification has formed the basis for generalizations of the resolution principle to second-order logic [21], [97] and general ω-order logic [51], [93], [98] (but see also [3]), the generalization of the method of matings [4] to higher-order [5], [6], [82], [94], higher-order logic programming in the language λProlog [85], [89], a means for providing flexible implementations of logical inference rules in theorem provers [29],[93], program synthesis, transformation, and development [55], [46], [47], [86], [96], and also has applications to type inferencing in polymorphic languages [95], computational linguistics [87], and certain problems in proof theory concerning the lengths of proofs [26]. Higher-order unification was studied by a number of researchers [21], [43], [44], [45], [97], [98] before Huet [53], [54] made a major contribution in showing that a restricted form of unification, called preunification, is sufficient for most refutation methods and in defining a method for solving this restricted problem which is used by most current higher-order systems.

General higher-order unification was reexamined in Snyder and Gallier [113] in terms of the method of transformations on systems. As we have stressed in previous sections, this method provides an abstract and mathematically elegant means of analyzing the invariant properties of unification in various settings by providing a clean separation of the logical issues from the specification of procedural information. The set of transformations for higher-order unification is developed from an analysis of the manner in which substitution and β-reduction make two terms identical, and shows clearly the relationship between first-order unification, higher-order preunification, and general higher-order unification. The use of this formalism yields a more direct proof of the completeness of a method for higher-order unification than has previously been available. In addition, this analysis provides another justification of the design of Huet's procedure, and shows how its basic principles work in a more general setting.

Anyone who has worked with the lambda calculus knows that there are many technical pitfalls, and that quite often, "obvious" properties true of first-order terms turn out to be false for lambda-terms. Thus, we feel that it is necessary, especially for the benefit of readers unfamiliar with the field, to provide a "crash course" in the simply-typed lambda calulus. This way, we will at least be able to state correct definitions and theorems! For a comprehensive treatment of the lambda calculus, the reader is referred to Barendregt [11], Hindley and Seldin [49], and Huet[54]. The main feature of the simply-typed lambda calculus is that terms are typed in such a way that application is restricted. In particular, self application is impossible. This is very important, because it allows us to work with normal forms of lambda terms.

Definition 5.1. Given a set $\mathcal{T}_0$ of *base types* (e.g., such as *int*, *bool*, etc.) we define the set of types $\mathcal{T}$ inductively as the smallest set containing $\mathcal{T}_0$ and such that if $\alpha, \beta \in \mathcal{T}$, then $(\alpha \to \beta) \in \mathcal{T}$.

Intuitively, the type $(\alpha \to \beta)$ is that of a function from objects of type α to objects of type β. We assume that the type constructor $\to$ associates to the right, and we shall often write type expressions such as $(\alpha_1 \to (\alpha_2 \to \ldots (\alpha_n \to \beta) \ldots))$ in the form $\alpha_1, \ldots, \alpha_n \to \beta$, with β an arbitrary type.

Definition 5.2. Let us assume given a set Σ of symbols, which we call *function constants*, each symbol f having a unique type $\tau(f)$ from $\mathcal{T}$. For each type $\alpha \in \mathcal{T}$, we assume given a countably infinite set of variables of that type, denoted V_α, and let $V = \bigcup_{\tau \in \mathcal{T}} V_\tau$. Furthermore, let the set of *atoms* $\mathcal{A}$ be defined as $V \cup \Sigma$. The set $\mathcal{L}$ of lambda-terms is inductively defined as the smallest set containing $\mathcal{A}$ and closed under the rules of function application and lambda-abstraction, namely,

(i) If $e_1 \in \mathcal{L}$ has type $\alpha \to \beta$, and $e_2 \in \mathcal{L}$ has type α, then $(e_1 e_2)$ is a member of $\mathcal{L}$ of type β.
(ii) If $e \in \mathcal{L}$ has type β and $x \in V_\alpha$ then $(\lambda x.\, e)$ is a member of $\mathcal{L}$ of type $\alpha \to \beta$.

We shall denote the type of a term e by $\tau(e)$.

By convention, application associates to the left, so that a term $(\ldots((e_1 e_2) e_3) \ldots e_n)$ may be represented as $(e_1 e_2 \ldots e_n)$. In general we represent a sequence of lambda abstractions $\lambda x_1.\, (\lambda x_2.\, (\ldots (\lambda x_n.\, e) \ldots))$ in the form $\lambda x_1 \ldots x_n.\, e$, where e is either an application or an atom. We shall often drop superfluous parentheses when there is no loss of clarity.

Definition 5.3. In a term $\lambda x_1 \ldots x_n.\, e$ where e is either an application or an atom, we call e the *matrix* of the term, the object $\lambda x_1 \ldots x_n$ is the *binder* of the term, and the occurrences of the variables are called *binding occurrences* of these variables. We define the *size* of a term u, denoted $|u|$, as the number of atomic subterms of u. A variable x occurs *bound* in a term e if e contains some subterm of the form $\lambda x.\, e'$, in which case the term e' is called the *scope* of this binding occurrence of x. A variable x occurs *free* in e if it is a subterm of e but does not occur in the scope of a binding occurrence of x. The set of free variables of a term e is denoted by $FV(e)$.

The "reduction rules" of the lambda calculus are as follows.

Definition 5.4. Let $u[t/x]$ denote the result of replacing each free occurrence of x in u by t, and $BV(t)$ be the set of bound variables in t. We have three rules of *lambda conversion.*

(i) (α-conversion) If $y \notin FV(t) \cup BV(t)$, then

$$(\lambda x.\, t) \succ_\alpha (\lambda y.\, (t[y/x])).$$

(ii) (β-conversion)

$$((\lambda x.\, s)\, t) \succ_\beta s[t/x].$$

(iii) (η-conversion)[16] If $x \notin FV(t)$, then

$$(\lambda x.\,(t\,x)) \succ_\eta t.$$

The term on the left side of each of these rules is called a *redex*. A term t which contains no β-redices is called a *β-normal form*, and η-normal forms and $\beta\eta$-normal forms are defined similarly. If we denote by $e[s]$ a lambda term with some distinguished occurrence of a subterm s, then let $e[t]$ denote the result of replacing this single subterm by the term t, where $\tau(s) = \tau(t)$. We define the relation $\longrightarrow_\alpha$ as

$$e[s] \longrightarrow_\alpha e[t] \quad \text{iff} \quad s \succ_\alpha t,$$

and similarly for $\longrightarrow_\beta$ and $\longrightarrow_\eta$. We define $\longrightarrow_{\beta\eta}$ as $\longrightarrow_\beta \cup \longrightarrow_\eta$. We also define the symmetric closure $\longleftrightarrow$, the transitive closure $\stackrel{+}{\longrightarrow}$, and the symmetric, reflexive, and transitive closure $\stackrel{*}{\longleftrightarrow}$ of each of these relations in the obvious fashion. The relations $\stackrel{*}{\longleftrightarrow}_\beta$, $\stackrel{*}{\longleftrightarrow}_\eta$, and $\stackrel{*}{\longleftrightarrow}_{\beta\eta}$ are called β-, η-, and *$\beta\eta$-equivalence* respectively.

It is easy to show that the type of a lambda term is preserved under these rules of lambda conversion. From now on, we will actually ignore α-conversion. Technically, this means that we will be dealing with α-equivalence classes of terms. In full rigor, it should be verified that the concepts we introduce are invariant under α-equivalence, but we will not bother to do so in this brief presentation.

In this section, we wish to present the fundamental logical issues as clearly as possible, and for this purpose, we feel it is sufficient to develop the notion of unification of terms in the typed $\beta\eta$-calculus. This is a natural assumption in practice, and all higher-order theorem proving systems known to the authors use this weak form of extensionality. The reader interested in the details of the non-extensional case may consult [54]. As we mentioned earlier, the fact that the terms under consideration are typed has some important consequences. Two of the major results concerning this calculus are the following. Proofs can be found in [11], Hindley and Seldin [49], and Huet[54].

Theorem 5.5. *(Strong Normalization) Every sequence of $\beta\eta$-reductions is finite.*

---

[16]This rule is a special case of the the axiom of extensionality, viz., $\forall f, g(\forall x(f(x) = g(x)) \Longrightarrow f = g)$, which asserts that two functions are equal if they behave the same on all arguments, regardless of their syntactic representation.

Theorem 5.6. *(Church-Rosser Theorem) If $s \overset{*}{\longleftrightarrow}_{\beta\eta} t$ for two lambda terms s and t, then there must exist some term u such that $s \overset{*}{\longrightarrow}_{\beta\eta} u \overset{*}{\longleftarrow}_{\beta\eta} t$.*

Each of these theorems remains true when restricted to just η-conversion or just β-conversion. One of the important consequences of these two results is that for each term t there exists a unique (up to α-conversion) term t' such that $t \overset{*}{\longrightarrow}_{\beta\eta} t'$ with t' in $\beta\eta$-normal form, and similarly for the restriction to just β- or just η-reduction. Another consequence is that the β-, η-, or $\beta\eta$-equivalence of two arbitrary terms may be decided by checking if the corresponding normal forms of the two terms are equal. For example, if we denote the unique β-normal form of a term t by $t{\downarrow}$, then $s \overset{*}{\longleftrightarrow}_{\beta} t$ iff $s{\downarrow} = t{\downarrow}$.

Terms in β-normal form have a pleasant structure. Indeed, a term in β-normal is of the form

$$\lambda x_1 \dots x_n (a\, e_1 \dots e_m),$$

where the *head* a is an atom, i.e., a is either a function constant, bound variable, or some variable free in this term, and the terms $e_1, \dots, e_m$ are in the same form. By analogy with first-order notation, such a term will be denoted $\lambda x_1 \dots x_n.\, a(e_1, \dots, e_m)$. As an abbreviation, we represent lambda terms using something like a "vector" notation for lists, so that $\lambda x_1 \dots x_n.\, e$ will be represented by $\lambda \overline{x_n}.\, e$. Furthermore, this principle will be extended to lists of terms, so that $\lambda \overline{x_n}.\, f(e_1, \dots, e_m)$ will be represented as $\lambda \overline{x_n}.\, f(\overline{e_m})$, and we shall even sometimes represent a term such as

$$\lambda \overline{x_k}.\, a(y_1(\overline{x_k}), \dots, y_n(\overline{x_k}))$$

in the form $\lambda \overline{x_k}.\, a(\overline{y_n(\overline{x_k})})$.

Definition 5.7. A term whose head is a function constant or a bound variable is called a *rigid* term; if the head is a free variable it will be called a *flexible* term. (For example, the term $\lambda x.\, F(\lambda y.\, y(x, a), c)$ is flexible, but both of its immediate subterms are rigid.)

One of the nice properties of the simply-typed lambda calculus under $\beta\eta$-conversion is that by modifying slightly the concept of normal form, it is possible to ignore the role of η-reduction, and yet exploit fully the type information. The formal justification for this is given by the following result known as postponement of η-reduction.

Lemma 5.8. *For any two terms s and t, we have $s \overset{*}{\longrightarrow}_{\beta\eta} t$ iff there exists a term u such that $s \overset{*}{\longrightarrow}_{\beta} u \overset{*}{\longrightarrow}_{\eta} t$.*

(For a proof see [11].) As a consequence, we can decide $\beta\eta$-equivalence by reducing terms to their β-normal forms, and then testing for η-equivalence, that is, $s \xleftrightarrow{*}_{\beta\eta} t$ iff $s\!\downarrow \xleftrightarrow{*}_{\eta} t\!\downarrow$. This allows us to 'factor out' η-conversion, by considering only η-equivalence classes of terms. We shall use the following means of representing such classes by canonical representatives (from [54]).

Definition 5.9. Let $e = \lambda x_1 \ldots x_n.\, a(e_1, \ldots, e_m)$ be a term in β-normal form of type $\alpha_1, \ldots, \alpha_n, \alpha_{n+1}, \ldots, \alpha_{n+k} \to \beta$, with $\beta \in \mathcal{T}_0$. The *η-expanded form* of e, denoted by $\eta[e]$, is produced by adding k new variables of the appropriate types to the binder and the matrix of the term, and (recursively) applying the same expansion to the subterms, to obtain

$$\lambda x_1 \ldots x_n x_{n+1} \ldots x_{n+k}.\, a(\eta[e_1], \ldots, \eta[e_m], \eta[x_{n+1}], \ldots, \eta[x_{n+k}]),$$

where $\tau(x_{n+i}) = \alpha_{n+i}$ for $1 \leq i \leq k$.

This normal form is more useful than the η-normal form because it makes the type of the term and all its subterms more explicit, and is therefore a convenient syntactic convention for representing the congruence class of all terms equal modulo the η−rule. It can be shown that we have a Church-Rosser theorem in the following form.

Theorem 5.10. *For every two terms s and t, we have $s \xleftrightarrow{*}_{\beta\eta} t$ iff $\eta[s\!\downarrow] = \eta[t\!\downarrow]$.*

From a technical point of view, in discussing higher-order unification modulo $\beta\eta$-conversion, it is convenient to work with the following languages $\mathcal{L}_{exp}$ and $\mathcal{L}_\eta$, which are subsets of $\mathcal{L}$, and deal primarily with terms in η-expanded form.

Definition 5.11. Let $\mathcal{L}_{exp}$ be defined as the set of all η-expanded forms, i.e., $\mathcal{L}_{exp} = \{\eta[e\!\downarrow] \mid e \in \mathcal{L}\}$. Define the set $\mathcal{L}_\eta$ as the smallest subset of $\mathcal{L}$ containing $\mathcal{L}_{exp}$ and closed under application and lambda abstraction, i.e., $(e_1 e_2)$ and $\lambda x.\, e_1$ are in $\mathcal{L}_\eta$ whenever $e_1 \in \mathcal{L}_\eta$ and $e_2 \in \mathcal{L}_\eta$.

The essential features of $\mathcal{L}_{exp}$ and $\mathcal{L}_\eta$ which will allow us to restrict our attention to η-expanded forms are proved in the next lemma, which is from [54].

Lemma 5.12. *For every variable x and every pair of terms e and e' of the appropriate types:*

(1) *$e, e' \in \mathcal{L}_{exp}$ implies that $(\lambda x.\, e) \in \mathcal{L}_{exp}$ and $(ee')\!\downarrow \in \mathcal{L}_{exp}$;*
(2) *$e \in \mathcal{L}_\eta$ implies that $e\!\downarrow \in \mathcal{L}_{exp}$;*
(3) *$e, e' \in \mathcal{L}_\eta$ implies that $(\lambda x.\, e) \in \mathcal{L}_\eta$ and $(ee') \in \mathcal{L}_\eta$;*
(4) *$e \in \mathcal{L}_\eta$ and $e \xrightarrow{*}_\beta e'$ implies that $e' \in \mathcal{L}_\eta$;*
(5) *$e, e' \in \mathcal{L}_\eta$ implies that $e'[e/x] \in \mathcal{L}_\eta$.*

These closure conditions for $\mathcal{L}_\eta$ (not all of which are satisfied by the set of η-normal forms) formally justify our leaving the η-rule implicit in the following sections by developing our method for higher-order unification in the language $\mathcal{L}_\eta$ and considering explicitly only β-conversion as a computation rule.[17] The reader interested in a more detailed treatment of these matters, including proofs of the previous results, is referred to [54] for details.

It remains to formalize the general notion of substitution of lambda terms for free variables in the $\beta\eta$-calculus, after which we show how this may be specialized to substitutions over $\mathcal{L}_{exp}$.

Definition 5.13. A *substitution* is any (total) function $\sigma : V \to \mathcal{L}$ such that $\sigma(x) \neq x$ for only finitely many $x \in V$ and for every $x \in V$ we have $\tau(\sigma(x)) = \tau(x)$. Given a substitution σ, the *support* (or *domain*) of σ is the set of variables $D(\sigma) = \{x \mid \sigma(x) \neq x\}$. A substitution whose support is empty is termed the *identity substitution*, and is denoted by Id. The set of variables *introduced by* σ is $I(\sigma) = \bigcup_{x \in D(\sigma)} FV(\sigma(x))$.

The application of a substitution to a term is defined recursively as follows.

Definition 5.14. Let σ_{-x} denote the substitution $\sigma|_{D(\sigma)-\{x\}}$. For any substitution σ,

$$\begin{aligned}
\widehat{\sigma}(x) &= \sigma(x) \text{ for } x \in V;\\
\widehat{\sigma}(a) &= a \text{ for } a \in \Sigma;\\
\widehat{\sigma}(\lambda x.\, e) &= \lambda x.\, \widehat{\sigma_{-x}}(e);\\
\widehat{\sigma}(\,(e_1\, e_2)\,) &= (\,\widehat{\sigma}(e_1)\, \widehat{\sigma}(e_2)\,)
\end{aligned}$$

[17] In fact, we shall depart from our convention in the interests of simplicity only when representing terms which are (up to η-conversion) variables, e.g., $\lambda xy.\, F(x, y)$. In some contexts, such as solved form systems, we wish to emphasize their character as variables, and will represent them as such, e.g., just F. In these cases, we shall be careful to say that 'F is (up to η-conversion) a variable,' etc.

Thus a substitution has an effect only on the *free* variables of a term.[18] In the sequel, we shall identify σ and its extension $\widehat{\sigma}$. It is easy to show that the type of a term is unchanged by application of an arbitrary substitution.

Remark: It is important to note that by $\sigma(e)$ we denote the result of applying the substitution σ to e *without* β-reducing the result; we shall denote by $\sigma(e)\downarrow$ the result of applying the substitution and then reducing the result to β-normal form. This rather non-standard separation we impose between substitution and the subsequent β-reduction is useful because we wish to examine closely the exact effect of substitution and β-reduction on lambda terms.

Definition 5.15. The *composition* of σ and θ is the substitution denoted by $\sigma\,;\theta$ such that for every variable x we have $\sigma\,;\theta(x) = \widehat{\theta}(\sigma(x))$.

As in the case of E-unification, we need to compare substitutions.

Definition 5.16. Given a set W of variables, we say that two substitutions σ and θ are *equal over* W, denoted $\sigma = \theta[W]$, iff $\forall x \in W$, $\sigma(x) = \theta(x)$. Two substitutions σ and θ are *β-equal over* W, denoted $\sigma =_\beta \theta[W]$ iff $\forall x \in W$, $\sigma(x) \overset{*}{\longleftrightarrow}_\beta \theta(x)$, or, equivalently, $\sigma(x)\downarrow = \theta(x)\downarrow$. The relations $=_\eta$ and $=_{\beta\eta}$ are defined in the same way but using $\overset{*}{\longleftrightarrow}_\eta$ and $\overset{*}{\longleftrightarrow}_{\beta\eta}$. We say that σ is *more general than* θ *over* W, denoted by $\sigma \leq \theta[W]$, iff there exists a substitution η such that $\theta = \sigma\,;\eta[W]$, and we have $\sigma \leq_\beta \theta[W]$ iff there exists some η' such that $\theta =_\beta \sigma\,;\eta'[W]$, and $\leq_\eta$ and $\leq_{\beta\eta}$ are defined analogously. When W is the set of all variables, we drop the notation $[W]$. If neither $\sigma \leq_{\beta\eta} \theta$ nor $\theta \leq_{\beta\eta} \sigma$ then σ and θ are said to be *independent*.

The comparison of substitutions modulo β-, η-, and $\beta\eta$-conversion is formally justified by the following lemma, which is easily proved by structural induction on terms:

Lemma 5.17. *If σ and θ are arbitrary substitutions such that either $\sigma =_\beta \theta$, $\sigma =_\eta \theta$, or $\sigma =_{\beta\eta} \theta$, then for any term u we have either $\sigma(u) \overset{*}{\longleftrightarrow}_\beta \theta(u)$, $\sigma(u) \overset{*}{\longleftrightarrow}_\eta \theta(u)$, or $\sigma(u) \overset{*}{\longleftrightarrow}_{\beta\eta} \theta(u)$, respectively.*

We now show that we can develop the notion of substitution wholly within the context of the language $\mathcal{L}_\eta$ developed above without loss of generality.

[18]Here, we are using fact that we are dealing with α-equivalence classes. We always α-rename bound variables to avoid capture.

Definition 5.18. A substitution θ is said to be *normalized* if $\theta(x) \in \mathcal{L}_{exp}$ for every variable $x \in D(\theta)$.

We can assume without loss of generality that no normalized substitution has a binding of the form $\eta[x]/x$ for some variable x. A normalized renaming substitution has the form $[\eta[y_1]/x_1, \ldots, \eta[y_n]/x_n]$; the effect of applying such a substitution and then β-reducing is to rename the variables $x_1, \ldots, x_n$ to $y_1, \ldots, y_n$. The justification for using normalized substitutions is given by the following corollary of Lemma 5.12.

Corollary If θ is a normalized substitution and $e \in \mathcal{L}_{exp}$, then $\theta(e) \in \mathcal{L}_\eta$ and $\theta(e)\!\downarrow\in \mathcal{L}_{exp}$.

It is easy to show that if σ and θ are normalized, then $\sigma =_{\beta\eta} \theta$ iff $\sigma = \theta$ and if θ' is the result of normalizing θ, then $\theta' =_{\beta\eta} \theta$.

In general, substitutions are assumed to be normalized in the rest of this section, allowing us to factor out η-equivalence in comparing substitutions, so that we may, e.g., use $\leq_\beta$ instead of $\leq_{\beta\eta}$. In fact, the composition of two normalized substitutions could be considered to be a normalized substitution as well, so that $\sigma \leq_\beta \theta$ iff $\sigma \leq \theta$, but this need *not* be assumed in what follows. For example, the composition $[\lambda x.\, G(a)/F]\,;[\lambda y.\, y/G]$ is defined as $[\lambda x.\, ((\lambda y.\, y)a)/F, \lambda y.\, y/G]$, *not* as $[\lambda x.\, a/F, \lambda y.\, y/G]$. We shall continue to use $=_\beta$ and $\leq_\beta$ to compare normalized substitutions, although strictly speaking the subscript could be omitted if no composition is involved.

Definition 5.19. A substitution σ is *idempotent* if $\sigma\,;\sigma =_{\beta\eta} \sigma$.

A sufficient condition for idempotency is given by[19]

Lemma 5.20. *A substitution σ is idempotent whenever $I(\sigma) \cap D(\sigma) = \emptyset$.*

It is shown in Snyder [112] that in most contexts, we may restrict our attention to idempotent substitutions without loss of generality. The net effect of these definitions, conventions, and results is that we can develop our method for unification of terms in the $\beta\eta$-calculus wholly within $\mathcal{L}_\eta$, leaving η-equivalence implicit in the form of the terms under consideration. We are now in a position where we can define higher-order unification.

[19]In the first-order case, this condition is necessary as well, but in our more general situation we have counter-examples such as $\sigma = [\lambda x.\, F(a)/F]$.

Definition 5.21. The notion of pairs and systems of terms carries over from the first-order case.[20] A substitution θ is a unifier of two lambda terms e_1 and e_2 iff $\theta(e_1)\overset{*}{\longleftrightarrow}_{\beta\eta}\theta(e_2)$.[21] A substitution is a unifier of a system S if it unifies each pair in S. The set of all unifiers of S is denoted $U(S)$ and if S consists of a single pair $\langle s,t\rangle$ then it is denoted $U(s,t)$.

This definition is more general than we shall need, in fact, since we shall develop our approach in $\mathcal{L}_\eta$ in order to factor out η-conversion, as explained earlier. Thus for two terms $s,t\in\mathcal{L}_\eta$, we say that a normalized substitution θ is in $U(s,t)$ iff $\theta(s)\overset{*}{\longleftrightarrow}_{\beta}\theta(t)$, or, alternately, if $\theta(s)\!\downarrow=\theta(t)\!\downarrow$.

A pair of terms is solved in a system S if it is in the form $\langle\eta[x],t\rangle$, for some variable x which occurs only once in S; a system is solved if each of its pairs is solved. Our only departure from the use of η-expanded form is that we shall represent pairs of the form $\langle\eta[x],t\rangle$ as $\langle x,t\rangle$ in order to emphasize their correspondence to bindings t/x in substitutions, as in the first-order case of the previous section.

Example 5.22. Let $S=\{\langle F(f(a)),f(F(a))\rangle\}$, where $F\colon int\to int$ (a variable). $\theta=[\lambda x.f(x)/F]$ is a unifier for S, since

$$\theta(F(f(a)))=(\lambda x.f(x))\,f(a)\longrightarrow_\beta f(f(a))$$
$$f(f(a))\longleftarrow_\beta f((\lambda x.f(x))a)=\theta(f(F(a))).$$

For all $n\geq 0$, $[\lambda x.f^n(x)/F]$ is also a unifier for S.

Example 5.23. If $u=f(a,g(\lambda x.\,G(\lambda y.\,x(b))))$ and $v=F(\lambda x.\,x(z))$, then $\theta=[\lambda x_2.\,f(a,g(x_2))/F,\lambda x_3.\,x_3(z_2)/G,b/z]$ is in $U(u,v)$, since $\theta(u)\!\downarrow=\theta(v)\!\downarrow$:

$$\begin{aligned}\theta(u)&=f(a,g(\lambda x.\,[(\lambda x_3.\,x_3(z_2))(\lambda y.\,x(b))]))\\&\longrightarrow_\beta f(a,g(\lambda x.\,[(\lambda y.\,x(b))z_2]))\\&\longrightarrow_\beta f(a,g(\lambda x.\,x(b)))\\&\longleftarrow_\beta (\lambda x_2.\,f(a,g(x_2)))(\lambda x.\,x(b))\;=\theta(v).\end{aligned}$$

As we know from section 3, the standard first-order unification problem is decidable. Unfortunately, this does not hold for higher-order unification. Indeed, it has been shown that the second-order unification problem is undecidable. This result was shown by Goldfarb [42] using a reduction from

[20] See definition 3.1.

[21] This is in the context of the $\beta\eta$-calculus; in the β-calculus the condition would be $\theta(e_1)\overset{*}{\longleftrightarrow}_{\beta}\theta(e_2)$.

Hilbert's Tenth Problem; previously, Huet [52] showed the undecidability of the third-order unification problem, using a reduction from the Post Correspondence Problem. These results show that there are second-order (and therefore arbitrarily higher-order) languages where unification is undecidable; but in fact there exist *particular* languages of arbitrarily high-order which have a decidable unification problem. Interestingly, Goldfarb's proof requires that the language to which the reduction is made contain at least one 2-place function constant. It has been shown in [27] that the unification problem for second-order monadic languages (i.e., no function constant has more than one argument place) is decidable, which has applications in certain decision problems concerning the lengths of proofs. A different approach to decidability is taken in [123], where decidable cases of the unification problem are found by showing that the search tree for some problems, although infinite, is regular, and that the set of unifiers can be represented by a regular expression. More generally, it has been shown by Statman [118] that the set of all decidable unification problems is polynomial-time decidable.

Besides the undecidability of higher-order unification, another problem is that *mgu*'s may no longer exist, a result first shown by [43]. For example, the two terms $F(a)$ and a have the unifiers $[\lambda x.\, a/F]$ and $[\lambda x.\, x/F]$, but there is no unifier more general than both of these. Thus, the situation is similar to that encountered in section 3 with E-unification. This leads us to extend the notion of *complete set* of unifiers to the higher-order case.

Definition 5.24. Given a system S and a finite set W of 'protected' variables, a set U of normalized substitutions is a *complete set of unifiers for S away from W* iff

(i) For all $\sigma \in U$, $D(\sigma) \subseteq FV(S)$ and $I(\sigma) \cap (W \cup D(\sigma)) = \emptyset$;

(ii) $U \subseteq U(S)$;

(iii) For every normalized $\theta \in U(S)$, there exists some $\sigma \in U$ such that $\sigma \leq_\beta \theta[FV(S)]$.

It can be shown that there is no loss of generality in considering only normalized substitutions. We now give a set of transformations first presented in Snyder and Gallier [113] and which turns out to be a complete set of transformations for higher-order unification.

Definition 5.25. (The set of transformations $\mathcal{HT}$)

$$\{\langle u, u\rangle\} \cup S \Longrightarrow S \tag{1}$$

$$\left\{\langle \lambda\overline{x_k}.\, a(\overline{u_n}), \lambda\overline{x_k}.\, a(\overline{v_n})\rangle\right\} \cup S \Longrightarrow \bigcup_{1\le i\le n} \left\{\langle \lambda\overline{x_k}.\, u_i,\ \lambda\overline{x_k}.\, v_i\rangle\right\} \cup S, \tag{2}$$

where a is an arbitrary atom.

$$\{\langle \lambda\overline{x_k}.\, F(\overline{x_k}), \lambda\overline{x_k}.\, v'\rangle\} \cup S \Longrightarrow \{\langle F, \lambda\overline{x_k}.\, v'\rangle\} \cup \sigma(S)\downarrow, \tag{3}$$

where $F \notin FV(\lambda\overline{x_k}.\, v')$, and $\sigma = [\lambda\overline{x_k}.\, v'/F]$.

$$\{\langle \lambda\overline{x_k}.\, F(\overline{u_n}),\ \lambda\overline{x_k}.\, a(\overline{v_m})\rangle\} \cup S \Longrightarrow \{\langle F,\ t\rangle,\ \langle \lambda\overline{x_k}.\, F(\overline{u_n}), \lambda\overline{x_k}.\, a(\overline{v_m})\rangle\} \cup S, \tag{4}$$

where, either

(a) $t = \lambda\overline{y_n}.\, a(\overline{\lambda\overline{z_{p_m}}.\, H_m(\overline{y_n}, \overline{z_{p_m}})})$, if either a is a constant, or a is a free variable $\neq F$ (this is called an imitation binding);

(b) $t = \lambda\overline{y_n}.\, y_i(\overline{\lambda\overline{z_{p_q}}.\, H_q(\overline{y_n}, \overline{z_{p_q}})})$, where a is some arbitrary atom. If $head(u_i)$ is a constant, then $a = head(u_i)$ (this is called a projection binding);

(c) $t = \lambda\overline{y_n}.\, b(\overline{\lambda\overline{z_{p_m}}.\, H_m(\overline{y_n}, \overline{z_{p_m}})})$, if a is a free variable, $a \neq F$, and $a \neq b$.

After (4), transformation (3) is immediately applied to the new pair $\langle F, t\rangle$.

Example 5.26.

$$\langle F(f(a)), f(F(a))\rangle$$

$$\Longrightarrow_{4a} \langle F, \lambda x.\, f(Y(x))\rangle,\ \langle \frac{(\lambda x.\, f(Y(x)))f(a)}{f(Y(f(a)))}, f(\frac{(\lambda x.\, f(Y(x)))a}{f(Y(a))})\rangle$$

$$\Longrightarrow_{2} \langle F, \lambda x.\, f(Y(x))\rangle,\ \langle Y(f(a))), f(Y(a))\rangle$$

$$\Longrightarrow_{4b} \langle F, \lambda x.\, f(\frac{(\lambda x.\, x)x}{x})\rangle,\ \langle Y, \lambda x.\, x\rangle,\ \langle \frac{(\lambda x.\, x)f(a)}{f(a)}, f(\frac{(\lambda x.\, x)a}{a})\rangle$$

$$\Longrightarrow_{1} \langle F, \lambda x.\, f(x)\rangle,\ \langle Y, \lambda x.\, x\rangle$$

The following theorem is proved in Snyder and Gallier [113].

Theorem 5.27. *The set $\mathcal{HT}$ is a complete set of transformations for higher-order unification.*

The set of transformations $\mathcal{HT}$ is complete for the problem of general higher-order unification. Unfortunately, the "don't know" non-determinism of this set causes severe implementation problems in the case of two flexible terms (case (4c)). This "guessing" of partial bindings cannot be avoided without sacrificing completeness, and so the search tree of all transformation sequences may be infinitely branching at certain nodes, causing a disastrous explosion in the size of the search space.

Huet's solution to this problem [53], [54] (a major breakthrough in higher-order unification theory) was to redefine the problem in such a way that such flexible-flexible pairs are considered to be already solved; this partial solution of the general higher-order unification problem turns out to be sufficient for refutation methods (see [51]), and this is the method used in most current systems. We show here how to explain this approach in terms of transformations on systems. The only changes have to do with redefining the notion of a solved system and restricting the set of transformations.

Definition 5.28. A pair of terms $\langle x, e\rangle$ is in *presolved form* in a system S if it is in solved form in S (as above) *or* if it is a pair consisting of two flexible terms. A system is in presolved form if each member is in presolved form. For a set S in presolved form, define the associated substitution σ_S as the *mgu* $\sigma_{S'}$ of the set S' of solved pairs of S.

Definition 5.29. Let $\cong$ be the least congruence relation on $\mathcal{L}$ containing the set of pairs $\{(u,v)\,|\,u,v \text{ are both flexible terms }\}$. A substitution θ is a *preunifier* of u and v if $\theta(u){\downarrow}\cong\theta(v){\downarrow}$.

The importance of pre-unifiers is shown by our next definition and lemma.

Definition 5.30. For every $\phi = \alpha_1, \ldots, \alpha_n \to \beta \in \mathcal{T}$, with $n \geq 0$, define a term

$$\widehat{e}_\phi \;=\; \lambda x_1 \ldots x_n.\, v,$$

where $\tau(x_i) = \alpha_i$ for $1 \leq i \leq n$ and $v \in V_\beta$ is a new variable which will never be used in any other term. Let ζ be an (infinite) set of bindings

$$\zeta \;=\; \{\widehat{e}_{\tau(x)}/x \,|\, x \in V\}.$$

Finally, if S' is a pre-solved system containing a set S'' of flexible-flexible pairs, then define the substitution

$$\zeta_{S'} \;=\; \zeta|_{FV(S'')}.$$

As in [54], it is easy to show this next result.

Lemma 5.31. *If S is a system in pre-solved form then the substitution $\sigma_S \cup \zeta_S$ is a unifier of S.*

This lemma asserts that pre-unifiers may always be extended to true unifiers by finding trivial unifiers for the flexible-flexible terms in the pre-solved system.

The set of transformations for finding preunifiers is a slightly restricted version of the set of transformations $\mathcal{HT}$. It was first presented in Snyder and Gallier [113].

Definition 5.32. (The set of transformations $\mathcal{PT}$) Let S be a system, possibly empty. To the transformations (1) and (3) from $\mathcal{HT}$ we add three (restricted) transformations:

$$\left\{\langle\lambda\overline{x_k}.\, a(\overline{u_n}), \lambda\overline{x_k}.\, a(\overline{v_n})\rangle\right\} \cup S \Longrightarrow \bigcup_{1\leq i\leq n} \left\{\langle\lambda\overline{x_k}.\, u_i,\, \lambda\overline{x_k}.\, v_i\rangle\right\} \cup S, \qquad (2')$$

where $a \in \Sigma$ or $a = x_j$ for some j, $1 \leq j \leq k$.

$$\{\langle\lambda\overline{x_k}.\, F(\overline{u_n}),\, \lambda\overline{x_k}.\, a(\overline{v_m})\rangle\} \cup S \;\Longrightarrow\; \{\langle F,\, t\rangle,\, \langle\lambda\overline{x_k}.\, F(\overline{u_n}), \lambda\overline{x_k}.\, a(\overline{v_m})\rangle\} \cup S, \qquad (4'a)$$

where $a \in \Sigma$ and $t = \lambda\overline{y_n}.\, a(\overline{\lambda\overline{z_{p_m}}.\, H_m(\overline{y_n}, \overline{z_{p_m}})})$,

$$\{\langle\lambda\overline{x_k}.\, F(\overline{u_n}),\, \lambda\overline{x_k}.\, a(\overline{v_m})\rangle\} \cup S \;\Longrightarrow\; \{\langle F,\, t\rangle,\, \langle\lambda\overline{x_k}.\, F(\overline{u_n}), \lambda\overline{x_k}.\, a(\overline{v_m})\rangle\} \cup S, \qquad (4'b)$$

where either $a \in \Sigma$ or $a = x_j$ for some j, $1 \leq j \leq k$, and

$$t = \lambda\overline{y_n}.\, y_i(\overline{\lambda\overline{z_{p_q}}.\, H_q(\overline{y_n}, \overline{z_{p_q}})}).$$

After each of (4′a) and (4′b), we apply transformation (3) to the new pair introduced.

We say that $\theta \in PreUnify(S)$ iff there exists a series of transformations from $\mathcal{PT}$

$$S = S_0 \Longrightarrow S_1 \Longrightarrow \ldots \Longrightarrow S_n,$$

with S_n in pre-solved form, and $\theta = \sigma_{S_n}|_{FV(S)}$.

The major results concerning this formulation of higher-order unification are from Snyder and Gallier [113]. They can be viewed as a reformulation of Huet's results [54], [53] in the framework of transformations.

Theorem 5.33. (Soundness) *If $S \overset{*}{\Longrightarrow} S'$, with S' in presolved form, then the substitution $\sigma_{S'}|_{FV(S)}$ is a preunifier of S.*

Theorem 5.34. *(Completeness) If θ is some preunifier of the system S, then there exists a sequence of transformations $S \stackrel{*}{\Longrightarrow} S'$, with S' in presolved form, such that*

$$\sigma_{S'}|_{FV(S)} \leq_\beta \theta.$$

The search tree for this method consists of all the possible sequences of systems created by transforming the original two terms. Leaves consist of pre-solved systems or systems where no transformation can be applied. These correspond to the **S** and **F** nodes in Huet's algorithm; in fact, the search trees generated are essentially the same as the *matching trees* defined in [53], except that here an explicit representation of the matching substitutions found so far is carried along in the system. The set of pre-unifiers potentially found by our procedure is the set of pre-solved leaves in the search tree.

A detailed comparaison of the procedure corresponding to the set $\mathcal{PT}$ and Huet's unification procedure [53] as well as a pseudo-code version of Huet's procedure (for the typed $\beta\eta$-calculus) are given in Snyder and Gallier [113].

6. Conclusion and Directions for Further Research

The method of transformations has been applied systematically to design procedures for a number of unification problems. The method was applied to E-unification in section 3, to rigid E-unification in section 4, and to higher-order unification in section 5. We claim that these example are excellent illustrations of the flexibility and power of the method of transformations for designing unification procedures. In the first and the second case, the method led to new results, and in the third case, to more insight into the nature of higher-order unification. Other approaches for studying unification problems have been proposed. We simply mention a categorical approach due to Goguen [40], and a natural deduction approach due to Le Chenadec [76], the latter in the case of standard unification.

We conclude with a few suggestions for further research. The next step after obtaining a complete set of transformations for a unification problem is to implement procedures based on these transformations, and look for optimizations. Basically, one needs to understand which data structures should be used, and what kind of control should be imposed. As far as we know, very little work has been done in this area (although Gallier and Snyder wrote a prototype procedure in the case of E-unification, and obtained satisfactory results on most examples appearing in the literature). Much more work remains to be done, and in particular, an implementation of rigid E-unification is necessary.

We believe that it would be interesting to investigate rigid E-unification modulo Associativity and Commutativity, and more generally modulo certain specific theories. In a different direction, the problem of combined higher-order and E-unification is also an interesting one. Some preliminary work has been accomplished by Wayne Snyder [115]. What would also be interesting (but probably technically rather involved) is to study the combination of higher-order and order-sorted E-unification.

References

1. Aït-Kaci, H. A lattice theoretic approach to computation based on a calculus of partially ordered type structures, Ph.D. thesis. Department of Computer and Information Science, University of Pennsylvania, PA (1984).
2. Aït-Kaci, H. An algebraic semantics approach to the effective resolution of type equations. *Theoretical Computer Science* 45, pp. 293-351 (1986).
3. Andrews, P.B., "Resolution in Type Theory," JSL 36:3 (1971) 414-432.
4. Andrews, P. Theorem Proving via General Matings. *J.ACM* 28(2), 193-214, 1981.
5. Andrews, P. *An Introduction to Mathematical Logic and Type Theory: To Truth Through Proof.* Academic Press, New York, 1986.
6. Andrews, P.B., D. Miller, E. Cohen, F. Pfenning, "Automating Higher-Order Logic," *Contemporary Mathematics* 29, 169-192, 1984.
7. Bachmair, L., *Proof Methods for Equational Theories*, Ph.D thesis, University of Illinois, Urbana Champaign, Illinois (1987).
8. Bachmair, L., *Canonical Equational Proofs*, Research Notes in Theoretical Computer Science, Wiley and Sons, 1989.
9. Bachmair, L., Dershowitz, N., and Plaisted, D., "Completion without Failure," *Resolution of Equations in Algebraic Structures*, Vol. 2, Aït-Kaci and Nivat, editors, Academic Press, 1-30 (1989).
10. Bachmair, L., Dershowitz, N., and Hsiang, J., "Orderings for Equational Proofs," In *Proc. Symp. Logic in Computer Science*, Boston, Mass. (1986) 346-357.
11. Barendregt, H.P., *The Lambda Calculus*, North-Holland (1984).
12. Bibel, W. Tautology Testing With a Generalized Matrix Reduction Method, *TCS* 8, pp. 31-44, 1979.
13. Bibel, W. On Matrices With Connections, *J.ACM* 28, pp. 633-645, 1981.
14. Bibel, W. *Automated Theorem Proving*. Friedr. Vieweg & Sohn, Braunschweig, 1982.
15. Boudet, A., Jouannaud, J.-P., and Schmidt-Schauss, M. Unification in Boolean Rings and Abelian Groups. *Journal of Symbolic Computation* 8(5), pp. 449-478 (1989).
16. Bürckert, H., Herold, A., and Schmidt-Schauss, M. On equational theories, unification, and (un)decidability. Special issue on Unification, Part II, *Journal of Symbolic Computation* 8(1 & 2), 3-50 (1989).
17. Bürckert, H., Matching–A Special Case of Unification? *Journal of Symbolic Computation* 8(5), pp. 523-536 (1989).
18. Choi, J., and Gallier, J.H. A simple algebraic proof of the NP-completeness of rigid E-unification, in preparation (1990).
19. Church, A., "A Formulation of the Simple Theory of Types," JSL 5 (1940) 56-68.
20. Comon, H. Unification and Disunification. Théorie et Applications, Thèse de l'Institut Polytechnique de Grenoble (1988).
21. Darlington, J.L., "A Partial Mechanization of Second-Order Logic," Machine Intelligence 6 (1971) 91-100.
22. Dershowitz, N,. "Termination of Rewriting," Journal of Symbolic Computation 3 (1987) 69-116.
23. Dougherty, D., and Johann, P., "An Improved General E-Unification Method," *CADE'90*, Kaiserslautern, Germany.
24. Elliot, C., and Pfenning, F., "A Family of Program Derivations for Higher-Order Unification," Ergo Report 87-045, CMU, November 1987.
25. Fages, F. and Huet, G., "Complete Sets of Unifiers and Matchers in Equational Theories," TCS 43 (1986) 189-200.
26. Farmer, W., *Length of Proofs and Unification Theory*, Ph.D. Thesis, University of Wisconsin—Madison (1984).
27. Farmer, W. "A Unification Algorithm for Second-Order Monadic Terms," Unpublished Technical Report, MITRE Corporation, Bedford, MA.

28. Fay, M., "First-order Unification in an Equational Theory," Proceedings of the 4th Workshop on Automated Deduction, Austin, Texas (1979).
29. Felty, A., and Miller, D., "Specifying Theorem Provers in a Higher-Order Logic Programming Language," Ninth International Conference on Automated Deduction, Argonne, Illinois (1988).
30. Gallier, J.H. *Logic for Computer Science: Foundations of Automatic Theorem Proving*, Harper and Row, New York (1986).
31. Gallier, J.H., Raatz, S., "Extending SLD-Resolution to Equational Horn Clauses Using E-unification", *Journal of Logic Programming* 6(1-2), 3-56 (1989).
32. Gallier, J.H., Raatz, S., and Snyder, W., "Theorem Proving using Rigid E-Unification: Equational Matings," *LICS'87*, Ithaca, New York (1987) 338-346.
33. Gallier, J.H., Narendran, P., Plaisted, D., Raatz, S., and Snyder, W., "Finding canonical rewriting systems equivalent to a finite set of ground equations in polynomial time," submitted to *J.ACM* (1987).
34. Gallier, J.H., Narendran, P., Plaisted, D., and Snyder, W., "Rigid E-unification is NP-complete," *LICS'88*, Edinburgh, Scotland, July 5-8, 1988, 218-227.
35. Gallier, J.H., and Snyder, W. Complete Sets of Transformations For General E-Unification. Special issue of *Theoretical Computer Science* 67(2-3), 203-260 (1989).
36. Gallier, J.H., Raatz, S, and Snyder, W. Rigid E-Unification and its Applications to Equational Matings. *Resolution of Equations in Algebraic Structures*, Vol. 1, Aït-Kaci and Nivat, editors, Academic Press, 151-216 (1989).
37. Gallier, J.H., Narendran, P., Plaisted, D., and Snyder, W. Rigid E-Unification: NP-completeness and Applications to Theorem Proving. To appear in a special issue of *Information and Computation*, pp. 64 (1990).
38. Gallier, J.H., Narendran, P., Raatz, S., and Snyder, W. Theorem Proving Using Equational Matings and Rigid E-Unification. To appear in *J.ACM*, pp. 62 (1990).
39. Girard, J.Y., "Linear Logic," *Theoretical Computer Science* 50:1 (1987) 1-102.
40. Goguen, J.A. What is Unification? A categorical View of Substitution, Equation and Solution. *Resolution of Equations in Algebraic Structures*, Vol. 1, Aït-Kaci and Nivat, editors, Academic Press, 217-261 (1989).
41. Goguen, J.A., and Meseguer, J., "Eqlog: Equality, Types, and Generic Modules for Logic Programming," in *Functional and Logic Programming*, Degroot, D. and Lindstrom, G., eds., Prentice-Hall (1985). Short version in Journal of Logic Programming 2 (1984) 179-210.
42. Goldfarb, W., "The Undecidability of the Second-Order Unification Problem," TCS 13:2 (1981) 225-230.
43. Gould, W.E., *A Matching Procedure for Omega-Order Logic*, Ph.D. Thesis, Princeton University, 1966.
44. Guard, J.R., "Automated Logic for Semi-Automated Mathematics," Scientific Report 1, AFCRL 64-411, Contract AF 19 (628)-3250 AD 602 710.
45. Guard, J., Oglesby, J., and Settle, L., "Semi-Automated Mathematics," JACM 16 (1969) 49-62.
46. Hannan, J. and Miller, D., "Enriching a Meta-Language with Higher-Order Features," Workshop on Meta-Programming in Logic Programming, Bristol (1988).
47. Hannan, J. and Miller, D., "Uses of Higher-Order Unification for Implementing Program Transformers," Fifth International Conference on Logic Programming, MIT Press (1988).
48. Herbrand, J., "Sur la Théorie de la Démonstration," in *Logical Writings*, W. Goldfarb, ed., Cambridge (1971).
49. Hindley, J., and Seldin, J., *Introduction to Combinators and Lambda Calculus*, Cambridge University Press (1986).

50. Huet, G. *Constrained Resolution: A Complete Method for Higher-Order Logic*, Ph.D. thesis, Case Western Reserve University (1972).
51. Huet, G., "A Mechanization of Type Theory," Proceedings of the Third International Joint Conference on Artificial Intelligence (1973) 139-146.
52. Huet, G., "The Undecidability of Unification in Third-Order Logic," Information and Control 22 (1973) 257-267.
53. Huet, G., "A Unification Algorithm for Typed λ-Calculus," TCS 1 (1975) 27-57.
54. Huet, G., *Résolution d'Equations dans les Langages d'Ordre* $1, 2, \ldots, \omega$, Thèse d'Etat, Université de Paris VII (1976).
55. Huet, G., and Lang, B., "Proving and Applying Program Transformations Expressed with Second-Order Patterns," Acta Informatica 11 (1978) 31-55.
56. Huet, G., "Confluent Reductions: Abstract Properties and Applications to Term Rewriting Systems," JACM 27:4 (1980) 797-821.
57. Hullot, J.-M., "Canonical Forms and Unification," CADE-5 (1980) 318-334.
58. Hussmann, H., "Unification in Conditional Equational Theories," Proceedings of the EUROCAL 1985, Springer Lecture Notes in Computer Science 204, p. 543-553.
59. Hsiang, J., and Jouannaud, J.-P. General E-unification revisited. In *Proceedings of UNIF'88*, C. Kirchner and G. Smolka, editors, CRIN Report 89R38, pp. 51 (1988).
60. Isakowitz, T. Theorem Proving Methods For Order-Sorted Logic, Ph.D. thesis. Department of Computer and Information Science, University of Pennsylvania (1989).
61. Jouannaud, J.-P., and Kirchner, C. Solving Equations in Abstract Algebras: A Rule-Based Survey of Unification. Technical Report, University of Paris Sud (1989).
62. Kaplan, S., "Fair Conditional Term Rewriting Systems: Unification, Termination, and Confluence," Technical Report 194, Universite de Paris-Sud, Centre D'Orsay, Laboratoire de Recherche en Informatique (1984).
63. Kirchner, C., "A New Equational Unification Method: A Generalization of Martelli-Montanari's Algorithm," CADE-7, Napa Valley (1984).
64. Kirchner, C., *Méthodes et Outils de Conception Systematique d'Algorithmes d'Unification dans les Theories Equationnelles*, Thèse d'Etat, Université de Nancy I (1985).
65. Kirchner, C., "Computing Unification Algorithms," LICS'86, Cambridge, Mass. (1986).
66. Kirchner, C. and Kirchner, H., *Contribution à la Resolution d'Equations dans les Algèbres Libres et les Varietés Equationnelles d'Algèbres*, Thèse de 3e cycle, Université de Nancy I (1982).
67. Kfoury, A.J., J. Tiuryn, and P. Urzyczyn, "An analysis of ML typability", submitted (a section of this paper will appear in the proceedings of CAAP 1990 under the title "ML typability is DEXPTIME-complete").
68. Kfoury, A.J., J. Tiuryn, and P. Urzyczyn, "The undecidability of the semi-unification problem", Proceedings of STOC (1990).
69. Kfoury, A.J., and J. Tiuryn, "Type Reconstruction in Finite-Rank Fragments of the Polymorphic Lambda Calculus," LICS'90, Philadelpha, PA.
70. Knight. K. "A Multidisciplinary Survey," *ACM Computing Surveys*, Vol. 21, No. 1, pp. 93-124 (1989).
71. Knuth, D.E. and Bendix, P.B., "Simple Word Problems in Univeral Algebras," in *Computational Problems in Abstract Algebra*, Leech, J., ed., Pergamon Press (1970).
72. Kozen, D., "Complexity of Finitely Presented Algebras," Technical Report TR 76-294, Department of Computer Science, Cornell University, Ithaca, New York (1976).
73. Kozen, D., "Positive First-Order Logic is NP-Complete," IBM Journal of Research and Development, 25:4 (1981) 327-332.
74. Lankford, D.S., "Canonical Inference," Report ATP-32, University of Texas (1975),
75. Lassez, J.-L., Maher, M., and Marriot, K. Unification Revisited. *Foundations of*

Deductive Databases and Logic Programming, J. Minker, editor, Morgan-Kaufman, pp. 587-625 (1988).

76. Le Chenadec, P. On the Logic of Unification. Special issue on Unification, Part II, *Journal of Symbolic Computation* 8(1 & 2), 141-200 (1989).
77. Lucchesi, C.L., "The Undecidability of the Unification Problem for Third Order Languages," Report CSRR 2059, Dept. of Applied Analysis and Computer Science, University of Waterloo (1972).
78. Machtey, M. and Young, P.R., *An Introduction to the General Theory of Algorithms*, Elsevier North-Holland, NY, 1977.
79. Martelli, A., Montanari, U., "An Efficient Unification Algorithm," ACM Transactions on Programming Languages and Systems, 4:2 (1982) 258-282.
80. Martelli, A., Rossi, G.F., and Moiso, C. Lazy Unification Algorithms for Canonical Rewrite Systems. *Resolution of Equations in Algebraic Structures*, Vol. 2, Aït-Kaci and Nivat, editors, Academic Press, 245-274 (1989).
81. Meseguer, J., Goguen, J. A., and Smolka, G. Order-Sorted Unification. *Journal of Symbolic Computation* 8(4), pp. 383-413 (1989).
82. Miller, D., *Proofs in Higher-Order Logic*, Ph.D. thesis, Carnegie-Mellon University, 1983.
83. Miller, D. A. Expansion Trees and Their Conversion to Natural Deduction Proofs. In *7th International Conference on Automated Deduction, Napa, CA*, edited by R.E. Shostak, L.N.C.S, No. 170, New York: Springer Verlag, 1984.
84. Miller, D. A compact Representation of Proofs. *Studia Logica* 4/87, pp. 347-370 (1987).
85. Miller, D., and Nadathur, G., "Higher-Order Logic Programming," Proceedings of the Third International Conference on Logic Programming, London (1986).
86. Miller, D., and Nadathur, G., "A Logic Programming Approach to Manipulating Formulas and Programs," IEEE Symposium on Logic Programming, San Franciso (1987).
87. Miller, D., and Nadathur, G., "Some Uses of Higher-Order Logic in Computational Linguistics," 24th Annual Meeting of the Association for Computational Linguistics (1986) 247—255.
88. Milner, R. A theory of type polymorphism in programming. *J. Comput. Sys. Sci.* 17, pp. 348-375 (1978).
89. Nadathur, G., *A Higher-Order Logic as the Basis for Logic Programming*, Ph.D. Dissertation, Department of Computer and Information Science, University of Pennsylvania (1986).
90. Nelson G. and Oppen, D. C. Fast Decision Procedures Based on Congruence Closure. *J. ACM* 27(2), 356-364, 1980.
91. Nutt, W., Réty, P., and Smolka, G., "Basic Narrowing Revisited," Special Issue on Unification, Part I *Journal of Symbolic Computation* 7(3 & 4), pp. 295-318 (1989).
92. Paterson, M.S., Wegman, M.N., "Linear Unification," Journal of Computer and System Sciences, 16 (1978) 158-167.
93. Paulson, L.C., "Natural Deduction as Higher-Order Resolution," Journal of Logic Programming 3:3 (1986) 237-258.
94. Pfenning, F., *Proof Transformations in Higher-Order Logic*, Ph.D. thesis, Department of Mathematics, Carnegie Mellon University, Pittsburgh, Pa. (1987).
95. Pfenning, F., "Partial Polymorphic Type Inference and Higher-Order Unification," in *Proceedings of the 1988 ACM Conference on Lisp and Functional Programming*, ACM, July 1988.
96. Pfenning, F., and Elliott, C., "Higher-Order Abstract Syntax," *Proceedings of the SIGPLAN '88 Symposium on Language Design and Implementation*, ACM, June 1988.

97. Pietrzykowski, T., "A Complete Mechanization of Second-Order Logic," JACM 20:2 (1971) 333-364.
98. Pietrzykowski, T., and Jensen, D., "Mechanizing ω-Order Type Theory Through Unification," TCS 3 (1976) 123-171.
99. Plotkin, G., "Building in Equational Theories," Machine Intelligence 7 (1972) 73-90.
100. Rémy, Didier, "Algèbres Touffues. Application au typage polymorphique des objects enregistrements dans les languages fonctionnels." Thèse, Université Paris VII, 1990.
101. Rémy, Didier, "Records and variants as a natural extension in ML," *Sixteenth ACM Annual Symposium on Principles of Programming Languages*, Austin Texas, 1989.
102. Rety, P., "Improving Basic Narrowing Techniques," Proceedings of the RTA, Bordeaux, France (1987).
103. Robinson, J.A., "A Machine Oriented Logic Based on the Resolution Principle," JACM 12 (1965) 23-41.
104. Robinson, J.A., "A Review on Automatic Theorem Proving," Annual Symposia in Applied Mathematics, 1-18 (1967).
105. Robinson, J.A., "Mechanizing Higher-Order Logic," Machine Intelligence 4 (1969) 151-170.
106. Siekmann, J. H. Unification Theory, Special Issue on Unification, Part I, *Journal of Symbolic Computation* 7(3 & 4), pp. 207-274 (1989).
107. Siekmann, J. H., and Szabo, P. Universal unification and classification of equational theories. *CADE'82*, Lecture Notes in Computer Science, No. 138 (1982).
108. Shieber, S. *An Introduction to Unification-Based Approaches to Grammar.* CSLI Lecture Notes Series, Center for the study of Language and Information, Stanford, CA (1986).
109. Slagle, J.R., "Automated Theorem Proving for Theories with Simplifiers, Commutativity, and Associativity," JACM 21 (1974) 622-642.
110. Schmidt-Schauss, M. Unification in Permutative Equational Theories is Undecidable. *Journal of Symbolic Computation* 8(4), pp. 415-422 (1989).
111. Schmidt-Schauss, M. Unification in a Combination of Arbitrary Disjoint Equational Theories. Special issue on Unification, Part II, *Journal of Symbolic Computation* 8(1 & 2), 51-99 (1989).
112. Snyder, W. Complete Sets of Transformations for General Unification, Ph.D. thesis. Department of Computer and Information Science, University of Pennsylvania, PA (1988).
113. Snyder, W., and Gallier, J.H., "Higher-Order Unification Revisited: Complete Sets of Transformations," Special issue on Unification, Part II, *Journal of Symbolic Computation* 8(1 & 2), 101-140 (1989).
114. Snyder, W., "Efficient Ground Completion: A Fast Algorithm for Generating Reduced Ground Rewriting Systems from a Set of Ground Equations," RTA'89, Chapel Hill, NC (journal version submitted for publication).
115. Snyder, W., "Higher-Order E-Unification," CADE'90, Kaiserslautern, Germany.
116. Snyder, W. *The Theory of General Unification.* Birkhauser Boston, Inc. (in preparation).
117. Snyder, W., and C. Lynch, "Goal–Directed Strategies for Paramodulation," Proceedings of the Fourth Conference on Rewrite Techniques and Applications, Como, Italy (1991).
118. Statman, R., "On the Existence of Closed Terms in the Typed λ-Calculus II: Transformations of Unification Problems," TCS 15:3 (1981) 329-338.
119. Szabo, P. *Unifikationstheorie erster Ordnung*, Ph.D. thesis, Universität Karlsruhe (1982).
120. Tiden, E., and Arnborg, S. Unification problems with one-sided distributivity. *Journal of Symbolic Computation* 3(1 & 2), pp. 183-202 (1987).

121. Winterstein, G., "Unification in Second-Order Logic," Electronische Informationsverarbeitung und Kybernetik 13 (1977) 399-411.
122. Yelick, K. Unification in combinations of collapse-free regular theories. *Journal of Symbolic Computation* 3(1 & 2), pp. 153-182 (1987).
123. Zaionc, Marek, "The Set of Unifiers in Typed λ-Calculus as a Regular Expression," Proceedings of the RTA 1985.

J. Gallier: Computer and Information Science Department, University of Pennsylvania, 200 South 33rd Street, Philadelphia, PA 19104

W. Snyder: Computer Science Department, Boston University, 111 Cummington Street, Boston, MA 02215

This research was partially supported by ONR under Grant No. N00014-88-K-0593.

NORMAL FORMS AND CUT-FREE PROOFS AS NATURAL TRANSFORMATIONS

JEAN-YVES GIRARD, ANDRE SCEDROV AND PHILIP J. SCOTT

Dedicated to the Memory of E.S. Bainbridge

ABSTRACT. What equations can we *guarantee* that simple functional programs must satisfy, irrespective of their obvious defining equations? Equivalently, what non-trivial identifications must hold between lambda terms, thought-of as encoding appropriate natural deduction proofs ? We show that the usual syntax guarantees that certain naturality equations from category theory are necessarily provable. At the same time, our categorical approach addresses an equational meaning of cut-elimination and asymmetrical interpretations of cut-free proofs. This viewpoint is connected to Reynolds' relational interpretation of parametricity ([27], [2]), and to the Kelly-Lambek-Mac Lane-Mints approach to coherence problems in category theory.

1. INTRODUCTION

In the past several years, there has been renewed interest and research into the interconnections of proof theory, typed lambda calculus (as a functional programming paradigm) and category theory. Some of these connections can be surprisingly subtle. Here we address some relationships of three fundamental notions:

- Cut-elimination from proof theory.
- Natural Transformations from category theory.
- Parametricity from the foundations of polymorphism.

Familiar work of Curry, Howard, Lambek and others [12, 15, 17] has shown how we may consider constructive proofs as programs. For example, Gentzen's intuitionistic sequents $A_1, \ldots, A_k \vdash B$ may be interpreted as functional programs mapping k inputs of types A_i, $1 \leq i \leq k$, to outputs of type B. More precisely, proofs are interpreted as certain terms of the typed lambda calculus.

The cut rule

$$\frac{\Gamma \vdash A \quad \Delta, A \vdash B}{\Gamma, \Delta \vdash B}$$

has a special status; the cut formula A appears simultaneously covariantly and contravariantly (i.e. to the right and the left of $\vdash$, resp.). In the functional formalism, cut corresponds to *composition*, i.e. substitution or "plugging together" of one program into another. One meaning of cut-elimination is that general substitution is already definable from the special instances implicit in the other Gentzen rules.

In this paper we show how to associate to natural deduction proofs in normal form certain families of normal lambda terms indexed by types. These families necessarily satisfy certain naturality conditions from category theory in any cartesian closed category. In the case of the syntax, we obtain as corollary that appropriate lambda terms *provably* satisfy, in addition to the usual beta eta equations, also these new equations.

We begin with two examples from simply typed lambda calculus. Recall, the types are built inductively from a countably infinite set of ground types (which can be considered as type variables) by the two type constructors $\Rightarrow$ and $\times$.

Example 1.1. Let r be any closed simple typed lambda term of type

$$\alpha \times \alpha \Rightarrow \alpha \times \alpha,$$

where α is a type variable. What can we say about such r in general? We shall show that for any two types A and B and any closed term f of type $A \Rightarrow B$ it must be the case that:

$$(f \times f) \circ r_A = r_B \circ (f \times f) \tag{1}$$

where r_A and r_B are instances of r, where $=$ is beta-eta conversion, and where $g \circ f$ denotes function composition.

The equation can be restated in cartesian closed categories (= ccc's). Recall that simply typed lambda calculi correspond to ccc's and that in any ccc morphisms $A \to B$ uniquely correspond to morphisms $1 \to A \Rightarrow B$ [23, 19]. We shall blur this latter distinction and abuse the notation accordingly. In any ccc, then, the above equation says that for any morphism $f : A \to B$ the following diagram commutes.

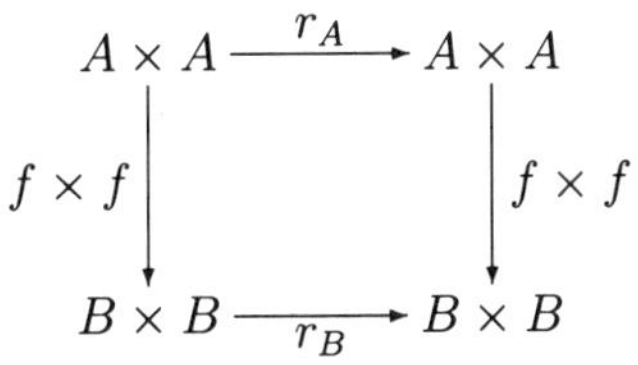

$$\begin{array}{ccc} A \times A & \xrightarrow{r_A} & A \times A \\ {\scriptstyle f \times f}\downarrow & & \downarrow{\scriptstyle f \times f} \\ B \times B & \xrightarrow[r_B]{} & B \times B \end{array}$$

This means that the instantiations of r at all values of type variable α form a natural transformation from the covariant endofunctor $F(A) = A \times A$ to itself.

In fact, in the example above there is a functional dependency on the map f, this dependency itself being natural. This is further explored in the following example.

Example 1.2. Let m be any closed simply typed lambda term of type

$$(\alpha \Rightarrow \beta) \Rightarrow (\alpha \times \alpha \Rightarrow \beta \times \beta),$$

where α and β are type variables. What can m possibly be? A first guess is that , for any argument f of appropriate type, $m(f) = f \times f$. We shall confirm this intuition by showing that for any two types A, B, and any closed term f of type $A \Rightarrow B$ it is the case that:

$$m_{AB}(f) = (f \times f) \circ m_{AA}(1_A) = m_{BB}(1_B) \circ (f \times f) \tag{2}$$

in the sense of beta-eta conversion. In other words, in any ccc, the equation holds for any morphism $f : A \longrightarrow B$. Thus $m_{AB}(f)$ "is" $f \times f$, up to composition with a distinguished endomorphism of the domain or codomain.

In the next section, these examples will be seen in their "natural" setting of multivariant functors and multivariant (dinatural) natural transformations.

Notation: In what follows, $\mathcal{C}$ denotes a category (usually cartesian closed). Objects of $(\mathcal{C}^\circ)^n \times \mathcal{C}^n$ are denoted $(\mathbf{A}; \mathbf{B})$, where $\mathbf{A}$, $\mathbf{B}$ are objects of $\mathcal{C}^n$ and semicolon separates the contravariant and covariant arguments. For notational ease, we often write the value of a functor $F : (\mathcal{C}^\circ)^n \times \mathcal{C}^n \longrightarrow \mathcal{C}$ at objects $(\mathbf{A}; \mathbf{B})$ by $F\mathbf{AB}$ rather than $F(\mathbf{A}; \mathbf{B})$.

For much of this paper, we purposely abuse notation between cartesian closed categories (= ccc's) and typed lambda calculus and blur, for example, the distinction (in the associated typed lambda calculus) between closed terms $m : A \Rightarrow B$, open terms $m(x) : B$ containing at most one free variable

$x : A$, and arrows $m : A \longrightarrow B$, ([19], pp. 75-78, and [23]). Substitution vs composition will be equally abused. Readers may easily restate results in the notation of their choice.

2. The Dinatural Calculus

The basic view that led to the work on *functorial polymorphism* in [2] is that we may interpret polymorphic type expressions $\sigma(\alpha_1, \ldots, \alpha_n)$, with type variables α_i, as certain kinds of multivariant "definable" functors

$$F : (\mathcal{C}^\circ)^n \times \mathcal{C}^n \longrightarrow \mathcal{C}$$

over an appropriate ccc $\mathcal{C}$; moreover, terms $t(x)$ of type τ, with variable x of type σ, are then interpreted as certain multivariant (= *dinatural*) transformations between (the interpretations of) the types. The resultant calculus has many interesting properties [2, 7], despite the fact that in general dinatural transformations do not compose. Previous studies [2, 3, 8] emphasized semantical (compositional) models for this calculus; in this paper we prove that the pure (cut-free) syntax itself admits a compositional dinatural interpretation.

We briefly review this functorial interpretation, following [2], Section 3. In this paper we emphasize the ccc connectives $\times$ and $\Rightarrow$; hence we omit the functorial interpretation of universal type quantification $\forall\alpha$, emphasized in [2]. In fact, we can now interpret the syntax over an *arbitrary* ccc, as we shall see.

Let $\mathcal{C}$ be any category, and $n \geq 1$ an integer.

Definition 2.1. A *dinatural transformation* between functors $F, G : (\mathcal{C}^\circ)^n \times \mathcal{C}^n \longrightarrow \mathcal{C}$ is a family of morphisms

$$\theta = \{\theta_{\mathbf{A}} : F\mathbf{AA} \longrightarrow G\mathbf{AA} \mid \mathbf{A} \in \mathcal{C}^n\}$$

satisfying the following: for any vector of arrows $\mathbf{f} : \mathbf{A} \longrightarrow \mathbf{B}$, the following diagram commutes:

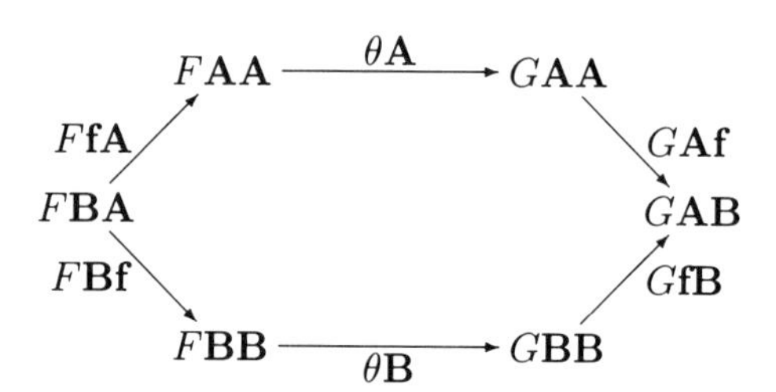

We write $\theta : F \longrightarrow G$ to denote that θ is a dinatural transformation from F to G, as above. Many examples of such transformations arise in practice, cf. [2, 20]; in particular, among other things, dinatural transformations include ordinary natural transformations as a special case. It is useful to note that the definition of dinaturality is an *equational* condition on the family θ in the category $\mathcal{C}$. Moreover, dinaturality is, on the face of it, a *semantical* condition: it must be verified for all objects $\mathbf{A}, \mathbf{B} \in \mathcal{C}^n$ and all morphisms $\mathbf{f} : \mathbf{A} \longrightarrow \mathbf{B}$.

Consider a simply typed lambda calculus whose types are freely generated from appropriate ground constants, including type variables, and whose terms are freely generated from infinitely many individual variables by the usual term-formation rules (appropriate to the ground types) [19, 12]. We impose familiar lambda calculus equations, including beta and eta, and product types with surjective pairing. [1]

Let $\mathcal{C}$ be any ccc. For each type $\sigma(\alpha_1, \dots, \alpha_n)$, with type variables α_i, we define its interpretation $\|\sigma\| : (\mathcal{C}^\circ)^n \times \mathcal{C}^n \longrightarrow \mathcal{C}$ by induction, as follows:

(1) If $\sigma = \alpha_i$, $\|\sigma\|(\mathbf{A}; \mathbf{B}) = B_i$, the projection onto the ith component of $\mathbf{B}$.
(2) If σ = C, a ground constant interpreted as an object of $\mathcal{C}$, then $\|\sigma\|(\mathbf{A}; \mathbf{B}) = K_C$, the constant functor with value C.
(3) (Product) If $\sigma = \tau_1 \times \tau_2$, then $\|\sigma\|(\mathbf{A}; \mathbf{B}) = \|\tau_1\|(\mathbf{A}; \mathbf{B}) \times \|\tau_2\|(\mathbf{A}; \mathbf{B})$.
(4) (Twisted Exponential) If $\sigma = \tau_1 \Rightarrow \tau_2$ then $\|\sigma\|(\mathbf{A}; \mathbf{B}) = \|\tau_1\|(\mathbf{B}; \mathbf{A}) \Rightarrow \|\tau_2\|(\mathbf{A}; \mathbf{B})$.

Having interpreted types as certain multivariant functors over an arbitrary ccc $\mathcal{C}$ we now sketch how to view the usual interpretation of typed lambda terms in ccc's as dinatural transformations.

Consider a legal typing judgement

$$x_1 : \sigma_1, \dots, x_k : \sigma_k \; t : \tau.$$

Let $\mathcal{C}$ be any ccc. As we've seen above, any assignment of all the type variables $\alpha_1, \dots, \alpha_n$ (in the typing judgement) to objects of $\mathcal{C}$ induces functors $\|\sigma_i\|, \|\tau\| : (\mathcal{C}^\circ)^n \times \mathcal{C}^n \longrightarrow \mathcal{C}$. We now show that the familiar interpretation of typed lambda terms in ccc's induces a family by substituting objects for type variables. Hence every term t induces a family

[1]This theory is actually a fragment of second order lambda calculus (e.g. [12]) in which types are (implicitly universally quantified) schema, closed under type substitution. The language is closely related to the "Core-ML" language of implicit polymorphism [24].

$\|t\|_{\mathbf{A}} : (\|\sigma_1\| \times \cdots \times \|\sigma_k\|)\, \mathbf{AA} \longrightarrow \|\tau\| \mathbf{AA}$, where $\mathbf{A} = (A_1, \ldots, A_n)$ and $A_i = \|\alpha_i\|$, as follows:

(1) If $t : \sigma_i$ is the variable x_i, then $\|t\| : \|\sigma_1\| \times \cdots \times \|\sigma_k\| \longrightarrow \|\sigma_i\|$ is the ith projection.

(2) If $t : \sigma \Rightarrow \tau$ is $\lambda x : \sigma.r$, where $\|r\| : \|\sigma_1\| \times \cdots \times \|\sigma_k\| \times \|\sigma\| \longrightarrow \|\tau\|$, then $\|t\| : \|\sigma_1\| \times \cdots \times \|\sigma_k\| \longrightarrow \|\sigma \Rightarrow \tau\|$ is defined by: $\|t\|\mathbf{A} = (\|r\|\mathbf{A})^*$, the curryfication (or exponential transpose) of the arrow $\|r\|\mathbf{A} : \|\sigma_1\|\mathbf{AA} \times \cdots \times \|\sigma_k\|\mathbf{AA} \times \|\sigma\|\mathbf{AA} \longrightarrow \|\tau\|\ \mathbf{AA}$ in the ccc $\mathcal{C}$ (cf. [19], p.61).

(3) If $t : \tau$ is ua, where $u : \sigma \Rightarrow \tau$ and $a : \sigma$, then $\|t\|$ is $ev\circ < \|u\|, \|a\| >$, where ev denotes the appropriate evaluation map and $\circ$ denotes composition in $\mathcal{C}$.

(4) If $t : \tau_1 \times \tau_2$ is $t =< t_1, t_2 >$ and $\|t_i\| : \|\sigma_1\| \times \cdots \times \|\sigma_k\| \longrightarrow \|\tau_i\|$ for $i = 1, 2$, then $\|t\|\mathbf{A} =< \|t_1\|\mathbf{A}, \|t_2\|\mathbf{A} >$.

(5) If $t : \tau_i$ is $\Pi_i(t')$ ($i = 1, 2$) where $\|t'\| : \|\sigma_1\| \times \cdots \times \|\sigma_k\| \longrightarrow \|\ \tau_1 \times \tau_2\ \|$ then $\|\Pi_i(t')\|$ is $\Pi_i\circ\|t'\|$, where Π_i is the ith projection.

One of our main points is that the above family $\|t\| = \{\|t\|_{\mathbf{A}} \mid \mathbf{A} \in \mathcal{C}^n\} : \|\sigma_1\| \times \cdots \times \|\sigma_k\| \longrightarrow \|\tau\|$,which arises from a single term by plugging in objects as values for type variables, is actually dinatural; moreover, all compositions of dinaturals in the interpretation are well-defined.

We summarize the results in the next two theorems, whose proof we shall discuss later.

Theorem 2.2. *Let $\mathcal{L}$ be simply typed lambda calculus with type variables, let $\mathcal{C}$ be any ccc, and let $\alpha_1, \ldots, \alpha_n$ be a list of type variables. Then any type $\tau(\alpha_1, \ldots, \alpha_n)$ with the indicated type variables induces a functor $\|\tau\| : (\mathcal{C}^\circ)^n \times \mathcal{C}^n \longrightarrow \mathcal{C}$. Furthermore, any legal typing judgement*

$$x_1 : \sigma_1, \ldots, x_k : \sigma_k\ t : \tau$$

induces a family of lambda terms $\|t\| = \{\|t\|_{\mathbf{A}} | \mathbf{A} \in \mathcal{C}^n\}$ which is actually a dinatural transformation $\|t\| : \|\sigma_1\| \times \cdots \times \|\sigma_k\| \longrightarrow \|\tau\|$. Beta-eta convertible lambda terms give the same dinatural family.

We shall call the multivariant functors and syntactic families in the above theorem *definable* functors and dinatural transformations, respectively.

Theorem 2.3. *Let $\mathcal{C}$ be any ccc. Then for each n, the definable functors $(\mathcal{C}^\circ)^n \times \mathcal{C}^n \longrightarrow \mathcal{C}$ and definable dinatural transformations between them* *form a ccc. The cartesian closed structure on definable functors is given by products and twisted exponentials, as in the discussion after Definition 2.1.*[2]

The nontrivial parts of the last two theorems are that syntactic families induced by terms actually yield dinatural transformations, and moreover that these definable dinatural transformations compose, to yield a category (cf. [2]). The proof proceeds through a rather subtle detour involving Gentzen's sequents and cut-elimination, which we begin in the next section. We also remark that a related semantical theorem (Theorem 2) in [8] is now only known to apply to the core-ML-fragment of polymorphic lambda calculus. The full second order version claimed in that paper is still open.

There is an immediate corollary of the above theorems. If we specialize $\mathcal{C}$ to be the ccc freely generated by $\mathcal{L}$, that is, the syntactical term model of typed lambda calculus with type variables, qua ccc (cf. [19],p. 77) then "equality" means "provable equality" in $\mathcal{L}$. In this setting, it follows that the syntactic families $\|t\|$ above are provably dinatural. Thus, at the linguisitic level, *arbitrary terms t, when validly typed, provably satisfy appropriate dinaturality equations in addition to any equations of the typed lambda calculus.*

For the remainder of this section, we reexamine and generalize Example 1.2 from the Introduction, in the light of the discussion above. Recall, from the typing of m we obtain the typing judgement

$$x : \alpha \Rightarrow \beta \quad m(x) : \alpha \times \alpha \Rightarrow \beta \times \beta \,.$$

Thus, in any ccc $\mathcal{C}$, the term m is interpreted as a definable dinatural transformation $F \longrightarrow G$ between certain functors $F, G : (\mathcal{C}^{\circ})^2 \times \mathcal{C}^2 \longrightarrow \mathcal{C}$ determined by the types. Indeed, from the discussion after **2.1**, F and G are given by $F = \|\alpha \Rightarrow \beta\|$ and $G = \|\alpha \times \alpha \Rightarrow \beta \times \beta\|$. It may be shown [3] that $F(A, B; A', B') = A \Rightarrow B'$, while $G(A, B; A', B') = A \times A \Rightarrow B' \times B'$, where $(A, B; A', B')$ denotes an object in $(\mathcal{C}^{\circ})^2 \times \mathcal{C}^2$. The dinaturality of $m (= m(x))$ says that for arbitrary morphisms $a : A \longrightarrow A'$ and $b : B \longrightarrow B'$ in $\mathcal{C}$, the following hexagon commutes:

[2]Indeed, we actually obtain a fibred ccc or hyperdoctrine, ignoring the indexed adjoints [25, 28]

[3]The type variables α and β correspond to covariant projections $P_i : (\mathcal{C}^{\circ})^2 \times \mathcal{C}^2 \longrightarrow \mathcal{C}$ where $P_i(\mathbf{C}; \mathbf{D}) = D_i, i = 1, 2$ (*resp.*). Thus $F(\mathbf{C}; \mathbf{D}) = P_1(\mathbf{D}; \mathbf{C}) \Rightarrow P_2(\mathbf{C}; \mathbf{D})$, and $G(\mathbf{C}; \mathbf{D}) = (P_1 \times P_1)(\mathbf{D}; \mathbf{C}) \Rightarrow (P_2 \times P_2)(\mathbf{C}; \mathbf{D})$.

$$\begin{array}{ccccc}
 & A \Rightarrow B & \xrightarrow{\;m_{AB}\;} & A \times A \Rightarrow B \times B & \\
{}_{-}\circ a \nearrow & & & & \searrow (b \times b)\circ{}_{-} \\
A' \Rightarrow B & & & & A \times A \Rightarrow B' \times B' \\
b\circ{}_{-} \searrow & & & & \nearrow {}_{-}\circ(a \times a) \\
 & A' \Rightarrow B' & \xrightarrow[\;m_{A'B'}\;]{} & A' \times A' \Rightarrow B' \times B' &
\end{array}$$

FIGURE 1

where $f\circ{}_{-}$ denotes the map $h \mapsto f\circ h$.

Letting $A' = B' = B, b = 1_B$, and $a = f$ in Figure 1, we obtain

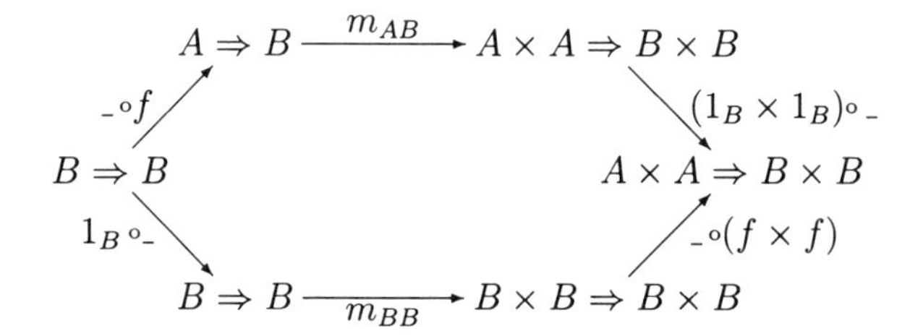

Equivalently, we obtain the following commutative square:

$$\begin{array}{ccc}
A \Rightarrow B & \xrightarrow{\;m_{AB}\;} & A \times A \Rightarrow B \times B \\
{}_{-}\circ f \uparrow & & \uparrow {}_{-}\circ(f \times f) \\
B \Rightarrow B & \xrightarrow[\;m_{BB}\;]{} & B \times B \Rightarrow B \times B
\end{array}$$

This says that for any $e : B \longrightarrow B$ it is the case that

(†)
$$m_{AB}(e\circ f) = m_{BB}(e)\circ(f \times f) \ ,$$

and so in particular for $e = 1_B$:

(2a)
$$m_{AB}(f) = m_{BB}(1_B)\circ(f \times f)$$

for any $f : A \longrightarrow B$.

A "dual" instance of the above equation is obtained in a similar manner by specializing in figure 1: $A' = B = A$ and $a = 1_A$, and then switching notation, by letting $B' = B$ and $b = f$. We obtain the following commutative square:

$$\begin{array}{ccc} A \Rightarrow A & \xrightarrow{m_{AA}} & A \times A \Rightarrow A \times A \\ {\scriptstyle f\circ\_}\downarrow & & \downarrow{\scriptstyle (f \times f)\circ\_} \\ A \Rightarrow B & \xrightarrow[m_{AB}]{} & A \times A \Rightarrow B \times B \end{array}$$

This says that for any $u : A \longrightarrow A$ it is the case that

(††)

$$m_{AB}(f \circ u) = (f \times f) \circ m_{AA}(u).$$

In particular, letting $u = 1_A$, we obtain

(2b)

$$m_{AB}(f) = (f \times f) \circ m_{AA}(1_A).$$

The equations (2a) and (2b), which together express equation 2 of Section 1, were first noted by Wadler [32].

Remark 2.4. The equations (†) and (††) are instances of natural transformations, as follows. From general considerations, dinatural transformations $m : F \longrightarrow G$, where $F, G : (\mathcal{C}^\circ) \times \mathcal{C} \longrightarrow \mathcal{C}$ essentially reduce to natural transformations if either the co- or the contravariant variable is dummy or a fixed constant. In the above discussion, consider the bifunctors $\| \alpha \Rightarrow \beta \|$ and $\| \alpha \times \alpha \Rightarrow \beta \times \beta \|$: $(\mathcal{C}^\circ)^2 \times \mathcal{C}^2 \longrightarrow \mathcal{C}$. Fix the covariant variable $\beta := B$ as a constant (so $\|\beta\| = K_B$). Then the induced composite functors $\|\alpha \Rightarrow B\|$ and $\|\alpha \times \alpha \Rightarrow B \times B\|$: $(\mathcal{C}^\circ) \times \mathcal{C} \longrightarrow \mathcal{C}$ may actually be considered as (inducing) contravariant functors $\mathcal{C}^\circ \longrightarrow \mathcal{C}$; the dinatural m then becomes a natural transformation $m_{\_B}$ between them. The commutative square above equation (†) expresses the special case of naturality along morphisms $f : A \longrightarrow B$.

The dual equation (††) is a special case of the fact that m also induces a natural transformation $m_{A\_}$ between $\|A \Rightarrow \beta\|$ and $\|A \times A \Rightarrow \beta \times \beta\|$, qua covariant functors $\mathcal{C} \longrightarrow \mathcal{C}$.

The above discussion centered on the type constructor $T(\alpha) = \alpha \times \alpha$, qua covariant definable functor $\mathcal{C} \longrightarrow \mathcal{C}$. For example, for this T, the just-mentioned dual equation (††) refers to a natural transformation $m_{A\_} : \|A \Rightarrow \beta\| \longrightarrow \| T(A) \Rightarrow T(\beta) \|$. We may generalize this discussion as follows.

Example 2.5. Consider m a closed simply typed lambda term, say

$$m : (\alpha \Rightarrow \beta) \Rightarrow (T(\alpha) \Rightarrow T(\beta)),$$

where T is any definable "pure" covariant functor (i.e. ignoring dummy arguments, consider T as simply a functor $\mathcal{C} \longrightarrow \mathcal{C}$). Then, according to Theorem 2.2 above, m yields a dinatural transformation from $() \Rightarrow ()$ to $T() \Rightarrow T()$, considered as functors $(\mathcal{C}^{\circ})^2 \times \mathcal{C}^2 \longrightarrow \mathcal{C}$ as in the previous discussion. Following the Remark above, if we fix the contravariant argument of these bifunctors to be $A \in \mathcal{C}$, m induces a natural transformation $m_{A\_}$ from the usual covariant functor $A \Rightarrow ()$ to $T(A) \Rightarrow T(\ )$, thought-of as functors $\mathcal{C} \longrightarrow \mathcal{C}$. Similarly, fixing the covariant argument to be B, m induces a natural transformation $m_{\_B}$ from the contravariant functor $() \Rightarrow B$ to $T() \Rightarrow T(B)$.

Thus, for any instantiations A of type variable α, B of type variable β, etc., and for all $f : B' \longrightarrow B$, the following square involving $m_{A\_}$ commutes:

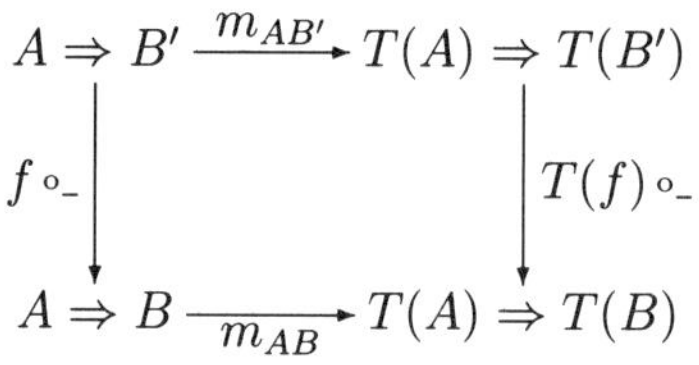

$$\begin{array}{ccc} A \Rightarrow B' & \xrightarrow{m_{AB'}} & T(A) \Rightarrow T(B') \\ {\scriptstyle f\circ\_}\downarrow & & \downarrow{\scriptstyle T(f)\circ\_} \\ A \Rightarrow B & \xrightarrow[m_{AB}]{} & T(A) \Rightarrow T(B) \end{array}$$

i.e. for all $g : A \Rightarrow B'$, we have $T(f) \circ m_{AB'}(g) = m_{AB}(f \circ g)$. Specializing to the case where $B' = A$ and $g = 1_A$, we thus have that m satisfies the equation

$$m_{AB}(f) = T(f) \circ m_{AA}(1_A).$$

for all $f : A \longrightarrow B$.

It is left to the reader to dualize the above discussion, using $m_{\_B}$ to obtain that m also satisfies

$$m_{AB}(f) = m_{BB}(1_B) \circ T(f).$$

for any $f : A \longrightarrow B$.

The last example, which includes all the previous ones for appropriate choices of T, may also be applied to other useful functorial type constructors, provided the ccc's $\mathcal{C}$ have the appropriate structure. As an example, we could introduce (as a definable functor) the covariant functor $T(\alpha) = \alpha - list$, and $T(f) = map(f)$ (as first described by Strachey, cf. [27]).

Remark 2.6. In the examples above, the two equations for $m_{AB}(f)$ are special cases of dinaturality, which is an equational condition. They were first observed by Wadler [32] from other considerations based on a universal Horn condition (see below). An interesting observation, due to Val Breazu-Tannen, is that these two special cases *actually imply* dinaturality. That is, following the discussion in the above example, if we have a family of morphisms

$$m = \{m_{AB} : F(AB;AB) \longrightarrow G(AB;AB) \mid AB \in \mathcal{C}^2\}$$

where $F, G : (\mathcal{C}^\circ)^2 \times \mathcal{C}^2 \longrightarrow \mathcal{C}$ are functors given by $F(AB;A'B') = A \Rightarrow B'$, while $G(AB;A'B') = T(A) \Rightarrow T(B')$, and if $m_{AB}(f)$ satisfies the following two equations (for any f of type $F(AB;AB)$):

$$m_{AB}(f) = T(f) \circ m_{AA}(1_A) = m_{BB}(1_B) \circ T(f)$$

then m actually is a dinatural transformation; i.e. for all arrows $a : A \longrightarrow A'$ and $b : B \longrightarrow B'$ the following diagram commutes:

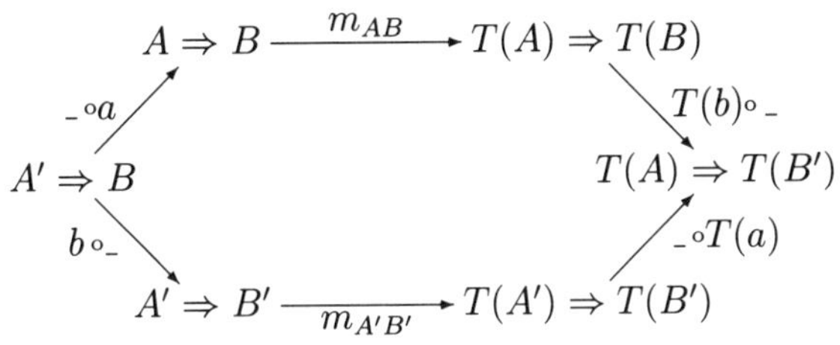

$$\begin{array}{ccccc}
 & A \Rightarrow B & \xrightarrow{m_{AB}} & T(A) \Rightarrow T(B) & \\
{\scriptstyle _\circ a} \nearrow & & & & \searrow {\scriptstyle T(b)\circ_} \\
A' \Rightarrow B & & & & T(A) \Rightarrow T(B') \\
{\scriptstyle b\circ_} \searrow & & & & \nearrow {\scriptstyle _\circ T(a)} \\
 & A' \Rightarrow B' & \xrightarrow[m_{A'B'}]{} & T(A') \Rightarrow T(B') &
\end{array}$$

Proof. $T(b) \circ m_{AB}(g \circ a) = T(b) \circ T(g \circ a) \circ m_{AA}(1_A) = T(b \circ g) \circ T(a) \circ m_{AA}(1_A) = T(b \circ g) \circ m_{AA'}(a) = T(b \circ g) \circ m_{A'A'}(1_{A'}) \circ T(a) = m_{A'B'}(b \circ g) \circ T(a)$. $\square$

This exercise is just a special case of a theorem that the universal Horn condition considered in Wadler [32] implies dinaturality, for definable functors. This general result, along with the categorical framework for this Horn condition (given by the notion of relator) is discussed in Abramsky, et al [1].

3. Gentzen Sequents and Natural Deduction

In what follows, we sketch the translation of Gentzen sequents into natural deduction calculi [12, 31, 33, 6, 22] which motivate our presentation.

3.1. Natural Deduction and Sequent Calculus. A deduction $\mathcal{D}$ of formula A (to be defined below) is a certain type of finite tree whose vertices are labelled by formulae and whose root is labelled by A. We denote such deductions by

$$\begin{array}{c}\mathcal{D}\\ A\end{array}$$

Certain occurrences of formulae among the leaves of $\mathcal{D}$ will be called *open assumptions* ; other leaves will be called *discharged* or *cancelled* assumptions.

We partition occurrences of assumption formulas into "assumption classes" ([31], [33]) or "parcels of hypotheses" ([12]), each class consisting of occurrences of a single formula. Note that a given formula may be in more than one class. We write

$$\begin{array}{c}[A_1]\cdots[A_k]\\ \mathcal{D}\\ B\end{array}$$

for a deduction $\mathcal{D}$ of formula B with certain distinguished assumption classes $[A_1]\cdots[A_k]$. We omit specifically denoting such classes where they are not relevant to the discussion.

Deductions are generated as follows:

(1) Basis. If A is a formula, the single node tree with label A is a deduction of A. A is an open assumption of this deduction; there are no cancelled assumptions, and there is one assumption class consisting of A itself.

(2) Inductive Step. New deductions are built from old ones by the following introduction and elimination rules.

$$\wedge\mathrm{I}\quad \frac{\begin{array}{cc}\mathcal{D}_1 & \mathcal{D}_2\\ A & B\end{array}}{A\wedge B} \qquad\qquad \wedge\mathrm{E}\quad \frac{\begin{array}{c}\mathcal{D}_1\\ A\wedge B\end{array}}{A} \qquad \frac{\begin{array}{c}\mathcal{D}_1\\ A\wedge B\end{array}}{B}$$

$$\Rightarrow\mathrm{I}\quad \frac{\begin{array}{c}[A]\\ \mathcal{D}_1\\ B\end{array}}{A\Rightarrow B} \qquad\qquad \Rightarrow\mathrm{E}\quad \frac{\begin{array}{cc}\mathcal{D}_1 & \mathcal{D}_2\\ A\Rightarrow B & A\end{array}}{B}$$

In $\Rightarrow$I, $[A]$ is a particular open assumption class, all of whose members (viz, occurrences of A) are simultaneously discharged. There may be other occurrences of A not discharged.

Now consider the following Gentzen rules for intuitionistic logic based on the connectives $\wedge, \Rightarrow$. Recall [12] sequents have the form $\Gamma \vdash A$, where Γ is a finite list of formulae and A is a single formula. These sequents satisfy the following axioms and rules (here Γ and Δ are finite lists of formulae):

Axiom: $A \vdash A$

Cut:
$$\frac{\Gamma \vdash A \quad \Delta, A \vdash B}{\Gamma, \Delta \vdash B}$$

Structural:

Exchange
$$\frac{\Gamma, A, B, \Delta \vdash C}{\Gamma, B, A, \Delta \vdash C}$$

Contraction
$$\frac{\Gamma, A, A \vdash C}{\Gamma, A \vdash C}$$

Weakening
$$\frac{\Gamma \vdash C}{\Gamma, A \vdash C}$$

Logical:

$$\wedge L_i \quad \frac{\Gamma, A_i \vdash C}{\Gamma, A_1 \wedge A_2 \vdash C}, \; i = 1, 2. \qquad \wedge R \quad \frac{\Gamma \vdash A \quad \Delta \vdash B}{\Gamma, \Delta \vdash A \wedge B}$$

$$\Rightarrow L \quad \frac{\Gamma \vdash A \quad \Delta, B \vdash C}{\Gamma, \Delta, A \Rightarrow B \vdash C} \qquad \Rightarrow R \quad \frac{\Gamma, A \vdash B}{\Gamma \vdash A \Rightarrow B}$$

The fundamental theorem of Gentzen (Cut-Elimination) states that the cut rule in the above formalism is redundant. Gentzen's theorem is closely related to normalization of lambda terms ([33, 12, 31]). One way to see this is to translate from sequent calculus to natural deduction (cf. Section 3.3 below), then use the so-called Curry-Howard isomorphism [12] to associate lambda terms to natural deduction proofs. This amounts to interpreting the sequent calculus directly into the typed lambda calculus, which we now describe.

3.2. From Sequents to Lambda Terms. Although the Curry-Howard procedure of associating typed lambda terms to natural deduction proofs (cf [12]) is by now quite familiar, a similar process applied to Gentzen sequent calculus appears less so, despite the related work of Lambek connected to categorical coherence theorems [17, 18].

One motivation of this term assignment is to think of an intuitionist sequent $\Gamma \vdash B$ (where $\Gamma = \{A_1, \ldots, A_k\}$) as an input/output device, accepting k inputs of types $A_1, \ldots, A_k$ and returning an output of type B. To this end, recall that our language of typed lambda calculus has types freely generated from type variables and constants under the usual type-forming operations $\times$ and $\Rightarrow$. We inductively define the assignment of lambda terms to proofs of sequents, using the special notation $x_1 : A_1, \ldots, x_k : A_k \; t[x_1, \ldots x_k] : B$, or sometimes for brevity the vector notation $\vec{x} : \Gamma \; t[\vec{x}] : B$. In this assignment, we assume *the variables* x_i *to the left of* *are distinct*, even if some of the formulae A_i are equal; moreover, the free variables of the term $t[x_1, \ldots, x_k]$ are contained in $\{x_1, \ldots, x_k\}$. At the same time, we shall also give the informal conversion equations that these terms satisfy, when appropriate. Notation: we write $t[a_1/x_1, \ldots, a_k/x_k]$ to denote the simultaneous substitution of a_i for x_i in term t.

Axiom: $x : A \; x : A$

Cut:
$$\frac{\vec{x} : \Gamma \; t[\vec{x}] : A \qquad \vec{y} : \Delta, z : A \; f[\vec{y}, z] : B}{\vec{x} : \Gamma, \vec{y} : \Delta \; f[\vec{y}, t[\vec{x}]/z] : B}$$

Structural:

Exchange
$$\frac{\vec{x} : \Gamma, y : A, z : B, \vec{w} : \Delta \; t[\vec{x}, y, z, \vec{w}] : C}{\vec{x} : \Gamma, z : B, y : A, \vec{w} : \Delta \; t^e[\vec{x}, z, y, \vec{w}] : C}$$
where $t^e[\vec{x}, z, y, \vec{w}] = t[\vec{x}, y, z, \vec{w}]$

Contraction
$$\frac{\vec{x} : \Gamma, y : A, z : A \; t[\vec{x}, y, z] : C}{\vec{x} : \Gamma, y : A \; t^c[\vec{x}, y] : C}$$
where $t^c[\vec{x}, y] = t[\vec{x}, y, y/z]$.

Weakening
$$\frac{\vec{x} : \Gamma \; t[\vec{x}] : C}{\vec{x} : \Gamma, y : A \; t^w[\vec{x}, y] : C}$$
where $t^w[\vec{x}, y] = t[\vec{x}]$.

Logical:

$\wedge L_i$, $i=1,2$
$$\frac{\vec{x} : \Gamma, y_i : A_i \; t[\vec{x}, y_i] : C}{\vec{x} : \Gamma, z : A_1 \wedge A_2 \; t^\wedge[\vec{x}, z] : C}$$
where $t^\wedge[\vec{x}, z] = t[\vec{x}, \pi_i(z)/y_i]$

$\wedge R$
$$\frac{\vec{x} : \Gamma \; s[\vec{x}] : A \qquad \vec{y} : \Delta \; t[\vec{y}] : B}{\vec{x} : \Gamma, \vec{y} : \Delta \; < s[\vec{x}], t[\vec{y}] >: A \wedge B}$$

$\Rightarrow L$
$$\frac{\vec{x} : \Gamma \; f[\vec{x}] : A \qquad \vec{y} : \Delta, z : B \; g[\vec{y}, z] : C}{\vec{x} : \Gamma \; , \vec{y} : \Delta, u : A \Rightarrow B \; g[\vec{y}, uf[\vec{x}]/z] : C}$$
where, if $u : A \Rightarrow B$ and $a : A$ then $ua : B$ denotes application of u to argument a.

$\Rightarrow R$
$$\frac{\vec{x} : \Gamma, y : A \; t[\vec{x}, y] : B}{\vec{x} : \Gamma \; \lambda y : A.t[\vec{x}, y] : A \Rightarrow B}$$

The inductive clauses in the term assignment above are *not* the same as the inductive clauses used in term judgements and term assignments of ordinary typed lambda calculus; however, the rules above are derivable from the usual term-formation rules. A cut-free version of the above calculus is obtained by omitting the cut rule.

From the point of view of the typed lambda calculus, the cut rule is a general substitution process obtained by plugging one term into another. In fact, the cut-elimination theorem can be restated as:

> *All general instances of term substitution arising from the cut rule are already derivable (up to equality of terms) from the special instances of substitution used in the other rules.*

Cut elimination permits "asymmetrical" interpretations of sequent calculi ([12], p.34). More precisely, suppose we assign to occurrences of atomic predicates a *signature* $+1$ or -1 depending upon whether they appear positively (resp. negatively) in a given sequent. (Obviously, this is closely related to representing type expressions as multivariant functors $(\mathcal{C}^{\circ})^n \times \mathcal{C}^n \longrightarrow \mathcal{C}$, up to rearrangement of factors). The central point is that the rules of the sequent calculus, *except for the identity and cut*, preserve the signature. What about identity and cut? If we think of term assignments $\vec{x} : \Gamma\ t(\vec{x}) : B$ as the functional process $\vec{x} \mapsto t(\vec{x})$ assigning "proofs" to formulae, the identity sequent maps (proofs of) negative occurrences, those with signature -1, to (proofs of) positive occurrences of the same formula. Conversely, cut substitutes a proof term whose type has positive signature for one of the same type, of negative signature; viz. in the cut rule above, in $f[\vec{y}, z] : B$ the term $t(\vec{x}) : A$ where A has positive signature is substituted into the variable slot $z : A$, where A has negative signature. In the absence of cut, the symmetry of the situation is broken: we may obtain asymmetrical interpretations of the sequents [12].

Let us give an example to illustrate the power of the method. In the paper [9] one associates to any lambda term an infinite series $\sum a_n \vec{e_n}$ where the coefficients a_n are cardinal numbers. The task is to prove that any normalisable lambda term is *weakly finite*, i.e. all a_n in its series are integers. To prove the statement (since the coefficients are invariant modulo β-conversion) one restricts to normal lambda terms. One introduces the auxiliary notion of being *strongly finite* (i.e. weakly finite plus almost all a_n null), and one checks that

- If $t[x]$ is weakly finite for any strongly finite x, then $\lambda x.t[x]$ is weakly finite.
- If t is strongly finite and u is weakly finite, then tu is strongly finite.

From this one easily proves that, for t normal

- If all variables in t are interpreted by strongly finite series then t becomes weakly finite.
- Furthermore, when t does not start with a λ, it is indeed strongly finite.

The method breaks down on cut, i.e. on general lambda terms, for which we would need the wrong symmetrical statement: strongly finite = weakly finite.

It is precisely the problem of linking positive and negative slots in type expressions which prevents, in general, the composing of dinaturals. Kelly [13], in his abstract treatment of coherence in categories, discusses this very situation in detail for a general calculus of transformations, while special cases of the general problem were already resolved in Eilenberg & Kelly [5]. Cut elimination theorems were successfully applied to coherence questions by Lambek [15, 16, 17, 18] and Mints [22]. Of course, for the simple types of this paper, normalization or cut elimination poses no problems. But even for these types, we shall obtain more: cut elimination implies that the internal language supports a compositional dinatural interpretation (between definable functors). In the language of Kelly-Mac Lane (applied to the case of ccc's), we shall show that every constructible arrow is dinatural.

As we have seen in the previous discussion, asymmetrical interpretation is one of the major approaches to cut elimination, closely related to dinaturality. For instance, dinaturals are like cut-free proofs and their compositionality is problematic just as cut-elimination is a highly non-trivial statement. But in some ways, until we can fully understand the second order (polymorphic) case from this viewpoint, we must honestly admit we are still far from understanding the main features of these fascinating questions.

3.3. From Sequents to Natural Deduction. The lambda term assignment occurs in two stages: the first stage is the translation of sequent calculus to natural deduction (see [12]); the second stage is the Curry-Howard isomorphism. Let us describe the first part. To a cut-free proof of a sequent $\Gamma \vdash A$ in sequent calculus one associates a natural deduction proof in normal form

$$\begin{array}{c}[\Gamma]\\ \mathcal{D}\\ A\end{array}$$

where $[\Gamma]$ denotes a collection of assumption classes built from formulas in Γ. The translation is reasonably well-known in proof theory ([12, 26, 31, 33]); we highlight the main points.

(1) To the axiom $A \vdash A$ associate the deduction A.
(2) The Exchange Rule is the identity on deductions (because the assumptions are not considered ordered).

(3) The Weakening Rule $\dfrac{\Gamma \vdash B}{\Gamma, A \vdash B}$ is interpreted as follows. Given the interpretation of the premiss $\begin{matrix}[\Gamma]\\ \mathcal{D}\\ B\end{matrix}$ we obtain $\begin{matrix}[\Gamma],[A]'\\ \mathcal{D}'\\ B\end{matrix}$ where $[A]'$ is an assumption class formed by *zero* occurrences of A, and $\mathcal{D}'$ is $\mathcal{D}$ with $[\Gamma]$ replaced by $[\Gamma],[A]'$.

(4) Contraction translates as follows: assuming we have interpreted the hypotheses, the conclusion is interpreted by forming the union of the assumption classes corresponding to the contraction formula A.

As for the logical rules,

(5) $\wedge L_i$ $\dfrac{\Gamma, A_i \vdash C}{\Gamma, A_1 \wedge A_2 \vdash C}$ translates to

$$\begin{matrix} & \dfrac{[A_1 \wedge A_2]}{A_i} \\ [\Gamma] & \vdots \\ & C \end{matrix}$$

That is, there is an operation ("projection") which allows us to go from $A_1 \wedge A_2$-assumption classes to A_i-assumption classes, $i = 1, 2$.

(6) $\wedge R$ $\dfrac{\Gamma \vdash A \quad \Delta \vdash B}{\Gamma, \Delta \vdash A \wedge B}$ translates as

$$\dfrac{\begin{matrix}[\Gamma] & [\Delta]\\ \vdots & \vdots \\ A & B\end{matrix}}{A \wedge B}$$

. In this case, we form the union of the two assumption classes.

(7) $\Rightarrow L$ $\dfrac{\Gamma \vdash A \quad \Delta, B \vdash C}{\Gamma\ , \Delta, A \Rightarrow B \vdash C}$ translates as follows: suppose

$$\begin{matrix}[\Gamma]\\ \mathcal{D}\\ A\end{matrix} \quad \text{and} \quad \begin{matrix}[\Delta],[B]\\ \mathcal{D}'\\ C\end{matrix}$$

interpret the premisses. Then

$$\begin{matrix} & \begin{matrix}[\Gamma]\\ \mathcal{D}\end{matrix} & \\ [\Delta] & \dfrac{A \quad [A \Rightarrow B]}{B} \\ & \mathcal{D}' \\ & C \end{matrix}$$

interprets the conclusion, where in $\mathcal{D}'$ we replace each occurrence of B

in the distinguished assumption class $[B]$ by the indicated proof with assumption class $[A \Rightarrow B]$. Proliferation of occurrences of formulas in Γ are taken into account by making a *new* assumption class, consisting of all the indicated occurrences of Γ so duplicated.

(8) $\Rightarrow$R $\dfrac{\Gamma, A \vdash B}{\Gamma \vdash A \Rightarrow B}$ translates as $\dfrac{\begin{matrix}[\Gamma], [A] \\ \mathcal{D} \\ B\end{matrix}}{A \Rightarrow B}$ where all formula occurrences in the assumption class $[A]$ of $\mathcal{D}$ are simultaneously discharged.

One subtlety of the natural deduction calculus is that natural deduction trees are closed under very general substitutions. For example, the Cut Rule $\dfrac{\Gamma \vdash A \quad \Delta, A \vdash B}{\Gamma, \Delta \vdash B}$ translates as an operation taking proofs of the premisses

$\begin{matrix}[\Gamma] \\ \vdots \\ A\end{matrix}$ and $\begin{matrix}[\Delta], [A] \\ \vdots \\ B\end{matrix}$ and obtaining a proof of the conclusion $\begin{matrix}[\Gamma], [\Delta] \\ \vdots \\ B\end{matrix}$

by making (possibly many) copies of the tree $\begin{matrix}[\Gamma] \\ \vdots \\ A\end{matrix}$, and grafting a copy onto each occurrence of formula A in the given proof of B. But in fact, only very special substitutions of natural deduction trees are required in this translation. We note that left sequent calculus rules translate as certain basic computations on assumption classes induced by these special substitutions.

4. Normal Forms and Cut-Free Proofs

In this section we shall prove the main theorems mentioned above. We shall also observe:

- How the process of cut-elimination, when interpreted by dinaturals, unlocks the clashes (i.e. dependencies) between positive and negative occurrences of formulas.
- How the kind of actions on assumption classes that arise in translating left sequent calculus rules into natural deduction are closely related to Reynolds' relational interpretation of parametric polymorphism [27].

The first observation above permits separate computations in the negative and positive parts of sequents, is related to the asymmetrical interpretation of cut-free proofs ([12]), and to the problem of composing dinaturals.

We shall show that the cut-free Gentzen sequent rules for the $\wedge, \Rightarrow$-fragment of intuitionist logic are interpretable as composable syntactic dinatural transformations between definable functors, as mentioned in Section 2 above. It follows that these transformations form a category. The fact that we use cut-free proofs obviates the need to consider the general problem of compositionality of dinaturals [2]. That is, we start with the translation of sequent calculus into natural deduction, from Section 3 above, then follow by the so-called Curry-Howard isomorphism [12] to obtain the associated lambda terms.

More precisely, lambda terms in normal form (arising from a valid typing judgement) correspond to normal natural deduction proofs of the formula represented by the types. Such normal deductions arise from (not necessarily unique) cut-free Gentzen sequent proofs. Cut-free Gentzen proofs without assumptions translate to certain closed normal lambda terms; these latter terms yield syntactic families which will be shown to be dinatural. Hence we prove the main result by induction on cut-free Gentzen sequent proofs (which is a kind of induction on head normal form [12], p.19).

To this end, we assume we already know how to interpret the formulas in a sequent as definable functors $(\mathcal{C}^{\circ})^n \times \mathcal{C}^n \longrightarrow \mathcal{C}$ as earlier. We shall interpret a sequent $\Gamma \vdash B$ as a dinatural transformation $\prod_{i \leq k} F_i(\mathbf{AA}) \longrightarrow G\mathbf{AA}$, where Γ is a finite list of formulae A_i, $F_i = \|A_i\|$ and $G = \|B\|$, by associating appropriate families of lambda terms to proofs.

The axiom is trivial to interpret: it is the identity transformation between functors. This is clearly a definable family, and moreover is dinatural over any ccc. For the remainder of the rules, we assume as inductive hypothesis that *we already know how to interpret the premisses of a rule as dinatural transformations*. We show how to interpret the conclusion of each rule. We look at the important cases:

(1) Contraction: We ignore the passive context Γ (which only adds free variable parameters to all terms). Suppose $F = \|A\|$ and $G = \|C\|$ are definable functors. Suppose the family of terms $d = \{d_{\mathbf{A}} : (F \times F)\mathbf{AA} \longrightarrow G\mathbf{AA} | \mathbf{A} \in \mathcal{C}^n\}$ interprets the premisses of Contraction. Recall that $(F \times F)\mathbf{AA} = F\mathbf{AA} \times F\mathbf{AA}$. Consider the family $d \circ \Delta = \{d_{\mathbf{A}} \circ \Delta_{F\mathbf{AA}} : \mathbf{A} \in \mathcal{C}^n\}$, given by the following diagram:

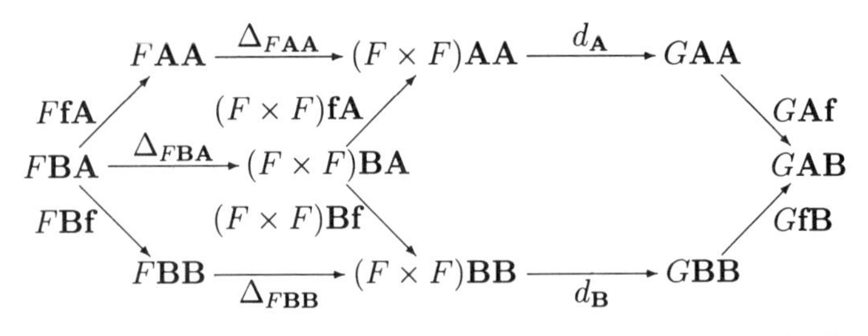

It is easily verified that the outer hexagon commutes. Thus the family $d \circ \Delta$ interprets the conclusion of contraction.

(2) Weakening: We may suppose without loss that the passive context Γ has only one formula. Suppose $F = \|\Gamma\|$, $G = \|A\|$ and $H = \|C\|$ are definable functors. Suppose the family of terms $d = \{d_{\mathbf{A}} : F\mathbf{AA} \longrightarrow H\mathbf{AA} | \mathbf{A} \in \mathcal{C}^n\}$ interprets the hypothesis of weakening. Consider the family $d \circ \Pi =$ $\{d_{\mathbf{A}} \circ \Pi_{F\mathbf{AA}} | \mathbf{A} \in \mathcal{C}^n\}$, given by the following diagram:

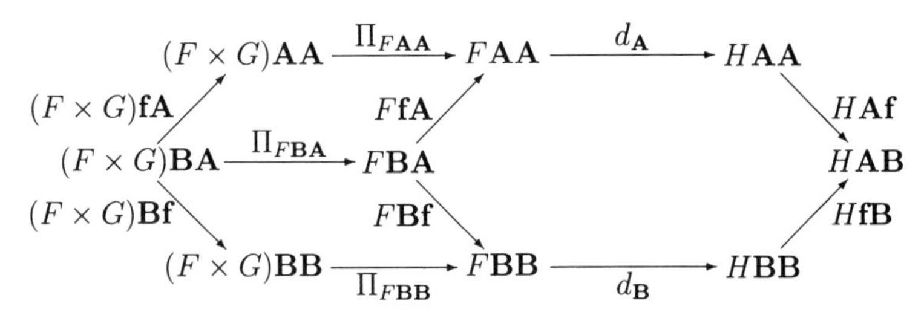

It is easily verified that the outer hexagon commutes. Thus the family $d \circ \Pi$ interprets the conclusion of weakening.

Although the left rules turn out to be more profound from the viewpoint of dinaturality, we shall first examine the right rules.

(3) $\wedge$R is interpreted essentially by taking the pointwise product of the two hexagons interpreting its premisses. That is, suppose (wolog) that Γ and Δ consist of only one formula, and are interpreted as definable functors $F = \|\Gamma\|$, $G = \|\Delta\|$, $H = \|C\|$, and $K = \|D\|$, resp. Suppose $s : F \longrightarrow H$ and $t : G \longrightarrow K$ interpret the premisses of $\wedge$R. Then the conclusion is interpreted by the family $d = \{d_{\mathbf{A}} \mid \mathbf{A} \in \mathcal{C}^n\} : F \times G \longrightarrow H \times K$, which is defined as follows: $d_{\mathbf{A}} = s_{\mathbf{A}} \times t_{\mathbf{A}} = \langle s_{\mathbf{A}} \circ \pi_1, t_{\mathbf{A}} \circ \pi_2 \rangle : (F \times G)\mathbf{AA} \longrightarrow (H \times K)\mathbf{AA}$. It is easily verified that the associated hexagon below commutes (where $\mathbf{f} : \mathbf{A} \longrightarrow \mathbf{B}$ and some arrows are omitted for clarity):

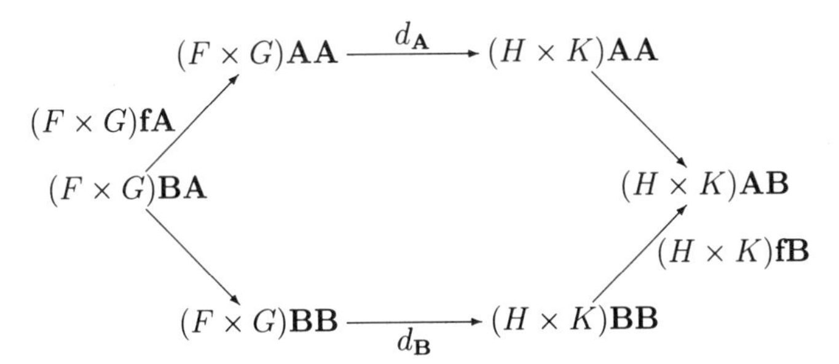

Note that each component of $d_{\mathbf{A}} = < s_{\mathbf{A}} \circ \pi_1, t_{\mathbf{A}} \circ \pi_2 >$ acts separately—there is no communication (cf. [11]).

(4) $\Rightarrow$R. Suppose $t : K \times F \longrightarrow G$ is a dinatural transformation interpreting the hypothesis of $\Rightarrow$R. Then the family $t^* = \{t^*_{\mathbf{A}} | t^*_{\mathbf{A}} : K\mathbf{AA} \longrightarrow G\mathbf{AA}^{F\mathbf{AA}}\}$ given by currying $t_{\mathbf{A}} : K\mathbf{AA} \times F\mathbf{AA} \longrightarrow G\mathbf{AA}$ is a dinatural transformation $t^* : K \longrightarrow G^F$. We leave the diagram and verification to the reader.

We now examine the left rules.

(5) $\wedge$L is interpreted almost identically to the case of Weakening above.

(6) $\Rightarrow$L. For ease of exposition, suppose the context Δ is empty. For notational convenience, we denote the functorial interpretations $\|\Gamma\|$, $\|C\|$,$\|D\|$, and $\|E\|$ by K,F, G, and H, resp. We are assuming dinatural transformations $w : K \longrightarrow F$ and $t : G \longrightarrow H$ which interpret the premisses. We wish to interpret the conclusion of $\Rightarrow$L ; i.e. (modulo exchange) we wish to find a dinatural transformation $d : G^F \times K \longrightarrow H$. We remind the reader that in functorial semantics, twisted exponentiation is defined by: $G^F\mathbf{BA} = G\mathbf{BA}^{F\mathbf{AB}}$. Behold the following diagram, where $\mathbf{f} : \mathbf{A} \longrightarrow \mathbf{B}$, $\theta_{\mathbf{A}} = 1 \times w_{\mathbf{A}}$ and $\theta_{\mathbf{B}} = 1 \times w_{\mathbf{B}}$.

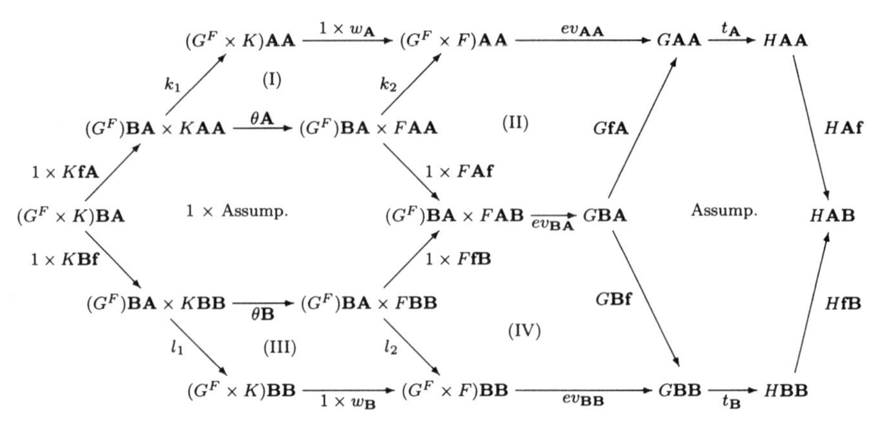

In this diagram, observe how the definition of twisted exponentials fits perfectly with the source and target of the middle horizontal evaluation map. Here, $k_1 = (G^F)\mathbf{fA} \times 1_{K\mathbf{AA}}$ and $k_2 = (G^F)\mathbf{fA} \times 1_{F\mathbf{AA}}$; Similarly for $l_i, i = 1, 2$ with appropriate $\mathbf{A}$ replaced by $\mathbf{B}$. The commutativity of (I) and (III) is trivial; the commutativity of (II) and (IV) comes from a direct calculation. The inductive assumption guarantees the commutativity of the indicated hexagons. Thus we obtain a dinatural transformation $G^F \times K \longrightarrow H$.

Note that in the figure above, (I) and (III) correspond to splitting the preparation for evaluation into " contravariant" and "covariant" channels, with the action separate in each part: there is no communication. This is precisely directed at the question of how positive and negative parts of formulas occur in cut-free proofs. Moreover, we emphasize again that in the case of the structural rules, as well as the left rules, the translation from Gentzen sequents to natural deduction translates to actions on assumption classes. These actions are given by special substitutions of lambda terms mentioned in Section **3** above. In our ccc context, these special substitutions are treated as precomposing by special families of morphisms, as in cases 1, 2 and the evaluation morphisms in case 6 above. The crucial point is that these special families satisfy strong naturality properties which have the shape

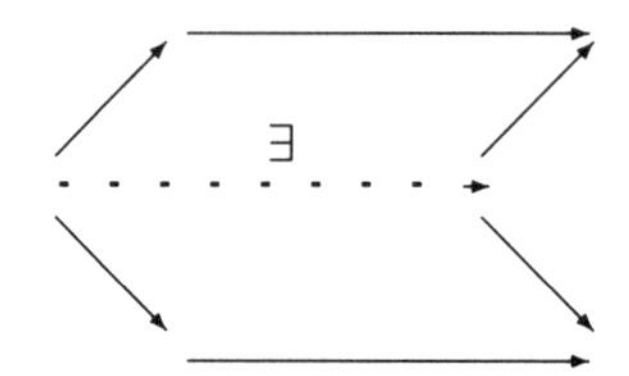

where the broken arrow indicates the existence of a factorization. Diagrams of this shape arise as Reynolds' invariance conditions in his approach to parametricity [27], [2], pp.53-55, as well as in Statman's theory of logical relations [30], and in Wadler's universal Horn conditions [32] which motivated this discussion. We emphasize, however, that the invariance conditions given by diagrams of this shape have deep roots in the left rules of Gentzen sequent calculus for propositional logic –prior to any considerations of second order type abstraction.

As a corollary, we obtain the following. Let $\mathcal{C}$ be the cartesian closed category generated by a simply typed lambda calculus $\mathcal{L}$ whose types are built inductively from some atomic types $\alpha, \beta, \ldots$, as in the Theorems of Section 2. Then

- Cut-free proofs are represented by closed terms in normal form, thus arrows in $\mathcal{C}$.
- Any arrow $m : A \longrightarrow B$ in $\mathcal{C}$ (qua closed normal form in $\mathcal{L}$ of type $A \Rightarrow B$) induces a dinatural transformation between definable functors $\|A\|$ and $\|B\|$. These normal forms compose by substitution.
- m provably satisfies the dinaturality equations.

We should remark that all this was done in the special case of intuitionist logic and cartesian closed categories. This analysis can be extended to the multiplicative fragment of linear logic (stemming from Barr's *-autonomous categories) and beyond.

Acknowledgments

A. Scedrov's research is supported by ONR contract N00014-88K0635, NSF Grant CCR-87-05596, and a Young Faculty Award of the Natural Sciences Association of the University of Pennsylvania. Scedrov would like to thank John C. Mitchell, the Computer Science Department, and the Center for the Study of Language and Information at Stanford University for their hospitality during his Sabbatical leave.

P.J. Scott's research is supported by an operating grant from the Natural Sciences and Engineering Research Council of Canada.

We thank Narciso Martí-Oliet and Robert Seely for helpful comments on a preliminary version of this work.

References

1. S. Abramsky, J. C. Mitchell, A. Scedrov, P. Wadler. Relators (to appear.)
2. E.S. Bainbridge, P. Freyd , A. Scedrov, and P. J. Scott. Functorial Polymorphism, *Theoretical Computer Science* **70** (1990), pp. 35–64.
3. E.S. Bainbridge, P. Freyd , A. Scedrov, and P. J. Scott. Functorial Polymorphism: Preliminary Report, *Logical Foundations of Functional Programming*, G. Huet, ed., Addison Wesley (1990), pp. 315–327.
4. M. Barr. **-Autonomous Categories*. Springer LNM 752, 1979.
5. S. Eilenberg and G. M Kelly. A generalization of the functorial calculus, *J. Algebra* **3** (1966), pp. 366–375.
6. A. Felty. A Logic Program for Transforming Sequent Proofs to Natural Deduction Proofs. In: *Proc. December 1989 Workshop on Extension of Logic Programming*, ed. by P. Schroeder-Heister, Springer LNCS, to appear.
7. P. Freyd. Structural Polymorphism, Manuscript, Univ. of Pennsylvania (1989), to appear in *Th. Comp. Science.*
8. P. Freyd, J-Y. Girard, A. Scedrov, and P. J. Scott. Semantic Parametricity in Polymorphic Lambda Calculus, *Proc. 3rd IEEE Symposium on Logic in Computer Science*, Edinburgh, 1988.
9. J-Y. Girard. Normal functors, power series, and λ-calculus, *Ann. Pure and Applied Logic* **37** (1986) pp. 129–177.
10. J-Y. Girard. The System **F** of Variables Types, Fifteen Years Later, *Theoretical Computer Science*, **45**, pp. 159–192.
11. J-Y. Girard. Linear Logic, *Theoretical Computer Science*, **50**, 1987, pp. 1–102.
12. J-Y. Girard, Y.Lafont, and P. Taylor. *Proofs and Types*, Cambridge Tracts in Theoretical Computer Science, **7**, Cambridge University Press, 1989.
13. G. M. Kelly. Many-variable functorial calculus.I, *Coherence in Categories* Springer LNM **281**, pp. 66–105.
14. G. M. Kelly and S. Mac Lane. Coherence in closed categories, *J. Pure Appl. Alg.* **1** (1971), pp. 97–140.
15. J. Lambek. Deductive Systems and Categories I, *J. Math. Systems Theory* **2** (1968), pp. 278–318.
16. J. Lambek. Deductive Systems and Categories II, *Springer LNM* **86** (1969), pp. 76–122.
17. J. Lambek. Multicategories Revisited, *Contemp. Math.* **92** (1989), pp. 217–239.
18. J. Lambek. Logic without structural rules, Manuscript. (McGill University), 1990.
19. J.Lambek and P. J. Scott. *Introduction to Higher Order Categorical Logic*, Cambridge Studies in Advanced Mathematics **7**, Cambridge University Press, 1986.
20. S. Mac Lane. *Categories for the Working Mathematician*, Springer Graduate Texts in Mathematics, Springer-Verlag, 1971.
21. S. Mac Lane. Why commutative diagrams coincide with equivalent proofs, *Contemp. Math.* **13** (1982), pp. 387–401.
22. G. E. Mints. Closed categories and the theory of proofs, *J. Soviet Math* **15** (1981), pp. 45–62.
23. J. C. Mitchell and P. J. Scott. Typed Lambda Models and Cartesian Closed Categories, *Contemp. Math.* **92**, pp. 301–316.
24. J.C. Mitchell and R. Harper, The Essence of ML, *Proc. 15th Annual ACM SIGACT-SIGPLAN Symposium on Principles of Programming Languages (POPL)*, San Diego, 1988, pp. 28–46.
25. A. Pitts. Polymorphism is set-theoretic, constructively, in: *Category Theory and Computer Science*, Springer LNCS 283 (D. H. Pitt, ed.) (1987) pp. 12–39.
26. D. Prawitz. *Natural Deduction*, Almquist & Wiksell, Stockhom, 1965.
27. J.C.Reynolds. Types, Abstraction, and Parametric Polymorphism, in: *Information*

Processing '83 , R. E. A. Mason, ed. North-Holland, 1983, pp. 513–523.
28. R. A. G. Seely. Categorical Semantics for Higher Order Polymorphic Lambda Calculus. *J. Symb. Logic* **52**(1987), pp. 969–989.
29. R. A. G. Seely. Linear Logic, *-Autonomous Categories, And Cofree Coalgebras. *Contemp. Math.* **92**(1989), pp. 371–382.
30. R. Statman. Logical Relations and the Typed Lambda Calculus. *Inf. and Control* **65**(1985), pp. 85–97.
31. A. S. Troelstra and D. van Dalen. *Constructivism in Mathematics*, Vols. I and II. North-Holland, 1988.
32. P. Wadler. Theorems for Free! *4th International Symposium on Functional Programming Languages and Computer Architecture*, Assn. Comp. Machinery, London, Sept. 1989.
33. J. Zucker. The Correspondence between Cut-Elimination and Normalization, *Ann. Math. Logic* **7** (1974) pp. 1–112.

JEAN-YVES GIRARD: ÉQUIPE DE LOGIQUE, UA 753 DU CNRS, MATHÉMATIQUES, T. 45-55, UNIV. DE PARIS 7, 2 PLACE JUSSIEU, 75251 PARIS CEDEX 05, FRANCE
Internet: GIRARD@MARGAUX.INRIA.FR

ANDRE SCEDROV: *On Sabbatical (1989-90)*: DEPT. OF COMPUTER SCIENCE, STANFORD UNIVERSITY, STANFORD, CA 94305-2140, USA
Permanent Address: DEPT. OF MATHEMATICS, UNIV. OF PENNSYLVANIA, 209 S. 33RD ST., PHILADELPHIA, PA 19104-6395, USA
Arpanet: ANDRE@CIS.UPENN.EDU

PHILIP J. SCOTT: DEPT. OF MATHEMATICS, UNIV. OF OTTAWA, 585 KING EDWARD, OTTAWA, ONT., CANADA K1N6N5
Bitnet: SCPSG@UOTTAWA.BITNET *Internet*: SCPSG@ACADVM1.UOTTAWA.CA

COMPUTER IMPLEMENTATION AND APPLICATIONS OF KLEENE'S S-M-N AND RECURSION THEOREMS

NEIL D. JONES

ABSTRACT. This overview paper describes new program constructions which at the same time satisfy Kleene's theorems and are efficient enough for practical use on the computer. The presentation is mathematically oriented, and along the way we mention some unsolved problems whose precise formulations and solutions might be of mathematical interest.

We identify a "programming language" with an acceptable enumeration of the recursive functions, and a program with an index. In computing terms, Kleene's s-m-n theorem says that programs can be *specialized* with respect to partially known arguments, and the second recursion theorem says that programs may without loss of computability be allowed to reference their own texts.

The theorems' classical proofs are constructive and so programmable—but in the case of the s-m-n theorem, the specialized programs are typically a bit slower than the original, and in the case of the recursion theorem, the programs constructed in the standard proofs are extremely inefficient. These results were thus of no computational interest until new methods were recently developed [12, 1], two of which are described here.

An important application: it was realized in [9, 21, 8] that one can, in principle, compile from one programming language **S** into another, **L**, by using the s-m-n theorem. Compiling is done by applying the s-m-n theorem to specialize an interpreter for **S** written in **L** to a fixed **S**-program. Further, compiler generation can be done by self-application: specializing an s-1-1 program to a fixed interpreter. A further use of self-application yields a compiler generator: a program that transforms interpreters into compilers.

1. INTRODUCTION

In this overview paper we describe some recent Computer Science research results closely related to two classical theorems from recursive function theory. The first concerns the s-m-n theorem and its now well-established applications to compiling and compiler generation [12]. In Computer Science a program with the s-m-n property is called a *partial evaluator*, perhaps better named a *program specializer*. Given a program and fixed values for some but not all of its inputs, a partial evaluator constructs a new program which, when applied to the values of the remaining inputs, yields the same result the original program would have given on all its inputs.

In the mid 1970's, Futamura in Japan and Turchin and Ershov in the Soviet Union [9, 21, 8] independently discovered the possibility of using partial evaluation for automatic compilation and for compiler generation as well, starting with a language definition in the form of an interpreter. This understanding was, however, only in principle since specialization as done in the classical proofs of the s-m-n theorem is of no computational interest due to its inefficiency. The first computer realization efficient enough to be of practical interest is reported in [12]; these and several subsequent papers have established the utility and conceptual simplicity of this approach.

We also describe a relatively efficient implementation of Kleene's second recursion theorem [1]. While this theorem has many applications in mathematics, it is not yet clear whether it will be as constructively useful in computer science as the s-m-n theorem has turned out to be.

1.1. The Problem of Compiler Generation. A compiler is a meaning preserving function from "source" programs (expressed in a programming language suitable for humans) into "target" programs (in a language suitable for computer execution). Correctness is of paramount importance: the compiler-generated code must faithfully realize the intended meaning of the source program being compiled. In practice, however, compiler construction is an arduous task, often involving many man-years of work, and it can be quite difficult to ensure correctness.

Program meanings can be defined precisely via denotational or operational semantics. A language's operational semantics can be given in the form of a program called an *interpreter*. In recursion theoretical terms, an interpreter for programming language **S** written in language **L** is an **L**-program computing a universal function for **S**; and a *compiler* from **S** to **L** is a meaning-preserving function from **S**-programs to **L**-programs (precise definitions appear later).

The compiler correctness problem naturally led to the goal of *semantics-directed compiler generation*: from a formal programming language definition automatically to derive a correct and efficient compiler for the defined language.

Once a correct compiler generator has been developed, every generated compiler will be faithful to the language definition from which it was derived. Such a system completely obviates the need for the difficult intellectual work involved in proving individual compilers correct [10, 16]. Ideally its role in the semantic part of practical compiler construction would be similar to that of YACC and other parser generators for syntax analysis.

Industrial-strength semantics-directed compiler generators do not yet exist, but noteworthy progress has been achieved in the past few years. Several systems based on the lambda calculus have been developed ([17, 18, 22], of which [22] is the fastest that has been used on large language definitions). However, all these systems are large and complex, and correctness proofs would be very hard to accomplish.

1.2. Partial Evaluation. A partial evaluator deserves its name because, when given a program and incomplete input, it will do part of the evaluation the program would do on complete input. A partial evaluator may be fruitfully thought of as a *program specializer*: given a program and the values of part of its input data, it yields another program which, given its remaining input, computes the same value the original program gives on all its input. In other words, a partial evaluator is an efficient computer realization of Kleene's s-m-n theorem [14].

Consider, for example, the following exponentiation program, written as a recursion equation:

```
p(n,x) =  if n=0      then 1 else
          if even(n) then p(n/2,x)**2 else x*p(n-1,x)
```

The s-m-n theorem is classically proved in a rather trivial way. For example if $\texttt{n} = 5$, a specialized version of the program above could be a system of two equations:

```
p5(x)  =  p(5,x)
p(n,x) =  if n=0      then 1 else
          if even(n) then p(n/2,x)**2 else x*p(n-1,x)
```

A better specialized exponentiation program for $\texttt{n} = 5$ can be got by unfolding applications of the function `p` and doing the computations involving `n`, yielding the residual program:

```
p5(x)  =  x*(x**2)**2
```

The traditional proofs of Kleene's s-m-n theorem do not take efficiency into account and so yield the equivalent of the first specialized program. Efficiency is very important in applications though, so partial evaluation may be regarded as the quest for efficient implementations of the s-m-n theorem.

1.3. Partial Evaluation and Compiling. We will see that compilation can be done by partial evaluation, and that compilers and even a compiler generator can be constructed by self-applying a partial evaluator, using an interpreter as input. These possibilities were foreseen in principle in [9, 21, 8], and first realized in practice as described in [12].

By this approach the partial evaluator and the language definition are the only programs involved in compiler generation. The resulting systems are much smaller than the above-mentioned semantics implementation systems. A side effect is that correctness is much easier to establish, e.g. see [11] for a complete correctness proof of a self-applicable partial evaluator for the untyped lambda calculus.

An interpreter for a programming language **S** usually has two inputs: a "source program" p to be executed, and its input data. Compiling from **S** is done by specializing the given **S**-interpreter with respect to p. The resulting "target program" is an optimized version of the interpreter, specialized so that it is always always applied to the same **S**-program p.

An *efficient* target program is obtained by performing at compile time all the interpreter's actions which depend only on source program p.

Compiler generation is more complex in that it involves *self-application.* One can generate a compiler by using the partial evaluator to specialize itself with respect to an interpreter as fixed input. Further, specializing the partial evaluator with respect to itself as fixed input yields a compiler generator: a program that transforms interpreters into compilers. These mind-boggling but practically useful applications of partial evaluation are explained more fully in section 2.

1.4. The Recursion Theorem. The constructions in the standard proofs of Kleene's recursion theorem, and Rogers' variant as well, were also programmed and found (as expected) to be far too inefficient for practical use. A new programming language was designed and implemented in which Kleene and Rogers "fixed-point" programs can be expressed elegantly and much more efficiently. Several examples have been programmed in an as yet incomplete attempt to find out for which sort of problems the recursion theorems are useful program generating tools.

1.5. Outline. In section 2 we review the definition of partial evaluation and state the "Futamura projections", which show in principle how partial evaluation can be used for compiling and compiler generation. In section 3 a concrete programming language is introduced and a self-interpreter for it is sketched. Section 4 begins with program running times and the computational overhead caused by interpretation. Efficiency is discussed in general terms, and we point out the desirability of a complexity theory for partial evaluation, analogous to but in some important respects different from traditional computational complexity theory. Some open problems are presented that appear susceptible to formulation and solution in such a new complexity theory. Finally, section 5 concerns efficient implementation of the Second Recursion Theorem.

2. Preliminaries

In both partial evaluation and the recursion theorem programs are treated as data objects, so it is natural to draw both programs and their data a single universal domain D. The traditional usage of natural numbers in recursive function theory is simple, abstract and elegant, but involves a high computational price: all program structures and nonnumeric data must be encoded by means of Gödel numbers, and operations must be done on encoded values.

Thus for practical reasons we diverge from the traditional recursion-theoretic use of the natural numbers. The implementations [12, 1] use a small Lisp-like language—especially suitable since Lisp programs *are* data values so programs and data have the same form. Ths avoids completely the need for the tedious coding of programs as Gödel numbers so familiar from number-based recursive function theory. A substantial advantage is that algorithms for the s-m-n and universal functions become much simpler and more efficient. For a concrete example, see section 3.2.

Our choice of D is thus the set of Lisp "S-expressions", defined as the smallest set satisfying:

- Any "atom" is in D, where an atom is a sequence of one or more letters, digits or characters, excluding (,) and blank.
- If $d_1, \dots, d_n \in D$ for n ≥ 0, then $(d_1 \dots d_n) \in D$.

2.1. Programming Languages. A *programming language* ϕ is a function $\phi : D \to (D \to D)$ that associates with each element $p \in D$ a partial function $\phi(p) : D \to D$. In accordance with standard notation in recursion theory, $\phi(p)$ will often be written φ_p. The n-ary function $\varphi_p^n(x_1, \dots, x_n)$ is defined to be $\lambda(x_1, \dots, x_n).\varphi_p((x_1 \dots x_n))$. The superscript n will generally be omitted.

Intuitively, a programming language is identified with its "semantic function" ϕ, so $\phi(p)$ is the partial input-output function denoted by program p. *All* elements p of D are regarded as programs; ill-formed programs can for example be mapped to the everywhere undefined function.

The following definition, originating from [19], captures properties sufficient for a development of (most of) the theory of computability independent of any particular model of computation.

Definition 2.1. The programming language ϕ is an *acceptable programming system* if it has the following properties:

(1) *Completeness property*: for any effectively computable function, $\psi : D \to D$ there exists a program $p \in D$ such that $\varphi_p = \psi$.
(2) *Universal function property*: there is a *universal function program* $up \in D$ such that for any program $p \in D$, $\varphi_{up}(p, x) = \varphi_p(x)$ for all $x \in D$.
(3) *S-m-n function property*: for any natural numbers m, n there exists a total recursive function $s_n^m : D^{m+1} \to D$ such that for any program $p \in D$ and any input $(x_1, \ldots, x_m, y_1, \ldots, y_n) \in D$

$$\varphi_p^{m+n}(x_1, \ldots, x_m, y_1, \ldots, y_n) = \varphi^n_{s_n^m(p, x_1, \ldots, x_m)}(y_1, \ldots, y_n)$$

s_n^m is called the *s-m-n function.*

The properties listed above actually correspond to quite familiar programming concepts. The completeness property states that the language is as strong as any other computing formalism. The universal function property amounts to the existence of a *self-* or *meta-circular* interpreter for the language which, when applied to the text of a program and its input, computes the same value that the program computes on the same input.

The s-1-1 property ensures the possibility of partial evaluation. Suppose, for example, that program $p \in D$ takes two inputs. When given its first input $d_1 \in D$, program p can be specialised to yield a so-called residual program $p' = s_1^1(p, d_1)$. When given the remaining input d_2, p' yields the same result as p applied to both inputs. This seemingly innocent property has become increasingly more important. Practical exploitation, however, requires non-trivial partial evaluators.

The s-1-1 definition is essentially the so-called *mix equation* central to our earlier articles. By completeness the function s_1^1 must be computed by some program, and such a program is traditionally called *mix* in the partial evaluation literature. Following are definitions of compilers, interpreters and a version of the *mix* equation.

Notational Variations. References [12, 11] and others write the semantic function in linear notation (to avoid deeply nested subscripts), associate function application from the left, and use as few parentheses as possible. For example, $(\phi(p))(d)$ would be written as $\phi\, p\, d$. Languages are usually denoted by roman letters, the ones we will use being **L** (the default "meta-" or "implementation" language), **S** (a "source language") to be interpreted or compiled, and **T** (a "target language"), the output of a compiler or partial evaluator.

2.2. Compilers, Interpreters and Partial Evaluators

Definition 2.2. Let **L**, **S** and **T** be programming languages.
Program *int* is an *interpreter for* ***S*** *written in* ***L*** if for all s, d ∈ D

$$\mathbf{S}\ s\ d = \mathbf{L}\ int\ (s,\ d)$$

Program *comp* is an ***S****-to-*___T___*-compiler written in* ***L*** if for all $s, d \in D$

$$\mathbf{S}\ s\ d = \mathbf{T}\ (\boldsymbol{L}\ comp\ s)\ d$$

Program *mix* is an ***S****-to-*___T___*-partial evaluator written in* ***L*** if for all $p, d_1, d_2 \in D$

$$\mathbf{L}\ p\ (d_1, d_2) = \mathbf{L}\ (\boldsymbol{L}\ mix\ (p, d_1))\ d_2$$

We will only use one-language partial evaluators, for which **L** = **S** = **T**. The universal program *up* is clearly an interpreter for **L** written in **L**; and *mix* is a program to compute s_1^1 from the previous definition.

2.3. The Futamura Projections for Compiling and Compiler Generation. Suppose we are given an interpreter *int* for some language **S**, written in **L**. Letting *source* be an **S**-program, a compiler from **S** to **L** will produce an **L**-program *target* such that $\mathbf{S}(source) : D \to D$ and $\mathbf{L}(target) : D \to D$ are the same input-output function.

The following three equations are the so-called "Futamura projections" [9, 8]. They assert that given a partial evaluator *mix* and an interpreter *int*, one may compile programs, generate compilers and and even generate a compiler generator.

$$\begin{array}{lcll} \mathbf{L}\ mix\ (int,\ source) & = & target & \\ \mathbf{L}\ mix\ (mix,\ int) & = & compiler & \\ \mathbf{L}\ mix\ (mix,\ mix) & = & cogen & \text{a compiler generator} \end{array}$$

Explanation. Program *source* from the *interpreted* language **S** has been translated to program *target*. This is expressed in the language **L** in which the interpreter is written (natural since *target* is a specialized version of **L**-program *int*). It is easy to verify that the target program is faithful to its source using the definitions of interpreters, compilers and the mix equation:

$$\begin{array}{lcl} output & = & \mathbf{S}\ source\ input \\ & = & \mathbf{L}\ int\ (source,\ input) \\ & = & \mathbf{L}\ (\boldsymbol{L}\ mix\ (int,\ source))\ input \\ & = & \mathbf{L}\ target\ input \end{array}$$

Verification that program *compiler* translates source programs into equivalent target programs is also straightforward:

$$\begin{array}{lcl} target & = & \mathbf{L}\ mix\ (int,\ source) \\ & = & \mathbf{L}\ (\boldsymbol{L}\ mix\ (mix,\ int))\ source \\ & = & \mathbf{L}\ compiler\ source \end{array}$$

Finally, we can see that *cogen* transforms interpreters into compilers by the following:

$$\begin{aligned} compiler &= \mathbf{L}\ mix\ (mix,\ int) \\ &= \mathbf{L}\ (\mathbf{L}\ mix\ (mix,\ mix))\ int \\ &= \mathbf{L}\ cogen\ int \end{aligned}$$

See [12] for a more detailed discussion.

2.4. Efficiency in Practice. A variety of partial evaluators satisfying all the above equations have been constructed ([12, 4, 13] contain more detailed discussions). Compilation, compiler generation and compiler generator generation can each be done in two ways, e.g.

$$target = \mathbf{L}\ mix\ (int,\ source) = \mathbf{L}\ compiler\ source$$

and similarly for generation of *compiler* and *cogen*. Although the exact timings vary according to the partial evaluator and the implementation language **L**, in our experience the second way is often about 10 times faster than the first (for all three cases). A less machine dependent and more intrinsic efficiency measure will be seen in section 4.

3. A Concrete Programming Language

We now describe the Lisp-like language used in our experiments with the s-m-n and recursion theorems, discuss program efficiency and outline a universal program.

3.1. The Language Mixwell. In Mixwell both programs and data are S-expressions, i.e. elements of D. A Mixwell program is the representation of a system of recursive equations as an element of D, with syntax

```
((f1 (x1 x2 ... xn) = expression1)
 (f2 (y1 y2 ... yp) = expression2)
  ...
((fm (z1 z2 ... zq) = expressionm))
```

Here `expression1`, `expression2`, etc. are constructed from variables (e.g. `y2`) and constants of form `(quote d)` where d is an element of D (for brevity, `(quote d)` is written as `'d` in examples). The operations allowed in expressions include application of base functions (arithmetic and comparison operations, constructors and destructors for D, etc.), conditionals `(if-then-else)` and calls to the user-defined functions `f1`, `f2`,.... As usual in Lisp a function call has form `(function-name argument...argument)`. For an example, consider a program to compute the function x^n (different from the one given before).

```
((p(n,x) = (if (zero? n) then '1 else
          (if (even? n) then (square (p (divide n '2) x))
           else
                                    (times x (p (subtract1 n) x)))))
```

Program semantics is as usual for statically scoped Lisp. Details are beyond the scope of this paper, but briefly: the program's meaning is the meaning of user-defined function `f1`; and call by value is used for all function calls (i.e. arguments of a call to `fi` are evaluated before the expression that defines `fi` is evaluated). It is easy to show that Mixwell is an acceptable programming system [1].

3.2. Outline of a Self-interpreter. The universal program *up* of Definition 2.1 is easy to program in Mixwell. Figure 1 contains a sketch of the natural self-interpreter:

```
 ((up (program data) =
     (eval (4th (1st program))        ; right side of the first
                                             equation
            (2nd (first program))     ; its list of formal
                                             parameter names
            (list data)               ; list of formal parameter
                                             values
            program)                  ; the program (used in
                                             function calls)
  )
  (eval (exp parnames values prog) =
     (if (quote? exp)
                 then (2nd exp) else
     (if (variable? exp)
                 then (lookupvalue exp parnames values) else
     (if (subtract1? exp)
                 then (subtract1 (eval (2nd exp) parnames
                            values program))
                 else
     (if ...     then ...
                 else (quote SYNTAX-ERROR) )))) )
  (lookupvalue (exp parnames values prog) =  ...) )
```

Figure 1: A Universal Program for Mixwell

Explanation. Here we have used base function `1st` to select element d_1

from the list $(d_1 d_2 \dots d_n)$, `2nd` to select element d_2, etc. (all can be expressed via Lisp's two primitives `car, cdr`). The interpreter's central function is `eval`, which is given as arguments: an expression to evaluate; the names of the parameters to the function currently being evaluated; the values of those parameters; and the text of the program being executed.

`eval` works by determining which of the allowable forms the expression has, and then taking the appropriate evaluation actions. If it has form `(quote d)`, then its value is `d`—the second component of `exp`, computed by `(2nd exp)`. If a variable, `eval` calls function `lookupvalue` to locate the variable in list `parnames`, and to extract the corresponding value from list `values`. If of form `(subtract1 e)`, then `eval` calls itself recursively to get the value of `e`, and the subtracts 1 from the result; and analogous computations are done for all the other cases.

The main function `up` simply calls `eval` with the appropriate arguments—the right side of the first equation, the name of its single argument, that argument's value `data`, and the whole program `prog`.

4. Techniques and Efficiency of Partial Evaluation

From the Futamura projections it is not at all clear how efficient we can expect *mix*-produced specialized programs to be, and it is also unclear how to measure the quality of specialized programs. For practical purposes, even a small speedup can be profitable if a program is to be run often. Further, it can be faster to compute even a single value $f(d_1, d_2)$ in two steps: first, specialize f's algorithm to d_1; and then run the specialized program on d_2. (A familiar example: compiling a Lisp program and then running the result is often faster than running Lisp interpretively.)

Since increased efficiency is the prime motivation for using partial evaluation on the computer, we now discuss techniques for gaining efficiency, and motivate the development of a more abstract and general understanding of efficiency.

4.1. Goals of Partial Evaluation. Let p be a program with two input parameters. Then $\mathbf{L}\ p\ (d_1, d_2)$ denotes the value gotten by running p on (d_1, d_2). If only d_1 is available, evaluation of $\mathbf{L}\ p\ (d_1, \_)$ does not make sense, as the result is likely to depend on d_2. However, d_1 might be used to perform *some* of the computations in p, yielding as result a transformed, optimized version of p.

The goal of a practical partial evaluator is thus: to analyze its subject program in order to find out which of its calculations may be performed on basis of the incomplete input data d_1; to perform them; and to construct a specialized program containing only those computations essentially dependent on d_2.

4.2. Some Techniques for Partial Evaluation. Successful self-applicable partial evaluators include [12], [5], [4] and [13]. The techniques used there include: applying base functions to known data; unfolding function calls; and creating versions of program functions which are specialized to data values computable from the known input d_1.

To illustrate these techniques, consider the well-known example of Ackermann's function (using the same informal syntax as in the introductory section):

```
a(m,n)  =  if m=0 then n+1 else
           if n=0 then a(m-1,1)
                  else a(m-1,a(m,n-1))
```

Computing `a(2,n)` involves recursive evaluations of `a(m,n')` for `m` $=$ 0, 1 and 2, and various values of `n'`. The partial evaluator can evaluate `m=0` and `m-1` for the needed values of `m`, and function calls of form `a(m-1,...)` can be unfolded (i.e. replaced by the right side of the recursive equation above, after the appropriate substitutions).

We can now specialize function `a` to the values of `m`, yielding the residual program:

```
a2(n)  =  if n=0 then 3 else a1(a2(n-1))
a1(n)  =  if n=0 then 2 else a1(n-1)+1
```

This program performs less than half as many arithmetic operations as the original since all tests on and computations involving `m` have been removed. The example is admittedly pointless for practical use due to the enormous growth rate of Ackermann's function, but it illustrates some simple and important optimization techniques.

4.3. Program Running Times. A reasonable approximation to program running times on the computer can be obtained by counting 1 for each of the following: constant reference, variable reference, test in a conditional or case, function parameter, and base or user-defined function call. Thus the exponentiation example has time estimate:

$$t(n,x) = \begin{cases} 4 & \text{if } n = 0\text{, else} \\ 14 + t(n/2, x) & \text{if } n \text{ even, else} \\ 14 + t(n-1, x) & \end{cases}$$

which can be shown to be of order $\log n$.

4.4. The Computational Overhead Caused by Interpretation. On the computer, interpreted programs are often observed to run slower than compiled ones. We now explain why, using the previously described self-interpreter as an example.

Let $t_p(d)$ denote the time required to calculate **L** p d. As seen in section 3.2 the basic cycle of the self-interpreter *up* is first syntax analysis: a series of tests to determine the main operator of the current expression to be evaluated; then evaluation of necessary subexpressions by recursive calls to `eval`; and finally, actions to perform the main operator, e.g. to subtract 1 or to look up the current value of a variable. It is straightforward to see that the running time of *up* on list $(p\,d)$ satisfies

$$at_p(d) \leq t_{up}((p\,d))$$

for all d, where a is a constant. In our experiments a is often around 10 for small source programs. (In this context "constant" means: a is independent of d, although it may depend on p.)

4.5. Efficiency of Partial Evaluation, Interpretation and Compilation. Program speedup is not a universal phenomenon, and one can only expect $s_1^1(p, d_1)$ to be significantly faster when p's control and subcomputations are largely determined by d_1. For example, specializing the exponentiation algorithm to its second argument would not decrease its running time. On the other hand for an interpreter, d_1 is the program being interpreted, and its structure strongly directs the computation.

The speedup of the previous section is typical in our experience: an interpreted program runs slower than one which is compiled (or executed directly, which amounts to being interpreted by hardware); and the difference is a linear factor, large enough to be worth reducing for practical reasons, and depending on the size of the program being interpreted. Further, clever use of data structures (hash tables, binary trees, etc) can make a grow slowly as a function of p's size.

A Goal for Compilation by Partial Evaluation. For practical purposes the trivial partial evaluation given by the traditional s-m-n construction (e.g. as illustrated in the introduction) is uninteresting; in effect it would yield a target program of the form "apply the interpreter to the source program text and its input data". Ideally, *mix* should remove *all computational overhead* caused by interpretation.

On the other hand, how can we meaningfully assert that a partial evaluator is "good enough"? Perhaps surprisingly, a machine-independent answer can be given. This answer involves the mix equation and the self-interpreter *up* seen earlier. For any program p

$$\begin{aligned} \mathbf{L}\ p\ d &= \mathbf{L}\ up\ (p,\ d) \\ &= \mathbf{L}\ (\mathbf{L}\ mix\ (up,\ p))\ d \end{aligned}$$

so $p' = \mathbf{L}\ mix\ (up,\ p)$ is an **L**-program equivalent to p. This suggests a natural goal: that p' be at least as efficient as p. Achieving this goal implies that *all computational overhead* caused by *up*'s interpretation has been removed by *mix*.

Definition 4.1. Mix is *optimal* on *up* provided $t_{p'}(d) \leq a + t_p(d)$ for some constant a and all $p, d \in D$, where $p' = \mathbf{L}\ mix\ (up,\ p)$.

We have satisfied this criterion on the computer for several partial evaluators for various languages, using self-interpreters such as the one sketched in section 3.2. For some of these p' is identical to p (up to variable renaming).

Discussion. *Mix* is a general specializer, applying uniform methods to all its program inputs. In practice the input program must be written in a straightforward manner, so *mix* can succeed in analyzing the way the control and computations of its program input p depend on the available data d_1. A special case: the universal program *up* must be "cleanly" written for *mix* to be optimal.

It can occur that $p' = \mathbf{L}\ mix\ (up,\ p)$ is *more* efficient than p, e.g. if p is written with redundant computations. This is the exception rather than the rule, though—it is unreasonable to expect speedup on all programs.

Question: How can this be precisely formulated and proven?

Unfortunately, one can "cheat" the optimality criterion above. For example, consider a *mix* algorithm that begins by testing its program input to see whether it is the specific universal program *up* from section 2.3. If so, then this version of *mix* just outputs its second argument without change; if not, it realizes a trivial partial evaluation. Such a partial evaluator would satisfy the letter of the above definition of optimality but not its spirit. (Disclaimer: our optimal *mix*es don't cheat.)

Open Problem: Find a better definition of optimality that rules out "cheating".

4.6. Desirable: A Complexity Theory for Partial Evaluation. Partial evaluation is chiefly concerned with *linear* time speedups—and the now well-developed theory of computational complexity [7] traditionally ignores linear factors (perhaps due to the abundance of unsolved open problems even bout larger increases in computing time). In any case, the open problems of complexity theory concern lower complexity bounds on specific problems, and these are especially difficult since a lower bound is less than or equal to the complexity of the best of *all possible* algorithms that solve the problem, regardless of the techniques employed.

On the other hand the s_1^1 function used for partial evaluation is a *uniform* program specializer that works on any program at all, and so is not problem-specific. Further, significant efficiency increases (and even optimality as defined above) can be accomplished with a rather limited set of program transformations. This gives some hope for a complexity theory of partial evaluation, since uniformity implies that the specializer is not expected to be sensitive to the perhaps mathematically sophisticated computation strategies used by the program being specialized.

We now give a simple question that we hope could be formulated precisely and solved, given a suitable complexity theory for partial evaluation. As before, let $t_p(d)$ be the time to compute $\mathbf{L}\ p\ d$, and let $p_{d_1} = \mathbf{L}\ mix\ p\, d_1$ be the result of specializing program p to known input d_1.

For a trivial partial evaluator as in the standard s-1-1 construction, one would expect

$$t_{p_{d_1}}(d_2) = t_p((d_1\, d_2)) + b$$

for constant $b \geq 0$, where b is the "set-up time" required to initialize p's first argument to d_1.

In our experiments (chiefly on interpreters, parsers and string matchers), p_{d_1} often runs substantially faster than p by factors ranging from 3 to 50, and speedups of over 200 have been reported in the literature.

Definition 4.2. We say that *mix* accomplishes *linear speedup* on p if for all $d_1 \in D$ there is a function $f(n) = an$ with $a > 1$ such that for all $d_2 \in D$

$$f(t_{p_{d_1}}(d_2)) \leq t_p((d_1\, d_2))$$

To clarify this definition, consider two examples. First, let p be an interpreter. We saw in the previous section that partial evaluation can sometimes eliminate all interpretation overhead, e.g. when applied to a self-interpreter. *A finer analysis*: the interpretation overhead involves two factors. One is independent of p and represents the time for the syntax analysis part of the "interpretation loop" together with the time to perform elementary operations. The other depends on p and is a (usually) slowly growing function representing the time required to fetch and/or store variables' values. Thus the exact speedup factor a will depend on the value of the program being interpreted, but the speedup is still linear by the definition above.

For a second example, let p be a string matching program that searches for an occurrence of pattern string d_1 in subject string d_2. By a naive algorithm this takes time $t_p((d_1\, d_2)) = O(mn)$ where m, n are the lengths of d_1 and d_2, respectively. The well-known Knuth-Morris-Pratt string matching algorithm yields an algorithm p_{d_1} that runs in time $O(n)$. In this case the matching time $t_{p_{d_1}}(d_2)$ is independent of d_1. A linear speedup with factor a as large as desired is thus obtainable by choosing d_1 to make m large enough.

Consel and Danvy have shown in [6] that partial evaluation, given a simple matching program p and pattern d_1, yields an algorithm that runs in time $O(n)$. Thus the speed of a sophisticated string matcher can be obtained by partially evaluating a fairly naive program, avoiding the need for the insight shown by Knuth, Morris and Pratt.

Open Question. If *mix* uses only the techniques given earlier in this section and preserves termination properties, do there exist programs p on which *mix* accomplishes superlinear speedups? *Discussion*: in our experience, speedups of interpreters and other programs are always linear in the above sense. Clearly, superlinear speedups can be obtained by changing p's computational algorithm, or by discarding unnecessary computations. On the other hand, algorithm changes would seem to involve nonuniform transformations; and discarding computations can change a nonterminating program into a terminating one. It thus remains unclear whether uniform techniques can make superlinear optimizations.

5. Efficient Realization of the Recursion Theorem

In programming terms, Kleene's "second recursion theorem" [14] says that programs may without loss of computability be allowed to reference their own texts. This powerful theorem has been used frequently in recursive function theory and in "machine-independent" computational complexity theory ([19, 15, 3]). Although its content is constructive, the theorem seems most often used negatively—as a tool for proving various hierarchies nontrivial by showing the existence of "pathological" functions and sets having great complexity or unexpected properties.

This section deals with a practical and, as it turns out, reasonably efficient implementation of the recursion theorem. We give a brief review of the fundamentals, then discuss our implementation of the theorem and report on two experiments with it; more may be seen in [1].

Theorem 5.1 (The second recursion theorem). *For any program $p \in D$ there is a program $e \in D$ such that $\varphi_p(e, x) = \varphi_e(x)$. We call such an e a* Kleene fixed-point *for p.*

Proof. By the s-m-n property for any program, $p \in D$

$$\varphi_p(y, x) = \varphi_{s_1^1(p,y)}(x)$$

It is evidently possible to construct a program $q \in D$ such that

$$\varphi_q(y, x) = \varphi_p(s_1^1(y, y), x)$$

Let e be the program $s_1^1(q, q)$. Then we have

$$\varphi_p(e, x) = \varphi_p(s_1^1(q, q), x) = \varphi_q(q, x) = \varphi_{s_1^1(q,q)}(x) = \varphi_e(x)$$

□

5.1. Some Applications. Following are some typical results which are easily proven using the recursion theorem, and which would seem rather difficult without it. All are language independent since they hold for any acceptable programming system.

Self-Reproduction. Let $r \in D$ be a program with $\varphi_r(p, x) = p$. According to theorem 5.1, there is a program e such that for all x

$$\varphi_e(x) = \varphi_r(e, x) = e$$

Thus e is a program that outputs its own text regardless of its input.

Eliminating Recursion. Consider the computable function

$$f(p, x) = [if\; x = 0\; then\; 1\; else\; x * \varphi_{up}(p, x - 1)]$$

By theorem 5.1, f has a fixed-point e with the property

$$\begin{gathered} \varphi_e(x) = f(e, x) = \\ [if\; x = 0\; then\; 1\; else\; x * \varphi_{up}(e, x - 1)] = \\ [if\; x = 0\; then\; 1\; else\; x * \varphi_e(x - 1)] \end{gathered}$$

Thus φ_e, which was found without explicit use of recursion, is the factorial function. It follows that any acceptable programming system is "closed under recursion".

Until now these are more or less the only ways the theorem has been used constructively to solve computationally interesting problems.

Unsolvability. The second recursion theorem can be used to give a very elegant proof (see [15] or [19]) of

Theorem 5.2 (Rice's theorem). *Let P be the set of partial recursive functions and let C be a proper, nonempty subset of P. Then it is undecidable whether, for a given program $p \in D$, φ_p belongs to C or not.*

Abstract Complexity Theory. The second recursion theorem has interesting (alas, also negative) applications in so-called abstract complexity theory, cf. Blum, [3]. In the following ψ is an *abstract resource measure*, $\psi_p(d)$ giving the resources used when running program p on input d, satisfying natural requirements stated in [3]. (Running time is one example, and memory usage is another.)

Theorem 5.3 (The speed-up theorem). *Let r be a total recursive function of 2 variables. Then there exists a total recursive function f taking values 0 and 1 with the property that to every program p computing f there exists another program q computing f such that*

$$r(n, \psi_q(n)) < \psi_p(n)$$

for almost all n.

A program computing f can be found constructively from one for r, but going from program p for f to the "faster" program q is a non-constructive operation. The theorem is therefore of limited interest for practitioners.

5.2. A Classical Implemention of the Recursion Theorem. Following the recipe implied by the following classical proof, we wrote a Mixwell program that when given another program as input, computes its Kleene fixed-point.

The method works on the computer but has some drawbacks. First, the fixed-point programs are rather large, since they contain both the whole input program text plus code (for the s-m-n function etc.) that in effect allows the fixed-point program to reference its own text. Second, applications often involve interpretation. i.e. applying the universal function. Thus *up* must be present in the fixed-point program, so a direct implementation quickly becomes inefficient since every nested call to *up* results in an extra level of interpretation.

For example running a fixed-point program to compute $n!$ as in the example given above results in n levels of interpretation. The run time is exponential in n, since each level slows down execution by a large constant factor (around 10 in the example above). Lesson: a too direct implementation is not practical.

5.3. The Reflect Language. To overcome these efficiency problems we devised a new language containing the essential features for expressing fixed-points directly. This language, Reflect, is just Mixwell extended with two new features:

(1) A built-in universal function (`univ` p i) that takes as arguments a program $p \in D$ and some input $i \in D$, and returns $\varphi_p(i)$.
(2) A special expression `*`, the value of which is the whole text of the program being executed.

Implementation. Each (`univ ...`) expression gives rise to a call of the same interpreter that is used to implement Reflect, and thus adds no extra levels of interpretation. The special expression `*` is quite naturally available, e.g. in section 3.2 it is the variable `prog`.

Reflect indeed has the power to express fixed-points, by the construction shown below. Its correctness proof (given in [1]) relies on a near-denotational semantics of the language, which is an almost trivial extension of the truly denotational semantics of Mixwell. It is an easy exercise to prove

Lemma 5.4. *Reflect is an acceptable programming system.*

Theorem 5.5. *Given a Reflect program p with the following structure:*

```
((p1 (f x) = expression1)
 (p2 (...) = ...  )
  ... )
```

The following Reflect program e is a Kleene fixed-point for p:

```
((fix (x) = (univ '((p1 (f x) = expression1)
                    (p2 (...) = ...  )
                     ... )
                   (list * x) )))
```

This avoids the nesting of interpretation levels (and the corresponding multiplication of running times) that occurred when applying the traditional Kleene construction. The reason is that by the way Reflect is implemented, applications of the universal function within p cause only a limited amount of extra overhead. Further, the time to run the program above on x is just a constant plus the time to run p on e and x.

Experiments with the Reflect System. Example 1. The very essence of the second recursion theorem is the ability for a program to reference its own text. Consequently, self-reproducing programs are very easily expressed in Reflect:

```
((selfreproduce (n) = *))
```

Running the program produces exactly the text shown above.

Example 2. The effect of recursion can also be obtained nonrecursively in Reflect:

```
((fact (n) = (if (= n '1)
               then '1 else (times n (univ * (list
                                 (difference n '1)))))) )
```

On the computer, the time to compute $n!$ grows linearly in n, so a major improvement has occurred over the direct implementation. The same holds if $n!$ is computed as in Theorem 5.5.

Remarks and Open Questions.

We designed a language Reflect that allows fixed-point programs to be expressed naturally and executed with reasonable efficiency, giving a framework for experiments with constructive applications of the second recursion theorem.

Is it cheating to define Reflect as a language with built-in self-reference? In our opinion the answer is no, for two reasons. First, Mixwell programs run just as fast under the Reflect interpreter as before; one would expect a slowdown if we had done something computationally unrealistic. (The reason is that the Reflect interpreter is simply a small extension of that for Mixwell.) Second, we can compute $n!$ and other reflexively defined functions *much* more efficiently in Reflect than by the classical construction from Kleene's theorem (either directly, as in the example above, or by the general construction of Theorem 5.5).

The most pressing open problem in this direction is to widen the spectrum of computationally interesting applications. Further, it seems likely that the second recursion theorem implies the existence of an efficiently implementable extensible programming language, but this possibility has not yet been realized in practice.

Acknowledgments

This paper owes much to the co-authors of [12] and [1], and to discussions and constructive criticism by the DIKU group, including Lars Ole Andersen, Anders Bondorf, Carsten Gomard and Torben Mogensen.

REFERENCES

1. Torben Amtoft, Thomas Nikolajsen, Jesper Larsson Träff, Neil D. Jones: Experiments with Implementations of two Theoretical Constructions. Logic at Botik, Lecture Notes in Computer Science,Springer-Verlag, vol. 363, pp. 119-133, 1989.
2. A. Appel: Semantics-Directed Code Generation. 12th ACM Symposium on Principles of Programming Languages, pp. 315-324,1985.
3. Manuel Blum: A Machine-Independent Theory of the Complexity of Recursive Functions. Journal of the Association for Computing Machinery, vol. 14, no. 2, April 1967, pp. 322-336.
4. Anders Bondorf: Automatic Autoprojection of Higher Order Recursive Equations. CAAP/ESOP1990 Proceedings, LNCS 432, Springer-Verlag, 1990.
5. A. Bondorf, O. Danvy: Automatic Autoprojection of Recursive Equations with Global Variables and Abstract Data Types. DIKU report 90-4, University of Copenhagen,1990.
6. C. Consel, O. Danvy: Partial Evaluation of Pattern Matching in Strings. Information Processing Letters, Vol. 30, No 2, pp 79-86, January 1989.
7. S. Cook: An Overview of Computational Complexity. Communications of the ACM 26, pp. 401-408,1983.
8. A. P. Ershov: Mixed Computation: Potential applications and problems for study. Theoretical Computer Science 18, pp. 41-67, 1982.
9. Y. Futamura: Partial Evaluation of Computation Process—an Approach to a Compiler-compiler. Systems, Computers, Controls, 2(5), pp. 45-50, 1971.
10. G. Goguen, J. W. Thatcher, E. G. Wagner, J. B. Wright: Initial Algebra Semantics and Continuous Algebras. Journal of the ACM vol. 24, pp. 68-95, 1977.
11. C. K. Gomard: Higher Order Partial Evaluation: H.O.P.E. for the Lambda Calculus. M. S. thesis, DIKU, University of Copenhagen, 1989.
12. Neil D. Jones, Peter Sestoft, Harald Søndergaard: MIX: A Self-applicable partial Evaluator for Experiments in Compiler Generation (Revised Version). Lisp and Symbolic Computation, issue 2, fall 1988.
13. N. D. Jones, C. K. Gomard, A. Bondorf, O. Danvy, T. Mogensen: A Self-applicable Partial Evaluator for the Lambda Calculus. IEEE Computer. Society 1990 International Conference on Computer Languages, pp. 49-58,1990.
14. Stephen Cole Kleene: Introduction to Metamathematics. North-Holland, 1952.
15. A. I. Mal'cev: Algorithms and recursive functions. McGraw-Hill, 1967.
16. F. L. Morris: Advice on Structuring Compilers and proving them correct. 1st ACM Symposium on Principles of Programming Languages, pp. 144-152, 1973.
17. P. Mosses: SIS—Semantics Implementation System, Reference Manual and User Guide. DAIMI Report MD-30, University of Aarhus, Denmark, 1979.
18. L. Paulson: A Semantics-Directed Compiler Generator. 9th ACM Symposium on Principles of Programming Languages, pp. 224-233, 1982. Wolters-Noordhoff Publishing, 1970.
19. Hartley Rogers: Theory of Recursive Functions and Effective Computability. McGraw-Hill, 1967.
20. D. Schmidt: Denotational Semantics: a Methodology for Language Development. Allyn and Bacon, 1986
21. V. F. Turchin: The Concept of a Supercompiler. ACM Transactions on Programming Languages and Systems, 8(3), pp. 292-325, 1986.

22. P. Weis: Le Systeme SAM: Metacompilation tres efficace a l'aide d'operateurs semantiques. Universite de Paris 7, 1987.

DIKU, DEPARTMENT OF COMPUTER SCIENCE, UNIVERSITY OF COPENHAGEN. UNIVERSITETSPARKEN 1, DK-2100 COPENHAGEN Ø, DENMARK
E-MAIL: NEIL@DIKU.DK

0-1 LAWS FOR FRAGMENTS OF SECOND-ORDER LOGIC: AN OVERVIEW

PHOKION G. KOLAITIS AND MOSHE Y. VARDI

ABSTRACT. The probability of a property on the collection of all finite relational structures is the limit as $n \to \infty$ of the fraction of structures with n elements satisfying the property, provided the limit exists. It is known that the 0-1 law holds for any property expressible in first-order logic, i.e., the probability of any such property exists and is either 0 or 1. Moreover, the associated decision problem for the probabilities is solvable.

We investigate here fragments of existential second-order logic in which we restrict the patterns of first-order quantifiers. We focus on fragments in which the first-order part belongs to a prefix class. We show that the classifications of prefix classes of first-order logic with equality according to the solvability of the finite satisfiability problem and according to the 0-1 law for the corresponding Σ_1^1 fragment are identical.

1. INTRODUCTION

In recent years a considerable amount of research activity has been devoted to the study of the model theory of finite structures. This theory has interesting applications to several other areas including database theory [CH82, Va82] and complexity theory [Aj83, Gu84, Im83, Im86]. One particular direction of research has focused on the asymptotic probabilities of properties expressible in different languages and the associated decision problem for the values of the probabilities [Co88].

In general, if C is a class of finite structures over some vocabulary and if P is a property of some structures in C, then *the asymptotic probability* $\mu(P)$ on C is the limit as $n \to \infty$ of the fraction of the structures in C with n elements which satisfy P, provided that the limit exists. We say that P *is almost surely true* on C in case $\mu(P)$ is equal to 1. Combinatorialists have studied extensively the asymptotic probabilities of interesting properties on the class G of all finite graphs. It is, for example, well known and easy to prove that μ(connectivity)=1, while μ(l-colorabilty)=0, for any l [Bo85]. A theorem of Pósa [Po76] implies that μ(Hamiltonicity)=1.

Glebskii et al. [GKLT69] and independently Fagin [Fa76] were the first to establish a fascinating connection between logical definability and asymptotic probabilities. More specifically, they showed that if C is the class of all finite structures over some relational vocabulary and if P is any property expressible in first-order logic (with equality), then $\mu(P)$ exists and is either 0 or 1. This result is known as *the 0-1 law for first-order logic*. The proof of the 0-1 law also implies that the decision problem for the value of the probabilities of first-order sentences is solvable. This should be contrasted with Trakhtenbrot's [Tr50] classical theorem to the effect that the set of first-order sentences which are true on *all* finite relational structures is unsolvable, assuming that the vocabulary contains at least one binary relation symbol.

It is well known that first-order logic has very limited expressive power on finite structures (cf. [AU79]). For this reason, one may want to investigate asymptotic probabilities for higher-order logics. Unfortunately, it is easy to see that the 0-1 law fails for second-order logic; for example, parity (i.e., the property that the domain contains an even number of elements) is definable by an existential second-order sentence. Moreover, the 0-1 laws fails even for existential monadic second-order logic [KS85, Ka87]. In view of this result, it is natural to ask: are there fragments of second-order logic for which a 0-1 law holds?

We note that the problem of identifying the second-order sentences that express properties with well-defined asymptotic probabilities is unsolvable. To see that, let $\varphi(\mathbf{S})$ be a first-order sentence without equality over the vocabulary $\mathbf{S}$ and let ψ be an existential second-order sentence over the empty vocabulary expressing the parity property. Then $\mu(\psi \vee \exists\mathbf{S}\varphi(\mathbf{S}))$ is well-defined (and equals 1) if and only if $\varphi(\mathbf{S})$ is finitely satisfiable. Finite satisfiability, however, is known to be unsolvable [Tr50]. Thus, we focus on syntactically defined fragments of second-order logic.

The simplest and most natural fragments of second-order logic are formed by considering second-order sentences with only existential second-order quantifiers or with only universal second-order quantifiers. These are the well known classes of Σ_1^1 and Π_1^1 sentences respectively. Fagin [Fa74] proved that a property is Σ_1^1 definable if and only if it is NP-computable. As we observed, the 0-1 law fails for Σ_1^1 in general (and consequently for Π_1^1 as well). Moreover, it is not hard to show that the Σ_1^1 sentences having probability 1 form an unsolvable set.

In view of these facts, we concentrate on fragments of Σ_1^1 sentences in which we restrict the pattern of the first-order quantifiers that occur in the sentence. If $\mathcal{F}$ is a class of first-order sentences, then we denote by $\Sigma_1^1(\mathcal{F})$ the class of all Σ_1^1 sentences whose first-order part is in $\mathcal{F}$. Two remarks are in order now. First, if $\mathcal{F}$ is the class of all $\exists^*\forall^*\exists^*$ first-order sentences (that is to say, first-order sentences whose quantifier prefix consists of a string of existential quantifiers, followed by a string of universal quantifiers, followed by a string of existential quantifiers), then $\Sigma_1^1(\mathcal{F})$ has the same expressive power as the full Σ_1^1. In other words, every Σ_1^1 formula is equivalent to one of the form

$$\exists \mathbf{S} \exists \mathbf{x} \forall \mathbf{y} \exists \mathbf{z} \theta(\mathbf{S}, \mathbf{x}, \mathbf{y}, \mathbf{z}),$$

where θ is a quantifier-free formula, $\mathbf{S}$ is a sequence of second-order relation variables and $\mathbf{x}, \mathbf{y}, \mathbf{z}$ are sequences of first-order variables (*Skolem normal form*). Second, if $\varphi(\mathbf{S})$ is a first-order sentence without equality over the vocabulary $\mathbf{S}$, then $\mu(\exists \mathbf{S} \varphi(\mathbf{S})) = 1$ if and only if $\varphi(\mathbf{S})$ is finitely satisfiable. Thus, for any first-order class $\mathcal{F}$ the decision problem for $\Sigma_1^1(\mathcal{F})$ sentences having probability 1 is at least as hard as the finite satisfiability problem for sentences in $\mathcal{F}$. The latter problem is known to be unsolvable even in the case where $\mathcal{F}$ is the class of $\exists^*\forall^*\exists^*$ sentences (cf. [Le79]). As a result, in order to pursue positive solvability results one has to consider fragments $\Sigma_1^1(\mathcal{F})$, where $\mathcal{F}$ is a class for which the finite satisfiability problem is solvable. Such classes $\mathcal{F}$ of first-order sentences are said to be *docile* [DG79].

There are exactly two docile *prefix classes*, i.e., classes of first-order sentences (with equality) defined by their quantifier prefix [DG79, Le79, Go84]:

- The *Bernays-Schönfinkel class*, which is the collection of all first-order sentences with prefixes of the form $\exists^*\forall^*$ (i.e., the existential quantifiers precede the universal quantifiers).
- The *Ackermann class*, which is the collection of all first-order sentences with prefixes of the form $\exists^*\forall\exists^*$ (i.e., the prefix contains a single universal quantifier).

These two classes are also the only prefix classes that have a solvable satisfiability problem [Go84, DG79, Le79].

We focus here on the question whether the 0-1 law holds for the Σ_1^1 fragments defined by first-order (with equality) prefix classes, and whether or not the associated decision problem for the probabilities is solvable. This can be viewed as a classification of the prefix classes according to whether the corresponding Σ_1^1 fragments have a 0-1 law. What we show is that the classifications of prefix classes according to the solvability of the finite satisfiability problem and according to the 0-1 law for the corresponding Σ_1^1 fragment are identical. Moreover, 0-1 laws in this classification are always accompanied by solvability of the decision problem for the probabilities. This is manifested by the positive results for the classes Σ_1^1(Bernays-Schönfinkel) and Σ_1^1(Ackermann), and the negative results for the other classes.

This paper is a survey that focuses on the overall picture rather than on technical details. The interested reader is referred to [KV87, KV90, PS89] for further details.

2. Random Structures

Let $\mathbf{R}$ be a vocabulary consisting of relation symbols *only* and let C be the collection of all finite relational structures over $\mathbf{R}$ whose universes are initial segments $\{1, 2, \ldots, n\}$ of the integers. If P is a property of (some) structures in C, then let $\mu_n(P)$ be the fraction of structures in C of cardinality n satisfying P. The *asymptotic probabilty* $\mu(P)$ on C is defined to be

$$\mu(P) = \lim_{n\to\infty} \mu_n(P),$$

provided this limit exists. In this probability space all structures in C with the same number of elements carry the same probability. An equivalent description of this space can be obtained by assigning truth values to tuples independently and with the same probability (cf. [Bo85]).

If L is a logic, we say that the 0-1 *law holds for* L *on* C in case $\mu(P)$ exists and is equal to 0 or 1 for every property P expressible in the logic L. We write $\Theta(L)$ for the collection of all sentences P in L with $\mu(P) = 1$. Notice that if L is first-order logic, then the existence of the 0-1 law is equivalent to stating that $\Theta(L)$ is a complete theory.

A standard method for establishing 0-1 laws, originating in Fagin [Fa76], is to prove that the following *transfer theorem* holds: there is an infinite structure $\mathbf{A}$ over the vocabulary $\mathbf{R}$ such that for any property P expressible in L we have:

$$\mathbf{A} \models P \Longleftrightarrow \mu(P) = 1.$$

It turns out that there is a unique (up to isomorphism) countable structure $\mathbf{A}$ that satisfies the above equivalence for first-order logic and for the fragments of second-order logic considered here. We call $\mathbf{A}$ the *countable random structure over the vocabulary* $\mathbf{R}$. The structure $\mathbf{A}$ is characterized by an infinite set of *extension axioms*, which, intuitively, assert that every *type* can be *extended* to any other possible type. More precisely, if $\mathbf{x} = (x_1, \ldots, x_n)$ is a sequence of variables, then a n-$\mathbf{R}$-*type* $t(\mathbf{x})$ *in the variables* $\mathbf{x}$ *over* $\mathbf{R}$ is a maximal consistent set of equality and negated equality formulas and atomic and negated atomic formulas from the vocabulary $\mathbf{R}$ in the variables $x_1, \ldots, x_n$. We say that a $(n+1)$-$\mathbf{R}$-*type* $s(\mathbf{x}, z)$ *extends the type* $t(\mathbf{x})$ if t is a subset of s. Every type $t(\mathbf{x})$ can be also viewed as a quantifier-free formula that is the conjunction of all members of $t(\mathbf{x})$. With each pair of types s and t such that s extends t we associate a first-order *extension axiom* τ which states that

$$(\forall \mathbf{x})(t(\mathbf{x}) \rightarrow (\exists z)s(\mathbf{x}, z)).$$

Let T be the set of all extension axioms. The theory T was studied by Gaifman [Ga64], who showed, using a *back and forth* argument, that any two countable models of T are isomorphic (*T is an ω-categorical theory*). The extension axioms can also be used to show that the unique (up to isomorphism) countable model $\mathbf{A}$ of T is *universal* for all countable structures over $\mathbf{R}$, i.e., if $\mathbf{B}$ is a countable structure over $\mathbf{R}$, then there is a substructure of $\mathbf{A}$ that is isomorhic to $\mathbf{B}$.

Fagin [Fa76] realized that the extension axioms are relevant to the study of probabilities on finite structures and proved that on the class C of all finite structures over a vocabulary $\mathbf{R}$

$$\mu(\tau) = 1$$

for any extension axiom τ. The 0-1 law for first-order logic and the transfer theorem between truth of first-order sentences on $\mathbf{A}$ and almost sure truth of such sentences on C follows from these results by a compactness argument. We should point out that there are different proofs of the 0-1 law for first-order logic, which have a more elementary character (cf. [GKLT69, Co88]). These proofs do not deploy infinite structures or the compactness theorem and they bypass the transfer theorem. In contrast, the proofs of the 0-1 laws for fragments of second-order logic that we present here seem to involve infinitistic methods in an essential way. In particular, we have no clue of how to obtain these 0-1 laws without establishing corresponding transfer theorems.

Since the set T of extension axioms is recursive, it also follows that $\Theta(L)$ is recursive, where L is first-order logic. In other words, there is an algorithm to decide the value (0 or 1) of the asymptotic probability of any first-order sentence. The computational complexity of this decision problem was investigated by Grandjean [Gr83], who showed that it is PSPACE-complete, when the underlying vocabulary $\mathbf{R}$ is assumed to be bounded (i.e., there is a some bound on the arity of the relation symbols in σ).

3. Existential and Universal Second-Order Sentences

The Σ_1^1 and Π_1^1 formulas form the syntactically simplest fragment of second-order logic. A Σ_1^1 formula over a vocabulary $\mathbf{R}$ is an expression of the form $(\exists \mathbf{S})\theta(\mathbf{S})$, where $\mathbf{S}$ is a sequence of relation symbols not in the vocabulary $\mathbf{R}$ and $\theta(\mathbf{S})$ is a first-order formula over the vocabulary $\mathbf{R} \cup \mathbf{S}$. A Π_1^1 formula is an expression of the form $(\forall \mathbf{S})\theta(\mathbf{S})$, where $\mathbf{S}$ and $\theta(\mathbf{S})$ are as above.

Both the 0-1 law and the transfer theorem fail for arbitrary Σ_1^1 and Π_1^1 sentences. Consider, for example, the statement "there is relation that is the graph of a permutation in which every element is of order 2". On finite structures this statement is true exactly when the universe of the structure has an even number of elements and, as a result, it has no asymptotic probability. This statement, however, is expressible by a Σ_1^1 sentence, which, moreover, is true on the countable random structure **A**. Similarly, the statement "there is a total order with no maximum element" is true on the countable random structure **A**, but is false on every finite structure. Notice that in the two preceding examples the transfer theorem for Σ_1^1 sentences fails in the direction from truth on the countable random structure **A** to almost sure truth on finite structures. In contrast, the following simple lemma shows that this direction of the transfer theorem holds for all Π_1^1 sentences.

Lemma 3.1. *Let* **A** *be the countable random structure over* **R** *and let* $(\forall \mathbf{S})\theta(\mathbf{S})$ *be an arbitrary* Π_1^1 *sentence. If* $\mathbf{A} \models (\forall \mathbf{S})\theta(\mathbf{S})$, *then there is a first-order sentence* ψ *over the vocabulary* σ *such that:*

$$\mu(\psi) = 1$$

and

$$\models \psi \to (\forall \mathbf{S})\theta(\mathbf{S}).$$

In particular, every Π_1^1 *sentence that is true on* **A** *has probability 1 on* C.

Proof Assume that the hypothesis of the lemma holds, i.e., $\mathbf{A} \models (\forall \mathbf{S})\theta(\mathbf{S})$. We claim that there is a finite subset T' of the set T of extension axioms such that

$$\models (\bigwedge T') \to (\forall \mathbf{S})\theta(\mathbf{S}).$$

Suppose to the contrary. Let

$$T^* = T \cup \{\neg\theta(\mathbf{S})\}.$$

By assumption the set $T' \cup \{\neg\theta(\mathbf{S})\}$ is satisfiable (over the expanded vocabulary $\mathbf{R}^* = \mathbf{R} \cup \{\mathbf{S}\}$) for any finite subset T' of T. By the compactness and the Löwenheim-Skolem theorems, the set T^* has a countable model **B** over $\mathbf{R}^*$. The reduct of **B** to the vocabulary **R** is a countable model of T and therefore it must be isomorphic to **A**, because T is an ω-categorical theory. But then $\mathbf{A} \models (\exists \mathbf{S})\neg\theta(\mathbf{S})$, where the symbols in **S** are interpreted by the additional predicates in **B**. □

Corollary 3.2. *Every* Σ_1^1 *sentence that is false on the countable random structure* **A** *has probability 0 on* C.

Corollary 3.3. *The set of* Π_1^1 *sentences that are true on* **A** *is recursively enumerable.*

We investigate here classes of Σ_1^1 and Π_1^1 sentences that are obtained by restricting appropriately the pattern of the first-order quantifiers in such sentences. If $\mathcal{F}$ is a class of first-order formulas, then we write $\Sigma_1^1(\mathcal{F})$ for the collection of all Σ_1^1 sentences whose first-order part is in $\mathcal{F}$.

The discussion in the introduction suggests that we consider prefix classes $\mathcal{F}$ that are docile, i.e., they have a solvable finite satisfiability problem. Thus, we focus on the following classes of existential second-order sentences:

- The class $\Sigma_1^1(\exists^*\forall^*)$ of Σ_1^1 sentences whose first-order part is a *Bernays-Schönfinkel* formula.
- The class $\Sigma_1^1(\exists^*\forall\exists^*)$ of Σ_1^1 sentences whose first-order is an *Ackermann* formula.

We also refer to the above as Σ_1^1(Bernays-Schönfinkel) class and Σ_1^1(Ackermann) class, respectively. Notice that the equality symbol $=$ is allowed in formulas in these classes.

Fagin [Fa74] showed that a class of finite structures over a vocabulary **R** is NP computable if and only if it is definable by a Σ_1^1 sentence over **R**. The restricted classes of Σ_1^1 sentences introduced above can not express all NP problems on finite structures. In spite of their syntactic simplicity, however, the classes $\Sigma_1^1(\exists^*\forall^*)$ and $\Sigma_1^1(\exists^*\forall\exists^*)$ can express natural NP-complete problems.

3.1. The Expressive Power of Σ_1^1(Bernays-Schönfinkel). If $k \geq 2$ is a fixed integer, then *k-COLORABILITY* is the problem:

- Given a finite graph $G = (V, E)$, can we assign one of k colors to every node of G in such a way that no two adjacent nodes have the same color?

It is easy to see that for each fixed $k \geq 2$ there is a $\Sigma_1^1(\forall\forall)$ sentence that expresses *k-COLORABILITY*. Moreover, it is well known that *k-COLORABILITY* is an NP-complete for each $k \geq 3$.

MONOCHROMATIC TRIANGLE is another example of an NP-complete problem on graphs that is expressible by a $\Sigma_1^1(\exists^*\forall^*)$ sentence, actually by a $\Sigma_1^1(\forall\forall\forall)$ sentence. This problem asks:

- Given a graph $G = (V, E)$, is there a partition of E to two disjoint sets E_1, E_2 such that neither $G = (V, E_1)$ nor $G = (V, E_2)$ contains a triangle? (cf. [GJ79]).

For an NP-complete problem of different flavor, consider the canonical problem of 3-*SATISFIABLITY*:

- Given a set of variables V and a collection K of clauses each containing exactly three variables, is there a satisfying assignment for K?

Let $\mathbf{R}$ be a vocabulary consisting of four ternary relation symbols P_i, where $i = 0, 1, 2, 3$. With every instance $I = (V, K)$ of 3-*SATISFIABILITY* we associate a structure $\mathbf{D}(I)$ over $\mathbf{R}$ by taking the universe of $\mathbf{D}(I)$ to be the set V of variables and encoding the clauses by the four relations P_i, as follows: a triple (u_1, u_2, u_3) of variables is in the relation P_i on $\mathbf{D}(I)$ if and only if there is a clause c in K such that precisely first i variables among u_1, u_2, u_3 occur positively in the clause c. Let Ψ be the following $\Sigma_1^1(\forall\forall\forall)$ sentence over $\mathbf{R}$:

$$(\exists S)(\forall x)(\forall y)(\forall z)[(P_0(x,y,z) \rightarrow \neg S(x) \vee \neg S(y) \vee \neg S(z)) \wedge \dots$$

$$\dots \wedge (P_3(x,y,z) \rightarrow S(x) \vee S(y) \vee S(z))].$$

It is easy to see that there is a satisfying assignment for an instance I of 3-*SATISFIABILITY* if and only if $\mathbf{D}(I) \models \Psi$.

It turns out that the class Σ_1^1(Bernays-Schönfinkel) can also capture interesting polynomial-time problems. For example, *DISCONNECTIVITY* of graphs is expressible by the $\Sigma_1^1(\exists\exists\forall\forall)$ sentence

$$(\exists S)(\exists x)(\exists y)(\forall z)(\forall w)[S(x) \wedge \neg S(y) \wedge (S(z) \wedge S(w) \rightarrow \neg E(z,w))].$$

3.2. The Expressive Power of Σ_1^1(Ackermann). The most well known example of an NP-complete problem is *SATISFIABILITY*:

- Given a set of variables V and a collection K of clauses over V, is there a satisfying truth assignment for K?

Consider now a vocabulary $\mathbf{R}$ consisting of a unary relation symbol V, and two binary relation symbols P and N. With every instance I of *SATISFIABILITY*, we associate a structure $\mathbf{D}(I)$ over $\mathbf{R}$ by taking the universe of $\mathbf{D}(I)$ to be the union $V \cup K$ of the set of variables and the set of clauses, and using the relation symbols P and N to code the positive and the negative occurrences respectively of the variables in the clauses (i.e., $P(c, v)$ means that the variable v occurs positively in the clause c, and analogously for N). This encoding of *SATISFIABILITY* appeared first in [Da84].

Let Ψ be the following Σ_1^1(Ackermann) sentence over $\mathbf{R}$:

$$(\exists S)(\forall y)(\exists z)\{[y \notin V] \rightarrow [(P(y,z) \wedge z \in S) \vee (N(y,z) \wedge z \notin S)]\},$$

where S is a unary relation variable. It is not hard to see that there is a satisfying assignment for I if and only if $\mathbf{D}(I) \models \Psi$. Thus, Ψ is a $\Sigma_1^1(\forall\exists)$ sentence that encodes *SATISFIABILITY* as a class of structures in a natural way. It can be shown that *SATISFIABILITY* cannot be expressed by a Σ_1^1(Bernays-Schönfinkel) sentence. Thus, Σ_1^1(Ackermann) is not less expressive than Σ_1^1(Bernays-Schönfinkel).

SET SPLITTING (problem [SP4] in [GJ79]) is another example of an NP-complete problem that can be expressed by a Σ_1^1(Ackermann) sentence, actually a $\Sigma_1^1(\forall\exists\exists)$ sentence. This problem asks:

- Given a collection K of subsets of a finite set S, is there a partition S into two subsets S_1 and S_2 such that no subset of S in K is entirely contained in either S_1 or S_2.

In terms of interesting polynomial-time properties, one can also show that the collection of all finite graphs containing a cycle (the *CYCLICITY* problem) is expressible by a $\Sigma_1^1(\exists\forall\exists)$ sentence.

4. The Class Σ_1^1(Bernays-Schönfinkel)

4.1. 0-1 Law. Lemma 3.1 and Corollary 3.2 reveal that in order to establish the 0-1 law for a class $\mathcal{F}$ of existential second-order sentences it is enough to show that if Ψ is a sentence in $\mathcal{F}$ that is true on the countable random structure $\mathbf{A}$, then $\mu(\Psi) = 1$. In this section we prove this to be the case for the class of Σ_1^1(Bernays-Schönfinkel) sentences.

Lemma 4.1. *Let $(\exists \mathbf{S})(\exists \mathbf{x})(\forall \mathbf{y})\theta(\mathbf{S}, \mathbf{x}, \mathbf{y})$ be a Σ_1^1(Bernays-Schönfinkel) sentence that is true on the countable random structure $\mathbf{A}$. Then there is a first-order sentence ψ over σ such that*

$$\mu(\psi) = 1$$

and

$$\models_{fin} \psi \rightarrow (\exists \mathbf{S})(\exists \mathbf{x})(\forall \mathbf{y})\theta(\mathbf{S}, \mathbf{x}, \mathbf{y}),$$

where $\models_{fin}$ denotes truth in all finite structures. In particular, if Ψ is a Σ_1^1(Bernays-Schönfinkel) sentence that is true on $\mathbf{A}$, then $\mu(\Psi) = 1$.

Proof Let $\mathbf{a} = (a_1, \ldots, a_n)$ be a sequence of elements of $\mathbf{A}$ that witness the first-order existential quantifiers $\mathbf{x}$ in $\mathbf{A}$. Let $\mathbf{A}_0$ be the finite substructure of $\mathbf{A}$ with universe $\{a_1, \ldots, a_n\}$. Then there is a first-order sentence ψ, which is the conjunction of a finite number of the extension axioms, having the property that any model of it contains a substructure isomorphic to $\mathbf{A}_0$. Now assume that $\mathbf{B}$ is a finite model of ψ. Using the extension axioms we can find a substructure $\mathbf{B}^*$ of the random structure $\mathbf{A}$ that contains $\mathbf{A}_0$ and is isomorphic to $\mathbf{B}$. Since universal statements are preserved under substructures, we conclude that

$$\mathbf{B} \models (\exists \mathbf{S})(\exists \mathbf{x})(\forall \mathbf{y})\theta(\mathbf{S}, \mathbf{x}, \mathbf{y}),$$

where $\mathbf{x}$ is interpreted by $\mathbf{a}$ and $\mathbf{S}$ is interpreted by the restriction to $\mathbf{B}$ of the relations on $\mathbf{A}$ that witness the existential second-order quantifiers. $\square$

From Lemmas 3.1 and 4.1 we infer immediately the 0-1 law and the transfer theorem for the class $\Sigma_1^1(\exists^*\forall^*)$.

Theorem 4.2. *The 0-1 law holds for Σ_1^1(Bernays-Schönfinkel) sentences on the class C of all finite structures over a relational vocabulary $\mathbf{R}$. Moreover, if $\mathbf{A}$ is the countable random structure and Ψ is a Σ_1^1(Bernays-Schönfinkel) sentence, then*

$$\mathbf{A} \models \Psi \Longleftrightarrow \mu(\Psi) = 1.$$

4.2. Solvability. As mentioned in Section 2, the proof of the 0-1 law for first-order logic showed also the solvability of the decision problem for the values (0 or 1) of the probabilities of first-order sentences. The preceding proof of the 0-1 law for Σ_1^1(Bernays-Schönfinkel) sentences does not yield a similar result for the associated decision problem for the probabilities of such sentences. Indeed, the only information one can extract from the proof is that the Σ_1^1(Bernays-Schönfinkel) sentences of probability 0 form a recursively enumerable set. We now show that the decision problem for the probabilities of sentences in the class Σ_1^1(Bernays-Schönfinkel) is solvable. We do this by proving that satisfiability of such sentences on $\mathbf{A}$ is equivalent to the existence of certain *canonical* models. For simplicity we present the argument for $\Sigma_1^1(\forall^*)$ sentences, i.e., sentences of the form

$$\exists S_1 ... \exists S_l \forall y_1 ... \forall y_m \theta(S_1, ..., S_l, y_1, ..., y_m).$$

Assume that the vocabulary σ consists of a sequence $\mathbf{R} = \langle R_i, i \in I \rangle$ of relation variables R_i. If B is a set and, for each $i \in I$, R_i^B is a relation on B of the same arity as that of R_i, then we write $\mathbf{R}^B$ for the sequence $\langle R_i^B, i \in I \rangle$. Let $<$ be a new binary relation symbol and consider structures $\mathbf{B} = (B, \mathbf{R}^B, <^B)$ in which $<^B$ is a *total ordering*. Let k be a positive integer. We say that $\mathbf{B}$ is *k-rich* if for any structure $\mathbf{D}$ with k elements over $\mathbf{R} \cup \{<\}$ (where $<$ is interpreted by a total ordering) there is a substructure $\mathbf{B}^*$ of $\mathbf{B}$ that is isomorphic to $\mathbf{D}$. Notice that the isomorphism takes into account *both* the relations R_i^D and the total ordering $<^D$.

Assume that S^B is an n-ary relation on B. We say that S^B is *canonical for the structure* $\mathbf{B} = (B, \mathbf{R}^B, <^B)$ if for any sequence $\mathbf{b} = (b_1, ..., b_n)$ of elements of B the truth value of $S^B(b_1, ..., b_n)$ depends *only* on the isomorphism type of the substructure of $\mathbf{B}$ with universe $\{b_1, ..., b_n\}$. An expanded structure $\mathbf{B}^* = (B, \mathbf{R}^B, <^B, S_1^B, ..., S_l^B)$ is *canonical* if every relation S_i^B, $1 \leq i \leq l$, is canonical on $\mathbf{B}$. The intuition behind canonical structures is that the relations S_i^B are determined completely by the $(\mathbf{R}, <)$-types.

We can state now the main technical result of this section.

Theorem 4.3. *Let* $\mathbf{A}$ *be the countable random structure over the vocabulary* σ *and let* Ψ *be a* Σ_1^1 *sentence of the form*

$$\exists S_1 ... \exists S_l \forall y_1 ... \forall y_m \theta(S_1, ..., S_l, y_1, ..., y_m).$$

Then the following are equivalent:

(1) $\mathbf{A} \models \Psi$.

(2) *There is a finite canonical structure* $\mathbf{B}^* = (B, \mathbf{R}^B, <^B, S_1^B, ..., S_l^B)$ *which is k-rich for every* $k \leq m$ *and such that*

$$\mathbf{B}^* \models \forall y_1 ... \forall y_m \theta(S_1, ..., S_l, y_1, ..., y_m).$$

In showing that (1) $\Longrightarrow$ (2) we will use certain Ramsey-type theorems which were proved by [NR77, NR83] and independently by [AH78]. We follow here the notation and terminology of [AH78] in stating these combinatorial results.

If X is any set and j is an integer, then $[X]^j$ is the collection of all subsets of X with j elements and $[X]^{\leq n} = \bigcup_{j \leq n} [X]^j$. A *system of colors* is a sequence $\mathbf{K} = (K_1, ..., K_n)$ of finite nonempty sets. A $\mathbf{K}$*-colored set* consists of a finite set X, a (total) ordering $<^X$ on X and a function $f : [X]^{\leq n} \mapsto K_1 \cup ... \cup K_n$ such that $f(Z) \in K_j$ for any $Z \in [X]^j$ and any $j \leq n$.

It is clear that every $\mathbf{K}$-colored set is isomorphic to a unique $\mathbf{K}$*-pattern*, that is a $\mathbf{K}$-colored set whose underlying set is an integer. If e, M are integers, $\mathbf{K}$ is a system of colors, P, Q are $\mathbf{K}$-patterns, then

$$Q \hookrightarrow (P)_M^{\leq e}$$

means that for every $\mathbf{K}$-colored set (X, f) of pattern Q and every partition $F : (X)^{\leq e} \mapsto M$ there is a subset Y of X such that $(Y, f |_Y)$ is of pattern P, and Y is *conditionally monochromatic* for F as $\mathbf{K}$-colored set, i.e. for $Z \in [Y]^j$ the value $F(Z)$ depends *only* on the $\mathbf{K}$-pattern of $(Z, f |_Z)$.

By iterated applications of Theorem 2.2 in [AH78], we can derive the following generalization of the classical Ramsey theorem [Ra28]:

Theorem 4.4. *For any integers* e, M*, any system of colors* $\mathbf{K}$*, and any* $\mathbf{K}$*-pattern* P*, there is a* $\mathbf{K}$*-pattern* Q *such that*

$$Q \hookrightarrow (P)_M^{\leq e}.$$

With every finite vocabulary $\mathbf{R}$ in which the maximum arity is n we can associate a system of colors $\mathbf{K}$ such that every finite structure $\mathbf{B} = (B, \mathbf{R}^B, <^B)$, where $<^B$ is a total ordering on B, can be *coded* by a $\mathbf{K}$-colored set $(B, <^B, f)$ with $f : [B]^{\leq n} \mapsto K_1 \cup ... \cup K_n$. For example, if σ consists of a single binary relation R, then K_1 has 2 elements, K_2 has 4 elements, and $f : [B]^{\leq 2} \mapsto K_1 \cup K_2$ is such that the value $f(\{x\})$ depends only on the truth value of $R^B(x,x)$, while the value $f(\{x,y\})$ depends only on the truth values of $R^B(min(x,y), max(x,y))$ and $R^B(max(x,y), min(x,y))$. Conversely, from any such **K**-pattern we can *decode* a finite structure **B**.

We now have all the combinatorial machinery needed to outline the ideas in the proof of Theorem 4.3.

Proof (1) $\Longrightarrow$ (2) Let Ψ be the Σ_1^1 sentence Ψ such that $\mathbf{A} \models \Psi$ and assume for simplicity that Ψ has only one ternary second-order existential variable S. We use the ternary relation S^A witnessing S on **A** to partition $[A]^{\leq 3}$ according to the *pure* S-type of a set $Z \in [A]^{\leq 3}$. This means that A is partitioned into two pieces defined by the truth value of $S^A(x,x,x)$, $[A]^2$ into $2^{2^3-2} = 64$ pieces defined by the truth values of $S^A(min(x,y), max(x,y), max(x,y))$, etc., and finally $[A]^3$ is partitioned into $2^{3!} = 64$ pieces defined by the truth values of $S^A(min(x,y,z), mid(x,y,z), max(x,y,z))$, etc.

Let $\mathbf{B} = (B, \mathbf{R}^B, <^B)$ be a finite structure which is k-rich for every $k \leq m$ and let P be a **K**-pattern which codes **B** in the way described above. We apply now Theorem 4.4 for $e = 3$, $M = 2 + 64 + 64 = 130$, and for P coding **B**. Let Q be a **K**-pattern such that $Q \hookrightarrow (P)_M^{\leq 3}$ and let **C** be the structure coded by Q. Using the extension axioms for the countable random structure **A** we can find in **A** a substructure $\mathbf{C}_1$ isomorphic to **C**. But now Theorem 4.4 guarantees that $\mathbf{C}_1$ contains a substructure $\mathbf{B}_1$ which is isomorphic to **B** and is conditionally monochromatic as a **K**-colored pattern. The structure $\mathbf{B}_1$ is k-rich for every $k \leq m$ and by taking the restriction of S^A on B_1 we can expand $\mathbf{B}_1$ to a canonical model $\mathbf{B}^*$ of $\forall y_1 \ldots \forall y_m \theta(S, y_1, \ldots, y_m)$, since universal sentences are preserved under substructures.

(2) $\Longrightarrow$ (1) From any canonical, k-rich model $(1 \leq k \leq m)$ of $\forall y_1 \ldots \forall y_m \theta$ $(S, y_1, \ldots, y_m)$ we can build a relation S^A witnessing the second-order existential quantifier in Ψ by assigning tuples to S^A according to their $(\mathbf{R}, <)$-type. □

Theorem 4.3 implies that the set of $\Sigma_1^1(\exists^*\forall^*)$ properties having probability 1 is recursively enumerable. On the other hand, Theorem 4.2 and Corollary 3.3 together imply that the complement of this set is also recursively enumerable. Thus, we have established that the decision problem for the probabilities of strict Σ_1^1 properties is solvable. This proof does not give, however, any complexity bounds for the problem. In [KV87] we analyzed the computational complexity of this decision problem and showed that it is NEXPTIME-complete for bounded vocabularies and 2NEXPTIME for unbounded vocabularies.

4.3. Expressiveness. The 0-1 law for Σ_1^1(Bernays-Schönfinkel) sentences has also some immediate applications to definability.

Corollary 4.5. *The collection of all finite structures over* $\mathbf{R}$ *whose universe has an even number of elements is not definable by any* Σ_1^1*(Bernays-Schönfinkel) sentence.*

As a further application of the 0-1 law for Σ_1^1(Bernays-Schönfinkel) sentences, we obtain definability results about *CONNECTIVITY* and *HAMILTONICITY* on the class $\mathcal{G}$ of finite (undirected) graphs.

Theorem 4.6. *Let* $\mathcal{G}$ *be the collection of all finite undirected graphs.*

(1) *There is no* Σ_1^1*(Bernays-Schönfinkel) sentence that defines the class of connected graphs on* $\mathcal{G}$*. As a result, the collection of* Σ_1^1*(Bernays-Schönfinkel) properties on* $\mathcal{G}$ *is not closed under negation.*
(2) *There is no* Σ_1^1*(Bernays-Schönfinkel) sentence that defines the class of Hamiltonian graphs on* $\mathcal{G}$*.*
(3) *There is no* Σ_1^1*(Bernays-Schönfinkel) sentence whose negation defines the class of Hamiltonian graphs on* $\mathcal{G}$*.*

Proof We sketch the proof for (1) and (3), the argument for (2) being similar to the one for (1). Since *CONNECTIVITY* has probability 1, if Ψ were a strict Σ_1^1 sentece defining it on $\mathcal{G}$, then Ψ would be true on the countable random countable graph $\mathbf{A}$. But then, the proof of Lemma 4.1 shows that there is a finite graph G such that every finite graph G* containing G as an induced subgraph is connected.

Assume that *HAMILTONICITY* is definable by a Π_1^1 sentence Φ whose negation is in the class Σ_1^1(Bernays-Schönfinkel). Since μ*(HAMILTONICITY)*$= 1$ [Po76], Theorem 4.2 above implies that $\mathbf{A} \models \Phi$. From Lemma 3.1 we infer that there is a first-order sentence ψ such that $\mu(\psi) = 1$ and $\models \psi \rightarrow \Phi$. In particular, $\models_{fin} \psi \rightarrow HAMILTONICITY$. This, however, violates a theorem in Blass and Harary [BH79] to the effect that there is no first-order property that has probability 1 and is a sufficient condition for a graph to be Hamiltonian. □

Notice that *HAMILTONICITY* is definable by a Σ_1^1 sentence on $\mathcal{G}$. On the other hand, by Fagin's characterization of NP [Fa74], proving that there is *no* Π_1^1 definition of *HAMILTONICITY* on $\mathcal{G}$ amounts to establishing that NP $\neq$ co-NP.

5. The Class Σ_1^1(Ackermann)

Our goal here is to establish the following:

Theorem 5.1. *Let* $\mathbf{A}$ *be the countable random structure over the vocabulary* $\mathbf{R}$ *and let* Ψ *be a* Σ_1^1*(Ackermann) sentence. If* $\mathbf{A} \models \Psi$, *then* $\mu(\Psi) = 1$.

This theorem will be obtained by combining three separate lemmas. Since the whole argument is rather involved, we start with a "high-level" description of the structure of the proof.

We first isolate a syntactic condition (condition (χ) below) for Σ_1^1-Ackermann sentences and in Lemma 5.2 we show that if Ψ is a Σ_1^1(Ackermann) sentence which is true on $\mathbf{A}$, then condition (χ) holds for Ψ. At the end, it will actually turn out that this condition (χ) is also sufficient for truth of Σ_1^1(Ackermann) sentences on the countable random structure $\mathbf{A}$. In Lemma 5.3, we isolate a "richness" property E_s, $s \geq 1$, of (some) finite structures over $\mathbf{R}$ and show that $\mu(E_s) = 1$ for every $s \geq 1$. The proof of this lemma requires certain asymptotic estimates from probability theory, due to Chernoff [Ch52]. Finally, in Lemma 5.4, we prove that if Ψ is a Σ_1^1(Ackermann) sentence for which condition (χ) holds, then for appropriately chosen s and for all large n the sentence Ψ is true on all finite structures of cardinality n over $\mathbf{R}$ that possess property E_s; consequently, $\mu(\Psi) = 1$. In this last lemma, the existence of the predicates $\mathbf{S}$ that witness Ψ is proved by a probabilistic argument, which in spirit is analogous to the technique used by Gurevich and Shelah [GS83] for showing the finite satisfiability property of first-order formulas in the Gödel class without equality.

Let $\mathbf{T}$ be a vocabulary, i.e. a set of relational symbols. Recall that, a *k-$\mathbf{T}$-type* $t(x_1, ..., x_k)$ is a maximal consistent set of equality, negated equality formulas, atomic and negated atomic formulas whose variables are among $x_1, ..., x_k$.

- If $t(x_1, ..., x_k)$ is a k-$\mathbf{T}$-type, then, for any m with $1 \leq m \leq k$, let $t(x_{i_1}, ..., x_{i_m})$ be the m-$\mathbf{T}$-type obtained by deleting from $t(x_1, ..., x_k)$ all formulas in which a variable $y \neq x_{i_1}, \dots, x_{i_m}$ occurs.
- If $\mathbf{S} \subseteq \mathbf{T}$, then the *restriction* of t to $\mathbf{S}$ is the k-$\mathbf{S}$-type obtained by deleting from t all formulas in which a predicate symbols in $\mathbf{T} - \mathbf{S}$ occurs.
- If $t(x_1, ..., x_k, x_{k+1})$ is a $(k+1)$-$\mathbf{T}$-type, and y is a variable different from all the x_i's, then $t(x_1, ..., x_k, x_{k+1}/y)$ is a $(k+1)$-$\mathbf{T}$-type obtained by replacing all occurrences of x_{k+1} by y.
- Let $t(x_1, ..., x_k)$ be a k-$\mathbf{T}$-type, and let $\varphi(x_1, ..., x_k)$ be a quantifier-free formula in the variables $x_1, ..., x_k$. We say that t *satisfies* φ if φ is true

under the truth assignment that assigns true to an atomic formula precisely when it is a member of t.

Let Ψ be a Σ^1_1(Ackermann) sentence of the form

$$(\exists \mathbf{S})(\exists x_1)\dots(\exists x_k)(\forall y)(\exists z_1)\dots(\exists z_l)\varphi(x_1,\dots,x_k,y,z_1,\dots,z_l,\mathbf{R},\mathbf{S}),$$

where φ is a quantifier-free formula over the vocabulary $(\mathbf{R},\mathbf{S}) = \mathbf{R}\cup\mathbf{S}$.

We say that *condition* (χ) *holds for* Ψ if there is k-$(\mathbf{R},\mathbf{S})$-type $t_0(x_1,...,x_k)$ and a set P of $(k+1)$-$(\mathbf{R},\mathbf{S})$-types $t(x_1,...,x_k,y)$ extending $t_0(x_1,...,x_k)$ such that the following are true:

(1) P contains as a member the $(k+1)$-$(\mathbf{R},\mathbf{S})$-type $t_0^{x_i}(x_1,\dots,x_k,y)$, for every $i=1\dots k$. Equivalently, for every i, $1\le i\le k$, there is a type $t_i(x_1,\dots,x_k,y)$ in P such that

$$t_i(x_1,\dots,x_k,y/x_i) = t_0(x_1,\dots,x_k).$$

(2) P is $\mathbf{R}$-*rich over* $t_0(x_1,...,x_k)$, i.e., every $(k+1)$-$\mathbf{R}$-type $t(x_1,...,x_k,y)$ extending the restriction of $t_0(x_1,...,x_k)$ to $\mathbf{R}$ is itself the restriction of some $(k+1)$-$(\mathbf{R},\mathbf{S})$-type in P to $\mathbf{R}$.

(3) For each $t(x_1,...,x_k,y)$ in P there is a $(k+l+1)$-$(\mathbf{R},\mathbf{S})$-type $t'(x_1,\dots,x_k,y,z_1,\dots,z_l)$ such that $t\subseteq t'$, t' satisfies $\varphi(x_1,\dots,x_k,y,z_1,\dots,z_l)$, and for each z_i, $1\le i\le l$, the $(k+1)$-$(\mathbf{R},\mathbf{S})$-type $t'(x_1,...,x_k,z_i/y)$ is in P.

Lemma 5.2. *Let* $\mathbf{A}$ *be the countable random structure over the vocabulary* $\mathbf{R}$ *and let* Ψ *be a* Σ^1_1*(Ackermann) sentence. If* $\mathbf{A}\models\Psi$*, then condition* (χ) *holds for* Ψ*.*

Proof The type t_0 and the set of types P required in condition (χ) are obtained from the relations on $\mathbf{A}$ and the elements of $\mathbf{A}$ that witness the existential second-order quantifiers $(\exists\mathbf{S})$ and the existential first-order quantifiers $(\exists x_1)\dots(\exists x_k)$ in Ψ respectively. To show that P is $\mathbf{R}$-rich, we use the fact that the countable random structure $\mathbf{A}$ satisfies the extension axioms, which in turn imply that the elements of $\mathbf{A}$ realize all possible $\mathbf{R}$-types. □

Let $\mathbf{D}$ be a structure over $\mathbf{R}$ and let $\mathbf{c}=(c_1,\dots,c_m)$ be a sequence of elements from $\mathbf{D}$. The *type* $t_{\mathbf{c}}$ of $\mathbf{c}$ on $\mathbf{D}$ is the unique m-$\mathbf{R}$-type $t(z_1,\dots,z_m)$ determined by the atomic and negated atomic formulas that the sequence $\mathbf{c}$ satisfies on $\mathbf{D}$, under the assignment $z_i\to c_i$, $1\le i\le m$. We say that a sequence $\mathbf{c}$ *realizes* a type t on a structure $\mathbf{D}$ if $t_{\mathbf{c}}=t$.

Let $s\ge 1$ be fixed. We say that a finite structure $\mathbf{D}$ over $\mathbf{R}$ with n elements *satisfies property* E_s if the following holds:

- For every number m with $1 \leq m \leq s$, every sequence $\mathbf{c} = (c_1, \ldots, c_m)$ from $\mathbf{D}$ and every $(m+1)$-$\mathbf{R}$-type $t(z_1, \ldots, z_m, z_{m+1})$ extending the type $t_{\mathbf{c}}$ of $\mathbf{c}$ on $\mathbf{D}$, there are at least $\sqrt{n}$ different elements d in $\mathbf{D}$ such that each sequence $(c_1, \ldots, c_m, d)$ realizes the type $t(z_1, \ldots, z_m, z_{m+1})$.

Lemma 5.3. *For every $s \geq 1$ there is a positive constant c and a natural number n_0 such that for any $n \geq n_0$*

$$\mu_n(E_s) \geq 1 - n^{s+1} e^{-cn}.$$

In particular, $\mu(E_s) = 1$, i.e. almost all structures over $\mathbf{R}$ satisfy property E_s, for every $s \geq 1$.

Proof The proof of this lemma uses an asymptotic bound on the probability in the tail of the binomial distribution, due to Chernoff [Ch52] (cf. also [Bo85]). We first fix a sequence $\mathbf{c}$ from $\mathbf{D}$ and a type t that extends $t_{\mathbf{c}}$, and apply this bound to the binomial distribution obtained by counting the number of elements d such that the sequence $(c_1, \ldots, c_m, d)$ realizes t. We then iterate through all types and all sequences $\mathbf{c} = (c_1, \ldots, c_m)$ for $1 \leq m \leq s$. □

The last lemma in this section provides the link between condition (χ), property E_s, $s \geq 1$, and satisfiability of Σ_1^1(Ackermann) sentences on finite structures over $\mathbf{R}$.

Lemma 5.4. *Let Ψ be a Σ_1^1(Ackermann) sentence of the form*

$$(\exists \mathbf{S})(\exists x_1) \ldots (\exists x_k)(\forall y)(\exists z_1) \ldots (\exists z_l)\varphi(x_1, \ldots, x_k, y, z_1, \ldots, z_l, \mathbf{R}, \mathbf{S})$$

for which condition (χ) holds. There is a natural number n_1 such that for every $n \geq n_1$, if $\mathbf{D}$ is a finite structure over $\mathbf{R}$ with n elements satisfying property E_{k+l+1}, then $\mathbf{D} \models \Psi$.

Proof The existence of the relations on $\mathbf{D}$ that witness the second-order quantifiers $(\exists \mathbf{S})$ in Ψ is proved with a probabilistic argument similar to the one employed by Gurevich and Shelah [GS83] for the finite satisfiability property of the Gödel class without equality. We use condition (χ) to impose on $\mathbf{D}$ a probability space of $\mathbf{S}$ predicates. The richness property E_{k+l+1} is then used to show that with nonzero probability (in this new space) the expansion of $\mathbf{D}$ with these predicates satisfies the sentence

$$(\exists x_1) \ldots (\exists x_k)(\forall y)(\exists z_l) \ldots (\exists z_l)\varphi(y, z_1, \ldots, z_l, \mathbf{S}).$$

□

This completes the outline of the proof of Theorem 5.1. Combining now this theorem with Lemma 3.1 we derive the main result of this section.

Theorem 5.5. *The 0-1 law holds for the Σ_1^1(Ackermann) class on the collection C of all finite structures over a vocabulary* **R**. *Moreover, if* **A** *is the countable random structure over* **R** *and Ψ is a Σ_1^1(Ackermann) sentence, then*

$$\mathbf{A} \models \Psi \Longleftrightarrow \mu(\Psi) = 1.$$

Notice that the preceding results also show that a Σ_1^1(Ackermann) sentence Ψ has probability 1 if and only if condition (χ) holds for Ψ. Since condition (χ) is clearly effective, it follows that the decision problem for the values of the probabilities of Σ_1^1(Ackermann) sentences is solvable. In [KV90] we analyzed the computational complexity of this decision problem and showed that it is NEXPTIME-complete for bounded vocabularies and Σ_2^{exp}-complete[1] for unbounded vocabularies.

The 0-1 law for Σ_1^1(Ackermann) sentences has also immediate applications to definability.

Corollary 5.6. *The collection of all finite structures over* **R** *whose universe has an even number of elements is not definable by any Σ_1^1(Ackermann) sentence.*

Using an argument analogous to the one in Theorem 4.6, we can also establish the following result.

Corollary 5.7. *There is no Σ_1^1(Ackermann) sentence whose negation defines HAMILTONICITY on the class $\mathcal{G}$ of finite undirected graphs.*

6. Negative Results

The Bernays-Schönfinkel and Ackermann classes are the only docile prefix classes (with equality), i.e., they are the only prefix classes for which the finite satisfiability problem is solvable [DG79, Le79, Go84]. A key role in this classification was played by the *Gödel class*, which is the class of first-order sentences with equality and with prefix of the form $\forall^2\exists^*$. In fact, the classification was completed only when Goldfarb [Go84] showed that the *minimal Gödel class* i.e., the class of first-order sentences with equality and with prefix of the form $\forall^2\exists$, is not docile. In this section we show that the same classification holds for the 0-1 law, namely, the 0-1 law holds for the Σ_1^1 fragments that correspond to docile prefix classes.

---

[1] Σ_2^{exp} is the second-level of the exponential hierarchy. It can be described as the class of languages accepted by alternating exponential-time Turing machines in two alternations where the machine start state is existential [CKS81] or as the class $\mathrm{NEXP}^{\mathrm{NP}}$ of languages accepted by nondeterministic exponential-time Turing machines with oracles from NP [HIS85].

It is easy to see that the 0-1 law does not hold for the Σ_1^1 fragment that correspond to the prefix classes $\forall\exists\forall$ and $\forall^3\exists$. For example, the property "there is an even number of elements" can be expressed by the following $\Sigma_1^1(\forall\exists\forall)$ sentence asserting that "there is a permutation in which every element is of order 2":

$$(\exists S)(\forall x)(\exists y)(\forall z)[S(x,y)\wedge(S(x,z)\rightarrow y=z)\wedge(S(x,z)\leftrightarrow S(z,x))\wedge\neg S(x,x)].$$

The statement "there is a permutation in which every element is of order 2" can also be expressed by the following $\Sigma_1^1(\forall\forall\forall\exists)$ sentence

$$(1)\quad (\exists S)(\forall x)(\forall y)(\forall z)(\exists w)[S(x,w)\wedge(S(x,y)\wedge S(x,z)\rightarrow y=z)\wedge (S(x,z)\leftrightarrow S(z,x))\wedge\neg S(x,x)].$$

Dealing with the class Σ_1^1(Gödel), i.e., the class $\Sigma_1^1(\forall^2\exists^*)$ is much harder. Pacholski and Szwast [PS89] established the following result.

Theorem 6.1. *There exists a $k \geq 1$ such that the 0-1 law fails for the class $\Sigma_1^1(\forall^2\exists^k)$.*

Proof Pacholski and Szwast construct a $\Sigma_1^1(\exists^*\forall\forall\exists^*)$ sentence Ψ over a certain vocabulary **R** such that a finite structure **D** over **R** satisfies Ψ if and only if the cardinality of the universe of **D** is of the form n^2+n for some integer n.

The construction of the above sentence Ψ uses ideas from Goldfarb's [Go84] proof of the unsolvability of the satisfiability problem for the Gödel class. The main technical innovation in that proof was the construction of a $\forall\forall\exists^*$ first-order sentence φ that is satisfiable, but has no finite models. The Σ_1^1(Gödel) sentence Ψ that has no asymptotic probability is obtained by modifying φ appropriately. □

Goldfarb [Go84] obtained the unsolvability results first for the Gödel class and then sharpened them to hold for the minimal Gödel class. Pacholski and Szwast showed that similar techniques can sharpen Theorem 6.1 [PS90].

Theorem 6.2. *The 0-1 law fails for the class $\Sigma_1^1(\forall^2\exists)$.*

Thus, we can conclude that the classifications of prefix classes according to the solvability of the finite satisfiability problem and according to the 0-1 law for the corresponding Σ_1^1 sentences are identical. This follows from the positive results for the classes Σ_1^1(Bernays-Schönfinkel) and Σ_1^1(Ackermann), and the negative results above.

Remark 6.3. The preceding results assume that the use of the equality symbol $=$ is allowed. At present, it is an open problem to determine whether or not the 0-1 law holds for the class Σ_1^1(Gödel) and for the class $\Sigma_1^1(\forall\exists\forall)$, if equality is precluded. On the other hand, we can show that the 0-1 law fails for the class $\Sigma_1^1(\forall^3\exists)$ even if equality is precluded [KV90].

7. 0-1 Laws and Finite Controllability

We showed here that the 0-1 law holds for the class $\Sigma_1^1(\mathcal{F})$ whenever $\mathcal{F}$ is a prefix docile class, i.e., a prefix class with a solvable finite satisfiability problem. It is known that there are nonprefix classes $\mathcal{F}$ that are docile. These classes are usually obtained by restricting appropriately the quantifier-free part of sentences in a prefix class with an unsolvable finite satisfiability problem. It is natural to ask if the docility of a class $\mathcal{F}$ is always accompanied by the 0-1 law for the class $\Sigma_1^1(\mathcal{F})$. It turns out, however, that this is not true.

A quantifier-free formula is said to be *Horn* if it is in conjuctive normal form and each conjunct has at most one (positive) atomic formula. A quantifier-free formula is *Krom* if it is in conjunctive normal form and each conjunct is the disjunction of two literals. The class $\forall\exists\forall$-Horn consists of all $\forall\exists\forall$ first-order sentences with a Horn matrix; similarly, the class $\forall\exists\forall$-Krom consists of all $\forall\exists\forall$ first-order sentences with a Krom matrix. It is known that both these classes have a solvable finite satisfiability problem (cf. [DG79]). We observe now that the 0-1 law fails for both the class $\Sigma_1^1(\forall\exists\forall\text{-Horn})$ and the class $\Sigma_1^1(\forall\exists\forall\text{-Krom})$. Indeed, in the previous section we gave a $\Sigma_1^1(\forall\exists\forall)$ sentence that has no asymptotic probability. An inspection of this sentence shows that its quantifier-free part is equivalent to both a Horn and a Krom formula.

Is there another property of a class $\mathcal{F}$ that goes hand on hand with the 0-1 law for the class $\Sigma_1^1(\mathcal{F})$? In what follows we give evidence suggesting that the real connection is between the *finite controllability* of $\mathcal{F}$ and the transfer theorem for $\Sigma_1^1(\mathcal{F})$.

A class $\mathcal{F}$ of first-order sentences is *finitely controllable* if satisfiability implies finite satisfiability for all sentences in $\mathcal{F}$. Notice that if $\mathcal{F}$ is finitely controllable, then the satisfiability problem for sentences in $\mathcal{F}$ is equivalent to the finite satisfiability problem for sentences in $\mathcal{F}$. Moreover, in this case both these problems are solvable. The solvability of both satisfiability and finite satisfiability for the Bernays-Schönfinkel and Ackermann classes is obtained by establishing that each of these classes is finitely controllable. On the other hand, the classes $\forall\exists\forall$-Horn and $\forall\exists\forall$-Krom have solvable satisfiability and finite satisfiability problems, but neither of them is finitely controllable.

We now show that the transfer theorem for a class $\Sigma_1^1(\mathcal{F})$ implies the finite controllability of the class $\mathcal{F}$. The proof make use of *relativization*. Let θ be a first-order sentence and let P be a unary predicate not occurring in θ. The relativization of θ to P, denoted θ^P, is the sentence obtained from θ by recursively replacing in θ all subformulas $(\exists x)\psi$ and $(\forall x)\psi$ by $(\exists x)(P(x) \wedge \psi)$ and $(\forall x)(P(x) \rightarrow \psi)$, respectively. We say that a class $\mathcal{F}$ of first-order sentences is *closed under relativization* if θ^P is in $\mathcal{F}$ whenever θ is in $\mathcal{F}$. Notice that every prefix class is closed under relativization.

Theorem 7.1. *Let* $\mathbf{A}$ *be the countable random structure over* $\mathbf{R}$, *and let* $\mathcal{F}$ *be a class of first-order sentences over* $\mathbf{R}$ *that is closed under relativization. If* $\mathbf{A} \models \Psi$ *precisely when* $\mu(\Psi) = 1$ *for each sentence* $\Psi \in \Sigma_1^1(\mathcal{F})$, *then the class* $\Sigma_1^1(\mathcal{F})$ *is finitely controllable.*

Proof Let Ψ be the sentence $(\exists \mathbf{S})\theta(\mathbf{R}, \mathbf{S})$. Suppose that Ψ is satisfiable. By the Löwenheim-Skolem theorem applied to θ, it follows that Ψ has a countable model $\mathbf{B}$. Since $\mathbf{A}$ is universal for all countable structures over $\mathbf{R}$, it must have a substructure $\mathbf{B}'$ isomorphic to $\mathbf{B}$. Let Ψ' be the sentence $(\exists P)(\exists \mathbf{S})\theta^P(\mathbf{R}, \mathbf{S})$. It is easy to see that $\mathbf{A} \models \Psi'$ (interpret P by the universe of $\mathbf{B}'$). Thus, $\mu(\Psi') = 1$ and, as a result, Ψ' is finitely satisfiable. Let $\mathbf{C}$ be a finite model of Ψ'. Then $\mathbf{C}$ must have a substructure $\mathbf{C}'$ that satisfies Ψ. Thus, Ψ is finitely satisfiable. □

The above Theorem 7.1 reveals that establishing the transfer theorem for a class $\Sigma_1^1(\mathcal{F})$ can be viewed as obtaining a strong version of the finite controllability for $\mathcal{F}$. The results presented here show that for prefix classes $\mathcal{F}$ the classification according to the finite controllability (and, hence, docility) of $\mathcal{F}$ and according to the transfer theorem (and, hence, the 0-1 law) for $\Sigma_1^1(\mathcal{F})$ are identical. It remains an interesting open problem to pursue transfer theorems and 0-1 laws for classes $\Sigma_1^1(\mathcal{F})$, where $\mathcal{F}$ is a nonprefix finitely controllable class.

Acknowledgments

We are grateful to Ron Fagin for many enlightening discussions.

References

[AH78] Abramson,F.D., Harrington, L.A.: Models without indiscernibles. *J. Symbolic Logic* 43(1978),pp. 572–600.

[Aj83] Ajtai, M.: Σ_1^1-formulas on finite structures. *Ann. of Pure and Applied Logic* 2443(1983),pp. 1–48.

[AU79] Aho, V.H., Ullman, J.D.: Universality of data retrieval languages. *Proceedings 6th ACM Symp. on Principles of Programming Languages*, 1979, pp. 110–117.

[BH79] Blass, A., Harary, F.: Properties of almost all graphs and complexes. *J. Graph Theory* 3(1979),pp. 225–240.

[Bo85] Bollobas, B: *Random Graphs.* Academic Press, 1985

[Ch52] Chernoff, H.: A Measure of Asymptotic Efficiency for Tests of a Hypothesis Based on the Sum of Observation. *Ann. Math. Stat.* 23(1952), pp. 493–509.

[CH82] Chandra, A., Harel, D.: Structure and Complexity of Relational Queries. *J. Computer and Systems Sciences* 25(1982), pp. 99–128.

[CKS81] Chandra, A., Kozen, D., Stockmeyer, L.: Alternation. *J. ACM* 28(1981), pp. 114–133.

[Co88] Compton, K.J.: 0-1 laws in logic and combinatorics, in *NATO Adv. Study Inst. on Algorithms and Order* (I. Rival, ed.), D. Reidel, 1988, pp. 353–383.

[Da84] Dalhlaus, E.: Reductions to NP-complete problems by interpretations. *Logic and Machines: Decision Problems and Complexity* (E. Börger et al., eds.), Springer–Verlag, Lecture Notes in Computer Science 171, 1984, pp. 357-365.

[DG79] Dreben, D., and Goldfarb, W.D.: *The Decision Problem: Solvable Classes of Quantificational Formulas.* Addison-Wesley,1979.

[Fa74] Fagin, R.: Generalized first–order spectra and polynomial time recognizable sets. *Complexity of Computations* (R. Karp, ed.), SIAM–AMS Proc. 7(1974), pp.43–73.

[Fa76] Fagin, R.: Probabilities on finite models. *J. Symbolic Logic* 41(1976), pp. 50–58.

[Ga64] Gaifman, H.: Concerning measures in first-order calculi. *Israel J. Math.* 2(1964). pp. 1–18.

[GJ79] Garey, M.R., Johnson, D.S.: *Computers and Intractability - A Guide to the Theory of NP-Completeness*, W.H. Freeman and Co., 1979.

[GKLT69] Glebskii, Y.V., Kogan, D.I., Liogonkii. M.I., Talanov, V.A.: Range and degree of realizability of formulas in the restricted predicate calculus. *Cybernetics* 5(1969), pp. 142–154.

[Go84] Goldfarb, W.D.: The Gödel class with equality is unsolvable. *Bull. Amer. Math. Soc. (New Series)* 10(1984), pp. 113-115.

[Gr83] Grandjean, E.: Complexity of the first–order theory of almost all structures. *Information and Control* 52(1983), pp. 180–204.

[Gu69] Gurevich, Y.: The decision problem for logic of predicates and operations. *Algebra and Logic* 8(1969), pp. 160–174.

[Gu76] Gurevich, Y.: The decision problem for standard classes. *J. Symbolic Logic* 41(1976), pp. 460–464.

[Gu84] Gurevich, Y.: Toward logic tailored for computational complexity. *Computation and Proof Theory* (M.M. Ricther et al., eds.), Springer–Verlag, Lecture Notes in Math. 1104, 1984, pp. 175–216.

[GS83] Gurevich, Y., and Shelah, S.: Random models and the Gödel case of the decision problem. *J. of Symbolic Logic* 48(1983), pp. 1120-1124.

[HIS85] Hartmanis, J., Immerman, N., Sewelson, J.: Sparse sets in NP-P – EXPTIME vs. NEXPTIME. *Information and Control* 65(1985), pp. 159–181.

[Im83] Immerman, N.: Languages which capture complexity classes. *Proc. 15th ACM Symp. on Theory of Computing*, Boston, 1983, pp. 347–354.

[Im86] Immerman, N.: Relational queries computable in polynomial time. *Information and Control* 68(1986), pp. 86–104.

[Ka87] Kaufmann, M.: *A counterexample to the 0-1 law for existential monadic second-order logic.* CLI Internal Note 32, Computational Logic Inc., Dec. 1987.

[KS85] Kauffman, M., Shelah, S: On random models of finite power and monadic logic. *Discrete Mathematics* 54(1985), pp. 285–293.

[KV87] Kolaitis, P., Vardi, M.Y.: The decision problem for the probabilities of higher-order properties. *Proc. 19th ACM Symp. on Theory of Computing*, New York, May 1987, pp. 425–435.

[KV90] Kolaitis, P., Vardi, M.Y.: 0-1 laws and decision problems for fragments of second-order logic. *Information and Computation*, in press.

[Le79] Lewis, H.R.: *Unsolvable Classes of Quantificational Formulas.* Addison-

Wesley, 1979.

[NR77] Nešetřil, J., Rödl, V.: Partitions of finite relational and set systems. *J. Combinatorial Theory A* 22(1977), pp. 289–312.

[NR83] Nešetřil, J., Rödl, V.: Ramsey classes of set systems. *J. Combinatorial Theory A* 34(1983), pp. 183–201.

[Po76] Pósa, L.: Hamiltonian circuits in random graphs. *Discrete Math.* 14(1976), pp. 359–364.

[PS89] Pacholski, L., Szwast, W.: The 0-1 law fails for the class of existential second-order Gödel sentences with equality. *Proc. 30th IEEE Symp. on Foundations of Computer Science*, 1989, pp. 160–163.

[PS90] Pacholski, L., Szwast, W.: A counterexample to the 0-1 law for existential second-order minimal Gödel sentences. Unpublished, 1990.

[Ra28] Ramsey, F.P.: On a problem in formal logic. *Proc. London Math. Soc.* 30(1928). pp. 264–286.

[Tr50] Trakhtenbrot, B.A.: The impossibilty of an algorithm for the decision problem for finite models. *Doklady Akademii Nauk SSR* 70(1950), PP. 569–572.

[Va82] Vardi. M.Y.: The complexity of relational query languages. *Proc. 14th ACM Symp. on Theory of Computing*, San Francisco, 1982, pp. 137–146.

P. G. Kolaitis: University of California, Santa Cruz

M. Y. Vardi: IBM Almaden Research Center

During the preparation of this paper the first author was partially supported by NSF Grant CCR-8905038.

NO COUNTER–EXAMPLE INTERPRETATION AND INTERACTIVE COMPUTATION

JAN KRAJÍČEK

ABSTRACT. No counter-example interpretation for bounded arithmetic is employed to derive recent witnessing theorem for S_2^{i+1}, functions $\square_{i+1}^p$-computable with counterexamples are shown to include all $\square_{i+2}^p$-functions, and two separation results for fragments of $S_2(\alpha)$ are proved.

Buss [1,2] has shown that functions $\exists\Sigma_{i+1}^b$-definable in S_2^{i+1} or T_2^i are precisely $\square_{i+1}^p$-functions. This was in [3] generalized in the following way.

Assume $T_2^i \vdash \exists x \forall y \exists z A(a,x,y,z)$, where A is a Σ_{i+1}^b-formula. Then function F assigning to a some b such that $\forall y \exists z A(a,b,y,z)$ is computable by a $\square_{i+1}^p$-algorithm which may ask constantly many times for counter-examples to $\forall y \exists z A(a,b,y,z)$ (i.e. for c such that $\neg \exists z A(a,b,c,z)$). In these questions b varies but a is fixed.

Pudlák [6] has recently proved similar theorem for S_2^{i+1}: the assumption $S_2^{i+1} \vdash \exists x \forall y \leq a A(a,x,y)$ (A again Σ_{i+1}^b) implies that function F assigning to a some b such that $\forall y \leq a A(a,b,y)$ is computable by a $\square_{i+1}^p$-algorithm which may ask for any (polynomial) number of counter-examples to $\forall y \leq a A(a,b,y)$. Here is a simple proof of this statement.

Extending the language of S_2^1 by some $\square_{i+1}^p$-functions and adding some universal axioms about them we may form theory $S_2^1(PV_{i+1})$, a conservative extension of S_2^{i+1}. We may also assume that A is existential.

As a formula implies its Herbrand's form, the assumption

$$S_2^{i+1} \vdash \exists x \forall y \leq A(a,x,y)$$

implies

$$S_2^1(PV_{i+1}, f) \vdash \exists x, f(a,x) \leq a \supset A(a,x,f(a,x)),$$

where f is a new function symbol.

By (relativization of) Buss's witnessing theorem there is functional $F(a, f)$ which satisfies:

$$f(a, F(a, f)) \leq a \supset A(a, F(a, f), f(a, F(a, f))),$$

and which is computable by a deterministic algorithm which may ask for values of some $\Box_{i+1}^{p}$-functions (from the language of $S_2^1(PV_{i+1})$) and for values of $f(a, x)$. Moreover, if f is of polynomial growth then the algorithm computing F runs in time polynomial in $|a|$. We may therefore call F a $\Box_{i+1}^{p}$-functional.

The algorithm computing F is the algorithm required in Pudlák's statement. This is because if f is a function computing some counter-examples:

$$f(a, b) := \begin{cases} \text{some } c \leq a, \text{ s.t. } \neg A(a, b, c) \\ a + 1, \text{ if } \forall y \leq a A(a, b, y) \end{cases}$$

then formula:

$$f(a, b) \leq a \supset A(a, b, f(a, b))$$

implies:

$$\forall y \leq a A(a, b, y).$$

(The additional property that a is fixed in the queries follows as we can treat a as a constant.)

The same argument works also if y in $\forall y$ is not bounded. But the run time of an algorithm computing F is then bounded only by a polynomial in $|a| + \sum_i |f(a, u_i)|$, where $f(a, u_i)$'s are all function values $f(a, x)$ asked for in the computation.

The statement clearly generalizes to arbitrary quantifier complexity: the assumption

$$S_2^{i+1} \vdash \exists x_1 \forall y_1 \ldots \exists x_k \forall y_k A(a, \vec{x}, \vec{y})$$

(bounds to y_j's implicitly in A) implies existence of $\Box_{i+1}^{p}$-functionals $F_1(a, \vec{f}), \ldots, F_k(a, \vec{f})$ such that for any a and $\vec{f}$ it holds:

$$A(a, \ldots, x_\ell / F_\ell(a, \vec{f}), \ldots, y_j / f_j(a, F_1(a, \vec{f}), \ldots, F_j(a, \vec{f})), \ldots).$$

The computation of F_ℓ's with particular f_j's computing counterexamples can be described again as an interactive computation.

The characterization of the witnessing functions in the T^i_2 case was in [3] used for a conditional separation of T^i_2 and S^{i+1}_2. The motivation for studying the S^{i+1}_2 case is the problem of separation of S^{i+1}_2 and T^{i+1}_2. Here however, a sceptical tone comes from the observation that any $\square^p_{i+2}$-function can be computed in the interactive manner associated with S^{i+1}_2, while Skolem functions for T^{i+1}_2 are also $\square^p_{i+2}$. This is seen as follows.

For M a deterministic oracle machine and $B(u) = \exists v C(u,v)$ a $\sum^p_{i+1}$-oracle take formula $D(a) = \exists x \forall \langle i,y\rangle A(a,x,\langle i,y\rangle)$ (bounds to v and $\langle i,y\rangle$ are implicitly in C and A resp.) where formula A is the conjunction of:

(i) $|x| \leq |a|^k$ ($|a|^k$ a time bound),

(ii) "x is a computation of M with some oracle",

(iii) "if the i-th step of computation x is a negative answer to an oracle query $[B(u_i)?]$ then either $\neg C(u_i,y)$ or ($\exists j < i$, "j-th step of x is also a negative answer to oracle query $[B(u_j)?]$ but $B(u_j)$ holds"),

(iv) "all positive answers to oracle queries are correct".

Formula A is clearly $\sum^b_{i+1}$.

Now consider the following algorithm. Take b_0 the computation of M on input a where we answer all oracle queries negatively, and ask for a counterexample to $\forall\langle i,y\rangle A(a,b_0,\langle i,y\rangle)$. If counterexample $\langle i_0,y_0\rangle$ is provided then the negative answer to oracle query $[B(u_{i_0})?]$ in step i_0 was the first incorrect one (and y_0 witnesses the positive answer). Construct computation b_1 identical with b_0 till step $i_0 - 1$, answering oracle query $[B(u_{i_0})?]$ positively and all later queries negatively. Then ask again for a counterexample to $\forall\langle i,y\rangle A(a,b_1,\langle i,y\rangle)$. If $\langle i_1,y_1\rangle$ is provided, step i_1 is the first incorrect one (a negative answer to an oracle query $[B(u_{i_1})?]$) and so take b_2 identical with b_1 till step $i_1 - 1$, answering $[B(u_{i_1})?]$ positively (with y_1 a witness to it), and all later queries, negatively.

In this way construct computations $b_0, b_1, b_2, \ldots$, with b_m correct at least till step m. Thus for $m := |a|^k$, b_m is the correct computation of M^B on input a. Output of M^B is read from b_m.

If unable to separate S^{i+1}_2 from T^{i+1}_2, a natural problem to look at is a separation of relativized versions of S^{i+1}_2 and T^{i+1}_2. Buss (unpublished) showed that $T^1_2(f)$ is not $\sum^b_1(f)$-conservative over $S^1_2(f)$ and Pudlák [6] employed his witnessing theorem to show that $S^{i+1}_2(\alpha) \neq T^{i+1}_2(\alpha)$, for $i = 0, 1$. Here I give an alternative proof of Buss's result and a strengthening of Pudlák's result for $i = 1$.

Theorem.

(a) *The following sequent is provable in* $T^1_2(\alpha, f)$ *but not in* $S^1_2(\alpha, f)$*:*

$$\alpha(0,0), \forall x, y \le a((\alpha(x,y) \wedge x \le y) \supset (\alpha(f(x,y), y+1) \wedge f(x,y) \le y+1)) \to \exists u \le a\ \alpha(u,a).$$

(b) *The following sequent is provable in* $T^2_2(\alpha)$ *but not in* $S^2_2(\alpha)$*:*

$$\forall u, v < a^2 \forall w < a(\alpha(u,w) \wedge \alpha(v,w) \supset u = v),$$
$$\forall u < a^2 \forall v, w < a(\alpha(u,v) \wedge \alpha(u,w) \supset v = w) \to \exists x < a^2 \forall y < a \neg\alpha(x,y).$$

Remark. The sequent from (b) is $\sum^b_2(\alpha)$ while $S^2_2(\alpha)$-axioms are $\sum^b_3(\alpha)$; in this respect (b) improves upon Pudlák's result.

Proof. In both cases we use a relativization of Buss's witnessing theorem.

(a) The sequent is clearly provable in $T^1_2(\alpha, f)$ by induction for formula $\exists u \le a\ \alpha(u,a)$. To show that the sequent is not provable in $S^1_2(\alpha, f)$ it is enough to show that for each polynomial time oracle machine $M^{\alpha,f}$ there exist $\alpha \subseteq \omega^2, a \in \omega$ and $f : \omega^2 \to \omega$ of polynomial growth such that $M^{\alpha,f}(a)$ does not witness the sequent.

Fix machine $M^{\alpha,f}$ and take $a \in \omega$ sufficiently large. We start the computation of $M^{\alpha,f}$ on a; when answering oracle queries we shall assign truth values (resp. values) to some $\alpha(x,y)$ (resp. some $f(x,y)$), for $x \le y \le a$.

(0) Assign to $\alpha(0,0)$ TRUE.

(i) Query $[\alpha(x,y)?]$: If for all $t \le y$, $t \ne x$ to $\alpha(t,y)$ is already assigned truth value FALSE, assign to $\alpha(x,y)$ TRUE and answer YES. Otherwise assign FALSE and answer NO.

(ii) Query $[f(x,y) =?]$: Consider three cases.

(1) To $\alpha(x,y)$ is assigned value TRUE. Choose some $t \le y+1$ such that to $\alpha(t, y+1)$ is assigned TRUE if it exists, or otherwise choose any $t \le y+1$ such that $\alpha(t,y+1)$ has not value assigned yet. Put $f(x,y) := t$ and assign to $\alpha(t,y+1)$ value TRUE.

(2) To $\alpha(x,y)$ is assigned value FALSE. Choose some $t \le y+1$ such that to $\alpha(t,y+1)$ is assigned FALSE if it exists, or otherwise choose any $t \le y+1$ such that $\alpha(t,y+1)$ has not value assigned yet. Put $f(x,y) := t$ and assign to $\alpha(t,y+1)$ value FALSE.

(3) $\alpha(x,y)$ has no value assigned yet. Assign to it some value according to (i) and then define $f(x,y)$ following (1) or (2) above.

The following claim is straightforward.

Claim. *To $\alpha(x, y)$ cannot be assigned value* TRUE *during answers to first y oracle queries.*

Hence obviously $M^{\alpha,f}(a)$ cannot find witness for the succedent. If the output is pair (x, y) such that $x \leq y \leq a$, $\alpha(x, y)$ is assigned TRUE then either $f(x, y)$ is correctly defined $\leq y+1$ and to $\alpha(f(x, y), y+1)$ is assigned TRUE too, or $f(x, y)$ is undefined. Then define it following (iii). Take α to be those pairs (x, y) such that to $\alpha(x, y)$ is assigned TRUE and f to be any extension of the partial function constructed during the computation. Clearly $M^{\alpha,f}(a)$ does not witness the sequent.

This proves clause (a).

(b) Assume $\alpha \subseteq a^2 \times a$ does not satisfy the sequent. Then α is a graph of a $1-1$ map from a^2 into a. In Paris-Wilkie-Woods [5, Thm. 1] it is proved in $I\Delta_0 + \Omega_1$ that there cannot be a Δ_0-definable, $1-1$ map from a^2 into a. Their proof readily formalizes in $S^3_2(\alpha)$ and hence in $T^2_2(\alpha)$ too. (I do not know if this remains true if we drop the second formula from the antecedent.)

To show that $S^2_2(\alpha)$ does not prove the sequent it is enough to show that for any polynomial time oracle machine M^B, and any $\sum^p_1(\alpha)$-predicate B there are $\alpha \subseteq \omega^2$ and $a \in \omega$ such that $M^B(a)$ does not witness the sequent.

Choose $a \in \omega$ sufficiently large. Assume that the $\sum^p_1(\alpha)$-oracle B has the form:

$$B(b) = \exists w \leq t(b)\ N^\alpha(w, b),$$

where $N^\alpha(w, b)$ formalizes

"w is an accepting computation of oracle machine N^α on input b".

We start the computation of M^B on a. During the computation we shall answer oracle queries and also construct partial approximations to $\alpha : \alpha^\pm_0 \subseteq \alpha^\pm_1 \subseteq \cdots \subseteq a^2 \times a$.

Put $\alpha^-_0 = \alpha^+_0 = \emptyset$. Let $[B(b_i)?]$ be the i-th oracle query. Consider two cases.

(i) There exist $\beta \subseteq a^2 \times a$ and $w \leq t(b_i)$ such that:

(1) $\beta \supseteq \alpha^+_{i-1}$ and $\beta \cap \alpha^-_{i-1} = \emptyset$,
(2) $N^\beta(w, b_i)$ holds,
(3) β is a graph of a partial $1-1$ function from a^2 to a.

Answer YES. Computation w contains at most $|a|^k$ oracle queries about β (some fixed $k \in \omega$). Add pairs (c, d) to α_{i-1}^+ resp. to α_{i-1}^- to form $\alpha_i^\pm$, according to whether the answer to oracle query $[\beta(c, d)?]$ in w was affirmative or negative.

In particular, $\mathrm{card}((\alpha_i^+ \cup \alpha_i^-) \backslash (\alpha_{i-1}^+ \cup \alpha_{i-1}^-)) \leq |a|^k$.

(ii) There are no such β and w. Answer NO and put $\alpha_i^\pm := \alpha_{i-1}^\pm$.

Put $\alpha := \bigcup_{i \leq |a|^\ell} \alpha_i^+$ where $|a|^\ell$ is the time bound of M^B. α satisfies the antecedent and so if $M^B(a)$ should witness the sequent it must output $x < a^2$ such that $\forall y < a \neg\alpha(x, y)$. But then we can always find $y < a$, $(x, y) \notin \bigcup_{i \leq |a|^\ell} \alpha_i^-$ and add pair (x, y) into α.

This proves clause (b). □

The sequent from clause (a) is a herbrandization of induction axiom for formula $\exists u \leq a\ \alpha(u, a)$. It would seem natural to conjecture that a herbrandization of induction axiom for $\sum_i^b(\alpha)$-formula

$$\exists x_1 \leq a \forall y_1 \leq a \ldots \alpha(\vec{x}, \vec{y}, a)$$

(i alternating quantifiers), namely:

$$\alpha(\vec{0}, \vec{0}, 0), \forall b, x_1, \ldots, x_m, t_1, \ldots, t_n \leq a[\hat{\alpha}(\vec{x}, y_j/f_j, b) \supset \hat{\alpha}(z_k/g_k, \vec{t}, b+1)] \rightarrow \exists u_1, \ldots, u_m \leq a\ \hat{\alpha}(\vec{u}, v_\ell/h_\ell, a),$$

where:

(0) $m = \frac{\ulcorner i \urcorner}{2}$, $n = \frac{i}{\llcorner 2 \lrcorner}$ and $\hat{\alpha}$ is the formula

$$(x_1 \leq b \wedge (y_1 \leq b \supset (\ldots(\alpha(\vec{x}, \vec{y}, b) \ldots),$$

(i) function f_j depends on $b, x_1, \ldots, x_j, t_1, \ldots, t_{j-1}$,
(ii) function g_k depends on $b, x_1, \ldots, x_k, t_1, \ldots, t_{k-1}$,
(iii) function h_ℓ depends on $a, u_1, \ldots, u_\ell$,

is not provable in $S_2^i(\alpha, \vec{f}, \vec{g}, \vec{h})$. However, this is not true. All these herbrandizations are provable already in $T_2^1(\alpha, \vec{f}, \vec{g}, \vec{h})$.

Remark. The problem whether S_2^{i+1} equals T_2^{i+1} was from a different perspective studied in [4]. Following that paper, Theorem above can be interpreted as results about structure of proofs in predicate calculus.

References

1. S. Buss, *Bounded Arithmetic*, Bibliopolis, Naples, 1986.
2. S. Buss, *Axiomatization and conservation results for fragments of bounded arithmetic*, Proc. Workshop in Logic and Computation, Contemporary Mathematics AMS (to appear).
3. J. Krajíček, P. Pudlák and G. Takeuti, *Bounded arithmetic and the polynomial hierarchy*, Annals of Pure and Applied Logic (to appear).
4. J. Krajíček and G. Takeuti, *On induction-free provability*, Discrete Applied Mathematics (to appear).
5. J. Paris, A. Wilkie and A. Woods, *Provability of the pigeon-hole principle and the existence of infinitely many primes*, Journal of Symbolic Logic **53** no. 4 (1988), 1235–1244.
6. P. Pudlák, *Some relations between subsystems of arithmetic and complexity of computations*, this volume.

Mathematical Institute, Czechoslovak Academy of Sciences, Žitná 25, 115 67, Prague-1

SEMANTIC CHARACTERIZATIONS OF NUMBER THEORIES

DANIEL LEIVANT

Dedicated to Leon Henkin
on the occasion of the 40th anniversary of his
Completeness in the Theory of Types

The first-order predicate calculus is complete for its intended semantics, by Gödel's Completeness Theorem. Type Theory, though not complete for its intended semantics, is complete for the more liberal yet natural semantics of Henkin-structures. We derive here similar semantic characterizations for number theories.[1] We show that Peano's Arithmetic is sound and complete for truth (with object variables interpreted as natural numbers) in Henkin-structures that are closed under abstract jump (Theorem 4.10). By "abstract jump" we mean here Barwise's strict-Π^1_1 definability, a notion that agrees with Turing jump over the natural numbers, and which we argue is of foundational significance. We also show that Σ_1-Arithmetic is sound and complete for validity in all Henkin-structures that contain their "abstract RE" (i.e. strict-Π^1_1 definable) sets (Theorem 4.11). The paper is a refinement (of both results and exposition) of [Lei90]. The author is grateful to P. Odifreddi, J. Sgall, W. Sieg, and S. Simpson for useful comments and conversations pertaining to this paper.

1. PRELIMINARIES

1.1. Number Theories. The vocabulary of First-Order Arithmetic, V_{FA}, has identifiers for zero and for all primitive recursive functions. The standard V_{FA}-structure has the set of natural numbers as universe, and the intended interpretations for 0 and for function identifiers.

First-Order Arithmetic, **FA**, is a V_{FA}-theory, with axioms for equality, primitive recursive definitions, and induction. The *equality axioms* are

[1]Of course, every first-order theory **T** has trivially a circular semantic characterization: **T** consists of the formulas valid in the models of **T**.

$\neg(0=1)$, $\forall x.\ x = x$, $\forall x, y, z.\ x = y \to x = z \to y = z$, and, for each function identifier f and each $i \leq r = arity(f)$,

$$\forall x_1, \ldots, x_r, y.\ x_i = y \to f(x_1, \ldots, x_r) = f(x_1, \ldots, x_{i-1}, y, x_{i+1}, \ldots, x_r).$$

The *induction axioms* of **FA** are the instances of induction for all (first-order) V_{FA}-formulas. The *primitive recursive definitions* of **FA** are the universal closures of defining equations for all primitive recursive functions.[2] We write **PR** for the set of these formulas. For a primitive recursive function f, we let *degree*(f) denote the length of the (shortest in **PR**) primitive recursive definition of f.

It is easy to see that **FA** is a conservative extension of Peano's Arithmetic. Note that Peano's third axiom follows from $\neg(0=1)$ by induction, and the fourth axiom follows from the defining equations for the predecessor function.

Σ_1-Arithmetic, $\mathbf{\Sigma_1 A}$, is like **FA** except that induction is restricted to existential V_{FA}-formulas. Computationally, $\mathbf{\Sigma_1 A}$ is the same as Primitive Recursive Arithmetic, **PRA** (which has induction only for V_{FA}-equations), since they prove the same Π^0_2 formulas, that is, they have the same provably recursive functions [Par77].

It will be useful to identify conditions for doing away with the axiom $\neg(0=1)$. For a number theory **A**, let $\mathbf{A}^-$ denote **A** without $\neg(0=1)$.

Lemma 1.1. *Let φ be a V_{FA}-formula in which equality has no negative occurrences, and let* **A** *be one of the number theories above. If* $\mathbf{A} \vdash \varphi$, *then* $\mathbf{A}^- \vdash \varphi$.

Proof. Troelstra observed [Tro73] that if a formula φ is provable in **A**, then the result of replacing (hereditarily) in φ every negated subformula $\neg\psi$ by $\psi \to 0=1$ is a theorem of $\mathbf{A}^-$. Let φ' be a prenex-disjunctive normal form for φ, so the equivalence $\varphi \leftrightarrow \varphi'$ is provable in first-order logic. Then $\mathbf{A} \vdash \varphi$ implies $\mathbf{A} \vdash \varphi'$, from which $\mathbf{A}^- \vdash \varphi'$, since φ' is without negation, whence $\mathbf{A}^- \vdash \varphi$. □

[2] Our development remains valid if we expand the set of functions to include all partial recursive functions, where the defining equations are Herbrand-Gödel functional programs, as in [Kle52], provably coherent in Primitive Recursive Arithmetic. A Herbrand-Gödel program $\mathcal{P}$ is *coherent* if its operational semantics generates a single-valued relation. Coherence is an undecidable property, but there is a collection C of functional programs such that membership in C is decidable in real time, and such that every partial recursive function has a program in C. The programs obtained from any one of the standard proofs of this fact (as e.g. in [Kle52]) are all provably coherent in Primitive Recursive Arithmetic.

Let *neq* be the primitive recursive characteristic function of inequality: $neq(x,y) = 0$ iff $x \neq y$. For a V_{FA}-formula φ, let $\tilde{\varphi}$ be the result of replacing in φ each negatively occurring equation, $t = s$, by $\neg(neq(t,s) = 0)$. Then $\tilde{\varphi}$ has no negative occurrences of equality, and we have, immediately from the definitions, $\mathbf{PRA} \vdash \varphi \leftrightarrow \tilde{\varphi}$.

Combined with Lemma 1.1 this implies

Lemma 1.2. *Let φ be a V_{FA}-formula, and let $\mathbf{A}$ be one of the number theories above. If $\mathbf{A} \vdash \varphi$ then $\mathbf{A} \vdash \varphi \leftrightarrow \tilde{\varphi}$ and $\mathbf{A}^- \vdash \tilde{\varphi}$.*

1.2. Henkin-Structures. The *language of second-order logic* is an extension of the language of first-order logic (with equality and function identifiers) with relational variables of all finite arities, and with quantification over them. Our basic proof-system for second-order logic, $\mathbf{SOL}_0$, is obtained from the first-order predicate calculus with equality by treating relational variables in par with object variables (without comprehension); see for example [Pra65, §V.1] for details. Given a class Φ of formulas, Φ-*Comprehension* is the schema $\exists R \forall \vec{x}\, (R(\vec{x}) \leftrightarrow \varphi)$, where R is a relational variable that does not occur free in φ, $arity(R) = arity(\vec{x})$, and $\varphi \in \Phi$. If Φ is a collection of second-order formulas, we write $\mathbf{SOL}(\Phi)$ for $\mathbf{SOL}_0$ augmented with Φ-comprehension. $\mathbf{SOL}$ will denote $\mathbf{SOL}_0$ with comprehension for all (second-order) formulas.

Since the collection of second-order formulas that are valid (under the standard interpretation of relational quantification) is not in the arithmetical hierarchy, let alone effectively enumerable, even $\mathbf{SOL}$ is incomplete for standard validity. However, second-order logic is complete for the broader class of Henkin-structures [Hen50]. A *Henkin-structure*, $\mathcal{H}$, consists of a first-order structure over some universe A, augmented with, for each $r \geq 1$, a collection $\mathcal{H}_r$ of r-ary relations over A. Semantic satisfaction, $\mathcal{H} \models \varphi$, is defined using $\mathcal{H}_r$ as the range of quantifiers over r-ary relations. If Φ is a class of second-order formulas, then $\mathcal{H}$ is *closed under* Φ if, for each $\varphi \equiv \varphi[\vec{x}, \vec{R}]$ in Φ, with free object variables $\vec{x} = (x_1 \ldots x_k)$, and free relational variables $\vec{R} = (R_1 \ldots R_l)$ (with $r_i = \text{arity}(R_i)$), and all $Q_1 \in \mathcal{H}_{r_1}, \ldots, Q_l \in \mathcal{H}_{r_l}$, the set $\{(a_1 \ldots a_k) \in A^k \mid \mathcal{H} \models \varphi[\vec{a}/\vec{x}, \vec{Q}/\vec{R}]\}$ is in $\mathcal{H}_k$.

The proof in [Hen50] establishes the following.

Theorem 1.3. [Henkin] *Let Φ be a class of second-order formulas. A second-order formula is valid in all Henkin-structures closed under Φ iff it is provable in $\mathbf{SOL}(\Phi)$.*

1.3. Computational Formulas. A second-order formula is *computational* if it is of the form $\forall\vec{R}\,\exists\vec{x}\,\psi$, where $\vec{R}$ are relational variables and ψ is quantifier-free, i.e. if it is *strict*-Π^1_1 in the sense of [Bar69, Bar75]. A computational formula without free relational variables is *relationally closed* (*r-closed* for short). **Comp** (respectively, $\mathbf{Comp_0}$) will denote the set of formulas equivalent in $\mathbf{SOL_0}$ to a computational formula (respectively, to an r-closed computational formula).

The term "computational" is induced by the fact that each computational formula describes a computation process, which becomes apparent when the formula is converted into an equivalent "computational normal formula," as follows. Let V be a vocabulary, and fix a tuple $\vec{R}$ of relational identifiers. Let χ be a syntactic parameter for quantifier-free V-formulas. A *computational normal formula* is a formula of the form

$$\varphi \;\equiv\; \forall R_1 \ldots R_r\, [\, \forall\vec{u}\,(\,\iota_1 \wedge \iota_2 \cdots \wedge \kappa_1 \wedge \cdots) \;\rightarrow\; \exists\vec{v}\,(\,\theta_1 \vee \cdots)\,],$$

where each ι_n is the disjunction of formulas of the form $\chi \rightarrow R_i(\vec{x})$, each κ_n is the disjunction of formulas of the form $\chi \wedge R_j(\vec{y}) \rightarrow R_i(\vec{x})$, and each θ_n is the conjunction of formulas of one the forms χ or $R(\vec{x})$. The formula φ states (about its free variables) that every process that initializes the values of $\vec{R}$ as prescribed by the ι_n's, and inductively closes these relations as prescribed by the κ_n's, will reach values that satisfy some "target condition" θ_n.

Each computational normal formula defines, uniformly for all V-structures, the operational semantics of a certain finite state machine (see [Lei89] for details). The connection with computational formulas is given by the straightforward observation that every computational formula is equivalent, in $\mathbf{SOL_0}$, to a computational normal formula.

The significance of computational formulas is further manifest in the following.

Theorem 1.4. [Kreisel] *Every computational V_{FA}-formula is equivalent in the standard V_{FA}-structure to an existential formula.*

Hence, every r-closed computational V_{FA}-formula defines in the standard V_{FA}-structure an RE set.

Proof. Using familiar sequence-coding, every computational formula is equivalent in the standard V_{FA}-structure to a computational formula of the form $\forall R\,\exists x\,\psi$, where R is unary, and $\psi \equiv \psi[R]$ is quantifier-free with no free variables other than x and u. Let

$$m_\psi(x,u) =_{df} \max\{\text{ value of } t[x,u] \mid R(t) \text{ is a subformula of } \psi\}.$$

Recall that König's Lemma states that every infinite finitely-branching tree has an infinite branch. This implies that for any formula $\varphi \equiv \varphi[R]$ with a single set variable R,

$$\forall R\, \exists m\, \varphi[R_{\leq m}] \;\rightarrow\; \exists h\, \forall R\, \exists m \leq h.\ \varphi[R_{\leq m}].$$

(Here $R_{\leq m} =_{df} R \cap \{0, \ldots, m\}$.)

We have

$$\begin{aligned} \forall R\, \exists x.\psi[R] \;&\leftrightarrow\; \forall R\, \exists x\, \exists m\, \psi[R_{\leq m_\psi(x,u)}] \\ &\leftrightarrow\; \forall R\, \exists m\, \exists x\, (\, m = m_\psi(x,u) \wedge \psi[R_{\leq m}]\,) \\ &\leftrightarrow\; \exists h\, \forall \sigma \subseteq \{0, \ldots, h\}\ \exists m \leq h\, \exists x\, (\, m = m_\psi(x,u) \wedge \psi[\sigma_{\leq m}]\,). \end{aligned}$$

The forward direction of the last equivalence holds by König's Lemma, and the backward direction is straightforward. The latter formula is equivalent in the standard structure to an existential formula, since the universal quantifier is bounded. □

The proof above of the equivalence of computational formulas to existential formulas clearly applies to any countable admissible structure [Bar69, Bar75 (Theorem VIII.3.1)]. However, this equivalence fails to hold in general for structures which do not contain a code for every completed computation over elements of the structure. For example, if $V_s = \{0, \mathbf{s}\}$, then it is easy to see that every RE set of natural numbers is defined in the standard V_s-structure[3] by a computational formula (compare Lemma 3.1 below), whereas the sets defined in the standard V_s-structure by existential formulas, or even by first-order formulas, are all recursive. (In fact, even for the vocabulary $V_+ = \{0, \mathbf{s}, +\}$, every set of natural numbers defined in the standard V_+-structure by a first-order formula is recursive, by [Pre30].) Thus, computational formulas might be regarded as the appropriate generalization to all structures of recursive enumerability; they reduce to existential formulas over structures in which computations are representable internally, but are in general stronger.

Of interest is also the strength of computational formulas as queries. A k-ary *query* (or *global relation*) over a class $\mathcal{C}$ of structures assigns to each structure $\mathcal{S} \in \mathcal{C}$ a k-ary relation over the universe $|\mathcal{S}|$ of $\mathcal{S}$. If $\mathcal{C}$ consists of V-structures, and φ is a V-formula whose free variables are among $u_1 \ldots u_k$, then $\lambda u_1 \ldots u_k.\ \varphi$ determines a query over $\mathcal{C}$, that assigns to $\mathcal{S} \in \mathcal{C}$ the relation $\{(a_1 \ldots a_k) \in |\mathcal{S}|^k \mid \mathcal{S} \models \varphi[\vec{a}/\vec{u}]\,\}$. Now, over ordered finite structures

[3] where 0 is interpreted as zero and $\mathbf{s}$ as the successor function

the queries defined by computational formulas are exactly the co-NP queries [Fag74, JS74], whereas the queries defined even by all first-order formulas, are a strict subclass of the queries computable in deterministic log-space[4]. Computational formulas as query definitions have found applications in Descriptive Computational Complexity [Lei89], in Finite Model Theory [KV87, KV88], and in Logics of Programs [Lei85, Lei85a].

Note that if a computational formula φ has free relational variables $\vec{Q}$, then it determines a computational process that uses $\vec{Q}$ as oracles, and is equivalent over countable admissible structures to an existential formula with $\vec{Q}$ free. Thus, φ defines an abstract notion of relative RE, that is — an abstract form of Kleene's jump.

2. Direct Second-Order Interpretation of Number Theories

In this section we define a second-order interpretation of V_{FA} which is direct, in the sense that the target formalism has defining equations for primitive recursive functions, in contrast to the "full" interpretation we define in the sequel. While the full interpretation is more logically prestine, the direct interpretation is easier to formulate and verify.

2.1. Definition of the Direct Interpretation. Let

$$N[x] \quad :\equiv \quad \forall R(\, Cl[R] \;\rightarrow R(x)\,),$$

where

$$Cl[R] \quad :\equiv \quad \forall u\,(R(u) \rightarrow R(\mathbf{s}(u)) \;\wedge\; R(0).$$

Note that $N \in \mathbf{Comp}_0$. If $\mathcal{S}$ is a structure that satisfies $\forall u\,\neg(\mathbf{s}(u)=0)$ and $\forall u, v\,(\mathbf{s}(u)=\mathbf{s}(v) \;\rightarrow\; u=v\,)$, then the extension of N in $\mathcal{S}$ is a copy of the natural numbers.

If $\vec{t} = (t_1 \ldots t_r)$ is a tuple of terms, we write $N[\vec{t}]$ for $N[t_1] \wedge \cdots \wedge N[t_r]$. If $\alpha_1, \ldots, \alpha_k$ are formulas or terms, we write $var(\alpha_1, \ldots, \alpha_k)$ for the set of variables that occur free in $\alpha_1, \ldots, \alpha_k$.

If φ is a V_{FA}-formula, then φ^N denotes φ with quantifiers relativized to N. Assuming $N[x]$ we get induction with respect to x for a formula φ, by instantiation of the universal relational quantifier to the relation $\lambda x.\varphi$:

$$\forall u\,(\varphi[u/x] \rightarrow \varphi[\mathbf{s}(u)/x]) \;\wedge\; \varphi[0/x] \;\rightarrow \varphi.$$

[4][Fag75] proves that graph connectivity is not a first-order definable query; an elegant simple proof of this can be found in [GV85]. [Imm87] observes that all first-order queries over finite ordered structures are computable in deterministic log-space.

This is legitimate by comprehension for φ. However, if φ is a first-order V_{FA}-formula, then φ^N is in general not first-order (because N is not), so first-order comprehension does not yield induction for the interpretation of first-order V_{FA}-formulas!

We define the *direct interpretation* of V_{FA} as having N as the formula defining the target universe, and with the V_{FA}-identifiers interpreted by themselves.

Given a formalism **C** for second-order (or higher-order) logic (with constant and function identifiers), the *directly-interpreted number theory of* **C** is

$$\begin{aligned}\mathbf{DNT}[\mathbf{C}] =_{df} & \{\varphi \mid \varphi \text{ is a closed } V_{FA}\text{-formula, and } \mathbf{C}, \mathbf{PR}, \neg(0=1) \vdash \varphi^N\} \\ = & \{\varphi \mid \varphi \text{ is a closed } V_{FA}\text{-formula, and } \mathbf{C}, \mathbf{PR} \vdash \tilde{\varphi}^N\}.\end{aligned}$$

It is not hard to delineate the direct number theory of (impredicative) Second-Order Logic, **SOL**. Recall that Impredicative Analysis, i.e. Second-Order Arithmetic, is the extension of **FA** with quantification over relations, with induction formulated as a single axiom, $\forall x.N[x]$, and with comprehension for all formulas in the language. The following is essentially due to Prawitz [Pra65].

Theorem 2.1. *A first-order V_{FA}-formula φ is a theorem of Impredicative Analysis iff φ^N is provable in* $\mathbf{SOL} + \mathbf{PR} + \neg(0{=}1)$, *i.e. iff $\tilde{\varphi}^N$ is provable in* $\mathbf{SOL} + \mathbf{PR}$.

2.2. Correctness of the Direct Interpretation. The following proposition states the correctness of the direct interpretation of V_{FA} in $\mathbf{SOL}(\mathbf{Comp}_0)$.

Proposition 2.2. *For every V_{FA}-identifier f,*

$$\mathbf{SOL}(\mathbf{Comp}_0) \vdash N[\vec{x}] \to N[f(\vec{x})]$$

(where $arity(\vec{x}) = arity(f)$).

Proof. By induction on $degree(f)$. If f is one of the initial functions, then the proposition is trivial. The case where f is defined by composition is straightforward.

Suppose f is defined by recurrence,

$$\begin{aligned} f(0, \vec{u}) &= g(\vec{u}) \\ f(\mathbf{s}(v), \vec{u}) &= h(v, \vec{u}, f(v, \vec{u})). \end{aligned}$$

Arguing within $\mathbf{SOL}(\mathbf{Comp}_0)$ we have, by induction assumption,

$$N[\vec{u}] \to N[g(\vec{u})] \tag{1}$$

and

$$\text{(2)} \qquad \forall w.\ N[v,\vec{u},w] \rightarrow N[h(v,\vec{u},w)].$$

Assume

$$\text{(3)} \qquad N[v,\vec{u}].$$

From $N[v] \equiv \forall R.Cl[R] \rightarrow R(v)$, instantiating $R(x)$ to the r-closed computational formula $N[f(x,\vec{u})]$, we get

$$\text{(4)} \qquad \forall x\,(N[f(x,\vec{u})] \rightarrow N[f(\mathbf{s}(x),\vec{u})]) \wedge N[f(0,\vec{u})] \ \rightarrow N[f(v,\vec{u})].$$

By (2) and (3) $N[f(x,\vec{u})]$ implies $N[h(x,\vec{u},f(x,\vec{u}))]$, which by the second defining equation for f yields $N[f(\mathbf{s}(x),\vec{u})]$. This proves the first conjunct in (4). The second conjunct is immediate from (1) and the first defining equation for f. Thus we get the conclusion of (4), $N[f(v,\vec{u})]$, based on assumption (3), which is precisely the statement of the proposition. □

2.3. Soundness of the direct interpretation. In this section we prove that the direct interpretation of **FA** is sound for **SOL**(**Comp**), that is, if **FA** $\vdash \varphi$ then **SOL**(**Comp**), **PR**, $\neg(0=1) \vdash \varphi$. Analogously, we show that the interpretation of $\mathbf{\Sigma_1 A}$ is sound for **SOL**($\mathbf{Comp_0}$).

Lemma 2.3. **Comp** *and* $\mathbf{Comp_0}$ *are closed under conjunction and disjunction.*

Proof. Trivial, by basic quantifier rules. □

Lemma 2.4. $\mathbf{SOL_0} \vdash Cl[N]$.

Proof. Straightforward. □

Lemma 2.5. *A formula φ of the form $\exists x^N\, \psi$, where ψ is [r-closed] quantifier-free, is equivalent in* **SOL**($\mathbf{Comp_0}$) *to an [r-closed] computational formula.*

Proof. We have

$$\varphi \ \equiv\ \exists x\,(\,\psi \wedge \forall R\,(\,Cl[R] \rightarrow R(x)\,)\,)$$

(assuming, without loss of generality, that R does not occur in ψ). We claim that this is equivalent to

$$\varphi' \ \equiv_{df}\ \forall R\, \exists x\,(\,\psi \wedge (\,Cl[R] \rightarrow R(x)\,)\,),$$

which is clearly an r-closed computational formula. φ implies φ' trivially. For the converse, instantiating R in φ' to N, yields $\exists x\,(\,\psi \wedge (\,Cl[N] \rightarrow N[x]\,)\,)$, which by Lemma 2.4 implies φ. □

Lemma 2.6. *For every V_{FA}-formula φ, comprehension for φ^N is provable in* **SOL(Comp)**.

Proof. By induction on φ. Without loss of generality, we assume that $\wedge$, $\neg$, and $\exists$ are the only logical constants. If φ is quantifier-free then it is computational, and the lemma is trivial.

If $\varphi \equiv \psi \wedge \chi$, then $\varphi^N \equiv \psi^N \wedge \chi^N$. By induction assumption **SOL(Comp)** proves $\exists P \forall \vec{u}\ (P(\vec{u}) \leftrightarrow \psi^N)$ and $\exists Q \forall \vec{v}\ (Q(\vec{v}) \leftrightarrow \chi^N)$, where $\vec{u}$ and $\vec{v}$ list $var(\psi)$ and $var(\chi)$, respectively. By comprehension for quantifier-free formulas

$$\forall P, Q\, \exists R\, \forall \vec{w}\ (\, R(\vec{w}) \leftrightarrow P(\vec{u}) \wedge Q(\vec{v})\,),$$

where $\vec{w}$ lists $var(\varphi) = \vec{u} \cup \vec{v}$. So

$$\mathbf{SOL(Comp)} \vdash \exists R\, \forall \vec{w}\, (\, R(\vec{w}) \leftrightarrow \varphi^N\,).$$

The case $\varphi \equiv \neg\psi$ is treated similarly.

Finally, if $\varphi \equiv \exists x\, \psi$, then $\varphi^N \equiv \exists x (N[x] \wedge \psi^N)$. By induction assumption, **SOL(Comp)** proves $\exists R\, \forall \vec{u}, x\, (\, R(\vec{u}, x) \leftrightarrow \psi^N\,)$, where $\vec{u}$ lists $var(\varphi)$. Hence, **SOL(Comp)** proves (using the same R), $\exists R\, \forall \vec{u}\, (\, (\exists x\, (N[x] \wedge R(\vec{u}, x))) \leftrightarrow \varphi^N)$. By Lemma 2.5 **SOL(Comp)** proves $\exists Q \forall \vec{u}\ (Q(\vec{u}) \leftrightarrow \exists x\, (\, N[x] \wedge R(\vec{u}, x))\,)$, i.e., $\exists Q \forall \vec{u}\, (\, Q(\vec{u}) \leftrightarrow \varphi^N\,)$. □

Lemma 2.7. *For each V_{FA}-term t,*

$$\mathbf{SOL(Comp_0)} \vdash N[var(t)] \rightarrow N[t].$$

Proof. By induction on (the structure of) t. The basis is trivial, and the induction step is Proposition 2.2. □

Lemma 2.8. *Let φ be a V_{FA}-formula.*

(1) *If* $\mathbf{FA}^- \vdash \varphi$, *then* $\mathbf{SOL(Comp)}, \mathbf{PR} \vdash N[var(\varphi)] \rightarrow \varphi^N$.
(2) *If* $\mathbf{\Sigma_1 A}^- \vdash \varphi$, *then* $\mathbf{SOL(Comp_0)}, \mathbf{PR} \vdash N[var(\varphi)] \rightarrow \varphi^N$.

Proof. By induction on proofs of $\mathbf{\Sigma_1 A}$ and **FA**, say in the Hilbert-style deductive calculus of [Kle52, §19] for the logical constants $\neg, \rightarrow$, and $\forall$.

The axioms are of the following types.

(1) φ is an instance of a propositional schema. Then φ^N is an instance of the same propositional schema.
(2) φ is of the form $\forall x \psi \rightarrow \psi[t/x]$. Then $N[var(\, \forall x \psi\,), t] \rightarrow \varphi^N$ outright. By Lemma 2.7, $N[var(t)] \rightarrow N[t]$. Since $var(t) \subseteq var(\varphi)$, we get $N[var(\varphi)] \rightarrow \varphi^N$.
(3) φ is one of the logical axioms for equality. Then φ is in $\mathbf{SOL}_0$.

(4) $\varphi \in \mathbf{PR}$, trivial.

(5) φ is an instance of Induction, $\forall x.\ Cl[\psi] \rightarrow \psi[x]$, with ψ existential for (2). Then $\varphi^N \equiv \forall x^N (Cl[\psi]^N \rightarrow \psi^N[x])$. We have comprehension for the formula $N[u] \wedge \psi^N[u]$: by Lemma 2.6 for (1), and by Lemma 2.5 for (2). Therefore, $N[x]$ implies

$$\forall u\, ((N[u] \wedge \psi^N[u]) \rightarrow (N[\mathbf{s}(u)] \wedge \psi^N[\mathbf{s}(u)]))\ \wedge\ (N[0] \wedge \psi^N[0])\ \rightarrow (N[x] \wedge \psi^N[x]),$$

which clearly implies

$$\forall u^N\, (\psi^N[u] \rightarrow \psi^N[\mathbf{s}(u)]) \wedge \psi^N[0]\ \rightarrow\ \psi^N[x],$$

i.e., $Cl[\psi]^N \rightarrow \psi^N[x]$. We have thus proved φ^N.

For the induction step of the proof, we check that the statement of the lemma is preserved under the two inference rules:

Detachment: φ is derived from $\chi \rightarrow \varphi$ and χ. By induction assumption, $N[var(\chi, \varphi)] \rightarrow (\chi^N \rightarrow \varphi^N)$ and $N[var(\chi)] \rightarrow \chi^N$ are both provable (in **SOL**(**Comp**) for part (1) of the lemma and in **SOL**($\mathbf{Comp_0}$) for (2)). Let χ_0 be χ with 0 substituted for all free occurrences of variables not free in φ. Then $N[var(\varphi)] \rightarrow (\chi_0^N \rightarrow \varphi^N)$ and $N[var(\varphi)] \rightarrow \chi_0^N$ are both provable. Thus $N[var(\varphi)] \rightarrow \varphi^N$ is provable, by Detachment.

Generalization: $\varphi \equiv \chi \rightarrow \forall x \psi$ is derived from $\chi \rightarrow \psi$. By induction assumption $N[var(\varphi), x] \rightarrow (\chi^N \rightarrow \psi^N)$ is provable. Therefore, $N[var(\varphi)] \rightarrow (\chi^N \rightarrow \forall x^N \psi^N)$ is provable, by Generalization, since x must not be free in χ.

This concludes the induction step and the proof. □

From Lemmas 1.2 and 2.8 we obtain

Theorem 2.9. *Let φ be a V_{FA}-formula. If* $\mathbf{FA} \vdash \varphi$, *then*

$$\mathbf{SOL}(\mathbf{Comp}),\ \mathbf{PR},\ \neg(0=1) \vdash N[var(\varphi)] \rightarrow \varphi^N, \text{ and}$$

$$\mathbf{SOL}(\mathbf{Comp}),\ \mathbf{PR} \vdash N[var(\varphi)] \rightarrow \tilde{\varphi}^N.$$

Similarly, if $\mathbf{\Sigma_1 A} \vdash \varphi$, *then*

$$\mathbf{SOL}(\mathbf{Comp_0}),\ \mathbf{PR},\ \neg(0=1) \vdash N[var(\varphi)] \rightarrow \varphi^N, \text{ and}$$

$$\mathbf{SOL}(\mathbf{Comp_0}),\ \mathbf{PR} \vdash N[var(\varphi)] \rightarrow \tilde{\varphi}^N.$$

3. Full Second-Order Interpretations of Number Theories

We define an interpretation of the vocabulary of **SOL** that has each V_{FA}-identifier interpreted by an r-closed computational formula that defines its graph. The target formalism of the interpretation cannot make do with no constants at all, since to interpret 0 and $\mathbf{s}$ in the absence of constants we would need second-order constant-free formulas, φ with only x free, and ψ with only x and y free, such that $\mathbf{SOL} \vdash \exists! x\, \varphi$ and $\mathbf{SOL} \vdash \forall y\,(M[y] \rightarrow \exists! x\, \psi)$ where M is a formula interpreting N. Clearly, no such formulas exist. We therefore assume that 0 and $\mathbf{s}$ are present in the target vocabulary.

3.1. Graphs of Primitive Recursive Functions. For each V_{FA}-identifier f, we define a formula G_f, by induction on *degree* (f), as follows.

- If f is the zero function, then $G_f[x, z] \equiv_{df} (z = 0)$.
- If f is the successor function, then $G_f[x, z] \equiv_{df} (z = \mathbf{s}(x))$.
- If f is the i'th-out-of-n projection function, then $G_f[x_1 \ldots x_n, z] \equiv_{df} (z = x_i)$.
- If f is defined by composition, $f(\vec{x}) = h(g_1(\vec{x}), \ldots, g_k(\vec{x}))$, then

$$G[\vec{x}, z] \equiv_{df} \exists y_1 \ldots y_k.\, G_{g_1}[\vec{x}, y_1] \wedge \cdots \wedge G_{g_k}[\vec{x}, y_k] \wedge G_h[\vec{y}, z].$$

- If f is defined by recurrence, $f(0, \vec{u}) = g(\vec{u})$, $f(\mathbf{s}(v), \vec{u}) = h(v, \vec{u}, f(v, \vec{u}))$, then

$$\begin{aligned} G[\vec{x}, z] \equiv_{df}\ & \forall Q\,(\forall \vec{u}, y\,(G_g[\vec{u}, y] \rightarrow Q(0, \vec{u}, y)) \\ & \wedge \forall \vec{u}, v, w, y\,(Q(v, \vec{u}, w) \wedge G_h[v, \vec{u}, w, y] \rightarrow \\ & \qquad Q(\mathbf{s}(v), \vec{u}, y)) \rightarrow Q(\vec{x}, z)) \end{aligned}$$

(where *arity* $(\vec{x})$ = *arity* $(\vec{u}) + 1$).

We need the following generalization of Lemma 2.5.

Lemma 3.1. *Let φ be a conjunction of formulas of the form $G_f[\vec{t}]$, formulas of the form $N[t]$, and quantifier-free [r-closed] formulas. Then $\exists \vec{x}\, \varphi$ is equivalent in* **SOL**($\mathbf{Comp_0}$) *to an [r-closed] computational formula.*

In particular, every formula of the form $G_f[\vec{t}]$ is equivalent in **SOL**($\mathbf{Comp_0}$) *to an r-closed computational formula.*

Proof. Suppose

$$\varphi \equiv G_{f_1}[\vec{t_1}, s_1] \wedge \cdots \wedge G_{f_k}[\vec{t_k}, s_k] \wedge N[q_1] \wedge \cdots \wedge N[q_l] \wedge \alpha,$$

where α is quantifier-free. We prove the lemma by main induction on the number l of conjuncts $N[q_i]$, secondary induction on $m = \max_{i \leq k}[degree(f_i)]$, and ternary induction on the number n of conjuncts G_{f_i} with $degree(f_i) = m$.

If $l = m = 0$ (i.e. $k = 0$), then the lemma is trivial.

If $l > 0$, then

$$\begin{array}{rrcl} & \varphi & \equiv & \exists x\,(\,\psi \wedge \forall R.\, Cl[R] \rightarrow R(q_l)\,) \\ \text{where} & \psi & \equiv & N[q_1] \wedge \cdots \wedge N[q_{l-1}] \wedge \alpha. \end{array}$$

As in the proof of Lemma 2.5, φ is equivalent to

$$\forall R\, \exists x\,(\,\psi \wedge (Cl[R] \rightarrow R(q_l))\,),$$

which is equivalent, by elementary quantifier rules, to

$$\varphi' \equiv_{df} \forall R\, \exists x, u\,(\,\psi \wedge \beta\,),$$

where u is fresh and β is the quantifier-free formula $(\,R(u) \rightarrow R(\mathbf{s}(u))\,) \rightarrow R(q_l)$. By induction assumption, φ' is equivalent in $\mathbf{SOL}(\mathbf{Comp_0})$ to an r-closed computational formula.

If $k > 0$ (and $l = 0$), then φ is of the form $\exists x\,(\,\psi \wedge G_f[\vec{t}, s]\,)$, where $degree(f) = m$. We proceed by cases for the definition of f. If f is initial, then $G_f[\vec{t}, s]$ is quantifier-free, and we are done by induction assumption. If f is defined by composition, then $G_f[\vec{t}, s]$ is of the form

$$\exists v_1 \ldots v_k\,(\,G_{g_1}[\vec{t}, v_1] \wedge \cdots \wedge G_{g_k}[\vec{t}, v_k] \wedge G_h[\vec{v}, s]\,),$$

with $degree(g_1), \ldots, degree(g_k), degree(h) < m$. So φ is equivalent to

$$\exists x, v_1 \ldots v_k\ (\,\psi \wedge G_{g_1}[\vec{t}, v_1] \wedge \cdots \wedge G_{g_k}[\vec{t}, v_k] \wedge G_h[\vec{v}, s]\,),$$

for which the lemma holds by induction assumption.

Suppose that f is defined by recurrence.

Claim 1. *$G_f[\vec{t}, s]$ is r-closed computational. By definition, $G_f[\vec{t}, s]$ is of the form $\forall Q.\, \chi$, where*

$$\begin{aligned} \chi \ \equiv\ & \forall \vec{u}, v\,(\,G_g[\vec{u}, v] \rightarrow Q(0, \vec{u}, v)\,) \\ & \quad \wedge\ \forall \vec{w}, y, z, a\,(\,Q(y, \vec{w}, z) \wedge G_h[y, \vec{w}, z, a] \rightarrow Q(\mathbf{s}(y), \vec{w}, a)\,) \\ & \quad \rightarrow Q(\vec{t}, s). \end{aligned}$$

The formula χ is equivalent to

$$\begin{aligned}\chi' \;\equiv_{df}\; & \exists \vec{u}, v\,(\,G_g[\vec{u}, v] \wedge \neg Q(0, \vec{u}, v)\,)\\ & \vee\, \exists \vec{w}, y, z, a\,(\,Q(y, \vec{w}, z) \wedge G_h[y, \vec{w}, z, a] \wedge \neg Q(\mathbf{s}(y), \vec{w}, a)\,)\\ & \vee\, Q(\vec{t}, s).\end{aligned}$$

Since $degree\,(g), degree\,(h) < m$, each one of the disjuncts of χ' is equivalent in **SOL**(**Comp**$_0$) to an r-closed computational formula, by induction assumption, and therefore, by Lemma 2.3, $G_f[\vec{t}, s]$ is also equivalent to an r-closed computational formula, proving Claim 1.

φ

Claim 2. *is equivalent, in* **SOL**(**Comp**$_0$), *to*

$$\varphi' \;\equiv_{df}\; \forall Q\, \exists x\,(\,\psi \wedge \chi'\,).$$

Clearly, $\varphi \equiv \exists x\,(\,\psi \wedge \forall Q\,\chi\,)$ *implies* $\forall Q\,\exists x\,(\,\psi \wedge \chi\,)$, *since* Q *is not free in* ψ. *This implies* $\forall Q\,\exists x\,(\,\psi \wedge \chi'\,)$, *i.e.* φ', *since* χ' *is equivalent to* χ.

For the converse, assume φ'. Then, by instantiating Q to $\lambda v, \vec{u}, w.\; G_f[v, \vec{u}, w]$, we get

$$\begin{aligned}\varphi_0 \;\equiv_{df}\; & \exists x\,(\,\psi \wedge \chi'[G_f\,/\,Q]\,)\\ \equiv\; & \exists x\,(\,\psi \wedge \chi[G_f\,/\,Q]\,).\end{aligned}$$

This instantiation is legitimate in **SOL**(**Comp**$_0$) by Claim 1. But the two premises of $\chi[G_f\,/\,Q]$, that is

$$\forall \vec{u}, v\,(\,G_g[\vec{u}, v] \rightarrow G_f[0, \vec{u}, v]\,)$$

and
$$\forall \vec{w}, y, z, a\,(\,G_f[y, \vec{w}, z] \wedge G_h[y, \vec{w}, z, a] \rightarrow G_f[\mathbf{s}(y), \vec{w}, a]\,)$$

are straightforward in **SOL**(**Comp**$_0$). Therefore, $\chi[G_f\,/\,Q]$ implies its antecedent, i.e. $G_f[\vec{t}, s]$. So φ_0 implies, in **SOL**(**Comp**$_0$), $\exists x\,(\,\psi \wedge G_f[\vec{t}, s]\,)$, i.e. φ. This proves Claim 2.

To prove the lemma it remains to show that $\varphi' \equiv \forall Q\,\exists x\,(\,\psi \wedge \chi'\,)$ is equivalent in **SOL**(**Comp**$_0$) to an r-closed computational formula. We have

$$\begin{aligned}\exists x\,(\,\psi \wedge \chi'\,) \equiv\; & \exists x, \vec{u}, v.\,(\,\psi \wedge G_g[\vec{u}, v] \wedge \neg Q(0, \vec{u}, v)\,)\\ & \vee\; \exists x, \vec{w}, y, z, a\,(\,\psi \wedge Q(y, \vec{w}, z) \wedge G_h[y, \vec{w}, z, a] \wedge \neg Q(\mathbf{s}(y), \vec{w}, a)\,)\\ & \vee\; \exists x\,(\,\psi \wedge Q(\vec{t}, s)\,).\end{aligned}$$

By induction assumption, each one of the disjuncts is equivalent in **SOL**(**Comp**$_0$) to an r-closed computational formula, so φ' is also equivalent to such a formula, by Lemma 2.3. □

3.2. Full Second-Order Interpretation of Arithmetic. The *full second-order interpretation of* V_{FA} has N as the formula that defines the target universe, with 0 interpreted by 0, $\mathbf{s}$ interpreted by $\mathbf{s}$, and every other V_{FA}-identifier f interpreted by the formula G_f, in the usual sense. The following is a more detailed description of the latter point.

Let us say that an equation is *simple* if it is of the form $f(\vec{u})=v$, where $\vec{u}, v$ are variables, and that a formula is *simple* if all equations therein are simple or are equations between atomic terms (i.e., variables or constants). For an equation E, let E^s be a simple formula, equivalent to E, obtained by hereditarily replacing equations by equivalent existential simple formulas. For example, $f(g(u))=v$ is replaced by $\exists w\,(\,g(u)=w \wedge f(w)=v\,)$.

For a V_{FA}-formula φ the *interpretation* φ^I *of* φ arises from φ by replacing each equation E (except simple equations and equations between atomic terms) by E^s, then replacing each simple equation $f(\vec{u}) = v$ by $G_f[\vec{u}, v]$, then relativizing quantifiers to N. Note that in the standard V_{FA}-structure $\forall \vec{x}, z\,(\,G_f[\vec{x}, z] \leftrightarrow f(\vec{x})=z\,)$, and so $\varphi \leftrightarrow \varphi^I$.

Given a second-order (or higher-order) formalism $\mathbf{C}$ (with constants 0 and $\mathbf{s}$), the *fully-interpreted number theory of* $\mathbf{C}$, $\mathbf{FNT}[\mathbf{C}]$, is

$$\begin{aligned}\mathbf{FNT}[\mathbf{C}] &=_{df} \{\varphi \mid \varphi \text{ is a closed } V_{FA}\text{-formula, and } \mathbf{C} \vdash \tilde{\varphi}^I\} \\ &= \{\varphi \mid \varphi \text{ is a closed } V_{FA}\text{-formula, and } \mathbf{C}, \neg(0=1) \vdash \varphi^I\}.\end{aligned}$$

3.3. Correctness of the Full Interpretation. We show that the full interpretation of V_{FA} in $\mathbf{SOL}(\mathbf{Comp})$, $\varphi \mapsto \varphi^I$, is correct, that is, that the interpretation of each V_{FA}-identifier is the graph of a function over the interpreted universe N, provably in $\mathbf{SOL}(\mathbf{Comp})$. We shall prove half of this already in $\mathbf{SOL}(\mathbf{Comp}_0)$, so our interpretation is "semi-correct" for $\mathbf{SOL}(\mathbf{Comp}_0)$.

Lemma 3.2. *If f is a V_{FA}-identifier, then*

$$\mathbf{SOL}(\mathbf{Comp}_0) \vdash \forall \vec{x}^N\, \exists z^N\, G_f[\vec{x}, z]$$

(where $arity(\vec{x}) = arity(f)$).

Proof. By induction on $degree(f)$. The induction basis is trivial, and the case where f is defined by composition is straightforward.

Suppose f is defined by recurrence, $f(0, \vec{u}) = g(\vec{u})$; $f(\mathbf{s}(v), \vec{u}) = h(v, \vec{u}, f(v, \vec{u}))$. By Lemma 3.1 the formula $\varphi[v] \equiv_{df} N[v] \wedge \exists z^N\, G_f[v, \vec{y}, z]$

is equivalent in $\mathbf{SOL}(\mathbf{Comp}_0)$ to an r-closed computational formula, and so $N[x]$ implies, in $\mathbf{SOL}(\mathbf{Comp}_0)$,

$$(\forall v\, \varphi[v] \to \varphi[\mathbf{s}(v)]\,) \wedge \varphi[0] \;\; \to \varphi[x].$$

By induction assumption applied to G_g, $N[\vec{y}] \to \exists z^N\, G_g[\vec{y}, z]$, and so $N[\vec{y}] \to \varphi[0]$, by the definition of G_f. By induction assumption applied to G_h we have

$$\forall w\, (\, N[v, \vec{y}, w] \to \exists z^N\, G_h[v, \vec{y}, w, z]\,),$$

so $N[\vec{y}]$ implies $\forall v\,(\,\varphi[v] \to \varphi[\mathbf{s}(v)]\,)$. We have proved $N[x, \vec{y}] \to \varphi[x]$, which trivially implies the statement of the lemma. □

Lemma 3.3. *If φ is a formula generated by propositional connectives and first-order quantifiers from atomic formulas, from formulas of the form $G_f[\vec{t}]$, and from formulas of the form $N[t]$, then comprehension for φ is provable in* $\mathbf{SOL}(\mathbf{Comp})$.

Proof. Analogous to the proof of Lemma 2.6, using Lemma 3.1 for the base case. □

Lemma 3.4. *Let f be a V_{FA}-identifier. Then*

$$\mathbf{SOL}(\mathbf{Comp}) \vdash \forall \vec{x}^N, z^N, s^N\, (\, G_f[\vec{x}, z] \wedge G_f[\vec{x}, s] \to z = s\,)$$

(where $arity(\vec{x}) = arity(f)$).

Proof. By induction on $degree(f)$. The cases for initial functions and for definition by composition are straightforward.

If f is defined by recurrence, then $G_f[y, \vec{x}, z]$ is of the form $\forall Q\, \chi$, where

$$\begin{aligned} \chi \;\equiv\;\; & (\forall \vec{u}, w\, (\, G_g[\vec{u}, w] \to Q(0, \vec{u}, w)\,) \\ & \wedge \forall v, \vec{u}, w, a\, (\, Q(v, \vec{u}, a) \wedge G_h[v, \vec{u}, a, w] \to Q(\mathbf{s}(v), \vec{u}, w)\,)\,) \\ & \quad \to Q(y, \vec{x}, z). \end{aligned}$$

Let

$$\psi[p, \vec{q}, r] \;\equiv_{df}\; G_f[p, \vec{q}, r] \wedge (\, N[p, \vec{q}, r] \to \forall s\, (\, G_f[p, \vec{q}, s] \to s = r\,)\,).$$

By Lemma 3.3 comprehension for ψ is provable in $\mathbf{SOL}(\mathbf{Comp})$, so $G_f[y, \vec{x}, z]$ implies $\chi[\psi/Q]$. The first conjunct of the premise of $\chi[\psi/Q]$ is provable in $\mathbf{SOL}(\mathbf{Comp})$, by induction assumption applied to the function g. Using $N[p]$ and induction assumption applied to the function h, the second conjunct in the premise of $\chi[\psi/Q]$ is also provable. Thus, the conclusion is provable, i.e. we have proved, under the assumption $G_f[x, \vec{y}, z]$, that

$$N[y, \vec{x}, z] \to \forall s\, (\, G_f[x, \vec{y}, s] \to s = z\,),$$

i.e. the statement of the lemma. □

From Lemmas 3.2 and 3.4 we have

Proposition 3.5. *The full interpretation of V_{FA} in* **SOL(Comp)** *is correct; that is, for every V_{FA}-identifier f,*

$$\mathbf{SOL(Comp)} \vdash \forall \vec{x}^N \, \exists! y^N \, G_f[\vec{x}, y].$$

3.4. Soundness of the full interpretation. We now show that the full interpretation of **FA** is sound for **SOL(Comp)**, and that the full interpretation of $\mathbf{\Sigma_1 A}$ is sound for $\mathbf{SOL(Comp_0)}$.

Lemma 3.6. *If $E \in$* **PR**, *then* $\mathbf{SOL(Comp_0)} \vdash N[var(E)] \to E^I$.

Proof. Without loss of generality, let E be the principal equation of a definition by recurrence, $\forall x, \vec{y}\, (\, f(\mathbf{s}(x), \vec{y}) = h(x, \vec{y}, f(x, \vec{y}))\,)$. Then

$$E^I \equiv \forall x^N, \vec{y}^N \, \exists u^N, v^N \, (\, G_f(x, \vec{y}, u) \wedge G_h(x, \vec{y}, u, v) \wedge G_f(\mathbf{s}(x), \vec{y}, v)\,).$$

This is provable in $\mathbf{SOL(Comp_0)}$ by Lemma 3.2 and the definition of G_f. □

Lemma 3.7. *Let f be a V_{FA}-identifier. If φ is an equality axiom of the form*

$$\varphi \equiv \forall x_1, \ldots, x_r, y\, (\, x_i = y \;\to\; f(x_1, \ldots, x_r) = f(x_1, \ldots, x_{i-1}, y, x_{i+1}, \ldots, x_r)\,),$$

then $\mathbf{SOL(Comp_0)} \vdash \varphi^I$.

Proof. We have

$$\varphi^I \equiv \forall x_1^N, \ldots, x_r^N, y^N \, (\, x_i = y \to \exists u^N \, G_f[x_1, \ldots, x_{i-1}, y, x_{i+1}, \ldots, x_r, u] \wedge G_f[x_1, \ldots, x_r, u]\,).$$

This is immediate by Lemma 3.2. □

Theorem 3.8. *Let φ be a V_{FA}-formula. If* $\mathbf{FA} \vdash \varphi$, *then*

$$\mathbf{SOL(Comp)}, \neg(0 = 1) \vdash N[var(\varphi)] \to \varphi^I.$$

and

$$\mathbf{SOL(Comp)} \vdash N[var(\varphi)] \to \tilde{\varphi}^I.$$

Proof. The proof of the first half of the proposition is essentially the same as for Proposition 2.8. Different are only the cases for equality axioms, which follow here from Lemmas 3.6 and 3.7.

Analogously, $\mathbf{FA}^- \vdash \varphi$ implies $\mathbf{SOL}(\mathbf{Comp}) \vdash N[var(\varphi)] \rightarrow \varphi^I$, which, combined with Lemma 1.1, yields the second half of the theorem. □

To prove an analogous statement for $\mathbf{\Sigma}_1\mathbf{A}$ and $\mathbf{SOL}(\mathbf{Comp}_0)$ we first make the following observation.

Lemma 3.9. *If φ is an existential V_{FA}-formula, then comprehension for φ^I is provable in* $\mathbf{SOL}(\mathbf{Comp}_0)$.

Proof. Immediate from Lemma 3.1. □

Theorem 3.10. *Let φ be a V_{FA}-formula. If $\mathbf{\Sigma}_1\mathbf{A} \vdash \varphi$, then*

$$\mathbf{SOL}(\mathbf{Comp}_0),\ \neg(0=1) \vdash N[var(\varphi)] \rightarrow \varphi^I.$$

and

$$\mathbf{SOL}(\mathbf{Comp}_0) \vdash N[var(\varphi)] \rightarrow \tilde{\varphi}^I.$$

Proof. Similar to the proof of Theorem 2.9, except that Lemma 3.9 is needed to justify the interpretation of existential instances of induction. □

4. Faithfulness of the Interpretations

In this section we prove the *faithfulness*, i.e. completeness for $\mathbf{SOL}(\mathbf{Comp})$, of the interpretations of **FA**, and the completeness of for $\mathbf{SOL}(\mathbf{Comp}_0)$ of the interpretations of $\mathbf{\Sigma}_1\mathbf{A}$. To do so we need to consider Theorem 1.4 more formally. Using a surjective coding of sequences by numbers (à la Kleene [Kle52]), we let

$$Init[Q,x] =_{df} \forall i{<}lth(x)\ (Q(i) \wedge (x)_i = 1)\ \vee\ (\neg Q(i) \wedge (x)_i = 0),$$

i.e., x is an initial segment of the characteristic function of Q. Put

$$\mathrm{WKL}_0 =_{df} \forall R\,(\,(\forall Q\,\exists x\ Init[Q,x] \wedge R(x)\,) \rightarrow \exists h\,\forall Q\,\exists x{<}h\ Init[Q,x] \wedge R(x)\,),$$

i.e., if R is a unary predicate over the universal binary tree, that contains an element on every branch, then there is a bound on the height of these elements.

Let **BT** be $\mathbf{SOL}_0$ with comprehension for quantifier-free formulas, quantifier-free induction, and **PR**. This is the same as the Base Theory of [Sie87], but formulated with relational variables. Let Σ^0_1 denote, as usual, the set of existential (first-order) V_{FA}-formulas, and $\tilde{\Sigma}^0_1$ the set of existential V_{FA}-formulas with relational parameters. A formalization of the proof of Theorem 1.4 establishes the following.

Lemma 4.1. *Every computational formula is equivalent, provably in* $\mathbf{BT}+\mathrm{WKL}_0$, *to a* $\tilde{\Sigma}^0_1$ *formula.*

Every r-closed computational formula is equivalent, provably in $\mathbf{BT}+\mathrm{WKL}_0$, *to a* Σ^0_1 *formula.*

We shall need the following two technical lemmas. If Φ is a collection of formulas, then $\mathbf{Ind}(\Phi)$ denotes induction for all $\varphi \in \Phi$.

Lemma 4.2. $\mathbf{BT}+\mathbf{SOL}(\tilde{\Sigma}^0_1)+WKL_0+\forall x N[x]$ *is conservative over* $\mathbf{FA}$.

Proof. Suppose

$$\mathbf{BT},\ \mathbf{SOL}(\tilde{\Sigma}^0_1),\ \mathrm{WKL}_0,\ \forall x\, N[x]\ \vdash\ \varphi,$$

where φ is first-order. Then

$$\mathbf{BT},\ \mathbf{SOL}(\tilde{\Sigma}^0_1),\ \mathrm{WKL}_0,\ \mathbf{Ind}(\tilde{\Sigma}^0_1)\ \vdash\ \varphi.$$

However, the straightforward proof of WKL_0 (in fact of full König's Lemma) is easily derivable in $\mathbf{BT}+\mathbf{SOL}(\tilde{\Sigma}^0_1)$ (compare [Fri69, Theorem 3]). So we get

$$\mathbf{BT},\ \mathbf{SOL}(\tilde{\Sigma}^0_1),\ \mathbf{Ind}(\tilde{\Sigma}^0_1)\ \vdash\ \varphi.$$

The latter theory is well-known to be conservative over $\mathbf{FA}$ (see e.g. [Tro73, §1.9.4]), concluding the proof. □

Proving the analogous statement for $\mathbf{\Sigma}_1\mathbf{A}$ requires a little more:

Lemma 4.3. $\mathbf{BT}+\mathbf{SOL}(\Sigma^0_1)+WKL_0+\forall x N[x]$ *is conservative over* $\mathbf{\Sigma}_1\mathbf{A}$.

Proof. Suppose

$$\mathbf{BT},\ \mathbf{SOL}(\Sigma^0_1),\ \mathrm{WKL}_0,\ \forall x\, N[x]\ \vdash\ \varphi,$$

where φ is first-order. Then, as in the previous proof,

$$\mathbf{BT},\ \mathbf{SOL}(\Sigma^0_1),\ \mathrm{WKL}_0,\ \mathbf{Ind}(\Sigma^0_1)\ \vdash\ \varphi,$$

i.e.,

$$\mathbf{BT},\ \mathrm{WKL}_0,\ \mathbf{Ind}(\Sigma^0_1)\ \vdash\ \chi \rightarrow \varphi,$$

where χ is the conjunction of instances of Σ^0_1-comprehension. Sieg [Sie87] showed that the latter theory is conservative over $\mathbf{BT}+\mathbf{Ind}(\Sigma^0_1)$ with respect to Π^1_1-sentences. (The formulation WKL used in [Sie87] for Weak König's

Lemma is easily seen to imply WKL_0 (in **BT**), so the result of [Sie87] applies to WKL_0.) Since χ is Σ^1_1, $\chi \to \varphi$ is Π^1_1, and so we get

$$\mathbf{BT},\ \mathbf{Ind}(\Sigma^0_1)\ \vdash\ \chi\ \to\ \varphi,$$

i.e.,

$$\mathbf{BT},\ \mathbf{SOL}(\Sigma^0_1),\ \mathbf{Ind}(\Sigma^0_1)\ \vdash\ \varphi.$$

Again, the latter theory is conservative over $\mathbf{\Sigma_1 A}$ (e.g. by the syntactic argument of [Tro73, §1.9.4]), yielding the lemma. □

Lemma 4.4. $\mathbf{SOL}_0, \mathbf{PR}\ \vdash\ N[\mathbf{s}(x)] \to N[x]$.

Proof. Let $\mathbf{p}$ denote the predecessor function, for which the defining equations are $\mathbf{p}(0) = 0,\ \mathbf{p}(\mathbf{s}(x)) = x$. By Lemma 2.2, $N[\mathbf{s}(x)]$ implies $N[\mathbf{p}(\mathbf{s}(x))]$, from which $N[x]$ by the definition of $\mathbf{p}$. □

Lemma 4.5. *For every V_{FA}-identifier f,*

$$\mathbf{BT},\ \mathbf{SOL}(\mathbf{Comp}) \vdash N[\vec{x}] \to (G_f[\vec{x}, z] \leftrightarrow f(\vec{x}) = z).$$

Proof. By induction on *degree*(f). The cases for initial functions are trivial. If f is defined by composition, $f(\vec{x}) = h(g_1(\vec{x}), \ldots, g_k(\vec{x}))$, then, in $\mathbf{BT} + \mathbf{SOL}(\mathbf{Comp}) + N[\vec{x}]$,

$$\begin{aligned}
G_f[\vec{x}, z] \ &\leftrightarrow\ \exists v_1 \ldots v_k.\ G_{g_1}(\vec{x}, v_1) \wedge \cdots \wedge G_{g_k}(\vec{x}, v_k) \wedge G_h(\vec{v}, z)\\
&\qquad\qquad \text{by the definition of } G_f\\
&\leftrightarrow\ \exists v_1^N \ldots v_k^N.\ G_{g_1}(\vec{x}, v_1) \wedge \cdots \wedge G_{g_k}(\vec{x}, v_k) \wedge G_h(\vec{v}, z)\\
&\qquad\qquad \text{by Lemma 3.5}\\
&\leftrightarrow\ \exists v_1^N \ldots v_k^N.\ g_1(\vec{x}) = v_1 \wedge \cdots \wedge g_k(\vec{x}) = v_k \wedge h(\vec{v}) = z\\
&\qquad\qquad \text{by induction assumption}\\
&\leftrightarrow\ f(\vec{x}) = z\\
&\qquad\qquad \text{by the definition of } f \text{ and Lemma 3.5.}
\end{aligned}$$

Suppose that f is defined by recurrence:

$$f(0, \vec{u}) = g(\vec{u}), f(\mathbf{s}(v), \vec{u}) = h(v, \vec{u}, f(v, \vec{u})).$$

Let $\varphi[v, \vec{u}, z]\ \equiv_{df}\ (N[v, \vec{u}] \to f(v, \vec{u}) = z)$. Assume $G_f[a, \vec{x}, z]$. Then, by Lemma 3.3,

$$\begin{aligned}
&\forall \vec{u}, y\, (\, G_g[\vec{u}, y] \to \varphi[0, \vec{u}, y]\,)\\
&\quad \wedge \forall \vec{u}, v, w, y\, (\, \varphi[v, \vec{u}, w] \wedge G_h[v, \vec{u}, w, y] \to \varphi[\mathbf{s}(v), \vec{u}, y]\,) \qquad (*)\\
&\quad \to \varphi[a, \vec{x}, z]\,).
\end{aligned}$$

Note that $G_g[\vec{u}, y]$ and $N[\vec{u}]$ imply, by induction assumption, $g(\vec{u}) = y$, from which $f(0, \vec{u}) = y$. Thus $G_g[\vec{u}, y] \to \varphi[0, \vec{u}, y]$, i.e. the first premise of (*).

Towards proving the second premise of (*) assume $\varphi[v, \vec{u}, w] \wedge G_h[v, \vec{u}, w, y] \wedge N[\mathbf{s}(v), \vec{u}]$. Then, by Lemma 4.4, $N[v, \vec{u}]$, so $f(v, \vec{u}) = w$ by $\varphi[v, \vec{u}, w]$, and so also $N[w]$, by Lemma 3.5. From $G_h[v, \vec{u}, w, y]$ we then have, by induction assumption, $h(v, \vec{u}, w) = y$, and so $f(\mathbf{s}(v), \vec{u}) = y$. This establishes the second conjunct in (*). Thus, (*) implies $\varphi[a, \vec{x}, z]$, proving $G_f[a, \vec{x}, z] \to f(a, \vec{x}) = z$.

To prove the converse, assume $f(a, \vec{x}) = z$, and assume

$$\forall \vec{u}, y \, (G_g[\vec{u}, y] \to Q(0, \vec{u}, y)) \tag{5}$$

and

$$\forall \vec{u}, v, w, y \, (Q(v, \vec{u}, w) \wedge G_h[v, \vec{u}, w, y] \to Q(\mathbf{s}(v), \vec{u}, y)). \tag{6}$$

Let

$$\psi[v] \equiv_{df} \forall y \, (N[v, \vec{u}] \wedge f(v, \vec{u}) = y \quad \to Q(v, \vec{u}, y)).$$

We have $\psi[0]$ from (5), the induction assumption for g, and the first equation for f. Similarly, $\psi[v] \to \psi[\mathbf{s}(v)]$ follows from (6) and the induction assumption for h. Thus $N[a]$ implies that (5) and (6) imply $\psi[a]$. Therefore, $N[a, \vec{x}]$ and $f(a, \vec{x}) = z$ imply that (5) and (6) imply $Q(a, \vec{x}, z)$, i.e. $G_f[a, \vec{x}, z]$. This concludes the backward direction for the case of definition by recurrence, and the proof of the lemma. □

Lemma 4.6. *For every V_{FA}-equation E, $E \leftrightarrow E^I$ is provable in* $\mathbf{BT} + \mathbf{SOL}(\mathbf{Comp}_0) + \forall x. N[x]$.

Proof. Straightforward, by Lemma 4.5. □

Lemma 4.7. *For every V_{FA}-formula φ, $\varphi \leftrightarrow \varphi^N$ and $\varphi \leftrightarrow \varphi^I$ are provable in* $\mathbf{BT} + \mathbf{SOL}(\mathbf{Comp}) + \forall x. N[x]$.

Proof. By induction on φ. The basis is trivial for the first equivalent, and is established by Lemma 4.6 for the second equivalence. The induction step is trivial. □

Theorem 4.8. $\mathbf{DNT}[\mathbf{SOL}(\mathbf{Comp})] = \mathbf{FNT}[\mathbf{SOL}(\mathbf{Comp})] = \mathbf{FA}$.

I.e., the following conditions are equivalent, for any closed V_{FA}-formula φ:

(1) $\mathbf{FA} \vdash \varphi$;
(2) $\mathbf{SOL}(\mathbf{Comp}), \mathbf{PR}, \neg(0{=}1) \vdash \varphi^N$;
(3) $\mathbf{SOL}(\mathbf{Comp}), \mathbf{PR} \vdash \tilde{\varphi}^N$;
(4) $\mathbf{SOL}(\mathbf{Comp}), \neg(0{=}1) \vdash \varphi^I$;
(5) $\mathbf{SOL}(\mathbf{Comp}) \vdash \tilde{\varphi}^I$.

Proof. The inclusions **FA** $\subseteq$ **DNT**[**SOL**(**Comp**)] and **FA** $\subseteq$ **FNT**[**SOL**(**Comp**)] are Theorems 2.9 and 3.8 above.

We prove the converse for the full interpretation (the case of the direct interpretation is identical). Suppose

$$\mathbf{SOL}(\mathbf{Comp}) \vdash \tilde{\varphi}^I.$$

Then, by Lemma 4.1,

$$\mathbf{BT},\ \mathbf{SOL}(\tilde{\Sigma}^0_1),\ \mathrm{WKL}_0 \vdash \tilde{\varphi}^I,$$

and so, by Lemmas 4.7 and 1.1,

$$\mathbf{BT},\ \mathbf{SOL}(\tilde{\Sigma}^0_1),\ \mathrm{WKL}_0,\ \forall x N[x] \vdash \varphi.$$

By Lemma 4.2 this implies **FA** $\vdash \varphi$. □

An analogous proof, using Lemma 4.3 in place of 4.2, yields:

Theorem 4.9. **DNT**[**SOL**($\mathbf{Comp}_0$)] $=$ **FNT**[**SOL**($\mathbf{Comp}_0$)] $= \mathbf{\Sigma}_1\mathbf{A}$.
I.e., the five conditions of Theorem 4.8 are equivalent, with $\mathbf{\Sigma}_1\mathbf{A}$ *in place of* **FA**, *and* $\mathbf{Comp}_0$ *in place of* **Comp**.

We summarize in the following theorems the semantic readings, based on Theorem 1.3, of Theorems 4.8 and 4.9. First, we have the following semantic characterizations of **FA**:

Theorem 4.10. *Let* φ *be a closed* V_{FA}*-formula. The following conditions are equivalent:*

(1) φ *is provable in* **FA**;
(2) $\tilde{\varphi}$ *[respectively,* φ*] is valid, as a statement about* N*, in all Henkin-models of* **PR** *that are closed under computational definitions (i.e. abstract jump) [and in which 0 and 1 are distinct];*
(3) $\tilde{\varphi}$ *[respectively,* φ*] is valid, as a statement about* N*, in all Henkin-structures that are closed under computational definitions [and in which 0 and 1 are distinct], and where the primitive recursive functions are defined by their graphs.*

Analogously, we have semantic characterizations of $\mathbf{\Sigma}_1\mathbf{A}$:

Theorem 4.11. *Let* φ *be a closed* V_{FA}*-formula. The conditions in Theorem 4.10 are equivalent, with* $\mathbf{\Sigma}_1\mathbf{A}$ *in place of* **FA**, *and with "r-closed computational" in place of "computational".*

References

[Bar69] Jon Barwise, *Applications of strict-Π^1_1 predicates to infinitary logic,* Journal of Symbolic Logic **34** (1969) 409–423.

[Bar75] Jon Barwise, "Admissible Sets and Structures", Springer-Verlag, Berlin and New York, 1975.

[Fag74] Ronald Fagin, *Generalized first order spectra and polynomial time recognizable sets,* in R. Karp (ed.). "Complexity of Computation", SIAM-AMS, 1974, 43–73.

[Fag75] Ronald Fagin, *Monadic generalized spectra,* Zeitschrift für mathematische Logik und Grundlagen der Mathematik **21** (1975) 89–96.

[Fri69] Harvey Friedman, *König's Lemma is weak*, Mimeographed note, Stanford University, 1969.

[GV85] Haim Gaifman and Moshe Vardi, *A simple proof that connectivity of finite graphs is not first-order definable,* Bulletin of the EATCS **26** (June 1985) 43–45.

[Hen50] Leon Henkin, *Completeness in the theory of types,* Journal of Symbolic Logic **15** (1950) 81–91.

[Imm87] Neil Immerman, *Languages which capture complexity classes*, SIAM Journal of Computing **16** (1987) 760–778.

[JS74] N.G. Jones and A.L. Selman, *Turing machines and the spectra of first-order formulas,* Journal of Symbolic Logic **39** (1974) 139–150.

[Kle52] S.C. Kleene, "Introduction to Metamathematics", Wolterns-Noordhof, Groningen, 1952.

[KV87] Phokion Kolaitis and Moshe Vardi, *The decision problem for the probabilities of higher-order properties,* Proceedings of the Nineteenth ACM Symposium on Theory of Computing, ACM, Providence, 1987, 425–435.

[KV88] Phokion Kolaitis and Moshe Vardi, *0-1 laws and the decision problem for fragments of Second-Order Logic,* Proceedings of the Third IEEE Symposium on Logic in Computer Science, IEEE, New York, 1988, 2–11.

[Lei85] Daniel Leivant, *Logical and mathematical reasoning about programs*, Conference Record of the Twelfth Annual Symposium on Principles of Programming Languages, ACM, New York, 1985, 132–140.

[Lei85a] Daniel Leivant, *Hoare's logic captures program semantics,* Manuscript, July 1985.

[Lei89] Daniel Leivant, *Descriptive characterizations of computational complexity*, Journal of Computer and System Sciences **39** (1989) 51–83.

[Lei90] Daniel Leivant, *Computationally based set existence principles*, in W. Sieg (ed.), "Logic and Computation", Contemporary Mathematics, volume 106, American Mathematical Society, Providence, R.I., 1990, pp. 197–212.

[Par77] Charles Parsons, *On a number-theoretic choice schema and its relation to induction,* in A. Kino et als. (eds.), "Intuitionism and Proof Theory", North-Holland, Amsterdam, 1977, 459–473.

[Pra65] Dag Prawitz, "Natural Deduction", Almqvist and Wiskel, Uppsala, 1965.

[Pre30] M. Presburger, *Über die Vollständigkeit eines gewissen Systems der Arithmetik ganzer Zahlen, in welchem die Addition als einzige Operation hervortritt,* Sprawozdanie z I Kongresu Matematików Krajów Słowiańskich, (Comptes-rendus du Ier Congrès des Mathématiciens des Pays Slaves), Warszawa 1929, Warsaw, 1930, pp. 92–101 and 395.

[Sie87] Wilfried Sieg, *Provably recursive functionals of theories with König's Lemma,* Rend. Sem. Mat. Univers. Politecn. Torino, Fasc. Specale: Logic and Computer Science (1986), 1987, 75–92.

[Tro73] Anne S. Troelstra, Metamathematical Investigation of Intuitionistic Arithmetic and Analysis, Springer-Verlag (LNM #344), Berlin, 1973.

SCHOOL OF COMPUTER SCIENCE
CARNEGIE MELLON UNIVERSITY

CONSTRUCTIVE KRIPKE SEMANTICS AND REALIZABILITY

JAMES LIPTON

ABSTRACT. What is the truth-value structure of realizability? How can realizability style models be integrated with forcing techniques from Kripke and Beth semantics, and conversely? These questions have received answers in Hyland's [33], Läuchli's [43] and in other, related or more syntactic developments cited below. Here we re-open the investigation with the aim of providing more constructive answers to both questions. A special, constructive class of so-called *fallible Beth models* has been shown *intuitionistically* to be complete for intuitionistic logic by Friedman, Veldman and others. Here we build such intuitionistic Beth models elementarily equivalent to a natural and broad class of realizabilities, thus showing that realizability interpretations correspond to the particularly effective kind of models yielded by the Veldman – Friedman – de Swart completeness theorem. In the other direction, an abstract realizability is shown constructively to be complete for intuitionistic logic. This extends earlier results along these lines due to Läuchli and others.

1. INTRODUCTION

Kripke models are a powerful metamathematical tool in constructive and computational mathematics. Often a simple diagram suffices to exhibit an intuitionistic counterexample. Such models, and their generalizations as toposes, Heyting-Valued sets, Beth models, or permutation models, also provide useful interpretations of constructive theories in terms of sets evolving through time, state-transitions or properties fixed by certain transformations.

These models were introduced and shown sound and complete by Kripke in 1963 [41], although similar ideas can be found in the work of Beth in the mid-fifties [3], in the topological models of Tarski and McKinsey in 1948 [53], and even in the early work of Jaskowsky (1936, [35]).

Kripke models are usually constructed in a classical metatheory. Most completeness proofs depend upon non-constructive principles such as the use of the Fan theorem, which is incompatible with 'strong' constructivity principles such as Church's thesis. A series of papers by Friedman [24, 23], Veldman [81], de Swart [72], Lopez-Escobar [50], Troelstra [77] and Van Dalen [76] showed that if one considers Beth –rather than Kripke– models, and relaxes the requirement that the set of formulas forced at a node of such a model be consistent, a completeness theorem can be established in a fully constructive metatheory. The arguments developed by these authors constitute a major untapped resource in theoretical computer science, despite their constructivity. They have played a major role in shaping the results in this paper.

Realizability, introduced by Kleene in 1945 [39], provides an inherently more constructive interpretation of intuitionistic reasoning, close in spirit to the propositions-as-types paradigm [13]. Propositions are witnessed by computations, which constitute constructive 'evidence' for assertions. Their use in constructive foundations and computer science ranges from extraction of computations from constructive specifications (McCarty [51], Beeson [1], Feferman [19], Hayashi [29]) to modelling the recursive 'universe' (e.g., McCarty [52], Hyland [33], Scedrov [67]) to furnishing models for computation and for consistency, conservativity and independence results for theories with a strong computational component (Goodman [27], Beeson [1, 2], McCarty [52], Mitchell-Moggi [55], Scedrov [66], Kreisel and Troelstra [40], Cook-Urquardt[12], Buss [6], Nerode-Remmel-Scedrov[59], Stein [70], Dragalin [15] and many others, see e.g. Lipton [46] for references).

The question of how to relate these two paradigms: truth-value semantics and evidence semantics has been addressed by a number of researchers. In 1982 Martin Hyland constructed a topos with the same semantics as Kreisel-Troesltra realizability ([76, 40]), extended to intuitionistic set theory [33]. This work was based on the Tripos-theory foundation worked out in Hyland, Johnstone and Pitts' 1980 paper [34]. A syntactic, first-order analogue of this is developed in the author's [46, 47], and a very similar Heyting-valued structure in Troesltra and van Dalen's [76]. Various authors (especially Rosolini [65]) have shown deep connections between Hyland's topos and "effective categories" studied by Ershov and Mulry (see Rosolini *op.cit* for references).

In the other direction, in 1970 [43], H. Läuchli showed how to build, out of a given Kripke model, a permutation model of (a variant of) the lambda calculus in which formulas true in the original Kripke structure are realized by certain fixpoints. In a classical metatheory he showed that one could restrict the class of fixpoints to those elements which were lambda-definable over a certain type theory. The arguments used, however, were nonconstructive.

A good understanding of the links between algebraic and realizability semantics for constructive formal systems is of great interest in designing semantics for analyzing the metamathematics of type theory, and theories based on the Curry-Howard isomorphism. Kripke models inspired by related work of H. Läuchli have been used in the study of logical relations and the semantics of programming languages by Mitchell and Moggi *op.cit.* and G. Plotkin [63]. Constructions similar to those below and in [47] have been used to model dependent types ([48]). The correspondences studied here are also of interest in automated deduction, in particular for obtaining the most effective possible information out of tableaux proofs.

In this paper we investigate both directions, endeavoring to remain as constructive as possible. In the first part, we show, in a constructive metatheory, that to each of a broad family of realizability interpretations, there corresponds an elementarily equivalent 'fallible' Beth model of the same computational complexity as the realizing theory, in the style of Troelstra-Van Dalen [76]. This result is achieved via a meta-theoretic translation of their proof of the intuitionistic completeness theorem for such models.

In the second part we give a constructive proof of a Läuchli-style converse. For each countable Kripke model we construct an elementarily equivalent realizability interpretation. The realizers are indices of functions partial recursive in the satisfaction predicate of the original Kripke model.

2. Constructive Beth Models for General Realizabilities

In this section we apply a translation to the (constructive) completeness theorem for Beth semantics, obtaining a functional version of this theorem. We are thus able to proceed from

$$\varphi \text{ is realized}$$

in a theory to

$$\varphi \text{ is everywhere forced in a constructive Beth model}$$

in a uniform constructive way.

We consider a broad notion of realizability. We start with any theory T extending a set of axioms formalizing the theory of partial application such as Beeson's $PCA+$, or EON [1], and we leave unspecified how realizability of atomic sentences is defined. We only require that the language over which the realized sentences are defined (the object language) be formalized in T.

Given this situation we define a constructive, fallible Beth Model which is elementarily equivalent to the given realizability interpretation. By constructive we mean:

(1) No appeal to the Fan theorem is made
(2) The construction uses intuitionistic reasoning for recursive theories T, and otherwise uses only decidability of T, i.e. is recursive in T.

Fallibility means $\perp$ (falsehood) may be forced at some nodes.

Our construction is a straightforward adaptation of the Friedman–Beth–Veldman–Lopez-Escobar–Troelstra–van Dalen intuitionistic completeness theorem as formulated in [76]. By interpreting 'and' as a suitable product $\times$, 'or' as a 'realizability' coproduct, implication as a function space, and so on, we obtain a metatheoretic translation of the proof of the completeness theorem itself in the spirit of the Curry-Howard isomorphism. While we do not wish to reproduce all the details of the translated argument here, it is important to spell out enough to show the kind of correspondence that mediates between Kripke semantics and realizability. Furthermore, our translation requires a more careful approach to some of the problems arising in the original construction. See [76], Veldman *op.cit.*, de Swart *op.cit.* and Troelstra [77] for details of the original completeness proof.

Definition 2.1. *([76])* A fallible, uniform (or strong) Beth Model $\mathcal{B} =< K,\ \preceq,\ \Vdash,\ D >$ is given by the following data

(1) A *fan* (see below) $< K, \preceq>$ where $\preceq$ is an ordering of the nodes by the initial segment relation.
(2) D, a domain function, which assigns inhabited sets to each $k \in K$ such that, if $k \preceq k'$, then $D(k) \subset D(k')$. Each constant symbol c in $\mathcal{L}$ has an interpretation c in every $D(k)$.
(3) A forcing relation $\Vdash$: a binary relation between nodes $k \in K$ and prime sentences $P(d_1, ..., d_n)$ over $\mathcal{L} \cup D(k)$ such that
(B1a) $[(\exists z \in N)(\forall k' \succeq_z k)(k' \Vdash P)] \ \Rightarrow k \Vdash P$ (covering).
(B1b) $k \Vdash P$ and $k' \succeq k \ \Rightarrow k' \Vdash P$ (monotonicity).

The notation $k' \succeq_z k$ means k' is a string extending k by length z, *i.e.*

$$k' \succeq_z k \equiv k' \succeq k \text{ and } \ lth(k') - lth(k) = z.$$

Definition 2.2. A fan T is a finitely branching tree, that is to say an inhabited, decidable set of finite sequences of natural numbers closed under initial segments in which each node has at least one successor and which is finitely branching. More formally:

(1) $<>\in T$, $\forall\sigma(\sigma \in T \vee \sigma \notin T)$ and $\forall\sigma\tau(\sigma \in T \& \tau < \sigma \rightarrow \tau \in T)$
(2) $\forall\sigma \in T \exists x \in N \ \ \sigma * < x >\in T$

(3) $\forall \sigma \in T \exists z \in N \ \forall x \in N(\sigma * < x > \in T \rightarrow x \leq z)$ where * denotes concatenation of strings.

We now define truth in a uniform Beth model.

Definition 2.3. For a node $k \in K$ we say a sentence is true at k, or is forced at k by the following inductive definition.

The prime case has already been specified.

$k \Vdash A \vee B \equiv \exists z \in N \ \forall k' \succeq_z k(k' \Vdash A \text{ or } k' \Vdash B)$

$k \Vdash \exists x A(x) \equiv \exists z \in N \ \forall k' \succeq_z k \ \exists d \in D(k')(k' \Vdash A(d))$

$k \Vdash A \& B \equiv k \Vdash A \text{ and } k \Vdash B$

$k \Vdash A \rightarrow B \equiv \forall k' \succeq k(k' \Vdash A \Rightarrow k' \Vdash B)$

$k \Vdash \forall x \ A(x) \equiv \forall k' \succeq k \ \forall d \in D(k')(k' \Vdash A(d))$

$k \Vdash \neg A \equiv k \Vdash A \rightarrow \perp$

In a fallible Beth model, falsity $\perp$ *may* be forced at a node (but not at every node) and the following condition must be met

if $k \Vdash \perp$ *then* for every sentence A, $k \Vdash$ A.

A few remarks are in order here about the definition just given. What gives a uniform, or strong Beth model its name is the fact that the usual definition of a *bar over a node* k in a Beth model, namely *a set of nodes which intersects every path through* k, is here made stronger: a *uniform bar* for k is a set of nodes which intersects every path through k and is *bounded.* Equivalently, as has been done here, it is simply defined as the set of all nodes of a given height n above k. The point is to *build compactness into the definition* so that no appeal to the Fan theorem is necessary. Such a definition of bars is just a special case of forcing with *covers* (e.g. Grayson's [28]). Grayson's cover axioms, essentially a reformulation of the definition–due to Joyal–of forcing over a Grothendieck topology (see the discussion of forcing over *sites* in [76]), guarantee soundness with respect to intuitionistic logic. These axioms, in our case, boil down to verifying the *monotonicity* and *covering* properties defined below. The proof that these properties guarantee soundness is routine (see, e.g., Grayson's paper, or Troesltra and Van Dalen *op.cit.*) and left to the reader.

Lemma 2.4. *In a uniform Beth model, for all sentences A*

(1) (monotonicity) $k \Vdash A$ *and* $k' \succeq k \Rightarrow k' \Vdash A$
(2) (covering) $(\exists z \in N)(\forall k' \succeq_z k)(k' \Vdash A) \to k \Vdash A$

Proof. By simultaneous induction on the structure of A: the details are straightforward. □

Realizability in Abstract Applicative Structures. We briefly sketch how realizability interpretations are defined in an abstract applicative theory. There are various versions of such a theory, APP introduced by Feferman in the mid 1970's (see [76]), the system of the same name formalized in set theory in McCarty's [51] and Beeson's EON [1]. In these theories the notions of convergence and partial application are usually taken as primitive. We adopt Beeson's EON here. Readers familiar with this theory may skip to the end of definition (2.9).

To avoid introducing two sets of variables, formal variables and metavariables (which range over terms built up using the former) EON adopts the convention that every variable converges, but does not allow substitution of terms for variables in a universally quantified statement unless said terms converge. Thus $(\forall x)x\downarrow$ is a theorem, but $t\downarrow$ is not, for an arbitrary term t since the axiom $\forall x A \,\&\, t\downarrow \to\ A[t/x]$ cannot be applied unless $t\downarrow$ has already been shown.

We now briefly describe the theory EON, as does Beeson, as an extension of a more basic theory PCA. We refer the reader to Beeson's book for the details.

The logic of partial terms (LPT):.This logic includes the usual rules for propositional logic plus the following rules of inference:

$$R\forall\ \frac{B \to A}{B \to \forall x A} \qquad R\exists\ \frac{A \to B}{\exists x A \to B} \quad (x \text{ not free in } B)$$

and the following axioms (note that A1, A2, A4, A5, A6 are axiom schemas using metavariables t, s, t_i for arbitrary terms and A7, A8 schemas for special terms.)

(A1) $\forall x A \,\&\, t\downarrow \to A[t/x]$
(A2) $A[t/x] \,\&\, t\downarrow \to \exists x A$

$\downarrow$ is a unary (post-fix) relation symbol in the language
$\simeq$ is a defined binary relation symbol

$$t \simeq s \equiv t\downarrow \vee s\downarrow \to\ \ t = s.$$

We have the following axioms governing $\downarrow$, $\simeq$ *and* $=$

(A3) $x = x \,\&\, (x = y \to y = x)$

(A4) $t \simeq s \,\&\, \varphi(t) \to \varphi(s)$

(A5) $t = s \to t \downarrow \,\&\, s \downarrow$

(A6) $R(t_1, ..., t_n) \to t_1 \downarrow \,\&\, ... \,\&\, t_n \downarrow$ for any atomic formula $R(t_1, ..., t_n)$ and any terms $t_1, ..., t_n$.

(A7) (i) For each constant symbol $c : c \downarrow$

(A8) (ii) For each variable $x : x \downarrow$

We note that (A5) is a special case of (A6). Another important special case of (A6) is

(A6') $f(t_1, t_2, ..., t_n) \downarrow \to t_1 \downarrow \,\&\, t_2 \downarrow \,\&\, ... \,\&\, t_n \downarrow$

N.B. (A6) does NOT imply that for any formula $\varphi, \varphi(t_1, ..., t_n) \to t_1 \downarrow \,\&\, ... \,\&\, t_n \downarrow$. Consider,e.g. $\neg t \downarrow \to t \downarrow$.

PCA.We now introduce the theory PCA over the logic of partial terms.

Language: Two constants, **k** and **s**.

A binary function symbol **Ap**

We will never explicitly write **Ap**. We use juxtaposition, (st) , or just st , to de note $Ap(s, t)$.

Axioms of PCA:

Those of LPT together with

(PCA1) $\mathbf{k}xy = x$

(PCA2) $\mathbf{s}xyz \simeq xz(yz) \,\&\, sxy \downarrow$

(PCA3) $\mathbf{k} \neq \mathbf{s}$

We now define **EON**, Beeson's *Elementary Theory of Operations and Numbers.* It is **PCA** together with

constants $\pi_0, \pi_1, p, d, S_N, P_N, 0$,and a predicate letter N, with axioms

EON1: $\mathrm{pxy} \downarrow \,\&\, \pi_0(pxy) = x \,\&\, \pi_1(pxy) = y$

EON2: $N(0) \,\&\, \forall x(N(x) \to [N(S_N(x)) \,\&\, P_N(S_N x) = x \,\&\, S_N x \neq 0])$

EON3: $\forall x(N(x) \,\&\, x \neq 0 \to N(P_N x) \,\&\, S_N(P_N x) = x)$

EON4: Definition by integer cases

$$N(a) \,\&\, N(b) \,\&\, a = b \to \quad d(a, b, x, y) = x$$

$$N(a) \,\&\, N(b) \,\&\, a \neq b \to \quad d(a, b, x, y) = y$$

EON5: Induction schema: for each formula φ

$$\varphi(0) \,\&\, \forall x[N(x) \,\&\, \varphi(x) \to \quad \varphi(S_N x)] \to \forall x(N(x) \to \quad \varphi(x)).$$

We will often write

if $a = b$ **then** x **else** y for $d(a, b, x, y)$

$\langle x, y \rangle$ *for* **p**xy.

Proofs of the following results about **EON** (and related systems) can be found in Beeson's book or in [76].

Theorem 2.5 (The Recursion Theorem). *There is a term R such that PCA proves*

$$Rf\downarrow \ \&\ [g = Rf \rightarrow \forall x(gx \simeq fgx)]$$

Theorem 2.6. *Let M be a model of EON. Then every partial recursive function is numerically representable in M.*

Theorem 2.7 (Numerical and Term Existence properties).
If EON $\vdash \exists x A$ *then there is a term t such that*

$$EON \vdash t\downarrow \ \&\ A(t).$$

If EON $\vdash \exists n(N(n)\ \&\ A(n))$ *there is a numeral* $\bar{m} = \underbrace{s(s...s(}_{m}0)...)$ *such that* $EON \vdash A(\bar{m})$.

The following definitions hold for EON as well as the enrichment by new constants EON$\underline{C}$ we will be considering below.

Definition 2.8. Let A, B be formulas in one free variable over the language of EON (with possibly a denumerable set of new constants added). Then

$(A \times B)(x) \equiv A(\pi_0 x)\ \&\ B(\pi_1 x)$

$(A + B)(x) \equiv N(\pi_0 x)\ \&\ (\pi_0 x = 0 \rightarrow |A|(\pi_1 x))$
$\&\ (\pi_0 x \neq 0 \rightarrow |B|(\pi_1 x))$

$(A \Rightarrow B)(x) \equiv \forall y[A(y) \rightarrow xy\downarrow \ \&\ B(xy)]$

Let $A(x, y)$ be a formula in two free variables. Then $(\Sigma_x A)$ and $(\Pi_x A)$ are formulas in one free variable given by

$$\begin{aligned}(\Sigma A)(z) &\equiv A(\pi_0 z, \pi_1 z)\\ (\Pi A)(z) &\equiv \forall y[(zy)\downarrow \ \&\ A(y, zy)]\end{aligned}$$

Definition 2.9. Let A, B be *sentences* over the language of EON (EON$\underline{C}$). Then we define inductively the *realizability formulas* $|A|$ in one free variable as follows:

If A is prime $|A|(x)$ is $A\ \&\ x\downarrow$

$|A\ \&\ B| \stackrel{def}{\equiv} |A| \times |B|$

$|A \vee B| \stackrel{def}{\equiv} |A| + |B|$

$|A \to B| \overset{def}{\equiv} |A| \Rightarrow |B|$
$|\exists y A(y)| \overset{def}{\equiv} (\Sigma_y |A(y)|)$, i.e. $|\exists y A(y)|(z) \equiv |A(\pi_0 z)|(\pi_1 z)$
$|\forall y A(y)| \overset{def}{\equiv} (\Pi_y |A(y)|)$,i.e. $|\forall y A(y)|(z) \equiv \forall y[(zy)\downarrow \ \& \ |A(y)|(zy)]$.
$|\neg A| \overset{def}{\equiv} |A \to \perp| \equiv \forall y \neg |A|(y)$

$|A|(x)$ is usually written $x \underset{\sim}{r} A$. Note that if A is a formula in n variables over EON (or EONC̲) then the above clauses define an associated realizability formula in $n+1$ variables. Note that if A is *prime*, A is logically equivalent to $|A|(x)$ for any variable x, since by the LPT conventions, $A \to A \,\&\, (x\downarrow)$. However $|A|[t/x]$ is not logically equivalent to A since $|A|[t/x]$ is $A \,\&\, (t\downarrow)$, which requires additional proof. The point of this definition is to guarantee the base case of the following lemma, which is easily proved by induction.

Lemma 2.10. *EON* $\vdash |A|(t) \to t\downarrow$ *for every sentence* A, *term* t.

Remarks, Conventions and Definitions:

We now make precise the assumptions on the theory T. Let $\mathcal{L}$ be a language. Let T be a theory extending EON with (names for) the constants a of $\mathcal{L}$ and the members of a denumerable set of fresh constants $C = \{c_i | i \in \omega\}$ together with axioms $a\downarrow$, $c_i\downarrow$, as well as the necessary extensions of the axiom schemas of EON to include the new terms so generated. We are distinguishing between the language of the sentences to be realized, $\mathcal{L} \cup C$, which will be called the *object language*, and the language $\mathcal{L}_T$ of the theory T, over which the realizability interpretation is taking place. $\mathcal{L}_T$ will have names e.g., for combinators and the usual vocabulary of the theory of applicative structures (in this paper Beeson's $PCA+$). The object language is formalized in T, i.e., we include in T a predicate $U(x)$ such that $U(c)$ is an axiom for each c of $\mathcal{L}$ or C.

We define realizability in T for $\mathcal{L} \cup C$ — sentences in the usual way, but with quantifiers over individuals relativized to the object language, U. We use the notation $|\varphi|(x)$ for the usual $x \underset{\sim}{r} \varphi$. Thus, for every sentence φ in the object language, $|\varphi|$ is a formula in one free variable in the language of the realizing theory.

For prime P, $|P|(x)$ will remain unspecified.

$|\varphi \& \psi|(x) \overset{def}{\equiv} (|\varphi| \times |\psi|)(x) \equiv |\varphi|(\pi_0 x) \& |\psi|(\pi_1 x)$
$|\varphi \vee \psi|(x) \overset{def}{\equiv} (|\varphi| + |\psi|)(x) \equiv$
$\quad N(\pi_0 x) \& (\pi_0 x = 0 \to |\varphi|(\pi_1 x)) \& (\pi_0 x \neq 0 \to |\psi|(\pi_1 x))$
$|\varphi \to \psi|(x) \overset{def}{\equiv} (|\varphi| \Rightarrow |\psi|(x) \equiv (\forall z)[|\varphi|(z) \to xz\downarrow \& |\psi|(xz)]$

$|\exists y\varphi(y)|(x) \stackrel{def}{\equiv} (\Sigma_{y\in U}|\varphi(y)|)(x) \equiv D(\pi_0 x)\&|\varphi(\pi_0 x)|(\pi_1 x)$
$|\forall y\varphi(y)|(x) \stackrel{def}{\equiv} (\Pi_{y\in U}|\varphi(y)|)(x) \equiv (\forall y)[D(y) \rightarrow xy\downarrow \&|\varphi(y)|(xy)]$

We will sometimes omit all reference to relativization of quantifiers to the domain U. In such cases, we will recall the dependency on U by writing $|\varphi|_U$. Notice that with this shorthand, e.g., $|\exists x\theta(x)| \equiv |\theta(\pi_0 x)|_U(\pi_1 x)$. Now, let Δ be the set of all variable-free terms over the language $\mathcal{L}_T$ of T.

Let $\{\varphi_n\}$ be an enumeration of all sentences of $\mathcal{L}\cup C$ (NOT of $\mathcal{L}_T$, but of the object language) with infinitely many repetitions. We now define a family of formulas of $\mathcal{L}_T$ in one free variable indexed by all finite binary strings $\{G_k | k \in 2^{<\omega}\}$.

Construction of G_k. $G_{<>}$ is the formula $x = 0$. Suppose the G_k have been defined for all $k \in 2^{<\omega}$ of length u. We now show how to define $G_{k*<0>}$ and $G_{k*<1>}$. There will be four cases. In all cases, if $k' \succeq k$ then $G_{k'} = G_k$ or $G_{k'} \equiv G_k \times B$ for some formula B (modulo associativity of the Cartesian product $\times$). Before proceeding with the construction we need one more bit of notation.

We write $T \vdash_u j : G_k \rightarrow A$ if, upon enumerating all proofs in the theory T of code $\leq u$, we find for some $i \in \Delta$ and some $k' \preceq k$, a proof of $T \vdash i : G_{k'} \rightarrow$ A. If $k' = k$ then j is i itself. If $k' \prec k$, then, by the remarks just made, G_k is either $G_{k'}$ or $G_{k'} \times B$ for some B. In the former case j is i, in the latter j is $i \circ \pi$ where π is the projection function such that $T \vdash \pi : G_k \rightarrow G_{k'}$. In this case, we have not actually witnessed a proof of $j : G_k \rightarrow A$ directly, but we already have evidence that such a proof will turn up (when the code is large enough to permit the proof of $\pi : G_k \rightarrow G_{k'}$ and the relevant composition and associativity facts to be combined with the proof of $i : G_{k'} \rightarrow A$). The point of the device just defined is to permit *looking back* at proofs about earlier formulas $G_{k'}$, so that once u is sufficiently large so that $T \vdash_u i : G_{k'} \rightarrow$ A, then for all $k'' \succeq k'$, $T \vdash_u j : G_{k''} \rightarrow A$ for some j. This property will only be needed in the last clause of the proof of Theorem 2.12.

Define $\mathcal{L}(G_k)$ to be the language $\mathcal{L}$ together with those constants $c \in C$ occurring in G_k.

Now, back to the construction. Recall, we have G_k defined for all k up to a certain length u, and we are about to define $G_{k*<n>}$ for $n = 0, 1$.

Case 1. $|\varphi_u| \notin \mathcal{L}(G_k)$; ($|\varphi_u| \in \mathcal{L}(G_k)$ just means that all fresh constants from C appearing in φ_u (or $|\varphi_u|$) already appear in G_k.) Then $G_{k*<j>} \equiv G_k$, for $j = 0, 1$.

Case 2. $|\varphi_u| \in \mathcal{L}(G_k)$, $\varphi_u = B \vee C$ (hence $|\varphi_u| = |B| + |C|$) and $(\exists i \in \Delta)T\vdash_u \forall x[G_k(x) \to ix \downarrow \&(|B| + |C|)(ix)]$. Then

$$G_{k*<0>} \equiv G_k \times |B|$$

$$G_{k*<1>} \equiv G_k \times |C|$$

Case 3. $|\varphi_u| \in \mathcal{L}(G_k)$, $\varphi_u = \exists x\theta(x)$ and $(\exists i \in \Delta)T\vdash_u \forall z[G_k(z) \to iz \downarrow \&|\exists x\theta(x)|(iz)]$. Then

$$G_{k*<j>} \equiv G_k \times |\theta(a)|_U \qquad j = 0, 1$$

where

$$a = \mu_c[c \in C\&c \notin \mathcal{L}(G_k \cup \{|\exists x\theta(x)|\})]$$

and, we recall, $|\theta(a)|_U$ is shorthand notation for $(U(a)\&|\theta(a)|)$.

Case 4. None of the above cases apply. Then

$$G_{k*<0>} \equiv G_k, \qquad G_{k*<1>} \equiv G_k \times |\varphi_u|$$

This completes the construction of the $\{G_k | k \in 2^{<\omega}\}$.

Lemma 2.11 (The Bar – condition lemma). *Let $k \in 2^{<\omega}$, $z \in N$, $|A| \in \mathcal{L}(G_k)$. (Recall that for k, $k' \in 2^{<\omega}$, $k' \succeq_z k$ means k' is a binary string extending k by precisely z bits.) Then:*

$$(1) \qquad (\exists i \in \Delta)T\vdash \forall x[G_k(x) \to ix \downarrow \&|A|(ix)]$$

iff

$$(2) \qquad \forall k' \succeq_z k \exists i \in \Delta \; T\vdash \forall x[G_{k'}(x) \to ix \downarrow \&|A|(ix)]$$

Notation. We will write $T\vdash i : G_k \to |A|$ for $T\vdash \forall x[G_k(x) \to ix \downarrow \&|A|(ix)]$.

Proof. By induction on z. The base case is

$$(\exists i \in \Delta)T\vdash i : G_k \to |A|$$

iff

$$(\exists i \in \Delta)T\vdash i : G_{k*<0>} \to |A| \text{ and } (\exists j \in \Delta)T\vdash j : G_{k*<1>} \to |A|$$

Notice that the $\Rightarrow$ direction is trivial, since in all cases $G_{k*<j>} \equiv G_k$ or $G_k \times |B|$ for some B. Thus $\pi_0 : G_k \times |B| \to G_k$, so if $i : G_k \to A$ then $i \circ \pi_0 : G_k \times |B| \to$ A.

We prove the $\Leftarrow$ direction of the base case. We investigate the four cases in the construction of the G_k.

Case 1. $G_{k*<j>} \equiv G_k$, trivial.

Case 2. Let $u = lth(k)$, and suppose $\varphi_u \equiv B_0 \vee B_1$, $|\varphi_u| \in \mathcal{L}(G_k)$ and

$$(\exists i \in \Delta) T \vdash_u i : G_k \rightarrow |B_0| + |B_1|.$$

Then $G_{k*<j>} \equiv G_k \times |B_j| \quad (j = 0, 1)$. By hypothesis, for some f_0, f_1 in Δ

$$T \vdash f_0 : G_{k*<0>} \rightarrow |A| \text{ and } \quad T \vdash f_1 : G_{k*<1>} \rightarrow |A|.$$

Then we have

$$T \vdash [\![f_0, f_1]\!] : (G_k \times |B_0|) + (G_k \times |B_1|) \rightarrow |A|$$

and

$$T \vdash \ll id, i \gg : G_k \rightarrow G_k \times (|B_0| + |B_1|)$$

where $id \equiv \lambda x \cdot x$, and where, for suitable f, g, $[\![f, g]\!]$ is the canonical map out of the coproduct, in this case $\lambda x \cdot$ **if** $x_0 = 0$ **then** $f x_1$ **else** $g x_1$, and $\ll f, g \gg$ is the canonical product map $\lambda x. < fx, gx >$. Then, the (self-inverse) map $h \equiv \lambda x \cdot < \pi_0(\pi_1(x)), < \pi_0 x, \pi_1 \pi_1 x >>$ gives

$$T \vdash (G_k \times |B_0|) + (G_k \times |B_1|) \overset{h}{\rightleftharpoons} G_k \times (|B_0| + |B_1|)$$

hence

$$T \vdash [f_0, f_1] \circ h \circ \ll id, i \gg : G_k \rightarrow |A|$$

i.e., $(\exists j \in \Delta)$, $T \vdash j : G_k \rightarrow |A|$.

Case 3. Suppose $\varphi_u = \exists x \theta(x)$, $|\varphi_u| \in \mathcal{L}(G_k)$ and $(\exists i \in \Delta) T \vdash_u i : G_k \rightarrow |\exists x \theta(x)|$, meaning $T \vdash \forall y [G_k(y) \rightarrow iy \downarrow \& |\theta(\pi_0(iy))|_U(\pi_1(iy))]$, hence

$$(3) \qquad T \vdash \forall y [G_k(y) \rightarrow \ll id, i \gg y \downarrow \& G_k(y) \& |\theta(\pi_0(iy))|_U(\pi_1(iy))].$$

By construction $G_{k*<j>} \equiv G_k \times |\theta(c)|_U$ for some new c, $(j = 0, 1)$, and

$$(4) \qquad (\exists f \in \Delta) T \vdash f : (G_k \times |\theta(c)|_U) \rightarrow |A|.$$

Now, by the definition of case 3 in the construction of the G_k, $c \notin \mathcal{L}(T \cup \{G_k\})$. By lambda abstraction it is easy to see that there is a c-free term f' such that $T \vdash f'c \simeq f$ in (4), hence, generalizing on c in (4),

$$(5) \qquad T \vdash (\forall z \in U) \forall y [(G_k \times |\theta(z)|)(y) \rightarrow (f'z) y \& |A|((f'z)y)].$$

Combining (3) and (5)

$$T \vdash \forall y [G_k(y) \rightarrow (f'(\pi_0(iy)))(< y, \pi_1(iy) >) \downarrow \& |A|((f'(\pi_0(iy)))(< y, \pi_1(iy) >))]$$

hence

$$(\exists g \in \Delta) T \vdash g : G_k \rightarrow \text{ A}.$$

Case 4. $G_{k*<0>} = G_k$. Then the result is immediate. Now for the inductive step: Assume the result is true for $x \leq n$. Then $k' \succeq_{n+1} k \Rightarrow \exists k''(k' \succeq_1 k'' \succeq_n k)$.

Apply the inductive hypothesis to get

$$(\exists i \in \Delta)T \vdash i : G_k \to |A| \text{ iff}$$

$$(\forall k'' \succeq_n k)(\exists i \in \Delta),\ T \vdash i : G_{k''} \to |A|$$

and use the same argument used in the base case to extend the result from k'' of length n greater than k to k'' of length $n+1$ over k. □

We are now in a position to define the uniform Beth model $\mathcal{B}$. The partially ordered set $< P, \preceq >$ is the full binary tree $2^{<\omega}$ with the initial segment order. The domain function is $D(k) \equiv$ *constants* of $\mathcal{L}(G_k)$, that is to say, all constants of $\mathcal{L}$ plus those fresh constants $c \in C$ which have shown up in G_k. The atomic forcing assignment is

$$k \Vdash P \text{ iff } \quad P \in \mathcal{L}(G_k) \text{ and } (\exists i \in \Delta)T \vdash \forall y[G_k(y) \to (iy)\downarrow \& |P|(iy)].$$

We have to show this satisfies monotonicity and covering (see the definition of uniform Beth model, conditions (B1a) and (B1b)).

Monotonicity: if $k' \succeq k$ and $(\exists i \in \Delta)T \vdash i : G_k \to |P|$ then, by construction $G_{k'} = G_k$ or $G_{k'} = G_k \times A$ for some T-formula A. (We are, of course, identifying $(((G_k \times A_1) \times A_2) \times \ldots \times A_n)$ with $G_k \times (A_1 \times \ldots \times A_n)$). So, for some projection function π (modulo associativity) $T \vdash \pi : G_{k'} \to G_k$, hence $T \vdash i \circ \pi : G_{k'} \to |P|$. Thus $k' \succeq k$ and $k \Vdash P \Rightarrow k' \Vdash P$. Covering follows immediately from Lemma 2.11, just proved:

$$(\exists z)(\forall k' \succeq_z k)k' \Vdash P \to k \Vdash P.$$

Theorem 2.12. *For every sentence $A \in \mathcal{L}(G_k)$,*

$$k \Vdash A \iff (\exists i \in \Delta)T \vdash i : G_k \to |A|$$

Proof. Atomic case: by definition of B.

and: Let $A \equiv \varphi \& \theta$, $A \in \mathcal{L}(G_k)$. Then φ, $\theta \in \mathcal{L}(G_k)$, and

$$k \Vdash \varphi \& \theta \Rightarrow k \Vdash \varphi \text{ and } \quad k \Vdash \theta$$

which implies, by inductive hypothesis

$$(\exists i, j \in \Delta),\ T \vdash i : G_k \to |\varphi| \text{ and } \quad T \vdash j : G_k \to |\theta|$$

hence

$$T \vdash \ll i, j \gg : G_k \to |\varphi| \times |\theta|.$$

Conversely: $T \vdash f : G_k \to |\varphi \& \theta| \Rightarrow T \vdash \pi_0 \circ f : G_k \to |\varphi|$ and $T \vdash \pi_1 \circ f : G_k \to |\theta|$. Hence, by inductive hypothesis

$$k \Vdash \varphi \text{ and } \quad k \Vdash \theta, \text{ so } \quad k \Vdash \varphi \& \theta.$$

<u>**or:**</u> Say $A \equiv \varphi \vee \theta,\ A \in \mathcal{L}(G_k)$. Then $k \Vdash \varphi \vee \theta \Rightarrow (\exists z) \forall k' \succeq_z k,\ k' \Vdash \theta$ or $k' \Vdash \varphi$. If θ, and φ are both in $\mathcal{L}(G_k)$ then θ and φ are in $\mathcal{L}(G_{k'})$ so by inductive hypothesis

$\forall k' \succeq_z k (\exists i \in \Delta) T \vdash i : G_{k'} \to |\varphi|$ or $(\exists i \in \Delta) T \vdash i : G_{k'} \to |\theta|$. For any $i \in \Delta$ witnessing the former case

$$T \vdash \lambda x \cdot < 0, i >: G_{k'} \to |\varphi| + |\theta|$$

and in the latter

$$T \vdash \lambda x \cdot < 1, i >: G_{k'} \to |\varphi| + |\theta|$$

hence, in all cases $\forall k' \succeq_z k (\exists h \in \Delta) T \vdash h : G_{k'} \to |\varphi| + |\theta|$. By Lemma 2.11

$$(\exists h \in \Delta) T \vdash h : G_k \to |\varphi| + |\theta|.$$

Conversely, suppose $T \vdash j : G_k \to |\varphi| + |\theta|$. Then, for some $z' \in N$,

$$T \vdash_{z'} j : G_k \to |\varphi| + |\theta|$$

(i.e. some number bounds the code of the proof). Let $z'' = \mu_y[y \geq z', lth(k)$ and $\varphi_y = \varphi \vee \theta]$ where, we recall that φ_y is the yth formula in our original enumeration of sentences over $\mathcal{L} \cup C$ with infinitely many repetitions. Then, letting $z = z'' - lth(k)$

$$\forall k' \succeq_z k, \exists j' \in \Delta, T \vdash_{z''} j' : G_{k'} \to |\varphi \vee \theta|,$$

where $\varphi \vee \theta$ is $\varphi_{z''}$, hence, by (case 2 of) the construction of the G_k

$$G_{k' * <0>} = G_{k'} \times |\varphi| \text{ and } \quad G_{k' * <1>} = G_k \times |\theta|$$

so for every k'' with $k'' \succeq_{z+1} k$ and for $i = \mathit{projection}$ onto second factor

$$T \vdash i : G_{k''} \to |\varphi|, \text{ or } \qquad T \vdash i : G_{k''} \to |\theta|$$

so, by the inductive hypothesis

$$(\forall k'' \succeq_{z+1} k) k'' \Vdash \varphi \text{ or } \quad k'' \Vdash \theta$$

hence $k \Vdash \varphi \vee \theta$ (definition of forcing $\vee$).

<u>**implies:**</u> Suppose $k \Vdash \psi \to \theta$ and $\psi \to \theta \in \mathcal{L}(G_k)$. Then $(\forall k' \succeq k) k' \Vdash \psi \Rightarrow k' \Vdash \theta$. Let $z' = \mu_y[y \geq lth(k)$ and $\varphi_y = \psi \vee \psi]$. (in other words, 'the disjunction of ψ and itself is the yth formula in our master enumeration $\{\varphi_y\}$). Now let $z = z' - lth(k)$. Then, for any $k' \succeq_z k$

$$(\exists i \in \Delta)T\vdash_z i : G_{k'} \to |\psi| + |\psi| or \neg(\exists i \in \Delta)T\vdash_z i : G_{k'} \to |\psi| + |\psi|$$

is recursively decidable in T. (In particular, if T is a axiomatizable, the argument that follows is taking place in a constructive metatheory). In the first case

$$G_{k*<0>} = G_{k*<1>} = G_{k'} \times |\psi|. \tag{6}$$

In the second

$$G_{k*<1>} = G_k \times (|\psi| + |\psi|). \tag{7}$$

Since $[\![id, id]\!] : |\psi| + |\psi| \to |\psi|$. We have, in all cases, for $k' \succeq_z k$,

$$(\exists h \in \Delta),\ T\vdash h : G_{k'*<1>} \to |\psi|,$$

and by the induction hypothesis $k'* < 1 > \Vdash \psi$. Now, by the assumption $k \Vdash \psi \to \theta$, we must have $k'* < 1 > \Vdash \theta$, so by the inductive hypothesis again

$$(\exists g \in \Delta)T\vdash g : G_{k'*<1>} \to |\theta|.$$

Depending on which case obtains above in (6,7), we have

$$(\exists g \in \Delta)T\vdash g : G_{k'} \times |\psi| \to |\theta| \tag{8}$$

or

$$(\exists g \in \Delta)T\vdash g : G_{k'} \times (|\psi| + |\psi|) \to |\theta|. \tag{9}$$

But, since $\lambda x \cdot < 0, x >: |\psi| \to |\psi| + |\psi|$, (9) implies (8), so (8) always obtains. Now, it is easy to show that a weak form of the cartesian closure axiom holds here:

$$(\exists g \in \Delta)T\vdash g : G_{k'} \times |\psi| \to |\theta|$$
$$\Rightarrow$$
$$(\exists h \in \Delta)T\vdash h : G_{k'} \to (|\psi| \Rightarrow |\theta|).$$

For each g in the first case take $h = \lambda x \lambda y \cdot g(< x, y >)$. Thus, we have established

$$\forall k' \succeq_z k(\exists h \in \Delta)T\vdash h : G_{k'} \to |\psi \to \theta|$$

which, by Lemma 2.11, gives the desired conclusion

$$(\exists h \in \Delta) T \vdash h : G_k \to |\psi \to \theta|.$$

Conversely, suppose $T \vdash h : G_k \to |\psi \to \theta|$ and $k' \succeq k$ with $k' \Vdash \psi$. Then, by the inductive hypothesis

$$(\exists f \in \Delta) T \vdash f : G_{k'} \to |\psi|.$$

Now, as observed earlier, for some projection function ρ, $T \vdash \rho : G_{k'} \to G_k$, so we have

$$(\exists g \in \Delta) T \vdash g : G_{k'} \to (|\psi| \Rightarrow |\theta|).$$

So if $w = \lambda x \cdot (gx)(fx)$ i.e., $w = sgf$, then $T \vdash w : G_{k'} \to |\psi|$, hence, by the induction hypothesis $k' \Vdash \theta$, which shows $k \Vdash \psi \to \theta$.

<u>exists:</u> Suppose $k \Vdash \exists x \theta(x)$ and $\exists x \theta(x) \in \mathcal{L}(G_k)$ Then $\exists y \in N \ \forall k' \succeq_y k$ $\exists c \in D(k') k' \Vdash \theta(c)$. Thus, for some $z \geq y$

$$\forall k' \succeq_z k, \exists c \in \mathcal{L}(G_{k'}) \ k' \Vdash \theta(c).$$

Therefore, by the inductive hypothesis

$$\forall k' \succeq_z k \ \exists c \in \mathcal{L}(G_{k'}) \ \exists f \in \Delta \ T \vdash f : G_{k'} \to |\theta(c)|.$$

But $c \in \mathcal{L}(G_{k'})$ means $c \in \mathcal{L} \cup C$, *i.e.* $U(c)$ is an axiom of T and hence, we have, above, $T \vdash f : G_{k'} \to U(c) \& |\theta(c)|$, or, using our shorthand:

$$\forall k' \succeq_z k \ \exists c \in \mathcal{L}(G_{k'}) \ \exists f \in \Delta \ T \vdash f : G_{k'} \to |\theta(c)|_U.$$

Let $h = \lambda x \cdot < c, fx >$. Then $T \vdash h : G_{k'} \to |\exists x \theta(x)|$. By Lemma 2.11

$$(\exists g \in \Delta) T \vdash g : G_k \to |\exists x \theta(x)|.$$

Conversely, suppose $T \vdash g : G_k \to |\exists x \theta(x)|$ with $|\exists x \theta(x)| \in \mathcal{L}(G_k)$. Then, for some $y \in N$, $T \vdash_y g : G_k \to |\exists x \theta(x)|$. Let $y' = \mu_w[w > y, lth(k)$ and $\varphi_w = \exists x \theta(x)]$ and let $z = y' - lth(k)$. Then, for $k' \succeq_z k$,

$$G_{k' * <j>} = G_{k'} \times |\theta(c)|_U \qquad j = 0, 1$$

for some $c \in \mathcal{L}(G_{k'*<j>}) \backslash \mathcal{L}(G_{k'})$, by construction of the G_k (case 3), hence

$$T \vdash \pi_1 : G_{k' * <j>} \to |\theta(c)|$$

and, by inductive hypothesis, $k' * < j > \Vdash \theta(c)$. So,

$$\forall k'' \succeq_{z+1} k \ \exists c \in D(k'') \ k'' \Vdash \theta(c), \text{ and hence } \quad k \Vdash \exists x \theta(x).$$

for all: Let $\forall x\theta(x) \in \mathcal{L}(G_k)$. Suppose $k \Vdash \forall x\theta(x)$. To begin with, we must show that we can find a sufficiently large $n \in N$ such that for some $c = c_{i(k')}$

$$\forall k' \succeq_n k \; c \in \mathcal{L}(G_{k'*<j>}) \backslash \mathcal{L}(G_{k'}),\; j = 0, 1.$$

The reason is that for some tautological existential sentence, e.g., $\exists x(P(x) \to P(x))$ for P a prime formula, some c is eventually used to Henkinize this sentence in the construction of the G_k (case 3), and this is done infinitely often. To be more precise, if P is a prime formula in one free variable, a a constant in the language $\mathcal{L}$, then

$$T \vdash \forall z[|P(a)|(z) \to (id)z \downarrow \& |P(a)|(z)]$$

hence $T \vdash |\exists x(P(x) \to P(x))|(< 0, id >)$, whence for

$$i = \lambda x \cdot if \;\; x = 0 \; then < 0, id > else \;\; x$$

we must have

$$T \vdash \forall y[y = 0 \to iy \downarrow \& |\exists x(P(x) \to P(x))|(iy)].$$

Thus $T \vdash i : G_{<>} \to |\exists x(P(x) \to P(x))|$ so for every G_k ,

$$T \vdash j : G_k \to |\exists x(P(x) \to P(x))|$$

for some j. Now let n be a number bounding a code for a proof of

$$T \vdash i : G_{<>} \to |\exists x(P(x) \to P(x))|.$$

Then, for each k, there is a $j \in \Delta$ such that $T \vdash_u j : G_k \to |\exists x(P(x) \to P(x))|$, j being the composition of i with the relevant projection function of $G_k \to G_{<>}$. Now, increase $lth\ k$ until we find a k' with $lth(k') \geq u$ and $lth(k') =$ some u' such that $\varphi_u = \exists x(P(x) \to P(x))$ in our enumeration of all $\mathcal{L} \cup C$ - sentences with infinitely many repetitions. Then for this k' we have, for some $j' \in \Delta$

$$T \vdash_{u'} j' : G_{k'} \to |\exists x(P(x) \to P(x))|$$

with $u' = lth(k')$ satisfying the requirements of case 3 of the construction of the $G_{k'}$. (This argument is the point of our somewhat bizarre definition of $\vdash_u$).

Thus, the next stage of the construction of the G_k required us to find a $c \notin \mathcal{L}(G_{k'})$ and put $G_{k'*<j>} = G_{k'} \times |P(c) \to P(c)|$. Thus $c \in \mathcal{L}(G_{k'*<j>}) \backslash \mathcal{L}(G_{k'})$ and, in particular, given our assumption $|\forall x\theta(x)| \in \mathcal{L}(G_k)$, we have $|\theta(c)| \in \mathcal{L}(G_{k'*<j>}) \backslash \mathcal{L}(G_{k'})$.

Since $k' \succeq k$ and $k \Vdash \forall x \theta(x)$ and $c \in D(k' * < j >)$ we may conclude $k' * < j > \Vdash \theta(c)$. The induction hypothesis then gives

$$(\exists f \in \Delta)\ T \vdash f : G_{k'*<j>} \rightarrow |\theta(c)| \qquad (j = 0, 1)$$

But, in fact,

$$(\exists g \in \Delta)\ T \vdash g : G_{k'} \rightarrow |\theta(c)|,$$

due to the special nature of $G_{k'*<j>} = G_{k'} \times |P(c) \rightarrow P(c)|$, as we now show. Since the identity realizes $|P(c) \rightarrow P(c)|$ we have

$$T \vdash (\lambda x \cdot id) : G_{k'} \rightarrow |P(c) \rightarrow P(c)|_U$$

hence

$$T \vdash \lambda x \cdot < id, \lambda x \cdot id >: G_{k'} \rightarrow G_{k'} \times |P(c) \rightarrow P(c)|$$

so

$$T \vdash f \circ (\lambda x \cdot < id, \lambda x \cdot id >) : G_{k'} \rightarrow |\theta(c)|.$$

By λ-abstraction (easy induction proof on the structure of terms in Δ) we can find an $h \in \Delta$ which is c-free (notation: $h \in \Delta/c$) such that

$$T \vdash hc \simeq f \circ (\lambda x \cdot < id, \lambda x \cdot id >).$$

Thus

$$\exists h \in \Delta/c,\ T \vdash hc : G_{k'} \rightarrow |\theta(c)|$$

with $c \notin \mathcal{L}(G_{k'} \cup \{\theta(x)\})$. Writing this out in full, we have

$$T \vdash G_{k'}(x) \rightarrow hcx \downarrow \& |\theta(c)|(hcx) \tag{10}$$

$$T + G_{k'}(x) \vdash hcx \downarrow \& |\theta(c)|(hcx). \tag{11}$$

Quantifying over c : (with a little care – we move logical axioms involving c, such as $U(c)$ [1] to the right of the $\vdash$ – symbol, generalize on c, obtaining universal quantification relativized to U)

$$T + G_{k'}(x) \vdash (\forall z \in U)(hzx \downarrow \& |\theta(z)|(hzx))$$

hence

$$T \vdash \forall x [G_{k'}(x) \rightarrow (\forall z \in U)(hzx \downarrow \& |\theta(z)|(hzx))].$$

[1]The finitely many sentences $\Gamma(c)$ involving c that might have occurred here are, as explained in the remarks after lemma (2.10), $U(c)$ and instances of the LPT-schemas (A1)-(A8), such as, e.g., $c\downarrow$, or $x \simeq c \,\&\, \varphi(x) \rightarrow \varphi(c)$. We form their conjunction $\Gamma_0(c)$, move them to the right of the turnstile, and generalize on c (which is now no longer present to the left of the $\vdash$), replacing it with e.g. the variable z. With the exception of $U(z)$ which is handled above, every conjunct of $\Gamma_0(z)$ is now an axiom of the c-free theory, and can therefore be dropped.

Now let $g = \lambda x \lambda z \cdot hzx$. By the combinatory completeness theorem for partial applicative structures (see, e.g., [1]) $gx\downarrow$ and

$$T \vdash \forall x[G_{k'}(x) \to gx \downarrow \& (\forall z \in U)(gzx \downarrow \& |\theta(z)|(gxz))]$$

which is to say,

$$T \vdash \forall x[G_{k'}(x) \to gx \downarrow \& (\forall z)(U(x) \to (gz)x \downarrow \& |\theta(z)|(gxz))]$$

i.e., $T \vdash g : G_{k'} \to |\forall x \theta(x)|$. This holds for all $k' \succeq k$ with $lth(k') = u'$, so by Lemma 2.11

$$(\exists g' \in \Delta)\ T \vdash g' : G_k \to |\forall x \theta(x)|.$$

Conversely, suppose $(\exists h \in \Delta)\ T \vdash h : G_k \to |\forall x \theta(x)|$, i.e.,

$$T \vdash \forall y[G_k(y) \to hy \downarrow \& \forall z(U(z) \to hyz \downarrow \& |\theta(z)|(hyz))] \tag{12}$$

Pick $k' \succeq k$ and $c \in \mathcal{L}(G_{k'})$, i.e., $c \in \theta(k')$. Then $U(c)$ is an axiom of T and can be dropped.

Now, by (12) and by monotonicity, i.e. the existence of a projection π such that

$$T \vdash \pi : G_{k'} \to G_k,$$

there is an $h' \in \Delta$ with

$$T \vdash \forall y[G_{k'}(y) \to h'y \downarrow \& h'yc \downarrow \& |\theta(c)|(hyc)].$$

Let $g = \lambda y \cdot h'yc$. Then $T \vdash \forall y[G_{k'}(y) \to gy \downarrow \& |\theta(c)|(gy)]$. So, by induction hypothesis, $k' \Vdash \theta(c)$. Thus $k \Vdash \forall x \theta(x)$. □

Corollary 2.13. *Let φ be a sentence over the language $\mathcal{L}$, and $\mathcal{B}$ the Beth model constructed above. Then*

$$\mathcal{B} \models \varphi \text{ iff } (\exists e \in \mathcal{L}(T)) T \vdash |\varphi|(e) \,\&\, e \downarrow$$

Proof. $B \models \varphi \Rightarrow <> \Vdash \varphi \Rightarrow (\exists h \in \Delta)\ T \vdash \forall x[G_{<>}(x) \to hx \downarrow \& |\varphi|(hx)]$
$\Rightarrow T \vdash \forall x(x = 0 \to hx \downarrow \& |\varphi|(hx))$
$\Rightarrow T \vdash h0 \downarrow \& |\varphi|(h0)$.
We have the problem that $h \in \Delta$ but h is not necessarily a term of T. As already discussed in the proof above, we repeatedly abstract out the finitely many parameters $c_1, ..., c_m$ of h which are not in the language of T, obtaining $h' \in T$ with $T \vdash h'c_1...c_m \simeq h$. Then universally quantify over these constants, and instantiate them to constants of T, obtaining, for some g in the language of T, $T \vdash g \downarrow \& |\varphi|(g)$.

Conversely, if $T \vdash e \downarrow \& |\varphi|(e)$ then $T \vdash ke0 \downarrow \& |\varphi|(ke0)$ hence $T \vdash ke : G_{<>} \to |\varphi|$. By the theorem, $<> \Vdash \varphi$ and $B \models \varphi$. $\square$

Thus every realizability notion formalized in some applicative theory T corresponds, uniformly to some elementarily equivalent *constructive* Beth model. The class of realizability interpretations so describable is quite broad: T may come equipped with a formalized diagram, which captures the theory of some model or of, say, fragments of analysis or set theory. We refer the reader to Beeson's detailed formulation (*op. cit.*) of EON**b**, where **b** is a member of 2^{ω}, satisfying some axiom $\varphi(\mathbf{b})$, and with reduction rules $b(n) \to \overline{b(n)}$. We do not require that φ be self-realizing (if it is, the interpretation will be sound with respect to EON**b** – deducibility). We may take the theory T to be EON**b** together with defining axioms φ, and use these special facts (i.e. the information supplied by φ) about **b** to define atomic realizability. Then the construction just given will produce a Beth model for this interpretation. A q – realizability variant of the above construction is discussed in [49].

3. A Realizability Interpretation Corresponding to Kripke Models

We now pursue the connections between Kripke and Kleene semantics in the other direction. We start with an arbitrary denumerable Kripke model $\mathcal{K}$ and construct a 'stratified' realizability interpretation which is elementarily equivalent to the original Kripke model. Informally, our aim is to build a nontrivial (all provably recursive functions are representable) category of realizers with the same semantics. We will construct a weakly cartesian closed category $\mathcal{C}$, with *weak* terminal object T such that $\mathcal{K} \models \varphi$ iff there is a 'realizing morphism' $e : T \to |\varphi|$ in $\mathcal{C}$.. By weakly cartesian closed we mean that exponents exist, but do not have a unique witnessing morphism. An object T of $\mathcal{C}$ is *weakly terminal* if for each object A of $\mathcal{C}$ there is a not necessarily unique morphism $A \to T$. These notions will be made precise below.

Let $\mathcal{L}$ be a countable language and $\mathcal{K} =< P, \leq, D, \Vdash >$ a countable Kripke model over $\mathcal{L}$.

We first need to make $\mathcal{K}$ into a fallible model, i.e., one with a top node * which forces falsehood $\perp$ *as* well as every other $\mathcal{L}$ - sentence. We simply add the relation $* \geq p$ for each node $p \in P$, and extend the D-function and the forcing relation as follows

$D(*) = \cup\{D(p) | p \in P\}$. For $p \in P$, $p \Vdash_* \varphi \equiv p \Vdash \varphi$, and

$* \Vdash \varphi$ for every atomic sentence over the language of $D(*)$.

The new model $\mathcal{K}* =< P*, \leq_*, D_*, \Vdash_* >$ (which will hereafter be identified with $\mathcal{K}$) is elementarily equivalent to $\mathcal{K}$. We leave the straightforward proof to the reader. We now drop the * — notation, but the reader should remember that the element * is now assumed to be in the underlying partial order of $\mathcal{K}$.

Now we need to formalize $\mathcal{K}$ within N, the natural numbers. Let $\mathcal{L}_k$ be the language of $\mathcal{L}$ together with all the individuals of $\mathcal{K}$ (i.e., $\mathcal{L}_k = \mathcal{L} \cup D(*)$). For each $p \in P$, (P now includes *) we have a Gödel number $\ulcorner p \urcorner \in N$. For each $c \in \mathcal{L}_k$, $\ulcorner c \urcorner \in N$. We also have the following functions and relations on N :

$P(\ulcorner p \urcorner) \iff p \in P$

$D(\ulcorner c \urcorner) \iff c \in D(*)$

$D(\ulcorner p \urcorner, \ulcorner c \urcorner) \iff c \in D(p)$ (we distinguish the binary D by writing D^2).

We Gödel-number all sentences over $\mathcal{L}_k$. Then define the following relations. For each prime $\mathcal{L}_k$-sentence φ we have

$ax(\ulcorner p \urcorner, \ulcorner \varphi \urcorner) \iff p \Vdash \varphi$ in $\mathcal{K}$ and also

$\mathcal{O}(\ulcorner p \urcorner, \ulcorner q \urcorner) \iff p \leq q$ in K.

We also have the full collection of Gödel-numbering auxiliary predicates and functions available, e.g.

$Sent(n) \iff n = \ulcorner \varphi \urcorner$ and φ is an $\mathcal{L}_k$-sentence

$Prime(n) \iff Sent(n) \,\&\, n = \ulcorner \varphi \urcorner \& \varphi$ is prime

$Sub(n, m, r) \iff r$ is the Gödel number of the formula resulting from the substitution of the term with Gödel number n for the variable whose Gödel number is m.

Let n be the Gödel number of an $\mathcal{L}_k$ – sentence φ. Then:

- $Dis(n)$ is true iff φ is a disjunction, in which case $l_\vee(n)$ and $r_\vee(n)$ give the Gödel numbers of its left and right disjuncts

- $Conj(n)$ is true iff φ is a conjunction, with $l_\&(n)$, $r_\&(n)$ the Gödel numbers of the conjuncts.

- $Imp(n)$ is true iff φ is an implication $\theta \rightarrow \psi$ with $\ulcorner \theta \urcorner = Ant(n)$, $\ulcorner \psi \urcorner = Cons(n)$

- $Ex(n)$ is true iff φ is an existential sentence $\exists x \theta(x)$ with $B_0var(n) = \ulcorner x \urcorner$ (the outermost bound variable) and $Pred(n) = \ulcorner \theta(x) \urcorner$.

Thus, if $\ulcorner c \urcorner = m$, $Sub(m, B_0var(n), Pred(n)) = \ulcorner \theta(c) \urcorner$.

- $Univ(n)$ is true iff φ is a universal sentence $\forall x \theta(x)$ with $B_0var(n) = \ulcorner x \urcorner$ and $Pred(n) = \ulcorner \theta(x) \urcorner$.

We now need the following special Skolem functions

$w : w(\ulcorner p \urcorner, \ulcorner \varphi \vee \psi \urcorner) = 0$ iff $p \Vdash \varphi$

$w(\ulcorner p \urcorner, \ulcorner \varphi \vee \psi \urcorner) = 1$ iff not $(p \Vdash \varphi)$ and $p \Vdash \psi$

$w(\ulcorner p \urcorner, \ulcorner \varphi \vee \psi \urcorner)$ is undefined otherwise

w is also a witness function for existential sentences

$$w(\ulcorner p \urcorner, \ulcorner \exists x \theta(x) \urcorner) = \mu m[m = \ulcorner c \urcorner \& D(p,c) \& p \Vdash \theta(c)]$$

if a witness exists.

We have in addition the following characteristic functions:

$$\chi_P, \chi_{D^1}, \chi_{D^2}, \chi_{Ax}, \chi_{\mathcal{O}}.$$

To summarize, we have added to N the following collection of functions

$$F_0 = \{w, \text{ and the characteristic functions of } \quad P,\ D(),\ D(\ ,\ ),\ \text{ax},\ \mathcal{O}\}$$

as well as the corresponding predicates. Formally none of these predicates are present: all information is carried by the functions, although the predicates will be used in the discussion below.

All other Gödel numbering predicates and functions can be effectively defined in terms of the functions in F_0.

Now let F be the set of functions partial recursive in F_0, i.e., the least class containing F_0, projections, constant functions, successor, and closed under primitive recursion and the μ-operator. Via the enumeration theorem (relativized to arbitrary functions or oracles, see, e.g. Kleene's textbook [38]), we can associate indices in N to each relativized algorithm computing functions in F, satisfying the (relativized) s-m-n and recursion theorems. We will informally write $e \in F$ to mean $\{e\}^{F_0}$, $e \in N$. We assume standard pairing and un-pairing. We write $e =< e_0, e_1 >$. For any term M, $\Lambda x \cdot M$ will be the usual F_0-recursive code whose existence is guaranteed by the s-m-n theorem.

For codes $e \in F$ we now define $\underset{\sim}{k}-$ realizability over N (not over the language of the original Kripke model) enriched with the predicates and functions given above, in the usual way:

Definition 3.1. If R is an atomic relation, e.g. $k = n+m$, $D(n,m)$, $\mathcal{O}(m,n)$, $P(n)$ etc., then

$$e \underset{\sim}{k} R \text{ iff } \quad R \text{ is true in } \quad N.$$

The inductive cases are as usual.

Now we introduce an (inductively) defined predicate $Sat(n, m)$ which formalizes satisfaction in the original Kripke model (not in N), i.e., $Sat(\ulcorner p \urcorner, \ulcorner \varphi \urcorner)$ formalizes $p \Vdash \varphi$ in $\mathcal{K}$.

$$
\begin{aligned}
& Sat(m, n) \equiv P(m) \,\&\, sent(n) \,\&\\
& \{ \quad [Prime(n) \,\&\, ax(m, n)]\\
& \quad \vee[Conj(n) \vee Sat(m, l_{\&}(n)) \,\&\, Sat(m, r_{\&}(n))]\\
& \quad \vee[Dis(n)\&[(w(m, n) = 0\& Sat(m, l_{\vee}(n))) \vee (w(m, n) = 1 \& Sat(m, r_{\vee}(n)))]]\\
& \quad \vee[Imp(n) \,\&\, \forall y(P(y) \,\&\, \mathcal{O}(m, y) \,\&\, Sat(y, Ant(n)) \rightarrow Sat(y, Cons(n)))]\\
& \quad \vee[Ex(n) \,\&\, D(m, w(m, n)) \,\&\, Sat[m, Sub(w(m, n), B_0 var(n), Pred(n))]]\\
& \quad \vee[Univ(n) \,\&\, \forall y \forall z[P(y) \,\&\, \mathcal{O}(m, y) \,\&\, D(y, z) \rightarrow\\
& \qquad\quad Sat[y, Sub(z, B_0 var(n), Pred(n))]]]\}
\end{aligned}
$$

Negation is handled by $p \Vdash \neg\varphi \equiv p \Vdash \varphi \rightarrow \perp$ where, we recall, at least one node, *, does force $\perp$.

Observe that realizability of $Sat(m, n)$ is already defined by induction, as it is already defined for the base case. Strictly for notational convenience we give a definition of $e \underset{\sim}{k} Sat(m, n)$ which 'skips' some of the trivial conjuncts (e.g. $Prime(n)$) which would otherwise lead to a proliferation of projections and subscripts. Such a definition will be legitimized by soundness of the interpretation, to be proven below.

Definition 3.2. $e \underset{\sim}{k} Sat(m, n) \equiv Sent(n) \,\&\, P(m) \,\&$
$\{[e_0 = 0 \,\&\, Prime(n) \,\&\, ax(m, n)] \vee$
$[e_0 = 1 \,\&\, Conj(n) \,\&\, e_{10} \underset{\sim}{k} Sat(m, l_{\&}(n)) \,\&\, e_{00} \underset{\sim}{k} Sat(m, r_{\&}(n))] \vee$
$[e_0 = 2 \,\&\, Dis(n) \,\&\, [(e_{10} = 0 \,\&\, e_{11} \underset{\sim}{k} Sat(m, l_{\vee}(n))) \vee$
$(e_{10} \neq 0 \,\&\, e_{11} \underset{\sim}{k} Sat(m, r_{\vee}(n)))] \vee$
$[e_0 = 3 \,\&\, Imp(n) \,\&\, \forall y \forall q$ [if $P(y) \,\&\, \mathcal{O}(m, y) \,\&\, q \underset{\sim}{k} Sat(y, Ant(n))$ then $e_1 y \downarrow \,\&\, e_1 yq \downarrow \,\&\, e_1 yq \underset{\sim}{k} Sat(y, \mathrm{Cons}(n))] \vee$
$[e_0 = 4 \,\&\, Ex(n) \,\&\, D(m, e_{10}) \,\&\, e_{11} \underset{\sim}{k} Sat(m, Sub(e_{10}, B_0 var(n), \mathrm{Pred}(n)))] \vee$
$[e_0 = 5 \,\&\, \mathrm{Univ}(n) \,\&\, \forall z \forall x[P(z) \,\&\, \mathcal{O}(m, z) \,\&\, D(z, x) \rightarrow e_1 z \downarrow \,\&\, e_1 zx \downarrow \,\&$
$e_1 zx \underset{\sim}{k} \mathrm{Sat}(z, \mathrm{Sub}(x, B_0 \mathrm{var}(n), \mathrm{Pred}(n)))]]\}$

Theorem 3.3. *Sat is self-realizing i.e.,*

(1) $\exists \hat{e} \in F$ *such that* $Sat(m, n) \rightarrow \hat{e}mn \underset{\sim}{k} Sat(m, n)$.
(2) $\forall q \in F\ (q \underset{\sim}{k} Sat(mSat(m, n))$.

$\hat{e}mn$ in (1) is Curried. It means $\{\{\hat{e}\}^{F_0}(m)\}^{F_0}(n)$.

Proof. By simultaneous induction on the length of sentences (e.g., define $lth(n) = lth(\varphi)$ if $n = \ulcorner\varphi\urcorner$, and 0 otherwise). First we give the definition of $\hat{e}$: By the (relativized) recursion theorem, pick $\hat{e}$ satisfying:

$\hat{e} = \Lambda m \cdot \Lambda n \cdot$**if** $\chi_P(m) = 1 \,\&\, \chi_{Sent}(n) = 1$ **then**

if $Prime(n) \,\&\, ax(m,n)$ **then** $< 0, 0 >$ **else**

if $Conj(n)$ **then** $< 1, < \hat{e}ml_{\&}(n), \hat{e}mr_{\&}(n) >>$ **else**

if $Dis(n)$ **then** $< 2, < w(m,n)$, **if** $w(m,n) = 0$ **then** $\hat{e}ml_{\vee}(n)$

else $\hat{e}mr_{\vee}(n) >>$ **else**

if $Imp(n)$ **then** $< 3, \Lambda y \Lambda q \cdot$ **if** $P(y) \,\&\, \mathcal{O}(m,y)$ **then** $\hat{e}yCons(n)$ >**else**

if $Ex(n)$ **then** $< 4, < w(m,n), \hat{e}mSub(w(m,n), B_0var(n), \mathrm{Pred}(n))$ >>**else**

if $\mathrm{Univ}(n)$ **then**

$< 5, \Lambda z \Lambda c \cdot$**if**$P(z) \,\&\, D(c,z) \,\&\, \mathcal{O}(m,z)$**then** $\hat{e}zSub(c, B_0var(n), Pred(n)) >$

Now for the proof: (1) and (2) will refer to the two conclusions of the theorem and the corresponding inductive hypothesis, here labelled $H(1)$ and $H(2)$.

prime case:

Suppose $P(m)$ and $Sent(n)$ and $Prime(n) \,\&\, ax(m,n)$.

(1) Then $\hat{e}mn =< 0, 0 >$, so $(\hat{e}mn)_0 = 0 \,\&\, Prime(n) \,\&\, ax(m,n)$ so $\hat{e}mn \underset{\sim}{k} Sat(m,n)$.

(2) Suppose $q \underset{\sim}{k} Sat(m,n)$,i.e. $q_0 = 0 \,\&\, Prime(n) \,\&\, ax(m,n)$. Then $Prime(n) \,\&\, \mathrm{ax}(m,n)$, i.e., $Sat(m,n)$.

and:

(1) Suppose $Conj(n) \,\&\, Sat(m, l_{\vee}(n)) \,\&\, Sat(m, r_{\&}(n))$. Then $(\hat{e}mn)_0 = 1$ and by $H(1)$ $(\hat{e}mn)_{10} \underset{\sim}{k} \mathrm{Sat}(m, l_{\vee}(n))$ and $(\hat{e}mn)_{11} \underset{\sim}{k} Sat(m, r_{\&}(n))$ so $\hat{e}mn \underset{\sim}{k} Sat(m,n)$.

(2) Suppose $q_0 = 1$ and $Conj(n) \& q_{10} \underset{\sim}{k} Sat(m, l_{\vee}(n)) \& q_{11} \underset{\sim}{k} Sat(m, r_{\&}(n))$. By $H(2) Conj(n) \,\&\, Sat(m, l_{\&}(n)) \,\&\, Sat(m, r_{\&}(n))$, so $Sat(m,n)$.

or:

(1) Suppose $Dis(n) \& [(w(m,n) = 0 \& Sat(m, l_{\vee}(n))) \vee (w(m,n) = 1 \& Sat(m, r_{\&}(n)))]$. Then
$(\hat{e}mn)_0 = 2$ and
$(\hat{e}mn)_1 =< w(m,n)$, **if** $w(m,n) = 0$ **then** $\hat{e}ml_{\vee}(n)$ **else** $\hat{e}mr_{\vee}(n) >$
so, by $H(1)$,

$[(\hat{e}mn)_{10} = 0 \ \& \ (\hat{e}mn)_{11} \underset{\sim}{k} Sat(m, l_\vee(n))] \vee [(\hat{e}mn)_{10} = 1 \ \& \ (\hat{e}mn)_{11}$
$\underset{\sim}{k} Sat(m,$
$r_\vee(n))]$.
Hence $\hat{e}mn \underset{\sim}{k} Sat(m,n)$.

(2) Suppose $q_0 = 2$ and

$$Dis(n)\&[q_{10} = 0\&q_{11} \underset{\sim}{k} Sat(m, l_\vee(n)))\vee(q_{10} \neq 0\&q_{11} \underset{\sim}{k} Sat(m, r_\vee(n)))].$$

By $H(2)$ and some elementary logic $q_{10} = 0 \vee q_{10} \neq 0$ and $\mathrm{Dis}(n)\&[Sat(m,$ $l_\vee(n)) \vee Sat(m, r_\vee(n))]$. Hence, by the definition of w,
$\mathrm{Dis}(n)\&[(w(m,n) = 0\&Sat(m, l_\vee(n))\vee(w(m,n) = 1\& \ Sat(m, r_\vee(n)))]$
so $Sat(m,n)$.

implies:

(1) Suppose $Sat(m,n)$ and $Imp(n)$.
Then $\forall y[P(y)\&\mathcal{O}(m,y)\&Sat(y, Ant(n)) \to Sat(y, Cons(n))]$, and
$(\hat{e}mn)_0 = 3$ and
$(\hat{e}mn)_1 = \Lambda y \Lambda q\cdot$ **if** $P(y) \ \& \ \mathcal{O}(m,y)$ **then** $\hat{e}yCons(n)$.
Now suppose y and q are given with $P(y)\&\mathcal{O}(m,y)\&q \underset{\sim}{k} Sat(y, Ant(n))$.
By $H(2)$, $Sat(y, Ant(n))$, hence $Sat(y, Cons(n))$.
By $H(1)$, $\hat{e}yCons(n) \underset{\sim}{k} Sat(y,\mathrm{Cons}(n))$ hence
$(\hat{e}mn)_1 y\downarrow$, $(\hat{e}mn)_1 yq\downarrow$ and $(\hat{e}mn)_1 yq \underset{\sim}{k} Sat(y, Cons(n)))$. Therefore
$\hat{e}mn \underset{\sim}{k} Sat(m,n)$.

(2) Suppose $q_0 = 3$ and
$\forall y \forall u[P(y) \ \& \ \mathcal{O}(m,y) \ \& \ u \underset{\sim}{k} Sat(y, Ant(n)) \to$
$q_1 y\downarrow \ \& \ q_1 yu\downarrow \ \& \ q_1 yu \underset{\sim}{k} Sat(y,\mathrm{Cons}(n))]$.
Further suppose y is given with $P(y) \ \& \ \mathcal{O}(m,y)$, and $Sat(y, Ant(n))$.
By $\mathbf{H}(1)$

$$\hat{e}y\mathrm{Ant}(n) \underset{\sim}{k} Sat(y, Ant(n)),$$

hence if $e' = \hat{e}yAnt(n)$ then

$$q_1 ye'\downarrow \ \& \ q_1 ye' \underset{\sim}{k} Sat(y, \mathrm{Cons}(n)).$$

By $H(2)Sat(y,\mathrm{Cons}(n))$. Hence $Sat(n,m)$.

The remaining cases are straightforward. See [46] for details. □

Stratified Realizability. Recall $\mathcal{L}_k$ is the original language for which $\mathcal{K}$ was a Kripke model, enriched with the constants c in $D(*) = \cup\{D(p) : p \in P\}$, where $*$ is the top node of $\mathcal{K}$, which forces all sentences over $\mathcal{L}_k$.

Let φ be a sentence over $\mathcal{L}_k$. We now define an associated realizability formula $|\varphi|$ in one free variable $|\varphi(x)| \equiv x \underset{\sim}{k} \varphi$.

Realizers are pairs $< e, p >$ with $e \in F$, $p \in P$. Formally, they are integers $< e, \ulcorner p \urcorner >$ where e is a code for an F_0-recursive function. Notationally, we will identify nodes p with their Gödel numbers. We define a truncated notion of application pq of node p to node q by

$$pq = if\ (p \leq q)\textbf{then}\ \ q\ \textbf{else}\ *. \tag{13}$$

Note that this is a 'degenerate supremum'. Our arguments work as well if we define $pq = if\ (p \leq q \text{ or } q \leq p)$ **then** $\max(p, q)$ **else** $*$. In fact, if $\mathcal{K}$ were sup-closed (as are many Kripke Models naturally associated with realizability, see [47]), then $pq = p \vee q$ would work.

We now define the realizability we will associate with the Kripke model K. For the sake of legibility, we write out the definitions of, e.g., $|\theta|$, as *sets* of realizers. Thus the following are all notational variants of the same notion:

$$x \in |\varphi|,\ |\varphi|(x),\ x \underset{\sim}{k} \varphi,\ x \underset{\sim}{r} \varphi$$

Definition 3.4. For φ atomic $|\varphi| = \{< n, p > | ax(p, \ulcorner\varphi\urcorner)\}$

$$|\varphi| \times |\psi| = \{< e, p > | < e_0, p > \in |\varphi| \ \& \ < e_1, p > \in |\psi|\}$$

$$|\varphi| + |\psi| = \{< e, p > | < (e_0 = 0 \ \& \ < e_1, p > \in |\varphi|) \vee (e_0 \neq 0 \ \& \ < e_1, p > \in |\psi|)\}$$

$$|\varphi| \Rightarrow |\psi| = \{< e, p > | (\forall < u, q > \in |\varphi|) e(< u, q >) \downarrow \ \& \ < e(< u, q >), pq > \in |\psi|\}$$

$$|\exists x \theta(x)| = \{< e, p > | < e, p > \in |\theta(e_0)| \ \& \ < e, p > \in D\}$$

$$|\forall x \theta(x)| = \{< e, p > | (\forall < c, z > \in D) e(< c, z >) \downarrow \ \& \ < e(< c, z >), pz > \in |\theta(c)|\}$$

Lemma 3.5. *For each $\mathcal{L}_k$-sentence φ there are functions $s_\varphi, t_\varphi \in F$ depending only on the parameter-free structure of φ such that*

(1) $< e, p > \in |\varphi| \Rightarrow t_\varphi(e, p) \underset{\sim}{k} Sat(p, \ulcorner\varphi\urcorner)$

(2) $e \underset{\sim}{k} Sat(p, \ulcorner\varphi\urcorner) \Rightarrow < s_\varphi(e, p), p > \in |\varphi|$

The functions can in fact be computed uniformly in a code for the parameter-free structure of φ.

Note: By parameter-free structure, we mean that φ can be replaced by $\dot{\varphi}$ which has variables in place of the parameters of φ. Thus, e.g.,

$$s_{\varphi(c)} = s_{\varphi(a)} \text{ for distinct parameters } \quad c \text{ and a.}$$

Proof. We sketch the routine simultaneous induction proof of (1), (2). The uniformity in φ will not be discussed, but is easy to see from the proof.

prime case:

(1) $< e, p > \in |\varphi| \Rightarrow ax(p, \ulcorner\varphi\urcorner) \Rightarrow Sat(p, \ulcorner\varphi\urcorner)$
$\Rightarrow < 0, e > \underset{\sim}{k}\, Sat(p, \ulcorner\varphi\urcorner)$

for any e. Hence $t_\varphi = \Lambda x \cdot < 0, x >$ will do.

(2) $e \ \underset{\sim}{k}\, Sat(p, \ulcorner\varphi\urcorner) \Rightarrow e_0 = 0 \,\&\, ax(p, \ulcorner\varphi\urcorner) \Rightarrow < x, p > \in |\varphi|$ for any $K \ \Rightarrow$ $< e, p > \in |\varphi|$ so $s_\varphi = \Lambda x \cdot x$ will work.

and:

(1) Suppose $< e, p > \in |\varphi| \times |\psi|$. Then $< e_0, p > \in |\varphi|$ and $< e_1, p > \in |\psi|$. By $H(1)$ there are codes t_φ, t_ψ such that $t_\varphi(p, e_0) \ \underset{\sim}{k}\, Sat(p, \ulcorner\varphi\urcorner)$ and $t_\psi(p, e_1)$ $\underset{\sim}{k}\, Sat(p, \ulcorner\psi\urcorner)$ hence
$< 1, < t_\varphi(p, e_0), t_\psi(p, e_1) >> \ \underset{\sim}{k}\, Sat(p, \varphi \vee \psi)$ and

$$t_{\varphi \,\&\, \psi} = \Lambda z \Lambda x \cdot < 1, \ < t_\varphi(z, x_0), t_\psi(z, x_1) >> .$$

(2) A similar argument shows we should take

$$s_{\varphi \,\&\, \psi} = \Lambda z \Lambda x \cdot < s_\varphi(z, x_{10}), s_\psi(z, x_{11}) > s_\varphi$$

or:

Take $t_{\varphi \vee \psi} = \Lambda z \Lambda x \cdot < 2, < x_0, \textbf{if } x_0 = 0 \textbf{ then } t_\varphi(z, x_1) \textbf{ else } t_\psi(z, x_1) >> .$
$s_{\varphi \vee \psi} = \Lambda z \Lambda x \cdot < x_{10}, \textbf{if } x_0 = 0 \textbf{ then } s_\varphi(z, x_{11}) \textbf{ else } s_\psi(z, x_{11}) >$

implies:

Suppose $< e, p > \in |\varphi| \Rightarrow |\psi|$ and $P(q)$, $p \leq q$, $u \ \underset{\sim}{k}\, Sat(q, \ulcorner\varphi\urcorner)$.
By $H(2)$, $< s_\varphi(u, q), q > \in |\varphi|$, hence $< e(< s_\varphi(u, q), q >), q > \in |\varphi|$.
By $H(1)$, $t_\psi(e(< s_\varphi(u, q), q >), q) \ \underset{\sim}{k}\, Sat(p, \ulcorner\varphi\urcorner)$ so
$< 3, \Lambda q \Lambda u \cdot \textbf{ if } \mathcal{O}(p, q) \textbf{ then } t_\psi(e(< s_\varphi(u, q), q >), q) > \ \underset{\sim}{k}\, Sat(p, \ulcorner\varphi \to \psi\urcorner)$
hence
$t_{\varphi \to \psi} = \Lambda w \Lambda p \cdot < 3, \Lambda q \Lambda u \cdot if \ \ \mathcal{O}(p, q) \textbf{ then } t_\psi(e(< s_\varphi(u, q), q >), q) > .$
A similar argument shows

$$s_{\varphi \to \psi} = \Lambda w \Lambda p \Lambda z \cdot \textbf{ if } \ \ \mathcal{O}(p, z_1) \textbf{ then } \ s_\psi(w(< t_\varphi(z_0, z_1), z_1 >), z_1) \textbf{ else } \ \ r_o.$$

where r_o denotes the (code of the) root node of $\mathcal{K}$.

exists: Take
$t_{\exists x\theta} = \Lambda w \Lambda p \cdot < 4, < w_0, t_\theta(w_1, p) >> \textit{ and } \ s_{\exists x\theta} = \Lambda w \Lambda p \cdot < w_{10}, s_\theta(w_{11}, p) >$

<u>**forall:**</u> Put $t_{\forall x\theta x} = \Lambda w \Lambda p\cdot < 5, \Lambda z \Lambda x$ **if** $p \leq z$ **then** $t_\theta(w(< x, z >), z)$, and let $s_{\forall x\theta} = \Lambda p \Lambda w \Lambda u\cdot$ **if** $(p \leq u)$ **then** $s_\theta(w_1 u_1 u_0, u_1)$ **else** 0. □

Corollary 3.6.

$$p \Vdash \varphi \iff (\exists x) < x, p > \in |\varphi|.$$

Proof. $p \Vdash \varphi \equiv Sat(\ulcorner p \urcorner, \ulcorner \varphi \urcorner) \Rightarrow$ by Theorem 3.3 $\exists e\ e \underset{\sim}{k} Sat(\ulcorner p \urcorner, \ulcorner \varphi \urcorner)$. By Lemma 3.5,

$$< s_\varphi(e, p), p > \in |\varphi|.$$

Conversely, suppose $< e, p > \in |\varphi|$. Then $t_\varphi(e, p) \underset{\sim}{k} Sat(\ulcorner p \urcorner, \ulcorner \varphi \urcorner)$. Hence, by Theorem 3.3, $Sat(\ulcorner p \urcorner, \ulcorner \varphi \urcorner)$, whence $p \Vdash \varphi$. □

We now briefly sketch how a realizing category can be built from the notions just defined, which is elementarily equivalent to the original Kripke model. We construct a weakly cartesian closed category $\mathcal{C}$ which is non-trivial in the sense that all partial recursive functions are representable by morphisms of $\mathcal{C}$ (in a way that will be made precise below), and define an abstract realizability in $\mathcal{C}$ as follows:

φ is *realizable* in $\mathcal{C}$ iff there is a morphism from the (weakly) terminal object T of $\mathcal{C}$ into the object $|\varphi|$ of $\mathcal{C}$ representing φ.

Local realizers of φ in C will be morphisms $e : A \to |\varphi|$ for arbitrary objects A of $\mathcal{C}$.

Definition 3.7. $\mathcal{C}_K$ is the category whose objects are *upward-closed* sets of pairs $< e, \ulcorner p \urcorner >$ such that $e \in N$ and $\ulcorner p \urcorner \in N$ is a code for a node p of the Kripke Model $\mathcal{K}$.

We will drop the brackets and identify p with its code. By A *upward closed*, we mean

$$< e, p > \in A \,\&\, r \geq p \Rightarrow < e, r > \in A.$$

(A somewhat more constructive category is obtained if we restrict $\mathcal{C}_K$ to objects generated by $\times, +, \Rightarrow, \Sigma, \Pi$ from the basic sets

$$|\varphi| = \{< n, p >: ax(p, \varphi),\ n \in N\}$$

and

$$Q_p = \{< n, r > | n \in N, r \geq p\}$$

where φ is atomic and $r \geq p$ is shorthand for $\mathcal{O}(\ulcorner p \urcorner, \ulcorner r \urcorner)$.)

Morphisms are triples (A, e, B) *where* $e \in N$ *and* e *denotes the function* $\{e\}^{F_0}$ whose action is given by

$$e| < n, p > := < e(< n, p >), p > .$$

This somewhat abusive notation means the following: the left hand side is defining a new *application* operator ' | ' , which is the application of the category $\mathcal{C}_K$ just defined. The right hand side is application as defined earlier in this section, to wit, $e(< n, p >)$ means $\{e\}^{F_0}(< n, p >))$. Furthermore, e must satisfy the requirement

$$< n, p > \in A \Rightarrow e(< n, p >) \downarrow \& e| < n, p > \in B.$$

As in the preceding section, morphisms (A, e, B) will just be referred to by the code e (i.e., domain and codomain will be tacit).

The identity id defined by $\pi_0 \equiv \Lambda x \cdot x_0$ satisfies $id| < n, p > = < n, p >$, so $1_A \equiv < A, \pi_0, A >$. Composition of two morphisms, f and e is denoted $f \| e$, and is given by $(f \| g)|x = f|(g|x)$. We define equality of morphisms $(A, e, B) = (A, f, B)$ by extensional equality, i.e.

$$< n, p > \in A \quad \Rightarrow \quad e| < n, p > = f| < n, p > .$$

We will not bother to distinguish notationally between e as a morphism in $\mathcal{C}_K$ and e as a map from N to N. The different application operators $e| < n, p >$ *and* $e(< n, p >)$ should make this distinction clear.

We recall that a category $\mathcal{C}$ has *weak exponents* if for every pair of objects A, B of $\mathcal{C}$ there is an object $(A \Rightarrow B)$ called the *weak exponent determined by* A *and* B and a morphism $app_{A,B}$ in $Hom_{\mathcal{C}}((A \Rightarrow B) \times A, B)$ such that for any object C and any morphism $f \in Hom_{\mathcal{C}}(C \times A, B)$ there is a not necessarily unique map Λf in $Hom_{\mathcal{C}}(C, A \Rightarrow B)$ making the diagram

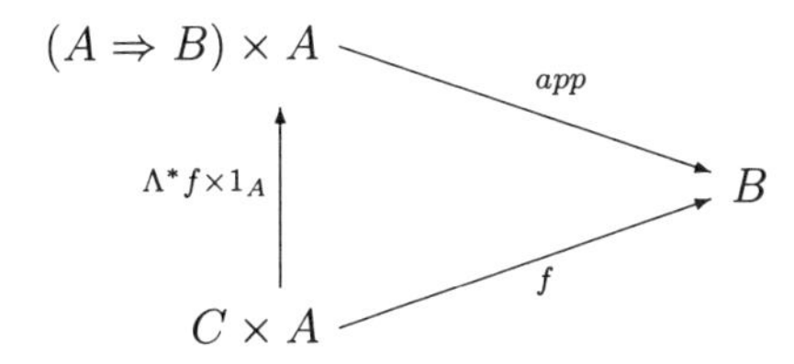

commute.

If the morphism Λf is unique for every such situation, we drop the word *weak* in the definition just given.

Lemma 3.8. *$\mathcal{C}_K$ is weakly cartesian closed. That is to say, it is closed under finite products, coproducts, and weak exponentiation, and has a weak terminal object T (there is a not necessarily unique morphism from every object A into T).*

Proof. We sketch the key definitions and leave the verifications to the reader. Products are given by

$$A \times B = \{<< e, f >, pq > \mid < e, p > \in A, < f, q > \in B\}$$

or, equivalently,

$$\{<< e, f >, p > \mid < e, p > \in A, < f, p > \in B\}.$$

The projection maps are ρ_0 and ρ_1, where

$$\rho_0 = \Lambda z \cdot z_{00} \ , \ \rho_1 = \Lambda z \cdot z_{01}.$$

Define

$$\ll e, f \gg \equiv \Lambda z \cdot < ez, fz > .$$

Then $\ll e, f \gg : C \to A \times B$, and $\rho_0 \| \ll e, f \gg = e$, and $\rho_1 \| \ll e, f \gg = f$ and the map is easily shown unique.

Coproducts $A + B$ are defined as in 3.4:

$$A + B = \{< e, p > \mid (e_0 = 0 \& < e_1, p > \in A) \vee (e_0 \neq 0 \& < e_1, p > \in B)\}$$

with the associated canonical inclusions:

$h_0 : A \to A + B$, given by $\Lambda z \cdot < 0, z_0 >$ and

$h_1 : B \to A + B$ by $\Lambda z \cdot < 1, z_0 >$.

If $e : A \to C$ and $f : B \to C$ in $\mathcal{C}_K$, then let $[\![e, f]\!] = \Lambda z \cdot$ **if** $z_{00} = 0$ then $e(< z_{01}, z_1 >$ *else* $f(< z_{01}, z_1 >)$. It is easily seen that $[\![e, f]\!] : A + B \to C$ is in $\mathcal{C}_K$ and

$$[\![e, f]\!] \| h_0 = e, \ [\![e, f]\!] \| h_1 = f.$$

Uniqueness is straightforward, and left to the reader.

Exponents: Define, for objects B, C of $\mathcal{C}_K$

$$B \Rightarrow C = \{< e, p > \mid (\forall < u, q > \in B) < e(< u, q >), pq > \in G\}.$$

Suppose $e : A \times B \to C$ is $\mathcal{C}_K$. Let $\Lambda^* e = \Lambda y \Lambda z \cdot e(<< y_0, z_0 >, y, z_1 >)$. Then, if $< w, s > \in A$.

$$\Lambda^* e| < w, s > = < \Lambda^* e(< w, s >), s >$$

$= < \Lambda z \cdot e(<< w, z_0 >, s z_1 >), s >$

We claim that $\Lambda^* e| < w, s > \in B \ \Rightarrow C$. For suppose $< v, q > \in B$. Then

$$< [\Lambda z \cdot e(<< w, z_0 >, sz_1 >)](< v, q >), sq >=< e(<< w, v >, sq >), sq >$$

is in C since $< w, s >\in A$, and $< r, q >\in B \Rightarrow << w, v >, sq >\in A \times B$, hence, by the assumption that $e : A \times B \to C$,

$$e| << w, v >, sq >=< e(<< w, v >, sq >), sq >\in C.$$

Thus, if $e : A \times B \to C$ then $\Lambda^* e : A \to (B \Rightarrow C)$.

Now suppose $e : A \to (B \Rightarrow C)$. In other words, $e \in \omega$ is a map of the type shown under $|$ — application. Then define the code $\Lambda_* e$ (of an F_0 – recursive function)

$$\Lambda_* e = \Lambda z \cdot (e(< z_{00}, z_1 >))(< z_{01}, z_1 >).$$

Then, under $|$ — application, $\Lambda_* e : A \times B \to C$ is easily checked.

Finally, observe that for r_o the root node of the Kripke model $\mathcal{K}$, the object

$$Q_{r_o} = \{< e, p > \mid p \geq r_o\} = N \times P$$

is weakly terminal: any object is included in it. □

Theorem 3.9. *Let $\mathcal{K}$ be a Kripke Model, formalized in N as described above, and let $\mathcal{C}_K$ be the category given in Definition 3.7 Then, for each node p, and $\mathcal{L}$-sentence φ*

$$p \Vdash \varphi \text{ iff there is an } \quad e : Q_p \to |\varphi| \text{ in } \quad \mathcal{C}_K$$

Proof. First observe that $< e, p >\in |\varphi| \iff \Lambda x \cdot e : Q_p \to |\varphi|$ in $\mathcal{C}_K$. This is almost by definition: if $< u, r >\in Q_p$, then $r \geq p$. Thus

$$(\Lambda x \cdot e)| < u, r >=< (\Lambda x \cdot e)(< u, r >), r >=< e, r >,$$

and $< e, r >\in |\varphi|$ if $< e, p >\in |\varphi|$. Conversely, if $\Lambda x \cdot e : Q_p \to |\varphi|$ then, since $< 0, p >\in Q_p$, we have $< e, p >\in |\varphi|$.

Now, by Corollary 3.6

$$p \Vdash \varphi \iff \exists x (< x, p >\in |\varphi|)$$

hence, $p \Vdash \varphi \iff$ there is some e such that $e : Q_p \to |\varphi|$ in $\mathcal{C}_K$ □

In particular, we have

Corollary 3.10. *φ is true in the model $\mathcal{K}$ iff it is realizable in the category $\mathcal{C}_K$.* □

To sum up: a straightforward generalization of the realizability categories one finds in ordinary practice gives a notion of realizability which is complete, and which can be obtained uniformly from a denumerable Kripke model. The interpretation can, of course be no more constructive than the Kripke model, but it will not be of any greater computational complexity.

If the model is "given" without any further information about how it was obtained, the realizability will be somewhat of a hybrid, since it must respect the truth-value structure imposed by the partial order of the original Kripke model. As can be seen in the Kripke models that naturally arise from realizability interpretations (e.g. for syntactic realizabilities, in Lipton [47], or, for Kreisel - Troelstra, and *semantic* Kleene realizability, in Hyland's [33], or in [76]), *the truth value structure imposed by realizability proper is essentially that of the degree of inhabitation* of sets of, or predicates on the realizers. This idea survives in the category $\mathcal{C}_K$ just described in the sense that there is a morphism $e : Q_p \to Q_r$ in $\mathcal{C}_K$ just in case $p \geq r$ in the original model $\mathcal{K}$. Thus $p \geq r$ corresponds to "Q_r is at least as inhabited as Q_p."

The realizers just constructed explicitly display more information than the nodes in the Kripke model do, in the sense that if $< e, p > \in |\varphi|$, we not only know that $p \Vdash \varphi$, but e tells us why. If φ is a conjunction, say, each conjunct of which is forced by p, the ordered pair structure of e will encode this. Of course the atomic information is simply copied from the diagram of $\mathcal{K}$.

If the Kripke structure itself has been constructively presented as is the case for Kripke models generated by Tableaux (as set forth in Fitting's [21], and in Nerode's [61]), the nodes in the category $\mathcal{C}_K$ can be replaced by descriptions of the algorithm that generates them. By so doing, one obtains a tableau refutation procedure, which produces "pure" realizability counterexamples. If a constructive metatheory is assumed, i.e. if it is not the case that for every p and φ $\;p \Vdash \varphi$ or not $p \Vdash \varphi$, then the positive and negative information produced by the tableau are not complements. Separate notions of realizability are required for the predicates $N(p, \varphi)$ and $F(p, \varphi)$ representing negative and positive information about what is forced. We briefly sketch the necessary definitions.

Realizability for Constructively Presented Kripke Models

Definition 3.11. The positive and negative satisfaction predicates, $F(p, \varphi)$ and $N(p, \varphi)$ are given by the following inductive definitions. We make use of the "truncated application" of nodes defined in (13) so as to build-in the applicative structure obtained in the preceding section.[2]

$$
\begin{aligned}
F(p,\varphi) &\stackrel{def}{\equiv} ax(p,\varphi) \qquad \varphi \text{ atomic}\\
F(p,\varphi \,\&\, \psi) &\stackrel{def}{\equiv} F(p,\varphi) \,\&\, F(p,\psi)\\
F(p,\varphi \vee \psi) &\stackrel{def}{\equiv} [(w(p,\varphi \vee \psi) = 0 \,\&\\
&\qquad F(p,\varphi)) \vee (w(p,\varphi \vee \psi) = 1 \,\&\, F(p,\psi))]\\
F(p,\varphi \to \psi) &\stackrel{def}{\equiv} \forall q[F(q,\varphi) \to F(pq,\psi)]\\
F(p,\exists x\theta(x)) &\stackrel{def}{\equiv} D(p, w(p,\exists x\theta(x))) \,\&\, F(p,\theta(w(p,\exists x\theta(x))))\\
F(p,\forall x\theta(x)) &\stackrel{def}{\equiv} \forall q \forall z[D(q,z) \to F(pq,\theta(z))]
\end{aligned}
$$

The **negative satisfaction** predicate $N(p,\varphi)$, which can be thought of a negatively signed tableau entry: "p does not force φ" is given by:

$$
\begin{aligned}
N(p,\varphi) &\stackrel{def}{\equiv} \neg ax(p,\varphi) \qquad \varphi \text{ atomic}\\
N(p,\varphi \,\&\, \psi) &\stackrel{def}{\equiv} [(w(p,\varphi \,\&\, \psi) = 0 \,\&\, N(p,\varphi)) \vee\\
&\qquad (w(p,\varphi \,\&\, \psi) = 1 \,\&\, N(p,\psi))]\\
N(p,\varphi \vee \psi) &\stackrel{def}{\equiv} N(p,\varphi) \,\&\, N(p,\psi)\\
N(p,\varphi \to \psi) &\stackrel{def}{\equiv} F(\nu(p,\varphi \to \psi),\varphi) \,\&\, N(\nu(p,\varphi \to \psi),\psi)\\
N(p,\exists x\theta(x)) &\stackrel{def}{\equiv} \forall z[D(p,z) \to N(p,\theta(z))]\\
N(p,\forall x\theta(x)) &\stackrel{def}{\equiv} D[\nu(p,\forall x\theta(x)), w(p,\forall x\theta(x))] \,\&\, N[\nu(p,\forall x\theta(x)), \theta(w(p,\\
&\qquad \forall x\theta(x)))]
\end{aligned}
$$

where:

(1) pq is the **application** of nodes: $p \cdot q =$ if $p \leq q$ then q else $*$.

(2) $w(p,\varphi)$ is the **witness** function, that returns 1 or 0 for left or right component if φ is a disjunction (positive forcing) or a conjunction (negative forcing), and returns an individual c in $D(p)$ if φ is existential, and positively forced. The negative forcing of implication and universal quantification require the introduction of a skolem function $\nu(p,\varphi)$ producing nodes, which we now define.

[2]Because the primary aim of these definitions is to develop realizability-tableaux, the induced applicative structure has been built into the satisfaction predicate. As can be seen from the use of the Skolem functions ν, the only "nodes" appearing in a tableau development are terms in ν, Gödel numbers of formulas, and the root node r_o. Notice that for ease of notation we have dropped Gödel-number brackets, and "typing-predicates" such as $D(p)$.

(3) $\nu(p,\varphi)$ is a node q above p providing a witness to the failure of node p to force φ if it is an implication, or, if φ is $\forall x\theta(x)$, of the failure of node p to force $\theta(w(p,\forall x\theta(x)))$.

Axioms for N,P and the witnessing functions:.

(1) $D(p, w(p, \exists x\theta(x)))$.
(2) $D(\nu(p,\forall x\theta(x)), w(p,\forall x\theta(x)))$.
(3) $p\cdot\nu(p,\varphi\to\psi)=\nu(p,\varphi\to\psi)$
(4) $p\cdot\nu(p,\forall x\theta(x))=\nu(p,\forall x\theta(x))$
(5) $N(p,\varphi)\ \&\ F(p,\varphi)\to\bot$

Definition 3.12. We now define **realizability** of the satisfaction predicates of definition (3.11) as follows:

Positive forcing:

$$
\begin{array}{lll}
e \underset{\sim}{k} F(p,\varphi) & \overset{def}{\equiv} & \\
\varphi \text{ atomic} & \Rightarrow & e_0 = 0\ \&\ F(p,\varphi) \\
\varphi\equiv\theta\ \&\ \psi & \Rightarrow & e_0 = 1\ \&\ e_{10} \underset{\sim}{k} F(p,\theta)\ \&\ e_{11} \underset{\sim}{k} F(p,\psi) \\
\varphi\equiv\theta\vee\psi & \Rightarrow & e_0 = 2\ \&\ (e_{10}=0\ \&\ e_{11} \underset{\sim}{k} F(p,\theta))\vee \\
 & & \qquad (e_{10}=1\ \&\ e_{11} \underset{\sim}{k} F(p,\psi)) \\
\varphi\equiv\theta\to\psi & \Rightarrow & e_0 = 3\ \&\ (\forall u)(\forall q)[u \underset{\sim}{k} F(q,\theta)\to \\
 & & \qquad e_1 q\downarrow\ \&\ (e_1 u)q\downarrow\ \&\ (e_1 u)q \underset{\sim}{k} F(pq,\psi) \\
\varphi\equiv\exists x\theta(x) & \Rightarrow & e_0 = 4\ \&\ D(p,e_{10})\ \&\ e_{11} \underset{\sim}{k} F(p,\theta(e_{10})) \\
\varphi\equiv\forall x\theta(x) & \Rightarrow & e_0 = 5\ \&\ \forall q\forall z[D(q,z)\to e_1 q\downarrow\ \&\ (e_1 q)z\downarrow\ \& \\
 & & \qquad e_1 qz \underset{\sim}{k} F(pq,\theta(z))]
\end{array}
$$

Negative forcing:

$$
\begin{array}{lll}
e \underset{\sim}{k} N(p,\varphi) & \overset{def}{\equiv} & \\
\varphi \text{ atomic} & \Rightarrow & e_0 = 0\ \&\ N(p,\varphi) \\
\varphi\equiv\theta\ \&\ \psi & \Rightarrow & e_0 = 1\ \&\ (e_{10}=0\ \&\ e_{11} \underset{\sim}{k} N(p,\theta))\vee \\
 & & \qquad (e_{10}=1\ \&\ e_{11} \underset{\sim}{k} N(p,\psi)) \\
\varphi\equiv\theta\vee\psi & \Rightarrow & e_0 = 2\ \&\ e_{10} \underset{\sim}{k} N(p,\theta)\ \&\ e_{11} \underset{\sim}{k} N(p,\psi) \\
\varphi\equiv\theta\to\psi & \Rightarrow & e_0 = 3\ \&\ P(e_{10})\ \&\ e_{110} \underset{\sim}{k} F(e_{10},\theta)\ \&\ e_{111} \underset{\sim}{k} N(e_{10},\psi) \\
\varphi\equiv\exists x\theta(x) & \Rightarrow & e_0 = 4\ \&\ \forall z[D(p,z)\to e_1 z\downarrow\ \&\ e_1 z \underset{\sim}{k} N(p,\theta(z))] \\
\varphi\equiv\forall x\theta(x) & \Rightarrow & e_0 = 5\ \&\ P(e_{100})\ \&\ D(e_{100},e_{101})\ \&\ e_{11} \underset{\sim}{k} N(e_{100},\theta(e_{101}))
\end{array}
$$

A similar proof to theorem 3.3 gives:

Theorem 3.13. *N and F are self-realizing*

With these definitions and a slight modification of the construction of the stratified realizability in definition (3.4) it is straightforward to give a realizability version of intuitionistic tableau proof development. This can be seen as a translation of Kripke-model based tableaux of, e.g., Fitting [21] or Nerode [60], along the lines of the preceding section.

From such realizers one is able to give highly effective counterexamples, when a tableau proof fails, and a Curry-Howard style proof term, if the tableau closes off. This constitutes an alternative constructive proof of the completeness theorem for intuitionistic logic. The details of these realizability tableaux are worked out in [49].

REFERENCES

1. Beeson, M. J. [1985], *Foundations of Constructive Mathematics*, Springer-Verlag, Berlin.
2. Beeson, M. J. [1982], "Recursive models of constructive set theories", Annals of Mathematical Logic **23**, 127–178.
3. Beth, E. W. [1956], "Semantic construction of intuitionistic logic", Koninklijke Nederlandse Akademie van Wetenschappen, Niewe Serie 19/11, 357–388.
4. Beth, E. W. [1947], "Semantical considerations on intuitionistic mathematics", Koninklijke Nederlandse Akademie van Wetenschappen, Proceedings of the Section of Sciences.
5. Boileau, A. and A. Joyal [1981], "La logique des topos", Journal of Symbolic Logic **46**, 6–16.
6. Buss, S. [1989], "On model theory for intuitionistic bounded arithmetic with applications to independence results", to appear.
7. Cohen, P. J. [1966], *Set Theory and the Continuum Hypothesis*, W. A. Benjamin Inc., New York.
8. Cohen, P. J. [1963], "The independence of continuum hypothesis", *Proceedings of the National Academy of Science, USA 50*, 1143–1148.
9. Constable, R. L., et al [1986], *Implementing Mathematics with the NUPRL Development System*, Prentice-Hall, N.J.
10. Constable, R. L. [1986], "The semantics of evidence", Cornell CS Technical Report, Ithaca, N.Y.
11. Constable, R. L. [1983], "Programs as proofs", Information Processing Letters, **16**(3), 105–112.
12. Cook, S. A. and Urquhart [1988], "Functional interpretations of feasible constructive arithmetic", Technical Report 210/88, Department of Computer Science, University of Toronto, June 1988.
13. Coquand, T. [1990], "On the analogy between propositions and types", in: *Logic Foundations of Functional Programming*, Addison-Wesley, Reading, MA.
14. Coquand, T. and G. Huet [1985], "Constructions: A higher order proof system for mechanizing mathematics", EUROCAL 85, Linz, Austria.
15. Dragalin, A. G. [1987], *Mathematical Intuitionism: Introduction to Proof Theory*, Translations of Mathematical Monographs **67**, AMS, Providence, R.I.
16. Dragalin, A. G. [1979], "New kinds of realizability", Sixth International Congress, Logic, Methodology and Philosophy of Science (Hannover, August 1979), Abstracts of Reports, Hannover.
17. Dummett, M. [1977], *Elements of Intuitionism*, Clarendon Press, Oxford.
18. Dyson, V. H. and G. Kreisel [1961], "Analysis of Beth's semantic construction of intuitionistic logic", Technical Report No. 3, Applied Mathematics and Statistics Laboratories, Stanford University.
19. Feferman, S. [1979], "Constructive theories of functions and classes", in Boffa, M., D. van Dalen and K. McAloon (eds.), *Logic Colloquium '78: Proceedings of the Logic Colloquium at Mons, 1978*, 159–224, North-Holland, Amsterdam.
20. Feferman, S. [1975], "A language and axioms for explicit mathematics", in: *Algebra and Logic*, Lecture Notes in Mathematics **450**, 87–139, Springer, Berlin.
21. Fitting, M. [1983], *Proof Methods for Modal and Intuitionistic Logics*, D. Reidel, Dordrecht, The Netherlands.
22. Fourman, M. P. and D. S. Scott [1979], "Sheaves and logic", in: Fourman, Mulvey and Scott, (eds.), *Applications of Sheaves*, Mathematical Lecture Notes **753**, 302–401, Springer-Verlag, Berlin.
23. Friedman, H. [1977], "The intuitionistic completeness of intuitionistic logic under Tarskian Semantics", unpublished abstract, State University of New York at Buffalo.

24. Friedman, H. [1975], "Intuitionistic completeness of Heyting's Predicate Calculus", *Notices of the American Mathematical Society* **22**, A–648.
25. Friedman, H. [1973], "Some applications of Kleene's methods for intuitionistic systems", in: A. R. D. Mathias (ed.), Cambridge Summer School in Mathematical Logic, Cambridge, Aug. 1971, Lecture Notes in Mathematics **337**, Springer-Verlag, Berlin.
26. Gabbay, D. G. [1981], *Semantical Investigations in Heyting's Intuitionistic Logic*, D. Reidel, Dordrecht, The Netherlands.
27. Goodman, N. [1978], "Relativized realizability intuitionistic arithmetic of all finite types", Journal of Symbolic Logic **43**, 23–44.
28. Grayson, R. J. [1983], "Forcing in intuitionistic systems without power set", Journal of Symbolic Logic **48**, 670–682.
29. Hayashi, S.and H. Nakano [1989], *PX: A Computational Logic*, The MIT Press, Cambridge.
30. Heyting, A. [1956], *Intuitionism: an Introduction*, North-Holland, Amsterdam.
31. Hyland, J. M. E., E. P. Robinson and G. Rosolini [1988], "The discrete objects in the effective topos", Manuscript.
32. Hyland, J. M. E. [1987], "A small complete category", Lecture delivered at the Conference "Church's Thesis after 50 years", to appear in Annals of Pure and Applied Logic.
33. Hyland, J. M. E. [1982], "The effective topos". in: Troelstra, A. S. and D. S. van Dalen (eds.), *L. E. J. Brouwer Centenary Symposium*, North-Holland, Amsterdam.
34. Hyland, J. M. E., P. T. Johnstone and A. M. Pitts [1980], "Tripos Theory", Math. Proceedings of the Cambridge Phil. Society **88**, 205–252.
35. Jaskowsky, S. [1936], "Recherches sur le systeme de la logique intuitioniste," in: *Actes du Congres International de Philosophie Scientifique, Paris*, Septembre, 1935, Vol. VI, 58–61, translation: *Studia Logica* **34**, (1975), 117–120.
36. Kleene, S. C. [1973], "Realizability: A retrospective survey", in: Mathias, A. R. D. (ed.), Cambridge Summer School in Mathematical Logic, Cambridge, Aug. 1971, Lecture Notes in Mathematics **337**, Springer-Verlag, Berlin.
37. Kleene, S. C. [1969], "Formalized recursive functionals and formalized realizability", Memoirs of the American Mathematical Society No. 89, American Mathematics Society, Providence, R.I.
38. Kleene, S. C. [1952], *Introduction to Metamathematics*, North-Holland (1971 edition), Amsterdam.
39. Kleene, S. C. [1945], "On the interpretation of intuitionistic number theory", Journal of Symbolic Logic **10**, 109–124.
40. Kreisel, G. and A. S. Troelstra [1970], "Formal systems for some branches of intuitionistic analysis", Annals of Mathematical Logic **1**, 229–387.
41. Kripke, S. [1965], "Semantical analysis of intuitionistic logic I", in: Crossley, J. N. and M. Dummett (eds.), *Formal Systems and Recursive Functions*, Proceedings of the Eighth Logic Colloquium, Oxford, 1963, North-Holland, Amsterdam, 92–130.
42. Lambek, J. and P. J. Scott [1986], *Introduction to higher order categorical logic*, Cambridge Studies in Advanced Mathematics **7**, Cambridge.
43. Läuchli, H. [1970], "An abstract notion of realizability for which predicate calculus is complete", in: Myhill, J., A. Kino, and R. E. Vesley (eds.), *Intuitionism and Proof Theory*, North-Holland, Amsterdam, 227–234.
44. Lawvere, W. [1971], "Quantifiers and sheaves", in: *Actes du Congres International des Mathematiciens, 1–10*, Septembre 1970, Nice, France, Vol. I., 329–334, Gauthier-Villars, Paris.
45. Leivant, D. [1976], "Failure of completeness properties of intuitionistic predicate logic for constructive models", Annales Scientifiques de l'Universite de Clermont No. 60 (Math. No. 13), 93–107.

46. Lipton, J. [1990], "Relating Kripke Models and realizability", Ph. D. dissertation, Cornell University.
47. Lipton, J. [1990], "Realizability and Kripke Forcing", to appear in the Annals of Mathematics and Artificial Intelligence, Vol. 4, North Holland, published in revised and expanded form as Technical Report TR 90-1163, Dept. of Computer Science, Cornell.
48. Lipton, J. [1990], *Some Kripke Models for "one-universe" Martin-Löf Type Theory*, Technical Report TR 90-1162, Dept. of Computer Science, Cornell University, Oct. 1990.
49. Lipton, J. [1991], *Constructive Kripke Models, Realizability and Deduction*, technical report, to appear.
50. Lopez-Escobar, E. G. K. and W. Veldman [1975], "Intuitionistic completeness of a restricted second-order logic", ISLIC Proof Theory Symposium (Kiel, 1974), Lecture Notes in Mathematics **500**, Springer-Verlag, 198–232.
51. McCarty, D. C. [1986], "Realizability and recursive set theory", Annals of Pure and Applied Logic **32**, 11–194.
52. McCarty, D. C. [1984], "Realizability and recursive mathematics", Doctoral Dissertation, Computer Science Department, Carnegie-Mellon University.
53. McKinsey, J. C. C. and A. Tarski [1948], "Some theorems on the sentential calculi of Lewis and Heyting", Journal of Symbolic Logic **13**, 1–15.
54. Makkai, M. and G. Reyes [1977], "First order categorical logic", Lecture Notes in Mathematics **611**, Springer-Verlag, Berlin.
55. Mitchell, J. and E. Moggi [1987], "Kripke-style models for typed lambda calculus", *Proceedings from Symposium on Logic in Computer Science*, Cornell University, June 1987, IEEE, Washington, D.C.
56. Mulry, P. S. [1985], "Adjointness in recursion", Annals of Pure and Applied Logic **32**, 281–289.
57. Mulry, P. S. [1980], "The topos of recursive sets", Doctoral Dissertation, SUNY-Buffalo.
58. Murthy, C., [1990], "Extracting constructive content from classical proofs: Compilation and the foundations of mathematics", Ph. D. dissertation, Cornell University.
59. Nerode, A., J. B. Remmel and A. Scedrov [1989], "Polynomially graded logic I: A graded version of System T", MSI Technical Report 89-21, Cornell University, Ithaca, N.Y, to appear in the proceedings of the Summer Workshop on Feasible Mathematics, Cornell.
60. Nerode, A. [1989], "Lectures on Intuitionistic Logic II", MSI Technical Report 89-56, Cornell University, Ithaca, N.Y.
61. Nerode, A. [1989], "Some lecures on Intuitionistic Logic I", Proceedings on Summer School on Logical Foundations of Computer Science, CIME, Montecatini, 1988, Springer-Verlag, Berlin.
62. Odifreddi, G. [1989], *Classical Recursion Theory*, North-Holland, Amsterdam.
63. Plotkin, G., [1980], "Lambda definability in the full type hierarchy", in : Seldin, J.P. and J. R. Hindley (eds.), *To H.B. Curry: Essays in Combinatory Logic, Lambda Calculus and Formalism*, Academic Press, New York.
64. Robinson, E. [1988], "How complete is PER?", Technical Report 88-229, Queen's University, Department of Computing and Information Science.
65. Rosolini, G., [1986], "Continuity and Effectiveness in Topoi", Ph. D. Dissertation, Oxford Univ.; also, report CMU-CS-86-123, department of Computer Science, Carnegie-Mellon University.
66. Scedrov, A. [1990], "Recursive realizability semantics for the calculus of constructions", in: *Logical Foundations of Functional Programming*, Addison-Wesley, Reading, MA.
67. Scedrov, A. [1985], "Intuitionistic Set Theory", in *Harvey Friedman's Research on the*

Foundations of Mathematics, North-Holland, Amsterdam.
68. Smorynski, C. A. [1973], "Applications of Kripke models", in: A. S. Troelstra (ed.), *Metamathematical Investigations of Intuitionistic Arithmetic and Analysis*, Lecture Notes in Math. **344**, Springer-Verlag, Amsterdam, 324–391.
69. Smullyan, R. M. [1968], *First-Order Logic*, Springer-Verlag, Berlin.
70. Stein, M. [1980], "Interpretation of Heyting's arithmetic—an analysis by means of a language with set symbols", Annals of Mathematical Logic **19**, North-Holland, Amsterdam.
71. Swart, H. C. M. de [1978], "First steps in intuitionistic model theory", Journal of Symbolic Logic **43**, 3–12.
72. Swart, H. C. M. de [1976], "Another intuitionistic completeness proof", Journal of Symbolic Logic **41**, 644–662.
73. Szabo, M. E. [1978], "Algebra of proofs", *Studies in Logic and the Foundations of Mathematics,* Vol. 88, North-Holland, Amsterdam.
74. Tait, W. W. [1975], "A realizability interpretation of the theory of species", Logic Colloquium (Boston, Mass., 1972/73), Lecture Notes in Mathematics **453**, Springer-Verlag, 240–251.
75. Tarski, A. [1956], *Sentenial Calculus and Topology in Logic, Semantics, and Metamathematics*, Clarendon Press, Oxford.
76. Troelstra, A. S. and D. van Dalen [1988], *Constructivism in Mathematics: An Introduction, Vol. II*, Studies in Logic and the Foundations of Mathematics, Vol. 123, North-Holland, Amsterdam.
77. Troelstra, A. S. [1977], *Choice Sequences: A Chapter of Intuitionistic Mathematics*, Clarendon Press, Oxford.
78. Troelstra, A. S. [1977], "Completeness and validity for intuitionistic predicate logic", Colloque International de Logique (Clermont-Ferrand, 1975), Colloques Internationaux du CNRS, no. 249, DNRS, Paris, 39–98.
79. Troelstra, A. S. [1973], *Metamathematical Investigation of Intuitionistic Arithmetic and Analysis*, Mathematical Lecture Notes **344**, Springer-Verlag, Berlin.
80. Troelstra, A. S. [1971], "Notions of realizability for intuitionistic arithmetic and intuitionistic arithmetic in all finite types", in: J. E. Fenstad (ed.), *Proceedings of the Second Scandinavian Logic Symposium*, Oslo, 1970, North-Holland, 369–405.
81. Veldman, W. [1976], "An intuitionistic completeness theorem for intuitionistic predicate logic", Journal of Symbolic Logic **41**, 159–166.

DEPT. OF COMPUTER SCIENCE, CORNELL UNIVERSITY
AND
DEPARTMENT OF MATHEMATICS, UNIVERSITY OF PENNSYLVANIA
AFTER SEPT. 1990: DEPT. OF MATHEMATICS, UNIVERSITY OF PENNSYLVANIA, PHILADELPHIA, PA 19104.

THIS WORK WAS PARTIALLY SUPPORTED BY A SLOAN DISSERTATION GRANT AND BY A FELLOWSHIP FROM THE MATH SCIENCES INSTITUTE, CORNELL

SPLITTING AND DENSITY FOR THE RECURSIVE SETS OF A FIXED TIME COMPLEXITY

WOLFGANG MAASS AND THEODORE A. SLAMAN

ABSTRACT. We analyze the fine structure of the time complexity classes induced by random access machines. Say that A and B have the same time complexity ($A+_C B$) if for all time constructible f $A \in DTIME_{RAM}(f) \iff B \in DTIME_{RAM}(f)$. The $=_C$-equivalence class of A is called it *complexity type*. We examine the set theoretic relationships between the sets in an arbitrary complexity type $\mathcal{C}$. For example, every recursive set X can be partitioned into two sets A and B such that $A =_C B =_C X$. Additionally, the ordering of $\mathcal{C}$ under $\subseteq^*$ (inclusion modulo finite sets) is dense.

1. INTRODUCTION

There has been a persistent intuition that the computational complexity of a set A of strings is related to the structure of the distribution of the elements of A. For example, Mahaney's [Ma] solution to the Berman-Hartmanis conjecture or Martin's [Mar] characterization of the Turing degrees of the maximal recursively enumerable sets grew out of the desire to test this intuition.

Although a large portion of current research in theoretical computer science is concerned with the investigation of problems that lie in P, very little is known about the relationship between the computational complexity and set theoretic aspects of sets in P. The results of this paper may be viewed as one step in this direction, since they provide nontrivial information about sets in P (for example about the structure of the class of sets of quadratic time complexity). We work in the context of a very fine scale for complexity and study the set theoretic properties among the recursive sets which are equally complex. Our results are all in the direction that there is a

---

The first author was partially supported by NSF-Grant CCR 8903398. The second author was partially supported by NSF-Grant DMS-8601856 and Presidential Young Investigator Award DMS-8451748 .

generous set theoretic variety among the sets of any particular time complexity. We show that for every recursive set X there are two sets of the same time complexity as X which partition the elements of X into disjoint pieces (Splitting Theorem); that if X and Y have the same complexity, $Y \subset X$ and $X - Y$ is infinite then there is an A such that $Y \subset A \subset X$ and both of $A - Y$ and $X - A$ are infinite (Density Theorem); and the countable atomless Boolean algebra can be embedded in the structure consisting of the collection of sets of the same complexity of X ordered by inclusion modulo the ideal of finite sets. The last result implies that the existential theory of this partial order is decidable.

In evaluating time complexity, we will use the random access machine (RAM) with uniform cost criterion (see [CR], [AHU], [MY], [P]) as our model for computation. This model is frequently adopted in considering the design of algorithms. It is also very sensitive to time complexity distinctions and allows sophisticated diagonalization constructions. It does not matter for the following which of the common versions of the RAM-model with instructions for ADDITION and SUBTRACTION of integers is chosen (note that it is common to exclude MULTIPLICATION of integers from the instruction set in order to ensure that the computation time of the RAM is polynomially related to time on a Turing machine). In order to be specific we consider the RAM model as it was defined by Cook and Reckhow [CR] (we use the common "uniform cost criterion" [AHU], i.e. $l(n) = 1$ in the notation of [CR]). In this model, a machine consists of a finite program, an infinite array $X_0, X_1, \ldots$ of registers (each capable of holding an arbitrary integer), and separate one-way input- and output-tapes. The program consists of instructions for ADDITION and SUBTRACTION of two register contents, the conditional jump "TRA m if $X_i > 0$" which causes control to be transferred to line m of the program if the current content of register X_i is positive, instructions for the transfer of register contents with indirect addressing, instructions for storing a constant, and the instruction "READ X_i" (transfer the content of the next input cell on the input-tape to register X_i) and "PRINT X_i" (print the content of register X_i on the next cell of the output-tape).

The relationship between computation time on RAM's and Turing machines is discussed in [CR] (Theorem 2), [AHU] (section 1.7), and [P] (chapter 3). It is obvious that a multi-tape Turing machine of time complexity $t(n)$ can be simulated by a RAM of time complexity $O(t(n))$. With a little bit more work (see [P]) one can construct a simulating RAM of time complexity $O(n + (t(n)/\log t(n)))$ (assuming that the output has length $O(n + (t(n)/\log t(n))))$. In addition, Cook and Reckhow demonstrate the existence of a universal RAM M^* such that for any RAM M there is a constant m such that M^* can simulate k steps in the execution of M using only $m \cdot k$ steps of its own. By a diagonal argument, they conclude a fine time hierarchy theorem for RAM's.

Say that a recursive function f is *time constructible* on a RAM if there is a RAM which can compute the function $1^n \mapsto 1^{f(n)}$ in $O(f(n))$ many steps. Let T be the set of recursive functions from $\mathbb{N}$ to $\mathbb{N}$ that are non-decreasing and time constructible on a RAM. We adopt T as our scale for measuring time complexity. Typically, when one calculates either an upper or lower bound on the running time of a computational procedure that bound is an element of T.

Suppose that f is in T. Let $DTIME(f)$ be the collection of recursive sets $A \subseteq \{0,1\}^*$ such that A can be computed by a RAM of time complexity $O(f)$.

Definition. (1) Say that A *has the same deterministic time complexity as* B (written $A =_C B$) if for all $f \in T$, A is in $DTIME(f)$ if and only if B is in $DTIME(f)$.

(2) A *complexity type* is an $=_C$-equivalence class.

We write $\mathbf{0}$ for $DTIME(n)$, which is the least complexity type. Note, for every complexity type $\mathcal{C}$ and every f in T either $\mathcal{C} \subseteq DTIME(f)$ or $\mathcal{C} \cap DTIME(f) = \emptyset$. We will investigate some basic properties of the partial order

$$PO(\mathcal{C}) = \langle \{X \mid X \in \mathcal{C}\}, \subseteq^* \rangle,$$

where $\mathcal{C}$ is an arbitrary complexity type and $\subseteq^*$ denotes inclusion modulo finite sets (i.e. $X \subseteq^* Y$ if and only if $X - Y$ is finite).

This paper is an expanded version of the expanded abstract [MS1] For a number of basic results about complexity types (and a list of open research problems) we refer to [MS2].

2. Splitting, Density and Embedding Theorems

We recall the following definition and results from [MS2], where it is shown that for each recursive set A there is a normal form for the presentation of the running times of the RAM's which compute A. To fix some notation, let $\{f\}$ denote the recursive function whose program on a RAM is coded by f.

Definition. $(t_i)_{i\in\mathbb{N}} \subseteq \mathbb{N}$ is called a *characteristic sequence* if $f : i \mapsto t_i$ is recursive and

(1) $(\forall i \in \mathbb{N})[\{t_i\} \in T$ and the program t_i is a witness for the time-constructibility of $\{t_i\}]$;
(2) $(\forall i, n \in \mathbb{N})[\{t_{i+1}\}(n) \leq \{t_i\}(n)]$.

Definition. Let A be a recursive subset of $\{0,1\}^*$ and let $\mathcal{C}$ be a complexity type. Then, $(t_i)_{i\in\mathbb{N}}$ is *characteristic for* A if $(t_i)_{i\in\mathbb{N}}$ is a characteristic sequence and

$$(\forall f \in T)\,[A \in DTIME(f) \iff (\exists i \in \mathbb{N})\,(f(n) = \Omega(\{t_i\}(n)))]\,.$$

Similarly, $(t_i)_{i\in\mathbb{N}}$ is *characteristic for* $\mathcal{C}$ if $(t_i)_{i\in\mathbb{N}}$ is characteristic for some $A \in \mathcal{C}$ (or equivalently, for all $A \in \mathcal{C}$).

In [MS2], it is shown that for every recursive set A there is a sequence $(t_i)_{i\in\mathbb{N}}$ that is characteristic for A and for every characteristic sequence there is a recursive set for which it is characteristic.

Theorem. (Splitting Theorem) *For every recursive set X, there are two disjoint recursive sets A and B such that $A \cup B = X$ and $A =_C B =_C X$.*

Proof. Let X be recursive, let $\mathcal{C}$ be the complexity type of X and let $(t_i)_{i\in\mathbb{N}}$ be a characteristic sequence for X. We build sets A and B so that the following conditions hold.

(1) A and B are disjoint and their union is equal to X.
(2) For every i, A and B are both elements of $DTIME(\{t_i\})$.
(3) If g is an element of T and one of A or B is an element of $DTIME(g)$ then X is an element of $DTIME(g)$.

The first condition states that A and B split X. The second and third conditions together imply that both A and B are elements of $\mathcal{C}$.

Condition (1) imposes a simple constraint on the construction: A and B must be constructed by dividing the elements of X into two disjoint sets. The further actions we take during the construction operate within this constraint.

There are three ingredients to the proof: the strategies to ensure conditions (2) and (3) and the mechanism by which the strategies are combined. The strategies for (2) and the global organization of the construction are taken from [MS2], where they were used to show that every characteristic sequence is characteristic for some recursive set.

We will eventually organize our construction as a stage by stage priority construction. We will assign the strategies a priority ranking. During stage s, we will decide the values of $A(\sigma)$ and $B(\sigma)$ for each string σ of length s by executing finitely many steps of finitely strategies with input σ. We adopt the values for $A(\sigma)$ and $B(\sigma)$ that satisfy the constraints imposed by the strategies of highest priority. If σ and τ are distinct strings of length s, it is reasonable to think of the values of $A(\sigma)$ and $B(\sigma)$ as being determined in parallel during stage s.

Time Control Strategies. The time control strategies are used to ensure that A and B are no more complex than X.

Suppose that f is in T. The time control strategy C_f associated with f ensures that A and B are in $DTIME(f)$. C_f will have a simulation constant m_f, fixed throughout the construction. It takes m_f many steps in the universal RAM to simulate 1 step in the execution of the RAM which computes f. C_f limits the total number of steps taken in the execution of strategies of lower priority during stage s to $O(f(s))$.

C_f uses the following mechanism to impose this constraint. For each string σ, C_f divides the execution of the strategies of lower priority into blocks of size b many steps, where b is less than $|\sigma|$. Each time that the construction executes b many steps for the sake of lower priority, C_f requires the construction to run $m_f \cdot b$ many steps in the simulation of the evaluation of of f at $|\sigma|$. If the computation of f at $|\sigma|$ converges then C_f constrains the strategies of lower priority to decide the values of A and B at σ immediately. Thus, within a constant factor, the attention of the global construction is equally shared between C_f and all of the strategies of lower priority. Since, for all σ, C_f calls a halt to the computations of $A(\sigma)$ and $B(\sigma)$ when it sees $f(|\sigma|)$ converge, the function mapping $|\sigma|$ to the total number of steps in the construction devoted to evaluating strategies of priority less than or equal to C_f on input σ is $O(f(|\sigma|))$.

C_f will ensure that A and B are in $DTIME(f)$ provided that there are only finitely many strings for which the strategies of higher priority than C_f choose values for A and B which disagree with those chosen by C_f and the strategies of lower priority.

The Complexity Strategies. The complexity strategies are used to ensure that A and B are at least as complex as X.

We describe the strategy R_M^A to ensure that if the RAM M computes A with run time function g, where g is in T, then X is in $DTIME(g)$. The strategy R_M^B is similar. Our approach is less direct than the one used in [MS2], where given $(t_i)_{i\in\mathbb{N}}$ we could directly construct a set so that $(t_i)_{i\in\mathbb{N}}$ was characteristic for that set. Here, we are additionally constrained by the fact that the only strings that we can put into A are those that already belong to X. Thus, if σ is not in X then we cannot diagonalize against a computation predicting that σ is not in A.

Let M and g be a RAM and a function as above. Given an input string σ, R_M^A acts as follows.

(1) First, R_M^A takes $|\sigma|$ many steps to look back and see whether an inequality between A and the set computed by M has already been established. If so then R_M^A halts its activity and does not impose any constraint on the values of $A(\sigma)$ or $B(\sigma)$. If not then R_M^A goes to step (2).

(2) R_M^A simulates the execution of M on input σ. If M halts then R_M^A goes to (3).

(3) If M returns value 1, then R_M^A constrains the construction from putting σ into A. If in fact $\sigma \in X$ then the constraint imposed by R_M^A implies that σ must be put into B. (*In this case, R_M^A has diagonalized A against M.*)

If M returns value 0, then R_M^A constrains the construction to put σ into A if σ is in X. (*In this case, either R_M^A has diagonalized A against M or we can infer that σ is not in X.*)

Suppose that f is a nonlinear element of T, c is a constant and M^X is a RAM that can be used to compute X in time $c \cdot f$. Suppose that we execute R_M^A within the constraint imposed by C_f. As noted earlier, there is a constant k_f such that the effect of C_f on R_M^A is to limit the execution of R_M^A on input σ to $k_f \cdot f(|\sigma|)$ many steps.

If A is not equal to the set computed by M, then there is a constant c_{finite} such that for all sufficiently long strings σ the evaluation of R_M^A takes only c_{finite} many steps after reading σ. Since f is not linear, $k_f \cdot f$ eventually dominates the number of steps it takes for R_M^A to look back and halt. Thus, C_f is essentially invisible to R_M^A in this case.

Now assume that M does compute A. In this case, if we reach (3) for a string σ it must be the case that M has output answer 0 and σ is not in X. Thus, the action of R_M^A is providing a subset of the complement of X that is computed in time less than or equal to $k_f \cdot f$. In fact, we show that X is in $DTIME(g)$. Suppose that σ is given. If $M(\sigma)$ converges in less than $k_f \cdot f(|\sigma|)$ many steps and σ is an element of X then we could keep σ out of A and diagonalize. Thus, if $g(|\sigma|)$ is less than $k_f \cdot f(|\sigma|)$ then $\sigma \notin X$. On the other hand, if $g(|\sigma|)$ is greater than $k_f \cdot f(|\sigma|)$ then we can evaluate $X(\sigma)$ in less than or equal to $c \cdot 1/k_f \cdot g(|\sigma|)$ many steps using M^X. Thus, there is a constant factor k such that the value of $X(\sigma)$ can be determined in $k \cdot g(|\sigma|)$ many steps, using whichever case occurs first.

Hence, if R_M^A is executed in the time control environment imposed by a function in the characteristic sequence for $\mathcal{C}$ then A will satisfy the associated complexity requirement.

Compatibility Between Complexity Strategies. We have already shown that R_M^A will satisfy its requirement if it is executed in a time control environment imposed by C_f and $X \in DTIME(f)$. In this context, we must show that the effect of R_M^A on the strategies of lower priority is essentially finite. If M does not agree with A then R_M^A's effect is explicitly finite, as established by the look back in step (1). Otherwise, there may be infinitely many strings σ such that R_M^A constrains the construction so that if $\sigma \in X$ then $\sigma \in A$. However, none of these strings can belong to X. Thus, any constraint imposed by R_M^A on σ is vacuous. Since any constraint imposed by a strategy $R_{M'}^A$ or $R_{M'}^B$ on σ will also be vacuous when $\sigma \notin X$, these constraints are compatible. Although the effect of R_M^A is not finite it can only contribute finitely many conflicts with the other strategies appearing in our construction.

Simultaneous Execution. The time complexity strategies impose a constraint on the way that we may distribute the computation steps in our construction. In particular, C_f requires that the total number of steps devoted to all the strategies of lower priority must be of the same order as the number of steps devoted to the analysis of f. We adapt a scheme from [MS2] by which we work within this constraint and still introduce infinitely many strategies of lower priority over the course of the construction. We give the reasoning behind the proof that this scheme works without reproducing all of the details found in [MS2].

Let M^* be a universal RAM. By the Cook-Reckhow theorem, for any strategy Q there is a constant q such that M^* can simulate n many steps in the execution of Q using only $q \cdot n$ many of its own steps. Further, by using distinct parts of memory, given a finite sequence of strategies $Q_1, \ldots, Q_k$ and the constants $q_1, \ldots, q_k$ associated with simulating these strategies and given numbers $n_1, \ldots, n_k$, M^* can simulate n_i many steps in each Q_i using only $\sum_{i=1}^{k} q_i \cdot n_i$ many of its own steps.

Consider the following pattern for dividing time resources between the strategies acting on the same input string σ. Let $Q_1, \ldots, Q_k$ be strategies and let $q_1, \ldots, q_k$ be their simulation constants as above. Let a *sweep* denote one implementation of the following recursion, beginning with Step(k) with s_k equal to 1.

Step(i) Execute s_i many steps of Q_i at σ. *Note that this takes* $q_i \cdot s_i$ *many steps for* M^*.

(a) If i is greater than 1 then let s_{i-1} equal $q_i \cdot s_i + s_i$ and go to Step($i-1$). s_{i-1} *is equal to the total number of steps that it would take* M^* *to run the sweep at* σ *through the execution of the strategies* $Q_k, \ldots, Q_i$.

(b) Otherwise, end.

There is a fixed number of steps to each sweep, depending only on the sequence $\vec{Q}$. In particular, the number of steps to a sweep does not depend on the length of the string which is taken as input for the strategies. Further, for each i, the number of steps assigned to Q_i during a sweep is exactly as many steps as are needed to simulate the total activity of all strategies of lower priority, independently of the number $k - i$ of strategies of lower priority that occur in the considered sweep. The first property ensures that for each sequence of strategies $\vec{Q}$ there is an s such that one sweep through the sequence $\vec{Q}$ takes less than s steps to execute. The second property shows that we can combine strategies in a global construction and respect the constraints imposed by the time control strategies.

The Global Construction. We build A and B by combining the strategies $C_{\{t_i\}}$ for t_i in $(t_i)_{i \in \mathbb{N}}$, the characteristic sequence for $\mathcal{C}$; the strategies $R^A_{M_i}$ for M_i a RAM, to ensure that if M_i computes A in time g_i then $X \in DTIME(g_i)$; and the symmetric strategies $R^B_{M_i}$. We give these strategies the following priority ordering.

$$C_{\{t_1\}}, R^A_{M_1}, R^B_{M_1}, C_{\{t_2\}}, R^A_{M_2}, R^B_{M_2}, \ldots$$

We construct A and B by stages. During stage s we decide the values of A and B on all strings σ of length s. Letting σ be a string of length s, the construction operates on σ as follows.

We alternate between two activities: executing sweeps through the sequence of strategies and updating sequence of active strategies and the tentative values to A and B at σ.

We let the initial sequence of active strategies be the initial segment of the priority ordering $\vec{Q}_0$ of length ℓ. We determine ℓ by executing the first s steps of the following iteration.

(1) Begin with ℓ equal to 1 and $\vec{Q}$ be $\langle C_{\{t_1\}} \rangle$.
(2) Given ℓ and $\vec{Q}$ of length ℓ, execute one sweep through $\vec{Q}$ at argument σ. Go to (3).
(3) Add 1 to ℓ and let $\vec{Q}$ be the initial segment of the priority list of length ℓ. Go to (2).

Let ℓ be the largest value for which the iteration completed a sweep before the $|\sigma|$ steps were completed. Given that $\vec{Q}$ is always taken to be an initial segment of the priority list, the number of steps needed to complete a sweep is determined solely by the length of $\vec{Q}$ and not by the argument σ. Thus the length of $\vec{Q}_0$ depends only on $|\sigma|$ and is a non-decreasing function of $|\sigma|$ with infinite limit.

Having determined $\vec{Q}_0$, we execute the following iteration, beginning with $\vec{Q}$ equal to $\vec{Q}_0$, $A(\sigma)$ equal to $X(\sigma)$ and $B(\sigma)$ equal to 0.

Given the sequence $\vec{Q}$ of currently active strategies, we iterate the execution of sweeps (continuing our simulations of $\{t_i\}(|\sigma|)$ and of $M_i(\sigma)$) through $\vec{Q}$ until a time control strategy $C_{\{t_i\}}$ computes the value of $\{t_i\}(|\sigma|)$ and calls a halt for the strategies of lower priority. We form the updated sequence of active strategies by omitting $C_{\{t_i\}}$ and all strategies of lower priority and retaining the others. If during the iteration of the sweeps, a strategy $R^A_{M_i}$ or $R^B_{M_i}$ imposed a constraint on the value of $A(\sigma)$ or $B(\sigma)$ then we adopt the updated values imposed by the strategy of highest priority as our tentative values. If i is greater than 1 then we continue as above with the truncated sequence of active strategies. Otherwise, we end the evaluation of the construction on σ when $C_{\{t_1\}}$ calls a halt, i.e. when our simulation of the evaluation of $\{t_1\}(|\sigma|)$ converges. We extend the definitions of A and B to σ by adopting the values current when $C_{\{t_1\}}$ calls a halt.

The time expended for the sake of determining A and B at σ is either spent in the initial setting up of $\vec{Q}_0$, in the constant number of steps spent in reading off the constraints imposed during the earlier iteration of sweeps and going to the next iteration or in executing a sweep. The first two involve only $O(|\sigma|)$ many steps in operational overhead.

By the observation that the length of a sweep does not depend on the stage when it is executed, we see that for every strategy S there is an s such that S is in $\vec{Q}_0$ as computed by every string of length greater than or equal to s. We have already argued that the complexity strategies only act in a way that is incompatible with the actions of lower priority strategies finitely often. This finite injury does not effect the ability of the lower priority strategies to satisfy their requirements. By operating with sweeps and an additional overhead of size $O(|\sigma|)$ and halting the actions of strategies of lower priority upon request, we have ensured that except for the finitely many exceptional strings mentioned above the construction respects every $C_{\{t_i\}}$. Thus, for all i A and B are in $DTIME(\{t_i\})$. Similarly, for each complexity strategy $R^A_{M_i}$ or $R^B_{M_i}$ and each argument σ of sufficient length, that strategy receives $O(\{t_i\}(|\sigma|)$ many steps during the series of sweeps at argument σ. Further, except for finitely many strings, all of its constraints are respected by the construction, either because it stops issuing constraints due to having found an inequality or because all of its constraints are vacuous. Thus, all of the complexity requirements are also satisfied. □

We can draw some corollaries from the splitting theorem and its proof.

Theorem. *For every non-trivial complexity type $\mathcal{C}$ the partial order $PO(\mathcal{C})$ of sets in $\mathcal{C}$ ordered by inclusion mod finite ($\subseteq^*$) has neither maximal nor minimal elements.*

Let $PO_{0,1}(\mathcal{C})$ be the partial order

$$\langle \{X \mid X \in \mathcal{C} \vee X = \{0,1\}^* \vee X = \emptyset\}, \subseteq^* \rangle .$$

$PO_{0,1}(\mathcal{C})$ is the result of adjoining a greatest and a least element to $PO(\mathcal{C})$.

Corollary. *For every complexity type $\mathcal{C}$ there is an embedding E from the countable atomless Boolean algebra (CBA), regarded as a partial order,* *into $PO_{0,1}(\mathcal{C})$. Further, this embedding associates the Boolean operations in CBA to the usual set theoretic ones.*

Proof. This corollary does not follow directly from the splitting theorem but rather from a modestly stronger version, which has almost the same proof. We first describe how to obtain the embedding E. Then, we read off the needed strong splitting theorem. Finally, we indicate how to alter the proof of the original splitting theorem to prove the stronger form.

There is a standard scheme to embed CBA into a partial order that satisfies the splitting theorem. We start with b_0 an intermediate element of CBA and X_0 an element of $\mathcal{C}$. We set $E(b_0)$ equal to X_0. Then we extend E to the Boolean closure of $\{b_0\}$ by mapping $\neg b_0$ to the complement of X_0, 0 to $\emptyset$ and 1 to $\{0,1\}^*$.

During the recursion step of our construction, we start with an isomorphism E between two finite Boolean algebras $\mathcal{B}$ contained in CBA and $E(\mathcal{B})$ contained in $PO_{0,1}(\mathcal{C})$. Given a new element b of CBA, let $\mathcal{B}_b$ be the finite subalgebra of CBA generated by $\mathcal{B}$ and $\{b\}$. Let $\{a_1, \ldots, a_k\}$ be the atoms of $\mathcal{B}$. The non-zero elements in $\mathcal{B}_b$ of the form $b \wedge a_i$ or $\neg b \wedge a_i$ generate $\mathcal{B}_b$ under join. Thus, it is enough to extend E to these elements and let the union operation determine E's extension to all of $\mathcal{B}_b$. Naively, for each i such that $b \wedge a_i \neq a_i$ and $b \wedge a_i \neq 0$, we could find images for $b \wedge a_i$ and $\neg b \wedge a_i$ by splitting $E(a_i)$ in $PO_{0,1}(\mathcal{C})$ into two sets $A_{i,0}$ and $A_{i,1}$. However, to generate E by taking unions we need to ensure that these sets have a stronger splitting property: that all finite unions of the $A_{i,j}$ belong to $\mathcal{C}$.

This conclusion will follow once we know the following. Every set X in $\mathcal{C}$ can be split into two sets X_0 and X_1 such that for every f in T, if there is a set U in $DTIME(f)$ with either $X \cap U = X_0$ or $X \cap U = X_1$ then X is in $DTIME(f)$. Symbolically, X_0 and X_1 satisfy the condition

$$(*) \qquad (\forall f \in T) \begin{bmatrix} (\exists U \in DTIME(f))\,(X \cap U = X_0 \vee X \cap U = X_1) \\ \Longrightarrow X \in DTIME(f) \end{bmatrix}.$$

Now suppose that for each i such that $b \wedge a_i \neq a_i$ and $b \wedge a_i \neq 0$, we produce sets $A_{i,0}$ and $A_{i,1}$ to split $E(a_i)$ and satisfy $(*)$ for $X = E(a_i)$, $X_0 = A_{i,0}$ and $X_1 = A_{i,1}$. If U is a finite union of the $A_{i,j}$ and U is not a element of $E(\mathcal{B})$ then there must be an i^* and a j^* such that $U \cap E(a_i) = A_{i^*,j^*}$. Applying $(*)$, for all $f \in T$ if $U \in DTIME(f)$ then $E(a_i) \in DTIME(f)$.

Thus, $E(a_i) \leq_C U$. Trivially, U is no more complex than $E(a_i)$ since U is a finite union of sets from $\mathcal{C}$. Consequently, U is in $\mathcal{C}$.

Thus, we have reduced the proof of the corollary to proving the strong splitting theorem. This form of the splitting theorem is proven using a slightly different version of the complexity strategies $R_M^{X_0}$ and $R_M^{X_1}$. Instead of simulating the output of M and attempting to diagonalize, we simulate the computation of U and attempt to differentiate $X \cap U$ from X_0 and X_1. Consider the case for X_0. If we are unable to complete the simulation evaluating $U(\sigma)$ then U is at least as complex as X at σ. If we see $\sigma \in U$, we issue the constraint that σ is not in X_0 and diagonalize between X_0 and U. (This is analogous to step (3) in R_M^A in step 3-case 1, when we ensured that M did not compute A.) When we see $\sigma \notin U$ we issue the constraint that $\sigma \in X_0$ if $\sigma \in X$ and ensure that if $X_0(\sigma) = U(\sigma)$ then U is at least as complicated as X at σ. (This is analogous to step 3-case 2 in R_M^A, when we ensured that either M did not compute A or we could compute X in order of the running time of M many steps.) If X_0 is produced by a construction that implements this strategy within a time constraint strategy C_f and U is equal to X_0 then, there is are constants c_1 and c_2 such that for every σ either the simulation of $U(\sigma)$ takes longer than $c_1 \cdot f(|\sigma|)$ many steps or we can compute that σ is not an element of X in c_2 times the running time to evaluate $U(\sigma)$. But then U is not a counter example to the strong splitting of X by X_0 and X_1.

These strong splitting strategies can be combined with the time control strategies as in the proof of the splitting theorem. □

Suppose that X and Y are subsets of $\{0,1\}^*$. Let $Y \underset{\infty}{\subset} X$ denote the condition that Y is a subset of X and that $X - Y$ is infinite.

Definition. We say that X is *at least as complex as* Y if for all f in T, $X \in DTIME(f)$ implies that $Y \in DTIME(f)$. In this case, we write $Y \leq_C X$.

Theorem. (Density Theorem) *Assume that* $Y \underset{\infty}{\subset} X$ *and* $Y \leq_C X$. *Then there is a set* A *such that* $A =_C X$ *and* $Y \underset{\infty}{\subset} A \underset{\infty}{\subset} X$.

Proof. Most of the ingredients in the proof of the density theorem are the same as in the splitting theorem, so we abbreviate our discussion.

For the sake of $Y \underset{\infty}{\subset} A \underset{\infty}{\subset} X$, we must put very element of Y into A and restrict the elements of A to come from X. Thus we cannot use strategies which issue constraints of the form $\sigma \notin A$. For these, we must substitute

$\sigma \notin Y \implies \sigma \notin A$. We analyze the resulting complexity strategy R_M^A in its outcome when M computes A. Given a string σ, there are three cases: we infer that $\sigma \in Y$ (and hence in X) from $M(\sigma) = 1$; we infer $\sigma \notin X$ from $M(\sigma) = 0$; or we compute X in less time than it takes to evaluate $M(\sigma)$ using the RAM associated with the time control strategy that caused R_M^A to halt. Consequently, we can design an algorithm to compute X that converges at least as fast as the one associated with the time control strategies of higher priority. □

Corollary. *For every complexity class $\mathcal{C}$, $PO_{0,1}(\mathcal{C})$ is dense.*

It is easy to see that $PO_{0,1}(\mathcal{C})$ is isomorphic to the countable atomless Boolean algebra if $\mathcal{C}$ is equal to $\mathbf{0}$. Furthermore, for every complexity type, CBA can be embedded in $PO_{0,1}(\mathcal{C})$. However, the following corollary suggests that the structure of $PO(\mathcal{C})$ is substantially more complicated than CBA when $\mathcal{C}$ is non-trivial.

Obviously, every complexity type is closed under complementation and so not closed under union or intersection. However, it could still be the case that any two sets A and B with an upper bound in $PO(\mathcal{C})$ have a least upper bound in $PO(\mathcal{C})$. This is ruled out by the following result.

Corollary. *Suppose that $\mathcal{C}$ is not equal to $\mathbf{0}$ and A and B belong to $\mathcal{C}$. Then, A and B have a least upper bound in $PO(\mathcal{C})$ if and only if $A \cup B \in \mathcal{C}$.*

Proof. Assume that $A \cup B \notin \mathcal{C}$ and that D is an upper bound both A and B in $PO(\mathcal{C})$. Then $A \cup B \underset{\infty}{\subset} D$ and $A \cup B \leq_{\mathcal{C}} D$. By the density theorem there is a set D^* such that $A \cup B \underset{\infty}{\subset} D^* \underset{\infty}{\subset} D$ and $D^* =_{\mathcal{C}} D$. Thus, D is not a least upper bound for $A \cup B$. □

A Question. We are left with an intriguing situation. Whether the union of two elements A and B from $\mathcal{C}$ is an element of $\mathcal{C}$ is a first order property of A and B in $PO(\mathcal{C})$. Thus, not all pairs from $\mathcal{C}$ are alike in $PO(\mathcal{C})$. Can this inhomogeneity in $PO(\mathcal{C})$ be used to show that the structure is complicated. In particular, is the first order theory of $PO(\mathcal{C})$ non-recursive?

Acknowledgment

We would like to thank Joel Berman for his acute and helpful comments.

References

[AHU] A.V. Aho, J.E. Hopcroft and J.D. Ullman, *The Design and Analysis of Computer Algorithms*, Addison-Wesley, Reading, 1974.

[CR] S.A. Cook and R.A. Reckhow, *Time-bounded random access machines*, J. Comp. Syst. Sc. **7** (1973), 354-375.

[MY] M. Machtey and P. Young, *An Introduction to the General Theory of Algorithms*, North-Holland, Amsterdam, 1978.

[Ma] S. Mahaney, *Sparse complete sets for NP: solution of a conjecture of Berman and Hartmanis*, J. Comp. Syst. Sc. **25** (1982), 130-143.

[Mar] D. A. Martin, *Classes of recursively enumerable sets and degrees of unsolvability*, Z. Math. Logik Grundlag. Math. **12** (1966).

[MS1] W. Maass and T. A. Slaman, *Extensional properties of sets of time bounded complexity (extended abstract)*, Proc. of the 7th Int. Conference on Fundamentals of Computation Theory, Lecture Notes in Computer Science, vol. 380, Springer, Berlin, 1989, pp. 318-326.

[MS2] ———, *The complexity types of computable sets (extended abstract)*, Proc. of the Structure in Complexity Theory Conference, 1989.

[P] W.J. Paul, *Komplexitaetstheorie*, Teubner, Stuttgart, 1978.

Department of Mathematics, Statistics and Computer Science; University of Illinois at Chicago; Chicago, IL 60680

Department of Mathematics; The University of Chicago; Chicago, IL 60637

REALS AND FORCING WITH AN ELEMENTARY TOPOS

SAUNDERS MAC LANE AND IEKE MOERDIJK

ABSTRACT. Certain special types of categories, called Toposes, can formulate basic facts about sheaf theory in topology and algebraic geometry and thus clarify the role of geometry in independence proofs by forcing. They also establish a remarkable connection with intuitionist logic. This paper will summarize these results.

1. TOPOS

In 1964, Lawvere observed that it should be possible to axiomatize set theory not in terms of set membership, but in terms of the composition of functions. His proposal for an "Elementary Theory of the Category of Sets" thus formulated as axioms the specific properties of the category *Sets*. Here, as usual, ([Mac1]) a *category* $\mathbb{C}$ is a collection of objects A, B, C and a collection of morphisms f, g, h; each morphism (or "arrow") $f : A \to B$ has a domain A and a codomain B. A composite gf is defined iff $g : B \to C$, and then $gf : A \to C$. The axioms then require that this composition be associative and have for each object A an identity morphism 1_A, with the evident property.

A category $\mathbb{C}$ has (all) finite limits when it has a terminal object 1 (for every A, a unique morphism $A \to 1$) and to each pair of coterminal arrows $f : A \to B$, $h : C \to B$ a pullback P; in particular this insures the presence of products $A \times C$ (take $B = 1$). A category has exponentials iff there is for each triple of objects A, C, D an object D^C and a natural bijection

$$\frac{A \times C \to D}{A \to D^C}. \tag{1}$$

A monomorphism $m : D \to F$ is a morphism such that $mf = mg$ implies $f = g$: in other words, a monomorphism m represents D as a subobject of F. Finally, a subobject *classifier* for $\mathbb{C}$ is an object Ω and a monomorphism

$1 \to \Omega$, called "true", such that every monomorphism m is a pullback of true along a unique arrow $x : F \to \Omega$, as in the pullback square

$$\begin{array}{ccc} D & \longrightarrow & 1 \\ {\scriptstyle m}\big\downarrow & & \big\downarrow{\scriptstyle \text{true}} \\ F & \underset{x}{\longrightarrow} & \Omega \end{array} \tag{2}$$

For example, in *Sets*, Ω is the two element set $\{0, 1\}$ and x is the characteristic function of the subset $D \subset E$.

A *topos* E is a category with all finite limits, with exponents, and with a subobject classifier. The category *Sets* of all sets is a topos, as is the category of all linear diagrams S of sets of the general form of "sets through (discrete) time", as in

$$S_0 \xrightarrow{s} S_1 \xrightarrow{s} S_2 \xrightarrow{s} S_3 \xrightarrow{s} \cdots. \tag{3}$$

In this case, a morphism f of this S into a similar T is simply a family of functions $f_n : S_n \to T_n$ with $sf_n = f_{n+1}s$ for every n. In any topos, the subobject classifier Ω acts as the "object of truth values". Using the intersection (i.e. the pullback) of two subsets D and D' of F in (2) one can define an intersection operator $\wedge : \Omega \times \Omega \to \Omega$ for Ω. Moreover, the other propositional operators "or", "implies", and "not" can be defined for Ω; this makes of Ω a Heyting algebra; more exactly, an internal Heyting algebra in which the usual identities for such an algebra are expressed ("internally") by way of commutative diagrams. Universal and existential quantifiers can also be introduced, in terms of suitable adjoints to projections.

The axioms for a topos are all first order, so they may serve to describe the "Elementary Theory of the Category of Sets". One may add further first order axioms, such as the requirement that the topos be Boolean (i.e. that the Heyting algebra Ω be Boolean) or that it satisfy the axiom of choice (say in the form that every epimorphism "splits"—i.e. has a section) or that it satisfy an axiom of infinity, in the form of the requirement that there be a "natural number object":

$$1 \xrightarrow{o} \mathbb{N} \xrightarrow{s} \mathbb{N}. \tag{4}$$

2. Posets and Sheaves

If P is a poset, consider functions $p \mapsto S_p$ from P to sets for which each $q \geq p$ gives a map $S_q \to S_p$ which has for $r \geq q \geq p$ the evident transitivity property for the composite

$$S_r \to S_q \to S_p. \tag{5}$$

All these functions S form an evident category, called $\mathrm{Sets}^{P^{op}}$, which is in fact an elementary topos (with a suitable Ω). Indeed, a special case of this topos is that of sets through time in (3), where P is the set of natural numbers with reverse ordering. Actually, a poset P can be considered to be a category, with exactly one morphism $q \to p$ when $q \leq p$. The object S considered in (5) can then be viewed as a (contravariant) functor $P^{op} \to \mathrm{Sets}$ (a functor is a map of categories which preserves identities and composition). More generally, for any category $\mathbb{C}$ the set of all contravariant functors $S : \mathbb{C}^{op} \to \mathrm{Sets}$ form a category $\mathrm{Sets}^{\mathbb{C}^{op}}$ which is in fact a topos.

In particular, if X is a topological space the open sets U of X form a poset, call it $\mathrm{Open}\, X$. Here a contravariant functor $(\mathrm{Open}\, X)^{op} \to \mathrm{Sets}$ assigns to each open U a set $S(U)$ such that $V \subset U$ gives a restriction map $S(U) \to S(V)$ with transitivity as in (5). Such a function $S : (\mathrm{Open}\, X)^{op} \to \mathrm{Sets}$ is called a *presheaf* on X. From analysis, we take the notion of a *sheaf*. It is a presheaf S with the following property: Given an open covering $U = \cup U_i$ and elements $x_i \in S(U_i)$ which *match* on every intersection $U_i \cap U_j$ (i.e. x_i restricted to $U_i \cap U_j$ equals x_j there) there is a unique $x \in S(U)$ which restricts to x_i on each U_i. In other words, this sheaf requirement states that objects x in $S(U)$ can be uniquely reconstructed (collated) from matching pieces x_i.

The category $Sh(X)$ of all sheaves on a topological space X is a topos; its subobject classifier is the sheaf Ω which assigns to each open U the set $\Omega(U)$ of all open subsets of U. This important topos is a subcategory of the presheaf topos:

$$Sh(X) \subset \mathrm{Presheaves}(X) = \mathrm{Sets}^{(\mathrm{Open})^{op}}. \tag{6}$$

Moreover, there is an inverse process: Each presheaf S has a sheaf $a(S)$ as "best approximation" sheaf: technically, "best" means that the passage $S \mapsto a(S)$ is left adjoint to the inclusion (6). Moreover, this left adjoint is *left exact*, in the sense that it preserves all finite limits.

This construction of sheaves from presheaves is an important way of passing from a given topos (that of presheaves) to a "better" one. It is called "sheafification". It was illustrated above for the ordinary notion of a topological space, where sheaves are defined in terms of coverings by open sets. In algebraic geometry, Grothendieck observed that one could use, instead of coverings by open sets $U \subset X$, suitable "coverings" by maps $Y \to X$ into the space X. For these Grothendieck topologies there is a different construction of a similar operation of sheafification. Finally, Lawvere and Tierney observed that such a notion of a covering could be described in an elementary topos by an operation $j : \Omega \to \Omega$ on the subobject classifier Ω of the topos, with suitable properties (7).

For a topological space X, a *sieve* D on an open set $U \subset X$ is a collection of subsets $V \subset U$ such that $W \subset V$ and $V \in D$ implies $W \in D$; in other words, if V "goes through" the sieve D, so does any smaller W; in other words, a sieve is "downward closed". The collection $\Omega(U)$ of all sieves on U is a functor of U, hence is a presheaf Ω. It is in fact the subobject classifier for the presheaf category. Now coverings enter: the sieve D *covers* W iff W is a union of sets V of D, thus the open sets covered by D form a larger sieve, call it $j(D)$, also on U. Thus everything about coverings is contained in the operation $D \mapsto j(D)$, written $j : \Omega \to \Omega$. This "modal" operator on truth values has three basic properties:

$$(7) \qquad j^2 = j, \quad j(\text{true}) = \text{true}, \quad j\wedge = \wedge(j \times j) : \Omega \times \Omega \to \Omega.$$

Now, mirabile dictu, any modal operator j in a topos E with these three properties suffices to define coverings and hence to define sheaves in the topos E. Moreover, all the sheaves for such a "Lawvere–Tierney" topology themselves form a subtopos of E

$$(7') \qquad Sh_j(E) \subset E.$$

Furthermore, every object A of E can be turned into a best approximation sheaf $a(A)$, and this operation a is a left exact left adjoint to the inclusion (7′). Thus the geometric operation of sheafification produces from j and E a new topos, and so from one model of a (generalized) set theory another one. This is the essence of forcing.

3. The Cohen Poset

Cohen's famous proof of the independence of the Continuum Hypothesis from the remaining axioms of Zermelo–Fraenkel set theory can be simplified and restated as an instance of this process of sheafification. In any topos one has for each object A its "power set" which is constructed as an exponential $PA = \Omega^A$. In the topos *Sets* consider some set B much larger than the power set $\wp(\mathbb{N})$ of the set $\mathbb{N}$ of all natural numbers. One wishes to "force" a monomorphism

$$g : B \rightarrowtail \wp(\mathbb{N}) = 2^{\mathbb{N}} \tag{8}$$

or instead the corresponding "transpose" under the bijection (adjunction) (2) defining the exponential

$$\check{g} = f : B \times \mathbb{N} \to 2 = \Omega. \tag{9}$$

As in Cohen's proof, consider instead all *finite* approximations to f,

$$p : F_p \to 2, \quad F_p \subset B \times \mathbb{N},$$

where F_p is a finite subset of the product $B \times \mathbb{N}$. If q is a second such, say that $q \leq p$ (q refines p) if $F_p \subset F_q$ and q restricted to F_p is exactly p. Thus $q \leq p$ means that q is a more complete approximation to $\check{g}$ than was p. The collection P of all these forcing conditions is then a poset, hence a category and one may form as before the topos

$$E = \mathrm{Sets}^{P^{op}}, \tag{10}$$

which is the collection of all contravariant functors $H : P^{op} \to \mathrm{Sets}$, with each $H(p)$ a set and each $H(p) \to H(q)$ for $q \leq p$ a map, with compositions $H(p) \to H(q) \to H(r)$ transitive.

But this E is not yet the desired model of set-theory because its "internal" logic is intuitionist and not classical. Specifically, its subobject classifier Ω is not Boolean, but is just an (internal) Heyting algebra. In particular, the negation operator $\neg : \Omega \to \Omega$ does not enjoy the classical property that $\neg\neg =$ identity, so that $\neg$ does not provide Boolean complements. Instead, one has only the weaker property

$$\neg\neg\neg = \neg : \Omega \to \Omega, \tag{11}$$

familiar from intuitionist logic. But this property implies that $\neg\neg\neg\neg = \neg\neg$, so that $j = \neg\neg$ is an idempotent operator. Moreover, $\neg\neg$ carries "true" into "true" and (by a modified de Morgan law) commutes with intersection.

In other words, double negation $j = \neg\neg$ is precisely a Lawvere–Tierney topology in the sense of (7) above. And this in turn means that we can construct the corresponding category of sheaves

$$E_o = Sh_{\neg\neg}(\mathrm{Sets}^{P^{op}}) \subset \mathrm{Sets}^{P^{op}} .$$

Moreover, this process has "forced" $\neg\neg$ to become the identity, so this new sheaf topos is indeed Boolean, in that its subobject classifier Ω_E is a Boolean algebra. It is likely to be a large Boolean algebra (by no means just two-valued). (This is readily adjusted—take a maximum filter (prime ideal) F in Ω_E and divide by F; one must divide not just Ω_E but the whole topos, by a quite standard colimiting process, to get E_o/F.)

Within this category E_o one can now actually see the desired monomorphism g of (8). In detail, the objects of E are not sets but functors $X : P^{op} \to \mathrm{Sets}$, while the objects of E_o are those functors X which are sheaves for the topology described by the operator $\neg\neg$. A map from one X to another Y is then a natural transformation $\alpha : X \to Y$ of functors, with $\alpha_p : X(p) \to Y(p)$ for each condition p. Each set A of the original model provides a constant functor ΔA with $(\Delta A)(p) = A$ for every p; this functor ΔA has a sheafification $\hat{A}$. In this context, one can "see" the desired mapping of (9). Namely, the nature of P allows us to define a subfunctor C of the constant functor $\Delta(B \times \mathbb{N})$ by setting

$$C(p) = \{(b, n) \mid p(b, n) = 0\} \tag{12}$$

—all those pairs (b, n) where p has already "decided" that $b \in f(n)$. This subfunctor C has a characteristic function to Ω. By properties of the operator $\neg\neg$, this characteristic function actually goes to the subobject classifier $\Omega_{\neg\neg}$ for sheaves, as

$$f' : \Delta(B \times \mathbb{N}) \to \Omega_{\neg\neg} \tag{13}$$

(compare (9) above). One then reverses the passage from (8) to (9) by taking the transpose of this f', written

$$g' : \Delta B \to \Omega_{\neg\neg}^{\Delta\mathbb{N}}. \tag{14}$$

It turns out that this transpose is a monomorphism, and that it induces a corresponding monomorphism for sheaves

$$\hat{g}' : \hat{B} \to \Omega_{\neg\neg}^{\hat{\mathbb{N}}}. \tag{15}$$

Here the object on the right is the power set in E_o of the natural numbers $\hat{\mathbb{N}}$ of E_o. So the original "large" set B now appears as a subset $\hat{B}$ of the new power set $\Omega_{\neg\neg}^{\hat{\mathbb{N}}}$—and this is possible because "set" now means "sheaf" (a suitable functor of the conditions p).

But B might be the whole of this new power set, we have only

$$\hat{B} \leq \Omega_{\neg\neg}^{\hat{\mathbb{N}}}. \tag{16}$$

What is needed is a proof that the construction of $\neg\neg$-sheaves from presheaves and hence from sets does preserve cardinal inequalities. This is proved much as in the original Cohen argument; the proof depends directly on the fact that the Cohen poset P satisfies the countable chain condition—any set of "pairwise disjoint" objects of P is at most countable. In the topos-theoretic version, the approach requires that one construct in any topos an object $\mathrm{Epi}(C, D) \subset D^C$ which acts as the "object of all epimorphisms" in the "function set" D^C. Thus, $\mathrm{Epi}(C, D) = 0$ will mean that $\operatorname{card} C < \operatorname{card} D$; one proves that $\mathrm{Epi}(C, D) = 0$ implies $\mathrm{Epi}(\hat{C}, \hat{D}) = 0$ for the corresponding sheaves $\hat{C}$, $\hat{D}$. In the original set theoretic model, one had

$$\mathbb{N} < 2^{\mathbb{N}} < B.$$

This preservation of inequalities and (16) then give in E_o

$$\hat{\mathbb{N}} < 2^{\hat{\mathbb{N}}} < \hat{B} \leq \Omega_{\neg\neg}^{\hat{\mathbb{N}}}. \tag{17}$$

Thus it is the sheafification of the "old" power set which has been "forced", via $\hat{B}$, to be in the middle, smaller than the new power set of the new natural numbers $\hat{\mathbb{N}}$.

4. Forcing and Languages

But, in this outline of the independence proof, "forcing" has not yet appeared. It will, and it is required in order to show that the axioms of set theory are indeed appropriately preserved in the passage to the new topos E_o/F.

For this purpose, the axioms are to be formulated in a certain "Mitchell–Bénabou" language associated to each topos E. Each object $X, Y, A, B, \ldots$ of E is to be regarded as a "type" $X, Y, A, \ldots$ of this language. For each type X take a stock of variables $x, x_1, \ldots$ of that type; with them, terms and formulas are to be constructed in the usual way, so that each term σ

of type, say, A and in the free variables $x, y, \ldots$ is to have an "interpretation" as an arrow $\sigma : X \times Y \to A$ of E; the formulas will be all those terms of type Ω (truth values). For example, if σ and τ are terms of the same type, $\sigma = \tau$ is interpreted as a suitable arrow to Ω, while a variable x is interpreted as the identity arrow $1 : X \to X$. In particular, the usual propositional connectives on formulas are interpreted by use of the corresponding operations already defined on the subobject classifier Ω of E, with an analogous treatment of the quantifiers. Thus if a formula $\varphi(x)$ in a variable x of type X is interpreted by an arrow to Ω the operation of pullback, as in the right hand square below, will produce that subobject of X usually written as $\{x \mid \varphi(x)\}$:

$$\begin{array}{ccccc} & & \{x \mid \varphi(x)\} & \longrightarrow & 1 \\ & \nearrow & \downarrow & & \downarrow \text{true} \\ U & \xrightarrow{\alpha} & X & \underset{\varphi(x)}{\longrightarrow} & \Omega \end{array} \tag{18}$$

This allows for the familiar set-like description of objects $\{x \mid \varphi(x)\}$ in the topos E.

The well-known semantics of Beth and Kripke for intuitionist and modal logics now leads via this language to the Kripke–Joyal semantics for a topos E. Instead of the usual "elements" of an object X of E, one considers "generalized elements" which are arrows $\alpha : U \to X$ from some starting object U. (If $E =$ Sets and $U = 1$, these are really the "ordinary" elements of the set X.) Then one defines "U forces $\varphi(x)$ at α", or $U \Vdash \varphi(\alpha)$, to mean that the arrow α factors through $\{x \mid \varphi(x)\}$, as displayed in the dotted slanting arrow on the left of (18) above. One then can derive the corresponding semantic rules for validity of $\varphi(x)$ (validity meaning that every α so factors). For example, given two formulas $\varphi(x)$ and $\psi(x)$ in a variable x of type X, one can show that

$$U \Vdash \varphi(\alpha) \vee \psi(\alpha) \tag{19}$$

iff there are arrows $p : V \to U$ and $q : W \to U$ such that $p+q : V+W \to U$ is an epimorphism from the coproduct $V + W$ and

$$V \Vdash \varphi(\alpha p) \text{ and } W \Vdash \psi(\alpha q). \tag{20}$$

(This is a familiar intuitionistic interpretation: For φ or ψ to hold, one must specify where φ holds and where ψ holds, this specification to cover everything ($p + q$ is an epimorphism).) With these and other appropriate semantical rules the treatment of the set-theoretical axioms in the above categorical model can be carried through. The process also uses Fourman's method [Fo] of constructing within a sufficiently complete elementary topos an imitation of the standard cumulative hierarchy of set theory.

5. The Freyd Topos for *AC*

The construction of topos models by means of presheaves and Lawvere–Tierney topologies is a flexible method. For example, it has been used by Bunge in [Bu] to reformulate the well-known Solovay–Tennenbaum proof of the independence of the Souslin hypothesis. Here we will summarize a surprising model of Peter Freyd [Fr] which gives a new proof that the axiom of choice is independent of the axioms of Zermelo–Fraenkel set theory.

Freyd's construction, unlike traditional forcing arguments, starts out not with a suitable poset P of conditions, but with an ingenious category $\mathbb{A}$. The objects of $\mathbb{A}$ are all finite sets of the form

$$n = \{0, 1, 2, \ldots, n\}$$

for a natural number n, while the morphism $f : n \to m$ in $\mathbb{A}$ are those functions f on n to m with $n \geq m$ and $f(i) = i$ for all $i \leq m$. This $\mathbb{A}$ could be described as the category of all the finite von Neumann ordinal numbers, with mappings restricted to the retractions of each n onto a smaller (or equal) m; thus if $n < m$ there are no morphisms from n to m.

From this category, one may construct, much as before, the topos $\mathbf{Sets}^{\mathbb{A}^{op}}$ of presheaves (i.e. functors on $\mathbb{A}^{op}$ to Sets), the double negation topology $\neg\neg$ in this topos, and the corresponding sheafification

$$a : \mathbf{Sets}^{\mathbb{A}^{op}} \to Sh_{\neg\neg}(\mathbf{Sets}^{\mathbb{A}^{op}}). \tag{21}$$

In the presheaf category, there is exactly one subfunctor of 1 for each natural number n (the functor U_n with $U_n(k) = 1$ iff $k \geq n$). Together with the empty subfunctor, these are all the subfunctors of 1. It follows from this that the sheaf category is two-valued; that is, that 1 has only two subsheaves. To see this, notice that a given subfunctor U_n must meet any other subfunctor U_k, since $U_n \cap U_k \supseteq U_{n+k}$. This means that every U_n is "dense" in the sense of the $\neg\neg$-topology. It follows that only two of these subfunctors, U_0 and 0, can be sheaves. This means that the sheaf category $\mathcal{J}$ is two-valued and therefore Boolean.

Now write H_n for the "hom functor" $\mathbb{A}(-, n) \to$ Sets, so that $H_n(m)$ is the set of all arrows $m \to n$ in $\mathbb{A}$. The sheaf category $\mathcal{J}$ must contain the sheafification $a(H_n) = F_n$ of each H_n. One can then readily prove that for each n the arrow $F_n \to 1$ to the terminal sheaf 1 is an epimorphism, and hence is non-zero—but that the infinite product $\prod_n F_n$ is the initial object 0. This demonstrates the failure in $\mathcal{J}$ of one familiar form of the (external) axiom of choice: The product of non-zero objects is non-zero. But to complete the independence proof one must show that the axiom of choice fails when formulated in the (internal) Mitchell–Bénabou language of $\mathcal{J}$: Every epimorphism $f : X \to Y$ has a section s. With f a variable of type Y^X and s one of type X^Y, this property is stated by the formula

$$\forall f [\forall y \, \exists x \, f(x) = y \Rightarrow \exists s \, \forall y \, fs(y) = y].$$

By a steady use of the semantics, one can show that this is equivalent to the requirement, for every object E of $\mathcal{J}$, that for each epimorphism $f : X \to Y$ the induced morphism $f^E : X^E \to Y^E$ is also an epimorphism.

This outlined proof again illustrates the flexibility of sheafification in the construction of models.

6. Equiconsistency

As this discussion indicates, the notion of a topos does provide a possible foundation for mathematics alternative to the conventional foundation in terms of sets—in which the usual primitive notion of membership in a set is replaced by that of composition of functions. With a general topos, this provides an intuitionistic type theory, as described in detail in the book by Lambek and Scott [LS]. To have a foundational system closer to classical set theory, one may use a *well-pointed* topos; that is, a topos which is non-degenerate ($0 \neq 1$) and in which the terminal object 1 is a generator (in the sense that for any two distinct arrows $f \neq g : A \to B$ there always exists an arrow $x : 1 \to A$ for which $fx \neq gx$). Such a well-pointed topos is necessarily Boolean (i.e., the subobject classifier Ω is a Boolean algebra). Thus, this notion of a well-pointed topos is finitely axiomatized and is substantially weaker than the standard Zermelo–Fraenkel axioms for set theory.

But these axioms for a well-pointed topos with choice are equiconsistent with the set-theoretic system "bounded Zermelo with choice" (BZC). The essential features of the latter are that there is no replacement axiom and that the usual comprehension axiom providing for the existence of sets $\{x \in A \mid \varphi(x)\}$ is restricted to those formulas $\varphi(x)$ of the language of set theory in which every quantifier is restricted (i.e., bounded) to some set b or c, as in $\forall y \in b \ldots$ or $\exists z \in c \ldots$. Thus, BZC consists of this plus the usual axioms: Extensionality, null set, pair, union, power set, foundation, axiom of infinity, axiom of choice and restricted comprehension. It is generally considered that this system BZC is adequate for a very large proportion of ordinary mathematics (where the quantifiers are indeed normally restricted). BZC does not provide for a proof that every well-ordered set is order isomorphic to an (von Neumann) ordinal number, and cannot prove that there is an infinite set of all whose members are infinite sets, no two of which have the same cardinal number (cf. Mathias [Ma]). The effectiveness of BZC for "ordinary" mathematics has been discussed in the first author's book on the nature of mathematics (MacLane [Mac2]).

To prove the equiconsistency of BZC and "well-pointed topos" one constructs from a model of each theory a model of the other theory. Starting with a set-model for BZC, the familiar construction of the corresponding category *Sets* of all these sets does at once provide a model of a well-pointed topos. For the converse construction one may use, following Mitchell [Mi], the familiar notion that a set may be regarded as a "tree" growing downward from the set S itself as a root, followed by nodes for the members of S, then for members of members of S and so on down. In the language of a well-pointed topos one can formulate a definition of such a tree as a reflexive partial order with a root U, well-founded up and down, in which the "up-segment" $\uparrow t$ above each node t is linearly ordered and which, moreover, is rigid (= has no non-trivial automorphisms). The formal proof that these trees, with an appropriate membership relation, do provide a model of BZC then makes direct use of the Mitchell–Bénabou language for the topos; indeed, William Mitchell first formulated this language exactly for this purpose [Mi].

Beyond equiconsistency, one can ask for an equivalence of theories, in the sense that two successive constructions of models (from sets to topos to sets, or vice versa) yield at the end the same (up to isomorphism) model of sets or of well-pointed topoi. To accomplish this, one must add two axioms to BZC: The axiom of transitive closure (every set is contained in a least transitive set) and a version of the Mostowski collapse lemma. Both of these added axioms are consequences of (but much weaker than) the usual replacement axiom of ZFC. Details are presented, for example, in Mitchell [Mi] and Johnstone [J].

Other related equiconsistency results (e.g. to aspects of Kripke–Platek set theory) are presented in a preprint of Mathias [Ma].

7. Brouwer's Theorem

In developing intuitionism L. E. J. Brouwer made the startling claim that every function (on and to the reals) is necessarily continuous. The realization of intuitionist logic in a topos now makes it possible to provide an explicit version of this result.

First observe that in a topos E with a natural numbers object $\mathbb{N}$ one can use this object to define the objects $\mathbb{Z}$ of integers, $\mathbb{Q}$ of rational numbers and $\mathbb{R}$ of Dedekind real numbers. Indeed, one can formulate the conventional definitions of these sets in the language of the topos, which then does provide the indicated objects $\mathbb{Z}$, $\mathbb{Q}$, and $\mathbb{R}$ there. For example, the definition of $\mathbb{R}$ uses the familiar description of a real number as a Dedekind cut in the ordered set $\mathbb{Q}$ of natural numbers. It then turns out that in the topos $Sh(X)$ of sheaves on a topological space X the object R_X of (Dedekind) reals is that sheaf on X which assigns to each open set U of X the set $C(U)$ of all continuous real-valued functions on U.

The definition of a continuous function on $\mathbb{R}$ and hence Brouwer's theorem about such functions can be formulated in the language of a topos E. One can then work in a large topos of "open cover" sheaves on subsets of Euclidean spaces $\mathbb{R}^n$—and in this context prove a form of Brouwer's theorem: Every function from the object of reals to itself is indeed continuous!

REFERENCES

[BW] M. Barr and C. F. Wells, *Triples, Topoi and Theories*, Springer–Verlag, Heidelberg, 1985.

[Be] J. L. Bell, *Toposes and Local Set Theories, an Introduction*, Clarendon Press, 1988, pp. 267.

[Bu] M. C. Bunge, *Topos theory and Souslin's hypothesis*, J. Pure and Applied Algebra **4** (1974), 159–187.

[C] P. J. Cohen, *The independence of the continuum hypothesis*, Proc. Nat. Acad. Sci. **50** (1963), 1143–1148 and **51** (1964), 105–110.

[Fo] M. P. Fourman, *Sheaf models for set theory*, J. Pure and Applied Algebra **19** (1980), 91–101.

[Fr] P. J. Freyd, *The axiom of choice*, J. Pure and Applied Algebra **19** (1980), 103–125.

[GV] A. Grothendieck and J. L. Verdier, *Theorie des Topos, SGA IV*, Springer Lecture Notes in Math., Vol. 269/270 (1972).

[J] P. T. Johnstone, *Topos Theory*, London Math. Soc. Monograph No. 10, Academic Press, New York, 1977, pp. 367.

[LS] J. Lambek and P. J. Scott, *Introduction to Higher Order Categorical Logic*, Cambridge Univ. Press, 1986, pp. 293.

[L] F. W. Lawvere, *An elementary theory of the category of sets*, Proc. Nat. Acad. Sci. USA **52** (1964), 1500–1511.

[Mac1] S. Mac Lane, *Categories for the Working Mathematician*, Graduate Texts in Mathematics No. 5, Springer–Verlag, Heidelberg, 1971.

[Mac2] ———, *Mathematics: Form and Function*, Springer–Verlag, New York/Heidelberg, 1986, 476 pages..

[MacMo] S. MacLane and I. Moerdijk, *A First Introduction to Topos Theory*, (Springer-Verlag; in preparation).

[Ma] A. R. D. Mathias, *Notes on MacLane set theory*, Preprint (1989), 11 pages..

[Mi] W. Mitchell, *Boolean topoi and the theory of sets*, J. Pure and Applied Algebra **2** (1972), 261–274.

[T] M. Tierney, *Sheaf theory and the continuum hypothesis*, in "Toposes, algebraic geometry, and logic," F. W. Lawvere (ed.), Springer Lecture Notes in Math 274 (1972), 13–42.

S. Mac Lane, Department of Mathematics, The University of Chicago, Chicago, IL 60637

I. Moerdijk, Department of Mathematics, The University of Utrecht, Utrecht, The Netherlands

Research at MSRI supported in part by NSF Grant DMS-8505550.

COMPLETENESS THEOREMS FOR LOGICS OF FEATURE STRUCTURES

LAWRENCE S. MOSS

ABSTRACT. We axiomatize the valid formulas of several logics of feature structures. The logics are similar to those studied by Johnson [2], Kasper and Rounds [3], Moshier and Rounds [4], and Rounds [7], and they include path equalities and set values. Our completeness proofs do not use reduction to normal forms or tableaux, in contrast to the papers cited. Instead they use structures built from maximal consistent sets, as in standard completeness arguments for modal logics. We also consider the forcing semantics of implication introduced to work on feature structure logics by Moshier and Rounds [4]. Our results here are based on, and strengthen, the connection between Kripke models and intuitionistic logic.

1. INTRODUCTION

When one uses mathematics in modeling, the choice of which objects to use can be critical. It affects the the basic vocabularies that one is likely to construct, and it even influences the observations that lie behind those theories. A case in point for this is linguistics. The use of trees as the bearers of linguistic structures suggests operations like transformations and concepts of hierarchical control. At the same time, it makes other ideas less appealing.

Several recent approaches in linguistics are moving in the direction of adopting a new primary mathematical object, the *feature structure*. The mathematical formulation of this is not yet fixed, but the idea is to have a class of objects which may or may not have certain pre-specified *features*. Moreover, the class of feature structures is organized hierarchically. And there is a notion of *informational extension* of feature structures.

To get an idea of all this, here is a feature structure, represented pictorially by an *attribute-value matrix*:

$$\begin{bmatrix} subj & \begin{bmatrix} agr & \begin{bmatrix} per & second \\ num & plur \\ gen & fem \end{bmatrix} \end{bmatrix} \\ pred & \begin{bmatrix} agr & \begin{bmatrix} num & sing \end{bmatrix} \\ verb & [\,] \end{bmatrix} \end{bmatrix}$$

The idea is that f might represent the information obtained at some stage in a parse of an English sentence S like *the girls stormed out* according to some grammar. The information about S represented in f is that the subject of S is a third person, female plural and that the predicate of S is plural. As a data structure, the feature structure f has two *attributes*: `subj` and `pred`. We think of these two as functions of f, whose values are themselves feature structures. The attributes may be composed. For example, the value of `subj : agr : num` on f is a feature structure `plur`. We can think of `plur` as an atomic feature structure; it makes no sense to apply an attribute like `verb` to it. So far, feature structures are little more than records, allowing for nesting of structures according to attributes (or *labels*, as we shall call them).

Feature structures have been put in great use in recent linguistic theories. In these approaches, a feature structure is a partial description of an object. Looking back at f, we see that `pred:verb` is a feature structure which is not an atom, and yet no attributes are defined on it. The idea is that f might represent the knowledge available at a certain moment of understanding. Later, we might learn that `pred:verb` has some properties, and we would want to update f.

In this way, feature structures are part of the basic vocabulary of a number of linguistic theories, including those which are most computationally oriented. The books by Johnson [2] and Pollard and Sag [5] contain substantial discussion of the use of feature structures in linguistic description and analysis. We shall not be concerned with particular grammars in this paper, but for readers unfamiliar with them, we stress that a feature structure like f could arise from an analysis of S according to a grammar in the spirit of [2] and [5].

The idea of giving logics for feature structures comes from the fact that statements of grammar can be viewed as formulas of an appropriate language. A very simple formulas of this type might be `num`: (`sing` $\vee$ `plur`). Sentences express intentions about feature structure, and once we have a formal language in which to make assertions, it is natural to ask for a semantics. There are other logical and computational connections as well, since the problem of parsing and interpretation according to grammars is reduced in the feature structure framework to problems of finding satisfying assignments to certain formulas.

The past few years has seen a number of proposals for logics of feature structures: Dawar and Vijay-Shanker [1], Johnson [2], Kasper and Rounds [3], Moshier and Rounds [4], Rounds [7], and others. These papers build on each other, by considering languages with more and more expressive power. They also differ a bit in what they see as the application of the structures which they propose. What these papers have in common, however, is that they consider languages, define semantics, and construct logical systems. The logical systems are either deductive systems to capture validity or tableau systems to characterize satisfiability. In a real sense, they are "logic from computational linguistics."

In reading these papers, we noticed that the completeness proofs involve syntactic methods, such as reduction to normal forms. The proofs seem more intricate than necessary if one is only interested in finding a completeness proof. The same is true for other logics; if one is interested in setting up a formal system, then it will be easier to prove the completeness of an axiomatic system rather than a technique like tableaux. On the other hand, if one is interested in actually finding formal proofs within a formal system, then it is easier to work with tableaux. In this paper, we present completeness proofs for several logics of feature structures. These systems are similar in most respects to those in the papers cited, with minor notational differences. But our proofs differ; they are model-theoretic. Our main purpose in writing this paper is to make the point that such proofs are available for logics of feature structures.

In addition, we give (possibly for the first time) axioms characterizing the validities in the logic with set values and path equations (Section 4), and for logics where the semantics is given in terms of Kripke models rather than individual structures (Section 5).

2. The Equational Proof System for Atoms and Labels

In this section, we want to consider the most basic language for feature structures. Further parts of this paper build on the language, semantics, and proof theory of this section.

Let A and L be arbitrary sets of *atoms* and *labels*, respectively. $\mathcal{L}$ is the smallest set containing 1 and each $a \in A$, and closed under the following formation rules: if $\phi, \psi \in \mathcal{L}$, then so are $\phi \wedge \psi$ and $\neg\phi$; if $\phi \in \mathcal{L}$ and $l \in L$, then $(l : \phi) \in \mathcal{L}$.

Definition. An **automaton (over** L **and** A**)** is a tuple $\mathcal{G} = \langle G, \delta, \alpha \rangle$ such that $\delta : G \times L \to G$ and $\alpha : G \to A$ are partial functions. Thinking of $\mathcal{G}$ as a kind of graph, we often refer to the members of G as **nodes**. For each $\mathcal{G}$, we define the **satisfaction relation** $\models_{\mathcal{G}}$ on $G \times \mathcal{L}$ by the following recursion:

$$\begin{array}{ll} n \models_{\mathcal{G}} 1 & \text{for all } n \\ n \models_{\mathcal{G}} \phi \wedge \psi & \text{if } n \models_{\mathcal{G}} \phi \text{ and } n \models_{\mathcal{G}} \psi \\ n \models_{\mathcal{G}} \neg\phi & \text{if } n \not\models_{\mathcal{G}} \phi \\ n \models_{\mathcal{G}} a & \text{if } \delta(n,l)\downarrow \text{ and } \delta(n,l) \models_{\mathcal{G}} \phi \\ n \models_{\mathcal{G}} l : \phi & \text{if } \alpha(n) \simeq a \end{array}$$

We write $S \models \phi$ if for all $\mathcal{G}$ and all $n \in G$, if $n \models_{\mathcal{G}} \psi$ for all $\psi \in S$, then $n \models_{\mathcal{G}} \phi$ also. Finally, we say ϕ is **valid** if $\models \phi$.

Remark. It should be mentioned that our semantics and model theory is not the only way to start out. It is possible to consider feature structures as automata with distinguished node, and then satisfaction is defined in terms of that node. The truth value of the $l : \phi$ is determined on the basis of the truth value of ϕ *in a different* structure. This alternative is closer to the hierarchical intuition behind feature structures. It tends to be a little more complicated notationally, and we prefer the version here.

Our first result is a characterization of validity in $\mathcal{L}$ using equational logic. We work in the signature containing constants 1,0, and the atoms $a \in A$; unary function symbols $\neg$ and 'l :' for each $l \in L$; and a binary function symbol $\wedge$. It should be noted that now formulas of $\mathcal{L}$ are terms, and we also add variables α_1, α_2, ... ranging over these terms. In the sequel, we use symbols ϕ, ψ,... to denote "formulas with variables." However, we are most interested in ground formulas, and when we write $\phi \in \mathcal{L}$, we still mean that ϕ contains no variables.

The equational proof systems deals exclusively with statements of the form $\phi = \psi$. The intended meaning of $\vdash \phi = \psi$, is that for all $\mathcal{G}$ and n, and all ground instances ϕ^* of ϕ and ψ^* of ψ, that $n \models_{\mathcal{G}} \phi^*$ iff $n \models_{\mathcal{G}} \psi^*$. In fact, for all our axiom systems, this soundness assertion will always be an easy induction on proofs. We will never write out such details.

Here are the laws which we take as axioms:

(A) The Boolean algebra laws using 1, 0, $\wedge$, and $\neg$.

(B) $(l : \phi) \wedge (l : \psi) \quad = \quad l : (\phi \wedge \psi)$

(C) $l : 1 \quad = \quad (l : \phi) \vee (l : \neg\phi)$

(D) $0 \quad = \quad l : 0$

(E) $\quad a \wedge b \quad = \quad 0 \qquad (a \neq b)$

(In (C), we are using $\vee$ as an abbreviation in the usual way.)

The only rules of proof are the reflexive, symmetric, and transitive laws of equality, and the rule of substitution of equals for equals. The substitution rule is: from $\vdash \phi = \psi$ deduce $\vdash \chi[\phi/\alpha] = \chi[\psi/\alpha]$. For example, $l : \phi = l : \psi$ is deducible from $\phi = \psi$. We write $\vdash \phi = \psi$ if $\phi = \psi$ can be proved from (A) – (E) using these rules. (However, in later sections we shall add equations without changing the provability symbol.)

We use $\phi \leq \psi$ as an abbreviation for $\phi \wedge \psi = \phi$. If S is a set of formulas (*not* equations) we overload our notation a bit to write $S \vdash \phi$ if there are $\psi_1, \ldots, \psi_n \in S$ such that $\vdash \psi_1 \wedge \cdots \wedge \psi_n \leq \phi$. (The empty conjunction is 1, so $\vdash \phi$ means $\vdash \phi = 1$.) A set $S \subseteq \mathcal{L}$ is **inconsistent** if $S \vdash 0$. Otherwise, S is **consistent**. S is **maximal consistent** if S is consistent but has no consistent supersets.

Let M be maximal consistent. By (A), if $M \vdash \phi$, then $\phi \in M$. Also, for all ϕ, M contains either ϕ or $\neg\phi$. The point of (B) – (D) is to insure that that if M is a maximal consistent set and $l \in L$, then the set

$$l^{-1}M \quad = \quad \{\phi \mid l : \phi \in M\}$$

is either empty or maximal consistent. Note also that $l^{-1}(M)$ is non-empty iff it contains 1. (E) insures that at most one of the atoms belongs to every consistent set.

Now we define an automaton $\mathcal{U}$. The node set U is the set of maximal consistent subsets of $\mathcal{L}$. Define

$$\delta(M, l) \quad \simeq \quad \begin{cases} l^{-1}(M) & \text{if } l : 1 \in M \\ \text{undefined} & \text{otherwise} \end{cases}$$

Finally, define $\alpha : U \to A$ by $\alpha(M) \simeq a$ iff $a \in M$.

Truth Lemma 1. *For all $M \in U$ and all $\phi \in \mathcal{L}$, $M \models_{\mathcal{U}} \phi$ iff $\phi \in M$.*

Proof. By induction on ϕ. If ϕ is 1 or an atom, then the statement is immediate. The induction steps for the the boolean connectives use the properties of maximal consistent sets. So we only check the induction step for prefixing by a label l. Suppose that $M \models_{\mathcal{U}} l : \phi$. Then $\delta(M, l)\downarrow$ and $\delta(M, l) \models_{\mathcal{U}} \phi$. The induction hypothesis tells us that $\phi \in \delta(M, l)$. Therefore, $l : \phi \in M$. Going the other way, if $l : \phi \in M$, then by (C), $l : 1 \in M$. Therefore $\delta(M, l)\downarrow$, and $\phi \in \delta(M, l)$. By induction hypothesis, $\delta(M, l) \models_{\mathcal{U}} \phi$, so $M \models_{\mathcal{U}} l : \phi$. $\square$

Completeness Theorem 2. *For all $S \subseteq \mathcal{L}$ and $\phi \in \mathcal{L}$, $S \models \phi$ iff $S \vdash \phi$.*

Proof. The soundness of deduction with S empty is an easy induction, and the general case follows from this by the finiteness of proofs.

Suppose that $S \not\vdash \phi$. Then $S \cup \{\neg\phi\}$ is consistent. Let $M \supseteq$ be maximal consistent. By Truth Lemma 1, $M \models_{\mathcal{U}} S \cup \{\neg\phi\}$. Therefore $S \not\models \phi$. $\square$

$\mathcal{L}$ is often used in situations where one also assumes that if for some $l \in L$, $\delta(n,l)\downarrow$, then $\alpha(n)\uparrow$. That is, the atoms are only defined on "end nodes" according to δ. For example, this is a standard assumption when dealing with attribute-value matrices. To get a complete system under this restriction, we need only add the equations

$$(l:1) \wedge a \quad = \quad 0 \qquad (a \in A,\, l \in L)$$

These laws are obviously sound, and adding them insures that in $\mathcal{U}$, $\alpha(M)\downarrow$ only when $\delta(M,l)\uparrow$ for all $l \in L$. Henceforth we shall not assume these laws, but they may be added to any of the remaining axioms systems if the desired semantics of atoms and labels is as above.

3. Adding Path Equations

We extend $\mathcal{L}$ to a more expresssive language by taking all of the expressions $(x \approx y)$ where x and y are finite strings from the set L of labels. (These pairs are sometimes written $x = y$, but this would be confusing here.) It is possible that x or y might be the empty string ϵ. We take these path equations as new atomic formulas. $\mathcal{L}_{\approx}$ contains all of the atomic formulas of $\mathcal{L}$ and is closed under the same operations.

This path equations are critical in the linguistic applications, because rules of grammar are expressed by equations of this form. For example, a hypothetical rule of a grammar might be: `subj:agr` $\approx$ `pred:agr`. In this rule, everything is a label. The intention is to specify that the agreement of the subject of a sentence is the same as the agreement of the verb of the verb phrase. In our setting, this would mean that the two are the very same node of the graph. In an automaton representing the attribute-value matrix f from the introduction, we would of course expect that this sentence fails.

We interpret the $\mathcal{L}_{\approx}$–formulas in automata. Extend δ from a partial function on $L \times G$ to a partial function on $L^* \times G$, by $\delta(n,\epsilon) \simeq n$, and $\delta(n,lw) \simeq \delta(\delta(n,l),w)$. Now we add to the inductive definition of satisfaction the clause

$$n \models_{\mathcal{G}} (x \approx y) \quad \text{if } \delta(n,x) \simeq \delta(n,y)$$

We use the same symbol $\models$ for this satisfaction relation; no confusion should result. For the proof theory, we take the equations (A) – (E), and we add the following:

(F) $(\epsilon \approx \epsilon) \quad = \quad 1$

(G) $(x \approx y) \quad \leq \quad (y \approx x)$

(H) $(x \approx y) \ \wedge \ (y \approx z) \quad \leq \quad (x \approx z)$

(I) $(x \approx y) \ \wedge \ x : \phi \quad \leq \quad y : \phi$

(J) $x : (y \approx z) \quad = \quad (xy \approx xz)$

(K) $(x \approx y) \ \wedge \ xz : 1 \quad \leq \quad (xz \approx yz)$

(L) $(x \approx y) \quad \leq \quad x : 1$

Here xy is the concatenation of x and y. The formula $x : \phi$ is defined by recursion on x in such a way that $\epsilon : \phi$ is ϕ, and $lx : \phi$ is $l : (x : \phi)$. This has the property that $xy : z$ is the same formula as $x : (y : z)$.

Remark. Note that x, y, and z are strings from L, so we have an infinite set of axioms here. It is possible to reformulate what we are doing so that x, y, and z become variables ranging over a new sort, say prefixes. Then the universe of prefixes is a monoid under concatenation with the labels in L as constants. We also could regard the prefixing operation : as a function symbol. Thus reinterpreted, (G) – (L) would be a finite list of equations with variables for prefixes.

Let $\mathcal{U}$ be the collection of all maximal consistent subsets of $\mathcal{L}_{\approx}$, made into a automaton according the definitions from the last section. It turns out that in order to get models of maximal consistent theories, it is not sufficient to consider $\mathcal{U}$. Let $\mathcal{V}$ be the **unfolding** of $\mathcal{U}$. That is, the nodes in V are the non-empty, finite, edge-labeled sequences

$$M_1 \overset{l_1}{\leadsto} M_2 \overset{l_2}{\leadsto} \cdots \overset{l_r}{\leadsto} M_{r+1}$$

through $\mathcal{U}$. This means that in $\mathcal{U}$, $\delta(M_i, l_i) \simeq M_{i+1}$ for $1 \leq i \leq r$. We call these sequences *paths*, and we use lower case letters like m to denote them. Also, define a map $\mathsf{Last} : V \to U$ by

$$\mathsf{Last}\,(m) \quad = \quad \text{the last node in the sequence } m\ .$$

Define $\delta_{\mathcal{V}}(m, l)$ when $\delta_{\mathcal{U}}(\,\mathsf{Last}\,(m), l)\downarrow$, and in this case,

$$\delta_{\mathcal{V}}(m, l) \quad \simeq \quad m \overset{l}{\leadsto} \delta_{\mathcal{U}}(\,\mathsf{Last}\,(m), l)\ .$$

This last path is the obvious extension of the path m. To complete the definition of $\mathcal{V}$ as an automaton, $\alpha_{\mathcal{V}}(m) \simeq \alpha_{\mathcal{U}}(\,\mathsf{Last}\,(m))$.

The Truth Lemma does not hold for $\mathcal{V}$. If $x \neq y$, then even if $(x \approx y) \in$ $\mathsf{Last}\ (n)$, $\delta(n, x)$ will not be the same sequence as $\delta(n, y)$. As a result, we must use a different structure.

Definition. $n\, E\, m$ iff there are $x, y \in L^*$ and $p \in V$ such that $\delta(p, x) = n$, $\delta(p, y) = y$, and $(x \approx y) \in$ $\mathsf{Last}\ (p)$.

Lemma 3. (a) *E is an equivalence relation.*
(b) *If $n\, E\, m$, then $\mathsf{Last}\ (n) = \mathsf{Last}\ (m)$.*
(c) *If $n\, E\, m$ and $\delta(n, l)\downarrow$, then $\delta(m, l)\downarrow$ and $\delta(n, l)\, E\, \delta(m, l)$.*

Proof. E is reflexive by (F), and symmetric by (G). For transitivity, suppose that $m\, E\, n$ and $n\, E\, p$. Let $a, b, c, d \in L^*$ and $r, s \in V$ be such that $\delta(r, a) = m$, $\delta(r, b) = n$, $\delta(s, c) = n$, $\delta(s, d) = p$, $(a \approx b) \in$ $\mathsf{Last}\ (r)$, and $(c \approx d) \in$ $\mathsf{Last}\ (s)$. The subpaths of n are linearly ordered by extension. Since r and s are subpaths of n, we may assume that r is a subpath of s. Let $e \in L^*$ be such that $\delta(r, e) = s$. Since $(c \approx d) \in$ $\mathsf{Last}\ (s)$, $(ec \approx ed) \in$ $\mathsf{Last}\ (r)$ by (J). But $ec = b$, so $(b \approx ed) \in$ $\mathsf{Last}\ (r)$. By (H), $(a \approx ed) \in$ $\mathsf{Last}\ (r)$. Finally, $\delta(r, ed) = \delta(s, d) = p$. Thus a, ed, and r show that $m\, E\, p$.

Part (b) is a consequence of (I), and part (c) uses part (b) and (K). □

Let $\mathcal{W}$ be the structure $\langle \mathcal{V}, E \rangle$. Now interpret $\mathcal{L}_{\approx}$ in $\mathcal{W}$ by changing the clause for the satisfaction of formulas $(x \approx y)$ to read:

$$n \models^*_{\mathcal{G}} (x \approx y) \quad \text{if } \delta(n, x) \text{ and } \delta(n, y) \text{ are defined, and } \delta(n, x)\, E\, \delta(n, y).$$

Note that we have used a different satisfaction symbol here, to emphasize that the semantics of the path equations uses E.

Truth Lemma 4. *For all $n \in W$ and all ϕ, $n \models^*_{\mathcal{W}} \phi$ iff $\phi \in$ $\mathsf{Last}\ (n)$.*

Proof. By induction on ϕ. The only case which needs an argument at this point is the case where ϕ is an atomic formula $x \approx y$. If $x \approx y$ belongs to $\mathsf{Last}\ (n)$, then by rule (L), both $\delta(n, x)$ and $\delta(n, y)$ are defined. It is immediate that $\delta(n, x)\, E\, \delta(n, y)$. Thus $n \models^*_{\mathcal{W}} x \approx y$.

Conversely, suppose that $\delta(n, x)\, E\, \delta(n, y)$. Let $a, b \in L^*$ and $m \in V$ be such that $\delta(m, a) = \delta(n, x)$, $\delta(m, b) = \delta(n, x)$, and $(a \approx b) \in$ $\mathsf{Last}\ (m)$. Now m and n are subpaths of $\delta(n, x)$, so one is a subpath of the other. Suppose first that n is a subpath of m, say $\delta(n, c) = m$. Then $ca = x$, and $cb = y$. By (J), $ca \approx cb \in$ $\mathsf{Last}\ (n)$; that is, $(x \approx y) \in$ $\mathsf{Last}\ (n)$ as desired. In the second case, suppose that m is a subpath of n. Then there is some $c \in L^*$ such that $a = cx$, $b = cy$, and $\delta(m, c) = n$. By (J) $c : (x \approx y) \in$ $\mathsf{Last}\ (m)$. So $(x \approx y) \in$ $\mathsf{Last}\ (n)$. □

Finally, we get a model which satisfies where the Truth Lemma holds, with the original semantics for $x \approx y$. Let $\mathcal{W}/E$ be the automaton whose nodes are the equivalence classes $[n]$ of nodes of $\mathcal{V}$ under E. Define $\delta_{\mathcal{W}/E}([n], l) \simeq [\delta_{\mathcal{W}}(n, l)]$ and $\alpha_{\mathcal{W}/E}([n]) \simeq \alpha_{\mathcal{W}}(n)$. These definitions do not depend on the choice of representatives, by parts (b) and (c) of Lemma 3.

Proposition 5. *For all $n \in W$, and all $\phi \in \mathcal{L}_{\approx}$, $[n] \models_{\mathcal{W}/E} \phi$ iff $n \models^*_{\mathcal{W}} \phi$.*

Completeness Theorem 6. *For all $S \subseteq \mathcal{L}_{\approx}$ and $\phi \in \mathcal{L}_{\approx}$, $S \models \phi$ iff $S \vdash \phi$.*

Proof. We only check that maximal consistent sets have models. Let M be maximal consistent, and consider M as a one point path. So $\mathsf{Last}\,(M) = M$, and by Truth Lemma 4 and Proposition 5, $[M] \models_{\mathcal{W}/E} M$. $\square$

4. Set Valued Feature Structures

Rounds [7] recently proposed a formalism which allows for the combination of set-theoretic and feature-theoretic ideas. The idea is to exploit the similarity of the graphical structure of sets with that already found in automata. The graph structure on sets is the one whereby $x \to y$ if $y \in x$. It plays a prominent role in non-well-founded set theory. (In fact, the anti-foundation-axiom (AFA) of Aczel is a formalization of the view that sets all possible graph structures give rise to sets, and two graphs picture the same set just in case they are bisimilar.) We shall not survey any of the work on set-valued feature structures. Rather, we present the appropriate formalism and show how our the earlier completeness theorems continue may be modified. The material of this section will not be used in Section 5, so the reader may skip this section.

Definition. The language $\mathcal{L}_{\Box}$ is obtained from $\mathcal{L}$ by adding a new unary operator $\Box$. A **graph automaton** is a tuple $\langle G, \delta, \alpha, R\rangle$ such that $\langle G, \delta, \alpha\rangle$ is an automaton and R is a binary relation on G. We define the satisfaction relation $\models_{\mathcal{G}}$ between nodes $n \in G$ and formulas $\phi \in \mathcal{L}_{\Box}$ by adding the clause

$$n \models_{\mathcal{G}} \Box\phi \quad \text{if for all } m \in G \text{ such that } n\,R\,m,\ m \models_{\mathcal{G}} \phi$$

to the earlier definition. (That is, we take G as a Kripke structure under R in order to interpret the modal operator.) We write $\Diamond\phi$ as an abbreviation for $\neg\Box\neg\phi$.

Now we extend our earlier axiom system by adding two additional laws:

$$\text{(M)} \qquad \Box\, 1 \;=\; 1.$$

$$\text{(N)} \qquad \Box\,(x \wedge y) \;=\; (\Box x) \,\wedge\, (\Box y).$$

In this section, $\vdash$ refers to the axiom system containing (M) and (N). We shall obtain completeness results for validity in $\mathcal{L}_\Box$ and in $\mathcal{L}_{\approx\Box}$. For $\mathcal{L}_\Box$, the next lemma is the key step. It is the heart of the completeness proof for the modal logic K using maximal consistent sets.

Lemma 7. *Let M be maximal consistent, and suppose that $\Diamond\psi \in M$, then there is a maximal consistent N such that $\psi \in N$ and whenever $\phi \in N$, $\Diamond\phi \in M$.*

Proof. By Zorn's Lemma, let N be maximal with such that $\psi \in N$, and if $\phi_1, \ldots, \phi_n \in N$, then $\Diamond(\phi_1 \wedge \cdots \wedge \phi_n) \in M$. We first check that N is consistent. If not, let $\chi_1, \ldots, \chi_n \in N$ be such that $\vdash \chi_1 \wedge \cdots \wedge \chi_n = 0$. Write $\chi_1 \wedge \cdots \wedge \chi_n$ as χ^*. Then $\Diamond\chi^* \in M$. But $\vdash \Diamond\chi^* = \Diamond 0 = 0$ by (M). So $0 \in M$. This contradiction shows that N is consistent.

To see that N is maximal consistent, suppose that neither ϕ not $\neg\phi$ belongs to N. Then by maximality, there are $\chi_1, \ldots, \chi_n \in N$ such that M contains both $\Box\neg(\chi^* \wedge \phi)$ and $\Box\neg(\chi^* \wedge \neg\phi)$, where $\chi^* = \chi_1 \wedge \cdots \wedge \chi_n$. By (A) $\vdash \neg(\chi^* \wedge \phi) \wedge \neg(\chi^* \wedge \neg\phi) = \neg\chi^*$. Applying $\Box$ to both sides and using (N),

$$\Box(\neg(\chi^*\wedge\phi)) \wedge \Box(\neg(\chi^*\wedge\neg\phi)) \quad = \quad \Box\{\neg(\chi^*\wedge\phi) \wedge \neg(\chi^*\wedge\neg\phi)\} \quad = \quad \Box\neg\chi^* .$$

Thus $\Box\neg\chi^* \in M$. But M therefore contains $\Box\neg\chi^* \wedge \Diamond\chi^* = 0$. Thus M is inconsistent. This contradiction shows that N is maximal consistent. $\Box$

Let $\mathcal{U}$ be the automaton of maximal consistent subsets of $\mathcal{L}_\approx$. We make $\mathcal{U}$ into a graph automaton by setting $M\,R\,N$ iff whenever $\phi \in N$, $\Diamond\phi \in M$.

Truth Lemma 8. *For all $M \in U$ and all $\phi \in \mathcal{L}_\Box$, $M \models_\mathcal{U} \phi$ iff $\phi \in \mathcal{U}$.*

The proof uses Lemma 7 in the induction step for $\Box$.

Completeness Theorem 9. *For all $S \subseteq \mathcal{L}_\Box$ and $\phi \in \mathcal{L}_\Box$, $S \models \phi$ iff $S \vdash \phi$.*

Once again, the axioms here are (A) – (E), (M), and (N).

We next add the path equalities to $\mathcal{L}_\Box$, and we call the result $\mathcal{L}_{\approx\Box}$. The semantics of the path equalities is as before, and we claim that laws (A) – (N) give a complete axiomatization of sentences valid on the class of graph automata. To prove this, it is sufficient to build models of maximal consistent sets. We do this following the construction in Section 3.

Let $\mathcal{U}$ be the automaton of maximal consistent subsets of $\mathcal{L}_{\approx\Box}$. Let V be the set of non-empty, finite, edge-labeled sequences

$$M_1 \overset{l_1}{\leadsto} M_2 \overset{l_2}{\leadsto} \cdots \overset{l_r}{\leadsto} M_{r+1}$$

where each $M_i \in U$, and each l_i either belongs to L or is the symbol $\Diamond$. If $l_i \in L$, then $\delta_{\mathcal{U}}(M_i, l_i) \simeq M_{i+1}$. If l_i is $\Diamond$, then $M_i \, R \, M_{i+1}$. We also define Last, δ, and α on V exactly as before. To complete the definition of a graph automaton $\mathcal{V}$, let R be the set of pairs of paths

$$\langle \; m \; , \; m \overset{\Diamond}{\rightsquigarrow} M \; \rangle$$

such that $\mathsf{Last}\,(m) \, R \, M$ in $\mathcal{U}$.

Consider the relation E defined on $\mathcal{V}$ exactly as before, by $n \, E \, m$ iff there are $x, y \in L^*$ and $p \in V$ such that $\delta(p, x) = n$, $\delta(p, y) = y$, and $(x \approx y) \in T(p)$. Lemma 3 holds for E on $\mathcal{V}$. Let $\mathcal{W} = \langle \mathcal{V}, \, E \, \rangle$, and use E to evaluate the path equations. The Truth Lemma holds for $\mathcal{W}$, by combining the proof of the Truth Lemma 4 with the proof of Lemma 7.

Finally, we need to take a quotient of $\mathcal{W}$ to get an analog of Proposition 5. The relation E is too small for this, so we enlarge it.

Definition. Let E' be the smallest relation on $\mathcal{W}$ such that

(a) $E \subseteq E'$.
(b) If $n \, E' \, m$, $n \, R \, n'$, $m \, R \, m'$, and $\mathsf{Last}\,(n') = \mathsf{Last}\,(m')$, then $n' \, E' \, m'$.
(c) If $n \, E' \, m$, and $(x \approx y) \in \mathsf{Last}\,(n)$, then $\delta(n, x) \, E' \, \delta(m, y)$.

Lemma 10. (a) *E' is an equivalence relation.*
(b) *If $n \, E' \, m$, then $\mathsf{Last}\,(n) = \mathsf{Last}\,(m)$.*
(c) *If $n \, E' \, m$ and $\delta(n, l)\!\downarrow$, then $\delta(m, l)\!\downarrow$, and $\delta(n, l) \, E' \, \delta(m, l)$.*
(d) *If $n \, E' \, m$ and $n \, R \, p$, then there is some q such that $m \, R \, q$ and $q \, E' \, p$.*
(e) *If $\delta(n, x) \, E' \, \delta(n, y)$, then already $\delta(n, x) \, E \, \delta(n, y)$.*

Proof. First, (c) and (d) are immediate. To prove the other parts, it is best to write E' as an infinite union, and check the properties by induction. Then (b) is easy, and the transitivity part of (a) is proved as in Lemma 3. The argument for part (e) is similar, and we omit the details. $\square$

We form the quotient $\mathcal{W}/\,E'$ as before, and we put in the relation R all the pairs $\langle [n], [m] \rangle$ such that $n \, E' \, m$ in $\mathcal{W}$. The analog of Proposition 5 holds, by an induction using Lemma 10. In this way, the equational system whose axioms are (A) – (N) characterizes the valid formulas of $\mathcal{L}_{\approx\Box}$.

Completeness Theorem 11. *For all $S \subseteq \mathcal{L}_{\approx\Box}$ and $\phi \in \mathcal{L}_{\approx\Box}$, $S \models \phi$ iff $S \vdash \phi$.*

5. Interpretations in Kripke Models

Feature structures are intended to be partial descriptions of attributed objects. Indeed, this is one of the main points in using them. Now wherever partial descriptions or partial objects are used, the status of negative information becomes problematic. For example, consider the feature structure f from the Introduction. According to our semantics f satisfies `pred`:`verb`:$\neg$`tense`:`past`. This is simply a consequence of the fact that the `tense` label is not defined on the trivial feature structure, and the fact that we are using the classical negation. However, if f represents information that is growing dynamically, then we might not want to use the classical negation.

The main point of Moshier and Rounds [4] is to motivate a different semantics, and to present a tableau proof system for satisfiability. Such a proof system also gives a procedure for finding models for satisfiable formulas. We study their semantics below, and we adapt our earlier proofs to axiomatize the valid formulas.

We should remark as well that there are other proposals for negation and implication in feature structures. For example, Dawar and Vijay-Shanker [1] suggest using partiality, in connection with Kleene's strong three-valued logic. Presumably every non-classical semantics of the logical connectives can be adapted to feature structures, and it will be necessary to be very clear about the intended applications in order to make the best choice.

Definition. A **Kripke model** for $\mathcal{L}$ is a pair $\mathsf{K}= \langle K, \sqsubseteq\rangle$ such that K is a set of automata, and $\sqsubseteq$ is a set of pairs $(\mathcal{G}, n)$ such that $n \in G$, and $\mathcal{G} \in K$. We require that $\sqsubseteq$ be reflexive and transitive. Assume that $(\mathcal{G}, n) \sqsubseteq (\mathcal{H}, m)$. We require that if $\alpha_{\mathcal{G}}(n)\downarrow$, then $\alpha_{\mathcal{H}}(m)\downarrow$ and $\alpha_{\mathcal{G}}(n) = \alpha_{\mathcal{H}}(m)$. We also require that if $\delta_{\mathcal{G}}(n, l)\downarrow$, then $\delta_{\mathcal{H}}(m, l)\downarrow$ and $(\mathcal{G}, \delta(n, l)) \sqsubseteq (\mathcal{H}, \delta(m, l))$.

Remark. This definition of Kripke model is not exactly equivalent to the one found in Moshier and Rounds. (They work with a different sort of automaton than we do, but this difference is unimportant.) In effect, they strengthen the definition above by requiring that the order $\sqsubseteq$ is maximal with respect to the other properties. The definition of forcing below is the same as theirs. We are confident that the completeness proofs below can be adapted the stronger definitions. We have not done this, because it is more difficult to construct models with the maximality property. At this point, it does not seem to be worth the effort since the intuitive motivation seems to be neutral on which definition to use.

The idea here is that the Kripke model might be something like the possible feature structures obtained in the course of a inference according to some grammar. The information states $(\mathcal{G}, n)$ are ordered by a relation of *information extension* which must be a preorder. The properties of $\sqsubseteq$ express our intuition about exactly what the informative part of a structure is. It demands that if $(\mathcal{H}, m)$ is more informative than $(\mathcal{G}, n)$, then all of the δ transitions available to n are also available to m. Think of $(\mathcal{G}, n) \sqsubseteq (\mathcal{H}, m)$ as meaning that the part of $\mathcal{H}$ below m is obtained from the part of $\mathcal{G}$ below n by adding positive information about δ. The condition on the atoms should be understood similarly.

In giving the definition below, we need to change the connectives of basic language $\mathcal{L}$. We now take them to be $\wedge$, $\vee$, and $\rightarrow$. We do not take $\neg$ as a primitive, but we add 0 as an atomic formula; in this way we may regard $\neg\phi$ as an abbreviation of $\phi \rightarrow 0$.

Definition. Let K be a Kripke model. Define the forcing relation $\Vdash_{\mathsf{K}}$ by recursion on formulas:

$(\mathcal{G}, n) \Vdash_{\mathsf{K}} 1$	always
$(\mathcal{G}, n) \Vdash_{\mathsf{K}} 0$	never
$(\mathcal{G}, n) \Vdash_{\mathsf{K}} a$	if $\alpha_G(n) \simeq a$.
$(\mathcal{G}, n) \Vdash_{\mathsf{K}} \phi \wedge \psi$	if $(\mathcal{G}, n) \Vdash_{\mathsf{K}} \phi$ and $(\mathcal{G}, n) \Vdash_{\mathsf{K}} \psi$
$(\mathcal{G}, n) \Vdash_{\mathsf{K}} \phi \vee \psi$	if $(\mathcal{G}, n) \Vdash_{\mathsf{K}} \phi$ or $(\mathcal{G}, n) \Vdash_{\mathsf{K}} \psi$
$(\mathcal{G}, n) \Vdash_{\mathsf{K}} \phi \rightarrow \psi$	if for all $(\mathcal{G}, n) \sqsubseteq (\mathcal{H}, m) \in K$, if $(\mathcal{H}, m) \Vdash_{\mathsf{K}} \phi$, then $(\mathcal{H}, m) \Vdash_{\mathsf{K}} \psi$
$(\mathcal{G}, n) \Vdash_{\mathsf{K}} l : \phi$	if $\delta(n, l)\downarrow$, and $(\mathcal{G}, \delta(n, l)) \Vdash_{\mathsf{K}} \phi$

Note that the clause for implication uses the Kripke structure in a way that the satisfaction relation did not. We say that $S \Vdash \phi$ if for all K, $\mathcal{G} \in K$ and $n \in G$, if $(\mathcal{G}, n) \Vdash_{\mathsf{K}} \psi$ for all $\psi \in S$, then $(\mathcal{G}, n) \Vdash_{\mathsf{K}} \phi$.

Perhaps the first thing to notice is that the law of the excluded middle, $\phi \vee \neg\phi = 1$, is not sound in this semantics. It is easy to construct Kripke models where neither $l : a$ nor $l : \neg a$ is forced by some particular $(\mathcal{G}, n)$. Also, the forcing relation is monotone in the sense that if $(\mathcal{G}, n) \Vdash_{\mathsf{K}} \phi$ and $(\mathcal{G}, n) \sqsubseteq (\mathcal{H}, m)$, then $(\mathcal{H}, n) \Vdash_{\mathsf{K}} \phi$. This goes along with the intuition that $\Vdash_{\mathsf{K}}$ means "positive information" and the order relation means "obtained by adding positive information".

We give an equational proof system for the valid formulas with the forcing semantics. Moshier and Rounds [4] speak of this logic as "intuitionistic" since the law of the excluded middle fails. Our axiomatization confirms this by taking as axioms

(O) The Heyting algebra laws using 1, 0, $\wedge$, $\vee$, and $\rightarrow$.

(Cf. Rasiowa and Sikorski [6], IV, 1.2. The axioms include the lattice laws of $\wedge$ and $\vee$, and they imply that $\phi \leq \psi$ iff $\phi \rightarrow \psi = 1$. The Heyting algebra laws are sound.) By incorporating these laws, we shall have the Deduction Theorem: $S \cup \{\phi\} \vdash \psi$ iff $S \vdash \phi \rightarrow \psi$. Concerning the laws for labels, (B), (D), and (E) are sound. But

(C) $\quad l : 1 \quad = \quad (l : \phi) \vee (l : \neg\phi)$

is not sound. This is the point of using the forcing semantics, since our intuition might be that $l : 1$ is forced by some $(\mathcal{G}, n)$, but neither $l : m : 1$ nor $l : \neg m : 1$ is forced there. Instead we take

(P) $\quad (l : \phi) \wedge l : (\phi \rightarrow \psi) \quad \leq \quad l : \psi$

(Q) $\quad l : (\phi \vee \psi) \quad = \quad (l : \phi) \vee (l : \psi)$

To summarize, we take as axioms (B), (D), (E), (O), (P), and (Q). We write $\vdash_I$ for the deduction relation using these axioms and the rules of equational logic. It should be mentioned that our interpretation of $\vdash \phi = \psi$ has changed. Now it is intended to mean that for all K, all $\mathcal{G} \in K$, all $n \in G$, and all ground substitution instances ϕ^* of ϕ and ψ^* of ψ, $(\mathcal{G}, n) \Vdash_{\mathsf{K}} \phi^*$ iff $(\mathcal{G}, n) \Vdash_{\mathsf{K}} \psi^*$.

Definition. A set S of formulas is **prime** if the following conditions hold:

(a) If $S \vdash_I \phi$, then $\phi \in S$.
(b) If $\phi \vee \psi \in S$, then either $\phi \in S$ or $\psi \in S$.
(c)) $0 \notin S$.

Chapter X of Rasiowa and Sikorski [6] has a great deal of information on prime theories and completeness proofs for intuitionistic logic.

Proposition 12. *Supppose that $S \not\vdash_I \phi$. Then there is a prime $T \supseteq S$ such that $T \not\vdash_I \phi$.*

Proof. This is a translation of a standard existence result for prime filters in distributive lattices (cf. Rasiowa and Sikorski [6] I.9.2.) By Zorn's Lemma, let T be maximal among supersets of S with the property that $T \not\vdash_I \phi$. The verifications of the primality properties are similar to the arguments of Lemma 7, except of course that only the Heyting algebra laws are used. □

Lemma 13. *If S is prime and $(l : 1) \in S$, then $\delta(S, l)$ is also prime.*

Proof. For (a), suppose that $\phi_1, \ldots, \phi_n \in \delta(S, l)$ and $\vdash_I \phi_1 \wedge \cdots \wedge \phi_n \leq \psi$. Let $\phi^* = \phi_1 \wedge \cdots \wedge \phi_n$; by (B), $\phi^* \in \delta(S, l)$. By (O), $\vdash_I \phi^* \rightarrow \psi = 1$, and thus

$\vdash_I l : (\phi^* \to \psi) = l : 1$. Since S is closed under deduction, $l : (\phi^* \to \psi) \in S$. By (P), $l : \psi \in S$. Thus $\psi \in \delta(S, l)$.

Part (b) is immediate from rule (Q), and (c) from (D). □

The collection of all prime sets is an automaton under the setwise definitions of δ and α from Section 3. Call this automaton $\mathcal{U} = \langle U, \delta_{\mathcal{U}}, \alpha_{\mathcal{U}} \rangle$. We are going to do is to turn $\mathcal{U}$ into a Kripke model. First, we specify the automata.

For each prime set S, let

$$S^* \quad = \quad \{\delta_{\mathcal{U}}(S, x) \ : \ x \in L^* \text{ and } \delta(S, x)\downarrow\} \ .$$

Let $\mathcal{S}^*$ be the automaton whose node set is S^* and such that $\delta_{\mathcal{S}^*}$ and $\alpha_{\mathcal{S}^*}$ are the restrictions of $\delta_{\mathcal{U}}$ and $\alpha_{\mathcal{U}}$. (We are using a subscript here because later we shall want a different structure defined from S, obtained by unfolding $\mathcal{S}^*$.)

Let P be the Kripke model defined in the following way. Let

$$P \quad = \quad \{\mathcal{S}^* \ : \ S \subseteq \mathcal{L} \text{ is prime}\} \ .$$

Let $(\mathcal{S}^*, X) \sqsubseteq (\mathcal{T}^*, Y)$ iff $X \subseteq Y$. This relation is a preorder, and we check that it has the right properties. Suppose that $(\mathcal{S}^*, X) \sqsubseteq (\mathcal{T}^*, Y)$. If $\delta_{\mathcal{S}^*}(X, l)\downarrow$, then $(l : 1) \in X$. Hence $(l : 1) \in Y$, so $\delta_{\mathcal{T}^*}(Y, l)\downarrow$. Similarly, if $\alpha_{\mathcal{S}^*}(X)\downarrow$, then $\alpha_{\mathcal{T}^*}(Y) \simeq \alpha_{\mathcal{S}^*}(X)$.

Truth Lemma 14. *For all $\mathcal{S}^* \in P$, $X \in S^*$, and ϕ, $(\mathcal{S}^*, X) \Vdash_{\mathsf{P}} \phi$ iff $\phi \in X$.*

Proof. By induction on ϕ. The steps for 1, 0, and the atoms are trivial, as is the conjunction step. The disjunction step is a direct application of primality.

For the implication step, fix $(\mathcal{S}^*, X)$, and suppose that $(\phi \to \psi) \in X$. Let $(\mathcal{S}^*, X) \sqsubseteq (\mathcal{T}^*, Y)$. Then $(\phi \to \psi) \in Y$. Suppose that $(\mathcal{T}^*, Y) \Vdash_{\mathsf{P}} \phi$. Then by induction hypothesis, $\phi \in Y$. As Y is closed under deduction, $\psi \in Y$. By induction hypothesis again, $(\mathcal{T}^*, Y) \Vdash_{\mathsf{P}} \psi$. This shows that $(\mathcal{S}^*, X) \Vdash_{\mathsf{P}} \phi \to \psi$.

Going the other way, suppose that $(\phi \to \psi) \notin X$. Then $X \cup \{\phi\} \nvdash_I \psi$. By Proposition 12, let Y be a prime set such that $X \cup \{\phi\} \subseteq Y \nvdash_I \psi$. So $(\mathcal{S}^*, X) \sqsubseteq (\mathcal{Y}^*, Y)$, where $\mathcal{Y}^*$ is the prime automaton determined by Y. By induction hypothesis, $(\mathcal{T}, T) \Vdash_{\mathsf{P}} \phi$, but $(\mathcal{Y}^*, Y) \nVdash_{\mathsf{P}} \psi$. Thus $(\mathcal{S}^*, X) \nVdash_{\mathsf{P}} \phi \to \psi$.

Finally, the induction step for prefixing by labels is trivial. □

Completeness Theorem 15. *For all $S \subseteq \mathcal{L}$ and $\phi \in \mathcal{L}$, $S \vdash_I \phi$ iff $S \Vdash \phi$.*

Proof. The soundness part is an easy induction. Going the other way, suppose that $S \nvdash_I \phi$. By Proposition 12, let $T \supseteq S$ be a prime set such that $T \nvdash_I \phi$. Consider the Kripke model P from above. By Truth Lemma 14, $(\mathcal{T}^*, T) \Vdash_{\mathsf{P}} S$ but $(\mathcal{T}^*, T) \nVdash_{\mathsf{P}} \phi$. Thus $S \nVdash \phi$. □

We conclude this paper with a discussion of validity in $\mathcal{L}_{\approx}$. We continue to interpret such formulas using Kripke models composed of automata. It is our intention to add to the definition of the forcing relation the clause

(1) $$(\mathcal{G}, n) \Vdash_{\mathsf{K}} (x \approx y) \quad \text{if} \quad \delta(n,x) \simeq \delta(n,y) \text{ in } \mathcal{G}$$

This seems to be the natural definition. In order to insure that the forcing relation is monotone, we add to the definition of Kripke models the condition that if $(\mathcal{G}, n) \sqsubseteq (\mathcal{H}, m)$ and $\delta_{\mathcal{G}}(n,x) \simeq \delta_{\mathcal{G}}(n,y)$, then $\delta_{\mathcal{H}}(m,x) \simeq \delta_{\mathcal{H}}(m,y)$.

Under this semantics, the laws of path equalities from Section 3 are valid. So we take as axioms of our proof system (F) – (L) together with the laws (B), (D), (E), (O), (P), and (Q) used in connection with $\mathcal{L}$.

For each prime subset $S \subseteq \mathcal{L}_{\approx}$, let $\mathcal{S}^{\dagger}$ be the following automaton. The nodes set $S^{\dagger}$ is the set of finite, edge-labeled paths $X_1 \overset{l_1}{\rightsquigarrow} X_2 \overset{l_2}{\rightsquigarrow} \cdots \overset{l_r}{\rightsquigarrow} X_{r+1}$ where each $X_i \in S^*$, $X_1 = S$, and $\delta_{\mathcal{S}^*}(X_i, l_i) \simeq X_{i+1}$ for $1 \leq i \leq r$. These paths are the nodes of $\mathcal{S}^{\dagger}$, and the functions $\delta_{\mathcal{S}^{\dagger}}$ and $\alpha_{\mathcal{S}^{\dagger}}$ are defined as in Section 3. Further, we want to consider $\mathcal{S}^{\ddagger} = \langle \mathcal{S}^{\dagger}, E_S \rangle$, where E_S is the relation on $S^{\dagger}$ with the same definition as the original E. The proofs of the last section only used (F) – (L) so they hold here as well. In particular, each E_S is an equivalence relation, and $n \, E_S \, m$ implies $\mathsf{Last}\ (n) = \mathsf{Last}\ (m)$.

Let K be the Kripke model whose structures are those of the form $\mathcal{S}^{\ddagger}$ for prime S. We set $(\mathcal{S}^{\ddagger}, m) \sqsubseteq (\mathcal{T}^{\ddagger}, n)$ iff $\mathsf{Last}\ (m) \subseteq \mathsf{Last}\ (m)$. It is easy to check that this definition satisfies all of the conditions on Kripke models.

We wish to study the forcing relation in K, and as with the structure $\mathcal{W}$ from Section 3, we want to use the relations E_S to interpret the path equality statments. So we define an auxilliary forcing notion $\Vdash^*_{\mathsf{K}}$, by replacing (1) by (2):

(2)
$$(\mathcal{G}, n) \Vdash^*_{\mathsf{K}} (x \approx y) \text{ iff } \delta(n,x) \text{ and } \delta(n,y) \text{ are defined, and } \delta(n,x) \, E \, \delta(n,y)$$

Combining the proofs of Truth Lemmas 4 and 14, we show that for all S, $n \in S^{\dagger}$, and ϕ, $(\mathcal{S}^{\ddagger}, n) \Vdash^*_{\mathsf{K}} \phi$ iff $\phi \in \mathsf{Last}\ (n)$.

Since we are interested in $\Vdash$ rather than $\Vdash^*$, we are not finished. For each $\mathcal{S}^{\ddagger}$, let $\mathcal{S}^{+}$ be the quotient $\mathcal{S}/E_S$. This is an automaton, and we use (1) to interpret the path equations. Define a Kripke model $\mathbf{J}$ by taking all these automata, and setting $(\mathcal{S}^{+}, [m]) \sqsubseteq (\mathcal{T}^{+}, [n])$ iff $\mathsf{Last}\ (n) \subseteq \mathsf{Last}\ (m)$. It is easy to now check that for all S, $n \in S^{\dagger}$, and ϕ that $(\mathcal{S}^{+}, [n]) \Vdash_{\mathsf{J}} \phi$ iff $(\mathcal{S}^{\ddagger}, n) \Vdash^*_{\mathsf{K}} \phi$.

Completeness Theorem 16. *For all $S \subseteq \mathcal{L}_{\approx}$ and $\phi \in \mathcal{L}_{\approx}$, $S \vdash_I \phi$ iff $S \Vdash \phi$.*

Proof. The soundness part is an easy induction. Going the other way, if $S \not\vdash_I \phi$, then let $T \supseteq S$ be a prime set such that $T \not\vdash_I \phi$. Consider the Kripke models K and $\mathbf{J}$. One of the automata in K is $\mathcal{T}^{\ddagger}$, and one of the nodes of $\mathcal{T}^{\ddagger}$ is the one-point path T. $\mathsf{Last}\ (T) = T$. So $(\mathcal{T}^{\ddagger}, T) \not\Vdash^{*}_{\mathsf{K}} \phi$, and thus $(\mathcal{T}^{+}, [T]) \not\Vdash_{\mathsf{J}} \phi$. But $(\mathcal{T}^{+}, [T]) \Vdash_{\mathsf{J}} S$. Hence $S \not\Vdash \phi$. $\square$

Acknowledgment

I would like to thank the Mathematical Sciences Department of the IBM T.J. Watson Research Center for their support during the writing of this paper.

References

1. Dawar, A. and K. Vijay-Shanker, *A three-valued interpretation of negation in feature structure descriptions,* Proc. 27th Meeting of the Association for Computational Linguistics (1989) 18–24.
2. Johnson, M., *Attribute-Value Logic and the Theory of Grammar*, CSLI Lecture Notes **16** (1988) University of Chicago Press.
3. Kasper, R. and W. C. Rounds, *A complete logical calculus for record structures representing linguistic information,* Proc. Symp. on Logic In Computer Science (1986) IEEE 38–48.
4. Moshier, M. D. and W. C. Rounds, *A logic for partially specified data structures,* Proc. 14th Annual ACM Symp. on Principles of Programming Languages, ACM (1986) 156–167.
5. Pollard, C. and I. A. Sag, *Information-Based Syntax and Semantics*, Vol. 1, CSLI Lecture Notes **13** (1987) University of Chicago Press.
6. Rasiowa, H. and R. Sikorski, *The Mathematics of Metamathematics*, Monographie Matematyczne tom **41** (1963) PWN–Polish Scientific Publishers, Warsaw.
7. Rounds, W. C., *Set values for unification based grammar formalisms and logic programming,* CSLI Tech. Report CSLI-88-129, Center for the Study of Language and Information, 1988.

Department of Mathematics, Indiana University, Bloomington, IN 47405

Mathematical Sciences Department, IBM Thomas J. Watson Research Center
P.O. Box 218, Yorktown Heights, NY 10510

CONCURRENT PROGRAMS AS STRATEGIES IN GAMES

ANIL NERODE, ALEXANDER YAKHNIS, AND VLADIMIR YAKHNIS

0. SUMMARY AND ABSTRACT

Running a newly written concurrent program feels, to the Programmer, like a contest with the Computer, in which the Computer is a devilish opponent trying its level best to execute the program in such a way that the program specification is violated. We make this into mathematics and into the basis of concurrent programming.

We show how to interpret each program as a strategy in a two-person game. The program specification becomes the condition that a play wins for player 0 ("Programmer"). In real programs, each program will denote a strategy equipped with an "internal state automaton" which guides the choice at each stage of the packet of instructions to be next submitted to the computer by player 0 (or "Programmer") for eventual execution by player 1 (or "Computer").

The principal new idea we contribute is to raise both Programmer and Computer to be first class citizens, independent Players in the game of computation ("computational game") played by Programmer against Computer. Unlike existing semantics, these plays explicitly list both Programmer's and Computer's previous moves. Positions, that is, sequences of past moves, entirely replace the conventional execution sequences of states (or sequences of executed instructions) used in previous semantics of concurrency. We contend that: 1) these plays are the "correct" way to represent computational behavior 2) that all informal reasoning about concurrency is more easily and more completely represented by using these plays rather than execution sequences 3) finally, that all reasoning about concurrency should be carried out in systems reflecting these plays rather than reflecting execution sequences.

We define several different strategy models, based on: abstract strategies, strategies induced by automata, strategies within one computational game tree, strategies on computational state trees, and strategies guided by the minimal state Nerode equivalence automaton. Strategies have a natural topology in which the strategy operations are continuous. The explicit state structure gives an additional "fine structure" not present in previous Scott domain semantics for concurrency. This fine structure allows Programmer to take into account the many possible behaviors of Computer, and this plays an essential role in applying our paradigm for program development.

Our paper is a leisurely tour of the basic notions. We discuss Park's example, and show how clear, and straightforwardly mathematical, the issues surrounding fairness become when Computer is assigned an independent role as player. We will treat in later papers the corresponding program logics, natural deduction systems, typed lambda calculi, and categories. We give a hint here of the new concurrent program development and verification methodology: see [A. Yakhnis, Ph.D. Diss., 1990], [V. Yakhnis, Ph.D. Diss., 1990], [A. and V. Yakhnis 1990, in prep.].

Outline

PART I: COMPUTATIONAL GAMES

1. INTRODUCTION

In contemporary computer science there is a thrust to develop principles of concurrent program development. We offer new tools, computational games and a calculus of strategies, which are a stengthened mathematical foundation and set of proof tools for dealing with this and other important issues in concurrency addressed in the pioneering papers of [Owicki and Gries 1976], [Lamport 1983], [Manna and Pnueli 1984, 1987], [Pnueli 1986], [deBakker and Zucker 1982], [Alpern and Schneider 1985, 1987], and [Chandra and Misra 1988]. We give a concise, easily manipulated, representation of instruction submission-execution mechanisms which we call "computational games". They yield a semantics for sequential and concurrent programs in which programs denote finite state strategies, the states of the associated automata being called the internal states of the program. This leads to a highly modular and verifiable concurrent programming paradigm based on construction rules for building finite state strategies, a calculus of strategies. This paradigm is easy to use at a high level. Our semantics of strategies is formulated to be applicable to all varieties of concurrent programs, whether they be terminating, infinitely prolonged, real time, or a mix of the above.

We begin by introducing games and strategies on trees (**Section 2**), define computational games (**Section 3**), give the connection with automata (**Section 4**), introduce computational state models (**Section 5**), introduce strategy models (**Section 6**), look at operations on strategies and continuity of strategy operations (**Section 7**), explain termination for Park's program (**Section 8**), and discuss the relation to other approaches and future work (**Section 9**). Other games we have seen in the literature are quite different from our computational games (**Section 9**). In games previously in the literature, processes play against processes or system plays against the environment, in contrast to our games, in which the submittor of program instructions, "Programmer", plays against the executor of those instructions, "Computer".

2. GAMES AND STRATEGIES

We will use the term "computational atom", or simply "atom" informally to mean an instruction or a test predicate. The usage of the term "atom" here is much like the usage of "atomic program" or "atomic test predicate" elsewhere; namely, one that we do not analyze further, an atom from which more complex programs are built. We separate program behavior in computation into two elements:

- Programmer, or the Submission Mechanism: a recipe for submitting packets of atoms to a pool of atoms intended for eventual execution, and
- Computer, or the Execution Mechanism: a recipe for executing atoms from the pool. See [Glushkov 65] for the origin of this approach.

A program P will be interpreted as denoting a strategy for selecting and submitting, at each stage of computation, a packet of atoms to the pool of as yet unexecuted atoms. Atoms in all submitted packets are put in a pool, and are potentially to be withdrawn from the pool and executed by Computer; but Computer may wait (possibly forever) before selecting and then executing any one of them. The recipe, or strategy, by which Programmer selects and submits packets of atoms to Computer should be based solely on the "finite past behavior" of Programmer and of Computer. Often in the literature a "finite past behavior" is defined as a finite sequence of "computation states". In many accounts of sequential computation, a "computation state" is determined by an assignment of values to program variables and a program counter value, an integer. But there is no generally accepted corresponding definition of program counter for concurrent computation. Our programs will denote strategies which come equipped with internal states of their own. These internal states reflect the intrinsic properties of strategies themselves rather than merely implementation properties according to Nerode equivalence (see **Section 4**). Such internal states can be used as implementation-independent concurrent analogues of sequential program counters, and clarify the meaning of sequential program counters as well.

So we do not define a "finite past behavior" as a sequence of computation states. Rather, we define it as a "play of a game", but then have to go to some pains to define what kind of game. The informal use above of the term "finite past behavior" is then formally replaced by the term "position in a game". The reason that we choose an abstract definition of computational game is that this provides a flexible modelling capability. It allows us to leave out information that we don't need, it allows us to include information that we do need which is not present in "computation states". We always include in our computational games information on which packets of atoms have been submitted for eventual execution by Programmer and which instructions have already been executed by Computer. We have included also information about unpredictable delays in execution of atoms, but not in numerical form. We could even introduce such information about numerical delays in real time systems, sequential or concurrent by simply choosing different definitions of what constitutes a position in one of our computational games. In our computational games, player 0 will be called "Programmer", player 1 will be called "Computer". This emphasizes that Programmer submits packets of atoms to Computer. But first,

What is a game tree?

It is very convenient to define the games that we need by using "game trees". If Σ is an alphabet, then let $\sum^*$ be the set of all finite words (finite sequences) from $\sum$, and let $\sum^\infty$ be the set of all infinite words (infinite sequences) from $\sum$, and finally let $\sum^* \cup \sum^\infty$ be the set of all words from $\sum$. If α is a finite word and β is a word, we write $\alpha \leq \beta$ if α is an initial segment of β, and call α a prefix of β and β an extension of α. For a finite word α and any word β, the concatenation $\alpha \cdot \beta$ of α in β is the sequence α followed by β. If A, B are sets of words, then $A \cdot B$ is the set of all concatenations $\alpha \cdot \beta$ such that $\alpha \in A$ and $\beta \in B$.

Definition. A tree T is a nonempty subset $T \subseteq \Sigma^*$ containing, with any word, all of its prefixes (including the empty, or null, word e). A position is an element of T. A word α in $\sum^* \cup \sum^\infty$ is a path of T if both

(1) every prefix of α is in T, and also
(2) α a finite word implies that no proper extension of α is in T.

Of course, any set A of words in $\sum^*$ is contained in a smallest subtree of $\sum^*$, obtained by adding into A all prefixes of words in A.

We reinterpret the paths of a tree as exactly the plays of a two-person game. Introduce the terminology of players 0 and 1. If Ω is one player, then $1 - \Omega$ denotes the other. The two players take alternate turns. Player 0 makes the zeroth move. Each move of either player consists of choosing a letter σ from Σ and appending it to the finite sequence handed over from the previous player at the end of the previous move, provided the resulting word is in T. Paths $\sigma_0 \cdots \sigma_n \ldots$ of T are interpreted as the plays of a game in which player 0 chooses σ_0 initially, and, afterwards, player Ω chooses σ_{n+1}

after player $1 - \Omega$ chooses σ_n. That is, player 0 makes even moves, player 1 makes odd moves. In a play, the finite prefixes of the play are defined to be the positions of that play, positions of even (odd) length are positions of player 0 (1) at which that player moves. If T is such a tree, let $Pos(\Omega, T)$ be the set of all positions of T at which player Ω moves. If T is known, we write $Pos(\Omega)$ instead of $Pos(\Omega, T)$. A tree T interpreted in this way we call a game tree.

- The paths of T are called the plays of T. They may be finite or infinite.
- The finite paths of T are called the terminal positions of the game.
- The set of all plays is denoted as Play(T).

Which player wins?

To specify a game based on a game tree T, we must specify a collection W of plays at which player Ω wins. This W is called the winning set for Ω. By convention every play is won either by player 0 or by player 1, we do not allow "draws". So (T, Ω, W) will describe a game.

Example. If W^c denotes the complement of W in the set of all plays Play(T), then the game $(T, 1-\Omega, W^c)$ has the same winning plays for the same players. This should be regarded as the same game as (T, Ω, W).

What is an end game?

Let $\Gamma = (T, \Omega, W)$ be a game. We introduce the customary notion of endgame following after a position. Namely, what happened up to that position is omitted.

(a) Suppose that p, q are positions in T, and that q is an extension of p. Then the residual position $x = q_p$ after p is defined as the x such that $q = p \cdot x$. If μ is a play and $p \leq \mu$, so $\mu = p \cdot \mu'$, then μ' is the residual play after p.

(b) Suppose p is a position and X is a subset of T. Then the residual set X_p is defined as the set of residuals after p of positions which are extension of p in X; in symbols, $X_p = [q_p : p \leq q \wedge q \in X]$.

(c) If p is an Ω-position, the residual set T_p is itself a game tree for a game (T_p, Ω, W_p), called the endgame after p, and W_p is the set of residual plays after p extending p in W. Note that the set of Ω-positions of the endgame tree T_p is the same as the set of residuals after p of the Ω-positions extending p, that is, $Pos(\Omega, T_p) = Pos(\Omega, T)_p$.

What is a strategy?

Let $\Gamma = (T, \Omega, W)$ be a game. Let X be a set of positions for player Ω. An Ω-strategy on X is a function f with domain a set of positions, defined on at least X, assigning to each position p in X a set of symbols $f(p)$ from $\sum$ such that

(*) for every position p in X and every symbol σ in $f(p), p \cdot \sigma$ is in T.

This definition allows $f(p)$ to be empty for some or all positions.

Here is another formulation of (*). Define the set p_T of children of position p. A child of p is an extension of p by one symbol, an extension which is also in T. Then condition * above can be written as $p \cdot f(p) \subseteq p_T$.

Example. The unrestricted (or vacuous) Ω-strategy $f = Vac(\Omega, T)$ is defined on $Pos(\Omega, T)$ by letting $f(p)$ be the set of all children of p. This is called unrestricted because it is no restriction on moves for Ω other than choosing a σ with $p\sigma$ on the game tree. (It is called a vacuous strategy in [A. and V. Yakhnis 1990] because it is a vacuous restriction.) We often write $Vac(\Omega)$. Note that $Vac(\Omega)$ is non-empty on the set of non-terminal positions of Ω.

Finally, we call f a strategy for player Ω, or equivalently an Ω-strategy, if f is defined on at least all positions for Ω.

Example. Here is the notion of an Ω-strategy induced in an end game. If f is an Ω-strategy in T, then the residual of f after p is the Ω-strategy f_p for the endgame (T_p, Ω, W_p) such that

$$f_p(q) = f(p \cdot q) \text{ for all } \Omega\text{-positions } q \text{ in } T_p.$$

Remark. Most of the literature going back to [Gale-Stewart 1953], including Büchi's fundamental papers, use deterministic strategies. [Gurevich and Harrington 1982] used nondeterministic strategies quite effectively. [Donald Martin 1985] mentions the possibility of using nondeterministic strategies, but himself uses only deterministic strategies in his famous Borel determinacy proof. He described nondeterministic strategies in terms of trees of consistent positions. The definition above is a modification of that of [Gurevich and Harrington 1982].

How to follow a strategy

For Ω to follow strategy f means that player Ω chooses moves in response to the Ω-position p only as allowed by the strategy, that is $\sigma \in f(p)$, as long as the strategy provides at least one choice, that is, as long as $f(p)$ is non-empty. But what positions result? Remember that, meanwhile, $1 - \Omega$ is free at $1 - \Omega$ positions to move to any child of p on the tree, which is $1 - \Omega's$ vacuous, or unrestricted strategy. We can specify the set U of possible positions resulting from $1 - \Omega$ making arbitrary moves, while Ω making moves according to f as a subtree of positions U associated with the Ω-strategy. U is defined as the smallest set U of positions such that

(a) U contains the null position.
(b) If p is an Ω-position in U, and σ is in $f(p)$, then $p \cdot \sigma$ is in U.
(c) If p is an $1 - \Omega$-position in U, then every child of p is in U.

We call the positions in U the positions consistent with f, writing $U = Con(f)$. We call a play in T consistent with f if all its positions are consistent with f; that is, if it is a path in the subtree U associated with the strategy.

Remark. Each subtree V of game tree T induces an Ω-strategy f on T. For any Ω-position p in T, let $f(p)$ be the set of all $p \cdot \sigma$ with $\sigma \in \sum$ and $p \cdot \sigma \in V$. This gives f trivial behavior for Ω-positions p in $T - V$, since then $f(p)$ is empty. Call an Ω-strategy f non-empty on set X of positions if f is defined on at least X and $f(p) \neq \Phi$ for all $p \in X$. The strategy f induced by a subtree V can only be non-empty on an $X \subseteq V$.

Now suppose the opponent $1 - \Omega$ is similarly restricted at $1 - \Omega$ moves to following a $1 - \Omega$ strategy g. Letting U play the role that T played before, we now get a subtree V of U consisting of all positions in U which can be reached in V by $1 - \Omega$ following g, that is the set of positions of U consistent with g. This we call the subtree $Con(f, g)$ of positions consistent with the pair of opposing strategies (f, g). Proceeding in the opposite order, restricting first by g, then by f, yields the same tree. This V can alternately be characterized by a), *b*) above, plus

(c') If p is a $1 - \Omega$-position in U, and σ is in $g(p)$, then $p \cdot \sigma$ is in U.

Every subtree V can be thought of as arising from a pair of opposing strategies f, g used at complementary positions of Ω, $1 - \Omega$. At a position p of Ω, $f(p) = [\sigma \in \sum : p \cdot \sigma \in V]$, at position p of $1 - \Omega$, $g(p) = [\sigma \in \sum : p \cdot \sigma \in V]$. Of course, if p is not in V, both are empty. Studying subtrees with the positions partioned into two classes and studying pairs of opposing strategies is the same thing.

Mutually perpetual strategies

Suppose that f is a strategy for Ω, and that g is a strategy for $1 - \Omega$. Imagine, as above, a game played by these opposing players according to these strategies, that is, on subtree V of positions consistent with (f, g) in tree T as defined above. In V it is perfectly possible for Ω, following f, to get stuck before the end of a play in T. That is, there may be an Ω-position p in V (that is, a result of the opponents playing according to (f, g)) with $f(p)$ empty, where p is not terminal in T and the game should be continued to extend the position to a play of T. So Ω cannot continue to follow the strategy further in the game; $1 - \Omega$ can get equally stuck because $g(p)$ is empty. We now give a name to the case when neither ever gets stuck before the end of a play of T.

Definition. Let T be a game tree. Let f be an Ω-strategy. Let g be a $(1 - \Omega)$-strategy. We say that f and g are mutually perpetual at p if whenever $q \geq p$ is not a terminal position in T and q is consistent with f starting at p and q is consistent with g starting at p, we have 1), 2).

1) If q is a position for Ω, then $f(q)$ is non-empty.
2) If q is a position for $1 - \Omega$, then $g(q)$ is non-empty.

Equivalently, this says that the plays (paths) of the subtree of positions extending p and consistent with (f, g) are all plays (paths) in T.

So positions where f, g are mutually perpetual are positions where f, g assign non-empty values enough to guarantee that, starting a game at p, the opponents always follow the strategies to produce finally plays in T.

Definition. The mutual perpetuity set $Q(f, g)$ of f and g is the set of all positions at which f, g are mutually perpetual. These are the positions p in T such that if the game is started from p, players Ω, $1 - \Omega$ following the respective strategies, then neither player ever gets stuck till the end of play.

Example. Consider the case when Ω-strategy f plays against an unrestrained opponent $1 - \Omega$ (the vacuous strategy). For Ω never to get stuck following f should mean that the pair $(f, Vac(1-\Omega))$ is perpetual. That is, call Ω-strategy f perpetual at p if f and the vacuous strategy are mutually perpetual at p; equivalently, if all plays (paths) in the subtree of positions extending p and consistent with f are also plays (paths) in T.

What is a winning strategy?

Let $(T,\ \Omega,\ W)$ be a game based on game tree T with winning set of plays W. Let f be an Ω-strategy. Let g be an $(1 - \Omega)$-strategy. We start by formalizing the idea of f winning against g starting at position p. The idea is simply that, starting in position p, if Ω follows f and $1 - \Omega$ follows g, and this produces a play (finite or infinite) of T, then that play is in the winning set. Formally,

Definition. A strategy f conditionally wins Γ at p against g if the winning set W contains every play μ of T such that $\mu \geq p$ and every prefix position of μ extending p is consistent with both f and g starting at p.

In this case, we do not regard strategy f as outright "winning" against strategy g starting at p if either f or g can get stuck at a non-terminal position q extending p due to $f(q)$ or $g(q)$ being empty and not supplying a move.

Definition. f wins Γ at p against g, if f and g are mutually perpetual at p, and f conditionally wins Γ at p against g.

This says that starting with p, if Ω plays according to f and $1 - \Omega$ plays according to g, neither gets stuck before producing a play in T, and this play will be in the winning set W.

Next, we specialize this concept to strategy f playing against the vacuous strategy for the opponent by giving the following definition.

Definition. f wins Γ at p if f wins Γ at p against the vacuous strategy. That is, if starting at p, no matter how an opponent chooses moves, Ω always has a move available from the strategy, until a play of T is produced, which is then in W.

Finally, game Γ is called determined if at least one of the players has a strategy that wins Γ at the empty position.

For convenience, in all game trees, we make an inessential change and identify the terminal positions (finite plays) by their last symbol. We therefore always suppose that there is a distinguished symbol " end " in $\sum$, and that a (finite) game position is terminal if and only if its last symbol is " end ".

Note. [Lachlan 1970] defined the notion of a "complete" deterministic strategy. This is, roughly, a strategy perpetual against the vacuous strategy. ("Mutual perpetuity" and "conditional win" were first introduced by [A. and V. Yakhnis 1989]. In [Gurevich and Harrington 1982], there are only strategies f with $f(p)$ non-empty for all p.)

Remark. Determinism and non-concurrency can be introduced and studied for strategies. Nonconcurrency for abstract strategies does not appear to have been mentioned before. Let T be a game tree. We call a strategy f for player Ω deterministic if:

for any position p of player Ω which is consistent with f,
the set $f(p)$ has at most one element.

That is, there are no real choices for player Ω. Player Ω has to "follow" player $1 - \Omega$. We call a strategy f for player Ω nonconcurrent if:

for any position p of $(1 - \Omega)$ with p consistent with f,
we have that p has at most one child.

That is, there are no real choices for player $1 - \Omega$ in responding to player Ω. Player $1 - \Omega$ has to "follow" player Ω.

The classes of nonconcurrent strategies and of deterministic strategies are overlapping, but neither is contained in the other. Our framework makes a clear distinction between the allowed nondeterminism and concurrency by associating them respectively with freedom of choice for players Ω and $1-\Omega$. In computational games as defined later, players 0 and 1 will be called "Programmer" and "Computer": and the definition of nonconcurrency will be slightly modified to allow Computer to be lazy-Computer can " wait " as well as have at most one choice of a real move, and still be called "nonconcurrent".

Scott and Boolean topologies

We associate with pairs $(f,\ g)$ of opposing strategies for Ω, $1-\Omega$ on game tree T Scott and Boolean topologies. Because this does not give results specific to programs, we do not go into this in depth. But it is a very useful point of view for placing the current work in the proper context of contemporary theoretical computer science and mathematics, and to make it easy to apply descriptive set theory. For simplicity, suppose that we are given a game tree T based on a countable total alphabet, and a pair $(f,\ g)$ of opposing strategies for Ω, $1 - \Omega$.

Here is the Scott topology connection. Let $V = Con(f,\ g)$ be the tree of all positions consistent with $(f,\ g)$, Let $\mathcal{T}(V)$, be the set of all the plays plus all the positions of V. Then $\mathcal{T}(V)$ is partially ordered by " $\leq$ " (extension), and is a CPO (Complete Partial Order). In this CPO, a play of V is exactly a maximal element, a set of all positions of a play μ of V is exactly a maximal linearly ordered subset of $\mathcal{T}(V)$. Then the Scott (compact T_0) topology declares that for each position p of $\mathcal{T}(V)$, the set $X(p) = [q \in \mathcal{T}(V) : q \geq p]$ is open. An arbitrary open set is a union of finite intersections of sets of this form. A position represents a past history of activity of each player, moving up in this partial order represents more and knowledge about that history. The inclusion relation of V into T is a 1-1 continuous map, but some plays in V may merely map into positions in T, not plays in T.

Let Play$(f,\ g)$ be the set of all plays of V. If the alphabet is finite, so the tree is finitely branching, Then Play$(f,\ g)$ is a separable Boolean space, that is, a totally disconnected compact Hausdorff space with a countable base. In case $(f,\ g)$ are mutually perpetual in T, so that plays of V map to plays of T, the inclusion map of V into T induces a 1-1 continuous map of Play$(f,\ g)$ to Play(T), where Play$(T) = Play(Vac(\Omega),\ Vac(1-\Omega))$. Since Play$(f,\ g)$ is compact and Play(T), is Hausdorff, $(f,\ g)$ mutually perpetual implies that the map of Play$(f,\ g)$ to Play(T) is a homeomorphism with closed image. To give a tree $V = Con(f,\ g)$ for a pair of opposing strategies and to give Play$(f,\ g)$ is the same thing; the plays have as their initial segments precisely all the positions which constitute V. So Play$(f,\ g)$ has all the information, and as a separable Boolean space can be investigated by known methods.

If we allow the alphabet to be countably infinite, Play$(f,\ g)$ is a complete metric space under the usual metric on countably branching trees.

3. Computational Games

Rules of the game

As seen above, perfectly general games have "clean" definitions of the concepts of position, plays, strategies, and wins. But in practice, when a particular game is defined by a game tree, the definitions of these same notions has a lot of necessary detail. Since we use computational games to represent programming, and programming has a lot of necessary detail, this is not surprising. We try to suppress that detail so far as possible, since we are not actually programming. A computational game should include at least:

- rules of the game, which tell which plays are legal,
- a winning condition.

Further, in practice, we restrict winning strategies to lie in a prescribed set C of strategies, those denoted by programs in a definite programming language L.

We give a very general definition of computational game. Our intention is that at even positions, Programmer moves by submitting a packet of atoms to a pool of all atoms which have been submitted in previous packets but have not yet been executed by Computer. At odd positions, Computer moves by either doing nothing (a " wait "); or alternately, by picking an executable atom from the pool, and executing that atom. In that case, the atom is removed from the aformentioned pool.

We suppose given two primitive notions, the non-empty set of instructions I and the non-empty set of test predicates TP. We define atoms as either instructions or pairs $(\varphi,\ x)$, where φ is a test predicate and x is t or f. That is, the set of atoms is $I \cup (TP \times \{t,\ f\})$. First, we discuss packets and symbols. We call a set A of atoms a packet provided that whenever A contains a $(\varphi,\ x)$ with φ a predicate, then A also contains $(\varphi,\ \neg x)$, where $\neg f = t$, $\neg t = f$. Among these packets is the empty packet " skip ". Let Programmer's set of symbols $\sum^P$ consist of all packets of atoms plus an extra symbol " end ". This " end " will be used solely as the last symbol of finite plays. Let the Computer's set of symbols $\sum^C$ consist of all atoms plus an extra symbol " wait ". This " wait " will be used as a Computer move at positions where the Computer does nothing. The full alphabet for the computational game is thus $\sum = \sum^P \cup \sum^C$.

Allowed positions. Let us refer to a choice of a set of instructions and a set of test predicates as a vocabulary. Let A be the set of atoms constructed from that vocabulary. We define the tree of allowed positions $T(A)$, based on this vocabulary A. An allowed position in $T(A)$ is a finite sequence ν_0, ..., ν_n of elements of the alphabet $\sum$, based on A, such that

- each even entry ν_{2i} is in $\sum^P$
- each odd entry ν_{2i+1} is in $\sum^C$
- only the last entry ν_n can be " end "

To repeat the first two, each ν_{2i} is either " end " or a packet (possibly empty) of atoms, and each ν_{2i+1} is either "wait" or an atom; and each atom is an instruction or a test (φ, f) or a test (φ, t). All computational game trees for the rest of the paper are subtrees of the tree $T(A)$ of allowed positions based on a vocabulary A, even length positions being called Programmer positions and odd length positions being called Computer positions.

We introduce the abstract notion of "executability of atoms at a Computer position". For each Computer position p assume as given a subset $E(p)$ of the set of atoms, called the set of "executable atoms at position p". The only assumption on these sets is that, for a test predicate φ, at most one of the two atoms (φ, t), (φ, f) is in the set- that is, executable at position p. Although in this section this is a purely abstract notion, in a computational state model (**Section 5**), there is a canonical definition of executability, inductively defined from the "execution functions" of atoms in such computational state models.

Certain subtrees of the tree of allowed positions will be called computational game trees. Here are the three requirements for the positions of a computational game tree T. We express these as "rules of the game". We assume that, before the game starts, and at the empty position e, the pool of instructions referred to in the first two rules is empty, and no atom has been executed. The pool referred to is, intuitively, the pool of all atoms in all packets that have been previously submitted but not yet executed. (In this paper we assume that all packets are finite. However, it is useful occasionally to allow infinite packets; we may wish to submit a "schema" of instructions whose instances are infinite in number.)

Rule 1. At an odd non-terminal position, Computer may choose as move either " wait ", or any single atom a from the pool, executable at that position. We remove a from the pool. In addition, if a is a test (φ, x), then $(\varphi, \neg x)$ is also removed from the pool. We say that a has been executed at the position.

Rule 2. At a non-terminal even position, Programmer may choose as move and also submit to the pool either " end ", or any packet of atoms disjoint from the pool. All atoms from that packet are added to the pool.

Rule 3. A position is terminal (a finite play) if and only if its last move is Programmer submitting " end ".

Remark. If all references to executability and execution are omitted in rule

1, we get the notion of the positions of a pure computational game tree. This is equivalent to declaring every atom executable at every Computer position, while dropping the requirement that at most one of (φ, t), (φ, f) is executable at p. So, in a computational game tree with a restricted notion of executability, at most one of (φ, t), (φ, f) in the pool can be used by Computer to create an extension of the position. But in a pure computational game tree, (φ, t), (φ, f) can both definitely be used to create two different extensions. For pure computational game trees, these three rules consitute an inductive definition of the set of all positions of the tree, by induction on the length of position. In playing a computational game, executability is used exclusively by Computer in making his moves, and is invisible to Programmer since positions in game trees do not mention executability.

To summarize: if we specify the vocabulary A of instructions and test predicates, rules 1), 2) 3) with reference to executability omitted define by induction the set of all allowed positions defining a computational game tree $T(A)$, the pure computational game tree based on this vocabulary. If we give a definition of executability by specifying $E(p)$ for all Computer positions p, the three rules then define inductively a subtree of executable positions of this pure computational tree. Such a family $[E(p) : p$ is a computer position$]$ will arise from each computational state model (**Section 5**) based on the same vocabulary. So each computational state model using the given vocabulary is represented by a computational game subtree of executable positions in the same pure computational tree for that vocabulary.

Remark. To make our model simple we have restricted Computer, unlike Programmer, to execute at most one atom per move. This is inessential. The Computer could execute a whole set of atoms from the pool at each computer move, instead of one atom, as long as he does not execute a test (φ, t) along with the test (φ, f), and most of this paper would be unchanged. A "packet execution" of a whole packet of atoms at one Computer move could be simulated by the set of all the positions in the computational game tree arising from executing all atoms in the packet in every possible order. Indeed, the "single instruction" moves for Computer are an important device for accurate modelling of interleaving for concurrent programs [Ben-Ari 1982], [Lamport, 1983].

A strategy f for Programmer (player 0) is any strategy with domain including the set of even positions p, and then $f(p)$ will be either a set of packets of atoms, or consist of " end " alone. Any member of $f(p)$ may be submitted by Programmer to Computer at position p. We require that if " end " is in $f(p)$, then no packet is in $f(p)$, so that there is no choice but to stop. A strategy g for Computer (player 1) is any strategy with domain including the set of odd positions, and then $g(p)$ will be " wait " or a single atom. A program specification will always be defined as a set of plays of the computational game.

What does it mean for a strategy for Programmer to win the game, or, equivalently, to satisfy the specification? Suppose that f is Programmer's strategy. For f to be a winning strategy means that no matter how Computer selects its moves, if Programmer follows strategy f, then the final play (finite or infinite) is in the prespecified winning set of plays.

Remark. Even if we prove mathematically that there exists a winning strategy for Programmer, this does not imply that there exists a strategy representing a real program. The strategy asserted to exist can be an arbitrary non-constructive function. For applications in recursion theory, that is, for proving results allowing computation with unbounded resources, we need only extract effective strategies as in [Lachlan 1970]. But for writing real programs we usually need to extract an explicit "finite state winning strategy" satisfying the specification, which then must be faithfully represented by a program in L denoting that strategy. Such a program is built up from finite state atomic programs by program constructs of L corresponding to operations on finite state strategies.

Remark. The present paper concentrates on the idea that programs denote Programmer strategies, with the Computer given an unrestricted strategy in a pure computational game tree. One can also allow Computer to denote a strategy, not necessarily unrestricted. Then we are working with a pair of opposing strategies $(f^P,\ f^C)$, where f^P is a Programmer strategy, and f^C is a Computer strategy. Such f^C can reflect the internal structure of Computer, its architecture and operating system. But for a fixed Computer strategy f^C, if V is the set of consistent positions for the pair $(Vac(P),\ f^C)$, we can analyze $(f^P,\ f^C)$ by obliterating f^C and working within the tree V of consistent moves for the pair $(f^P|V,\ Vac(C))$. That amounts to using Programmer strategies while always giving Computer the unrestricted strategy, which is what we do here.

Convention. From now on, every Programmer strategy is regarded as assigning the empty set of moves to every position outside its subtree of consistent positions, the Computer strategy is the unrestricted strategy in this subtree.

Where is non-determinism? Where is concurrency?

If $f : Pos(0) \rightarrow \mathcal{P}(\sum)$ is a strategy for Programmer, we regard non-determinism as a reflection of the fact that for each position p of Programmer (that is, for p a position of player 0), such an $f(p)$ can have more than one element. We regard concurrency as a reflection of the fact that at a position p of Computer, p has more than one move available besides " wait ", a choice of execution of one of possibly several atoms from the pool. Our model will divide the responsibility for the existence of multiple executions of a single program between two agents; one is the Programmer, or equivalently the atom submission mechanism, which is responsible for non-determinism; the other the Computer, or equivalently the atom execution mechanism, which is responsible for concurrency.

Suppose that T is a computational game tree, and $\mathcal{S}$ is a Programmer strategy.

I. Let the tree T^C of "Solely Computer" positions consist of all sequences p' of Computer moves such that there is a position p in T which has p' as the subsequence of all its Computer moves. If $\mathcal{S}$ is deterministic, then $\mathcal{S}$ induces a "sequential non-deterministic" strategy $\mathcal{S}^C$ on T^C. Strategy $\mathcal{S}^C$ reflects perfectly the concurrency choices for Computer, but obiterates any trace of Programmer since it is on a "single player" tree. If Programmer is absent as an independent player, concurrent behavior of $\mathcal{S}$ on T collapses to nondeterministic sequential behavior of $\mathcal{S}^C$ on T^C. Current semantics of concurrent programs using as execution sequences solely Computer moves, such as the record of executions of instructions or machine or computational states, work implicitly on T^C and encounter difficulties due to not being able to express relations to Programmer.

II. Let the tree T^P of "Solely Programmer" positions consist of all sequences p' of Programmer moves such that there is a position p in T which has p' as the subsequence of all its Programmer moves. If $\mathcal{S}$ is nonconcurrent, then $\mathcal{S}$ induces a "non deterministic sequential" strategy $\mathcal{S}^P$ on T^P. Strategy $\mathcal{S}^P$

reflects perfectly the nondeterminacy choices for Programmer, but obliterates any trace of Computer So if Computer is absent as an independent player, nonconcurrent behavior on T collapses to nondeterministic sequential behavior on the single-player tree T^P. Semantics of programs, such as the flowchart approach, using as execution sequences solely the record of submissions of instructions and "waits" to Computer implictly work on T^P and have difficulties due to not being able to express relations to Computer.

We believe that many definitions needed for concurrency, including those connected with fairness, can only be fully defined and proved on two-player computational game trees, with independent roles for Programmer and Computer.

4. Automata and Strategies

What is an automaton?

For automata, see [Hopcroft and Ullman 1979]. Here are brief definitions. A non-deterministic automaton $\mathcal{A}$ is a 5-tuple $(\sum, S, M, S_{in}, Q)$ such that $\sum$ is a non-empty set (the alphabet), S is a nonempty set (the states), M is a function $M : S \times \Sigma \to \mathcal{P}(S)$, (the state transition function), S_{in} is a non-empty subset of S (the initial states), and Q is a subset of S (the, possibly empty, set of terminal states). In case $M(\sigma, s)$ consists of one element s', we abuse notation and write $M(\sigma, s) = s'$. If every $M(\sigma, s)$ consists of one element, we call $\mathcal{A}$ deterministic, and then usually we arrange it so that $S_{in} = \{s_{in}\}$ consists of one initial state.

There is no assumed finiteness or constructiveness in this general definition. That is, S, S_{in}, Q are arbitrary sets, M is an arbitrary function. Of course, if S and $\sum$ are finite, we then have a nondeterministic finite automaton $\mathcal{A}$ in the standard sense. Very often such automata are used as devices for accepting (finite) input words from $\sum^*$ by the automaton being in a state in Q at the end of processing an input word from $\sum^*$. In this paper, such acceptance states are used only to identify termination of finite plays, while we also allow infinite plays.

A (positional) run of $\mathcal{A}$ on word $\sigma_0 \cdots \sigma_n ...$ from $\sum^* \cup \sum^\infty$ is a word $s_0 \cdots s_n ...$ from $S^* \cup S^\infty$ such that $s_0 \in S_{in}$ and for all i with σ_i defined, $s_{i+1} \in M(s_i, \sigma_i)$. So as a play $\sigma_0 \cdots \sigma_n ...$ develops its positions $\sigma_0 \cdots \sigma_n$, the automaton with this as input responds by developing non-deterministically a run $s_0 \cdots s_{n+1}$, which yields possibly an infinite run on an infinite input word $\sigma_0 \cdots \sigma_n \ldots$. The automaton may have many runs on a given position or play.

An automaton $\mathcal{A}$, run on positions from a game tree T, should obviously be called deterministic if $\mathcal{A}$ has at most one run for each input word from T.

To describe strategies, we use automata $\mathcal{A}$ equipped with an additional output function $\check{f} : S \to \mathcal{P}(\sum)$ producing from any state s a possibly empty subset $\check{f}(s)$ of $\sum$. An automaton with nondeterministic output function is a pair $(\mathcal{A}, \check{f})$ such that $\mathcal{A}$ is a non-deterministic automaton and $\check{f} : S \to \mathcal{P}(\sum)$. How does $(\mathcal{A}, \check{f})$ operate?

What is a state strategy?

Suppose we are given player Ω on game tree T. We introduce the class of "state strategies" for Ω, given by automata with output function on the same alphabet as the tree T, Strategies on T, as we defined them in the last section, base decisions on moves solely on position in T. To the contrary, state Ω-strategies will choose Ω-moves on additional information stored outside the position, namely stored in the current state of a non-deterministic automaton. The automaton state can reflect global features of the tree T not reflected by single positions in T. Here is how a state strategy is intended to guide Ω using $(\mathcal{A}, \check{f})$, a nondeteministic output automaton $\mathcal{A} = (\sum, S, M, S_{in}, Q)$ and a nondeterministic output function $\check{f}$.

(1) Before the play begins, Ω chooses an arbitrary element s_0 from S_{in}, designating s_0 as the state at the initial position.
(2) At any turn of Ω at a position $p\sigma$, Ω uses the state s at p to compute $M(s, \sigma)$, and chooses an element of $M(s, \sigma)$ as the state at position $p\sigma$.
(3) At an Ω-position p in state s at p, Ω chooses any $\sigma \in \check{f}(s)$ with $p\sigma$ on T.

This is a bit austere. Here is a more anthropomorphic explanation. Think of the automaton $(\mathcal{A}, \check{f})$ for an Ω-state strategy as an independent observer, reacting only to the successive symbols moves in the developing play. The automaton "wears his heart upon his sleeve" in that he advertises his current state s and the current value of $\check{f}(s)$, but never has any idea who is making use of that advertisement; in particular, he has no idea, ever, whose turn it is. But player Ω, at his turns, notes the value $\check{f}(s)$ advertised by the automaton and restricts his move to a σ from $\check{f}(s)$. To the contrary, $1-\Omega$, at his turns, simply ignores $\check{f}(s)$, and, free of restrictions, makes any legal move. How does the automaton change state? The automaton, reacting only to the succession of symbols which are the moves, observes the symbol moved, and changes state accordingly. (If, in addition, we wanted $\mathcal{A}$ to convey whose turn it is, we would double the number of states by using $S \times \{0, 1\}$, with $(s, 0)$ a state for Ω-positions, and $(s, 1)$ a state for $1-\Omega$-positions. This is natural, but is unnecessary for our purpose here.)

Terminal states and " end ". A terminal state for the automaton is, by definition, a state in Q. We require that Q consist of all states s such that $\check{f}(s) = \{ \text{ end } \}$. This and our earlier definition of permitted Programmer's strategies imply that if s is not in Q, then $\check{f}$ does not contain " end ". After submitting an " end ", Programmer does not need any further value of $\check{f}(s)$, so it is not necessary to define the transition function M at (s, σ) for which $s \in Q$.

Auxiliary positions. By putting enough information into states, we usually define $\check{f}$ so that $p\sigma$ is on T for all $\sigma \in \check{f}(s)$, where s is any possible current state at p. State strategies, as described informally above, limit their use of information about the position in T to information coded in current state at that position, forgetting the past history of the position except as it is reflected in the current state. If we wish to represent this behavior as a strategy based on position alone in the sense of the last section, we can introduce an alphabet $S \times (\sum \cup \{\tau\})$, consisting of pairs (s, σ) consisting of a state s of the automaton and a symbol σ of $\sum$ or an extra symbol τ. Define the tree of auxiliary positions as follows. It uses the initial states and transition table, but not the output function.

(1) For any s_0 of S_{in}, (s_0, τ) is an auxiliary position and the current state is s_0.

(2) If $(s_0, \tau) \cdot (s_1, \sigma_1) \cdots (s_n, \sigma_n)$ is an auxiliary position, and $\sigma_1 \cdots \sigma_n \cdot \sigma_{n+1}$ is a position of T and $s_{n+1} \in M(s_n, \sigma_{n+1})$, then

$$(s_0, \tau) \cdot (s_1, \sigma_1) \cdots (s_n, \sigma_n) \cdot (s_{n+1}, \sigma_{n+1})$$

is an auxiliary position.

Thus, each run of the nondeterministic automaton on a tree position in T is combined with that position to become a position in an auxiliary tree.

But the auxiliary tree is only a tree, it is not a strategy. There is a natural projection of the auxiliary tree onto a subtree of T. Namely, any $(s_0, \tau)\cdot(s_1, \sigma_1)\cdots(s_n, \sigma_n)$ projects to $\sigma_1 \cdots \sigma_n$. But since any $M(s, \sigma)$ may be empty, the projection can be much smaller than T because appropriate runs do not exist. To put this another way, for any position of T, there will be a largest prefix (possibly empty) which is in the projection, which may be smaller than the position. On the other hand, if the automaton has $M(s, \sigma)$ always non-empty, the projection is all of T. Infinite plays in the auxiliary tree project to infinite plays in T. But there may be infinite plays in T which cannot be lifted to the auxiliary tree even though each finite position of the play can be lifted. (But if every $M(s, \sigma)$ is a finite set, *König's* Lemma implies that every infinite path in T can be lifted, provided that every finite prefix can be lifted.)

Warning: in contrast to T, because of introducing τ, on the auxiliary tree the positions of player 0 are odd, the positions of player 1 are even.

We use the output function $\check{f}$ to get a (positional) Ω-strategy on this auxiliary tree. Namely, if $(s_0, \tau)\cdot(s_1, \sigma_1)\cdots(s_n, \sigma_n)$ is any Ω-position, define strategy f mapping the Ω-positions of the auxiliary tree to $\mathcal{P}(S\times(\sum\cup\{\tau\}))$ by

$$f((s_0, \tau)\cdot(s_1, \sigma_1)\cdots(s_n, \sigma_n))$$
$$= [(s_{n+1}, \sigma_{n+1}) : s_{n+1} \in M(s_n, \sigma_{n+1}) \wedge \sigma_{n+1} \in \check{f}(s_{n+1})]$$

The strategy above on the auxiliary tree is based on the last move of the position, ignoring the earlier part of the position. This strategy can be used to guide Ω in playing at Ω-positions on T, as described intuitively before, relative to a corresponding run of the automaton as parameter. This proof is in the spirit of Martin's covering trees [Martin 1985]. If the automaton used is a finite state automaton, we call this a finite state Ω-strategy. We use primarily finite state strategies induced by deterministic finite automata with a nondeterministic output function. This is a wide enough class to serve as denotations for all concurrent programs in the current standard concurrent programming languages. The unrestricted strategies for both Programmer and Computer that define the game tree T itself are of this kind. All non-determinacy is represented though the choices available to Programmer implicit in the nondeterministic output function $\check{f}$. All concurrency is represented through the choices of atoms available to Computer at Computer's moves.

Remark. We can investigate strategies induced by other classes of automata than finite automata, such as recursive, polynomial time, NP, exponential, pushdown. See [Condon 1989] for applicable methods. This would be useful for applications to strategic games in economics. This does not appear to have been done.

What is a Büchi Strategy?

Deterministic strategies can be thought of as denoting state strategies governed by a deterministic automaton $\mathcal{B} = (\sum^C,\ S_B,\ M_B,\ s_{inB},\ Q_B)$, where

$$M_B : S_B \times \sum^{C} \to S_B,\ Q_B \subseteq S_B,\ s_{inB} \in S_B,$$

together with an output function $\check{f}_B : S_B \to \sum^P$.

This $\check{f}_B$ produces Programmer's move $\check{f}_B(s)$ by acting on a state $s \in S_B$, a state which is found by letting the automaton run on the sequence of past Computer moves. Strategies equipped with a deterministic automaton used in this way are called Büchi strategies. Suppose that a Büchi strategy is given with automaton $\mathcal{B} = (\sum^C,\ S_B,\ M_B,\ s_{inB},\ Q_B)$ and output function $\check{f}_B$. Then one can construct an automaton $\mathcal{A} = (\sum,\ S,\ M,\ s_{in},\ Q)$, $\sum = \sum^P \cup \sum^C$ and output function $\check{f}$ inducing the same (positional) strategy as the Büchi automaton. We simply let $S = S_B$ and define $M : S \times \sum \to S$ by

$$M(s,\ \sigma) = M_B(s,\ \sigma) \quad \text{for } \sigma \in \sum^{C},$$
$$M(s,\ \sigma) = s \quad \text{for } \sigma \in \check{f}_B(s),$$

define $\check{f}$ by $\check{f}(s) = \check{f}_B(s)$, $Q = Q_B$, $s_{in} = s_{inB}$.
There is also a simple converse.

Proposition. *If $F = (\mathcal{A},\ \check{f})$ is a state Ω-strategy induced by a deterministic automaton $\mathcal{A}$ with deterministic output function $\check{f}$, its behavior on an Ω-position p is determined by the sequence of previous moves of $1-\Omega$, that is, by a Büchi state strategy.*

Proof. By induction on the length of Ω-positions in T.

Remark. Automata were originally used as devices for accepting sets of finite words in an alphabet. Büchi introduced the use of deterministic automata with deterministic output functions on infinite words to guide strategies, on Computer moves. [Alpern and Schneider, 1985], [Vardi and Wolper, 1986], [Vardi, 1987] use Büchi automata in this class to recognize infinite execution sequences, as do others in the references. We use the larger class of deterministic automata with non-deterministic output functions on game trees, which have very strong closure properties, and seem to be the largest class needed for concurrent programming.

Nerode equivalence

It may seem infelicitous that we first defined strategies f on positions, then independently defined state strategies using deterministic automata with non-deterministic output functions. We came to this latter class experimentally because finite state strategies in this class were precisely the strategies that showed up naturally when we wrote out denotations for concurrent programs. But afterwards we observed, using the appropriate Nerode equivalence, that every (positional) Ω-strategy f arises in a natural way from a state Ω-strategy. In fact, continuous maps induced by state automata were the basis of [Nerode, 1958], see also [Nerode 1957, 1959].

So arbitrary (positional) strategies are precisely the strategies that result from omitting mention of the states for a state strategy based on a deterministic state automaton with nondeterministic output function. This is why automata-theoretic ideas are relevant to game theory, and game theory to automata.

The state strategy, which we are able to associate canonically with an arbitrary (positional) strategy in general, is an automaton with countably many states, even when the alphabet $\sum$ is finite. The automaton is no more constructive than the strategy it was obtained from, but of course we will be primarily interested in the finite state case where this is not relevant. To state the equivalence result precisely, we need tree runs, rather than the previous (position) runs. Such runs were used heavily in [Rabin 1969], and therefore in all the literature connected with S2S.

Define a tree run of automaton $\mathcal{A} = (\sum, S, M, s_{in}, Q)$ as any map $r : \sum^* \to S$ such that

(1) $r(e) = s_{in}$, where s_{in} is the initial state of $\mathcal{A}$.
(2) whenever $\alpha \in \sum^*$ and $\sigma \in \sum$, we have that $r(\alpha \cdot \sigma) \in M(r(\alpha), \sigma)$.

Note that, by an induction on length of positions, a deterministic automaton with a total transition function has exactly one tree run.

Suppose we are given a state Ω-strategy given by $(\mathcal{A}, \check{f})$, with deterministic transition function M and nondeterministic output function $\check{f}$. Then

Proposition. *For any state Ω-strategy $(\mathcal{A}, \check{f})$ with M a deterministic transition function and $\check{f}$ a nondeterministic output function, there is a corresponding Ω-strategy f defined on T such that $f(p) = \check{f}(r(p))$, where r is the tree run of $\mathcal{A}$.*

There is a converse.

Proposition. *For each Ω-strategy f on T, there exists a state Ω-strategy on T, given by an automaton $(\mathcal{A}, \check{f})$ with deterministic total transition function M and nondeterministic output function $\check{f}$ such that if r is the run of $\mathcal{A}$, for all $p \in T$, $f(p) = \check{f}(r(p))$.*

To prove this proposition, we construct from Ω-strategy f a suitable deterministic automaton $\mathcal{A} = (\sum, S, M, s_{in}, Q)$ such that $M : S \times \sum \to S$ is a total function and a nondeterministic output function $\check{f} : S \to \mathcal{P}(\sum)$. Let V be the subtree of all positions of T consistent with f as defined earlier. For α, $\beta \in \sum^*$, define an equivalence relation, called for the last thirty years the Nerode equivalence relation [Nerode 1958], by $\alpha \sim \beta$ if and only if for all $\gamma \in \sum^*$,

$$\alpha \cdot \gamma \in V \text{ iff } \quad \beta \cdot \gamma \in V.$$

Of course, $\sim$ *is* right-invariant, namely $\alpha \sim \beta$ implies $\alpha \cdot \gamma \sim \beta \cdot \gamma$.

Let the desired set of states S be the set of all equivalence classes $[[\alpha]]$ of words in $\sum^*$. Since $\sim$ *is* right invariant, we may define the desired transition function $M : S \times \sum \to S$ by $M([[\alpha]], \sigma) = [[\alpha \cdot \sigma]]$ for $s \in S$, $\sigma \in \sum$. The desired initial state s_{in} is the equivalence class of the empty word. The set Q' of acceptance states for the tree V is the set of $[[\alpha]]$ with $\alpha \in V$. Finally, define $\check{f} : S \to \mathcal{P}(\sum)$ by $\check{f}(s) = [\sigma \in \sum : M(s, \sigma) \in Q']$. Note that for $s \in S - Q'$, we have that $\check{f}(s)$ is empty. The set of terminal states is $Q = [s \in Q' : \check{f}(s) = \{ \text{ end } \}]$.

We call the automaton above the intrinsic automaton associated with the strategy, its states we call the intrinsic states of the strategy. By the usual Nerode equivalence argument, this automaton is a minimal state realization of the desired behavior for strategy f. This means there is a finite state automaton with the above behavior for f if and only if the Nerode equivalence relation has a finite number of equivalence classes. Finite state strategies are exactly the denotations we generally intend for programs. When one writes a program, it will denote an intended finite state strategy with explicit associated automaton. But this may not be the intrinsic automaton of the strategy. This is because in actual programming we seldom reduce to minimal state automata. But we may wish to minimize resource use. The standard operations we introduce on state strategies equipped with automata do not generally lead from minimal state automata to minimal state automata.

Example. For deterministic strategies in the sense defined earlier the intrinsic states give a deterministic output function.

Remark. Since V is closed under prefixes, its complement forms an equivalence class which we may call $\perp$, or "sink", and $S = Q' \cup \{\perp\}$. A (positional) run passing through $\perp$ *stays* there.

5. Computational States

Up to this point we have modelled only the submission-execution mechanism for atoms, but not the dynamics of execution. We have spoken of the Computer as "executing an atom a" whenever Computer picks an atom from the pool. But execution might as well have meant that the atom was dumped into a black hole and forgotten. We model as "computational states", the states roughly of all the storage variables used in a program. The definition we use here is an abstract form of shared memory models; programs "pass data" to one another through a global computational state. Other models of computation, such as communicating channels, can be coded into shared memory models, but this introduces an artificiality. So we leave modelling other forms of communication by computational games to later papers, where we will have space to deal with a variety of Computer strategies other than the unrestricted strategies discussed in this paper.

Suppose a set of instructions and test predicates is given. A computational state model Π consists of a set Π (the set of computational states) and a denotation for each instruction or test predicate as follows.

I. Each instruction i denotes a partial function $\pi \to \pi' = i\pi$, whose domain is a subset of Π, and whose values are in Π. This function is called the execution function for atom i. Then instruction i is said to be executable at state π if $i\pi$ is defined.

II. Each test predicate φ denotes a partial function $\pi \to \varphi\pi \in \{t, f\}$ defined on a subset of Π. This function is called the execution function for φ. Then φ is called executable at state π if $\varphi\pi$ is defined. Also, an atom (φ, x) is called executable at state π if $\varphi\pi$ is defined and $x = \varphi\pi$.

Remark. No change, except for notation, is needed to use non-deterministic denotations $i\pi \subseteq \Pi$ for instructions i. But atomic constructs denote deterministic state transitions in contemporary programming languages, so we do not bother.

Suppose that A is the set of atoms based on a given vocabulary of instructions and test predicates. For each state π_0, define a subtree $T(A, \pi_0)$ of $T(A)$, called the computational state subtree based on π_0, as follows. This will not be a pure game tree, because the freedom of branching at tests is eliminated. Relative to the choice of initial state π_0, we define, by induction on length of positions

- the positions of $T(A, \pi_0)$,
- the computational state at any position,
- the pool at any position,
- the set $E(p)$ of executable atoms at any Computer position,
- the atom executed at any Computer position, if any.

I. The empty position $e \in T(A, \pi_0)$. The computational state at the empty position p (before the zeroth move of programmer) is π_0. The pool U at p is empty.

II. Suppose that p is a position of Programmer in $T(A, \pi_0)$. Suppose that $p' = p \cdot \sigma$ and σ is a packet of atoms, none in the pool U at p. Suppose that the computational state at p is π. Then $p' \in T(A, \pi_0)$, the computational state at p' is π, and the pool U' at p' is $U' = U \cup \sigma$.

III. Suppose that p is a position of Computer in $T(A, \pi_0)$, $p' = p \cdot \sigma$ and σ is " wait ". Suppose that the computational state at p is π. Then p' is in $T(A, \pi_0)$, the computational state π' at p' is π, and the pool U' at p' is the same as the pool U at p. No atom is executed at p'.

IV. Suppose that p is a position of Computer in $T(A, \pi_0)$, $p' = p \cdot \sigma$, and that σ is an instruction i. Suppose that the computational state at p is π, the pool at p is U, and that σ is executable at p (that is, $i\pi$ is defined), and σ is in the pool U. Then p' is in $T(A, \pi_0)$, the computational state π' at p' is $i\pi$, the pool U' at p' is $U - \{i\}$, and i is executed at p'.

V. Suppose that p is a position of Computer in $T(A, \pi_0)$, $p' = p \cdot \sigma$, $\sigma = (\varphi, x)$, (φ, x) is executable at p; that is, $\varphi(\pi)$ is defined and $\varphi(\pi) = x$), and (φ, x) is in the pool at p. Suppose the computational state at p is π. Then p' is in $T(A, \pi_0)$, the computational state at p' is π, the pool U' at p' is $U - [(\varphi, x), (\varphi, \neg x)]$, and (φ, x) is executed at p.

VI. Suppose that p is a position of Programmer in $T(A, \pi_0)$, $p' = p \cdot \sigma$, $\sigma =$ " end ". Suppose the computational state at p is π. Then p' is in $T(A, \pi_0)$, the computational state at p' is π, the pool at p' is the pool at p, and no atom is executed at p.

Example. We can define, for each Computer position p, the set $E(p)$ of atoms executable at p. If p is in $T(A, \pi_0)$ and π is the computational state at p, then let $E(p)$ be the set of all atoms executable at state π. Otherwise, let $E(p)$ be empty. It is routine to show that $[E(p) : p$ is a computer position$]$, together with rules 1), 2), 3) for computational games (**Section 3**) define the same tree $T(A, \pi_0)$. Then the set $[E(p) : p$ is a Computer position$]$ has the following property.

(**) If p is a position of Computer and σ is a legal move at p and σ is not an instruction, and σ' is a legal move at $p\sigma$, then $E(p) = E(p \cdot \sigma \cdot \sigma')$.

Is the converse true? If we attach abstractly to each computer position an arbitrary set $E(p)$ of atoms satisfying (**), is there a computational state model in which $E(p)$ is precisely the set of executable atoms at position p for all computer positions p? The answer is an easy yes.
We can rewrite (**) as follows. For any position $p = \sigma_0, \ldots, \sigma_n$, let p^I be the sequence obtained from p by deleting from p every symbol except those moves of Computer which are instructions. A proof by induction shows that (**) is equivalent to the following statement (***).

(***) If p, q are positions of Computer and $p^I = q^I$, then $E(p) = E(q)$.

Supposing that the collection $[E(p) : p$ is a Computer position$]$ satisfies (**) and therefore (***), we can define a computational state model as follows.

Π is the set of all p^I such that p is a Computer position.

Instruction i denotes the partial function $i(\pi) = \pi \cdot i$
if, and only if, there exists a Computer position p with

$\pi = p^I$ and $i \in E(p)$.

Test predicate φ denotes the partial function φ such that $\varphi(\pi) = x$ if, and only if, for some position p of Computer, $\pi = p^I$ and $(\varphi,\, x) \in E(p)$.

It is easy to show, by induction on length of positions, that this computational state model induces the given collection $[E(p) : p$ is a Computer position$]$.

Computational game trees and SYSTEM

The Computational state subtree of a pure computational game tree based on a computational state model of execution can be characterized as the tree of positions consistent with a pair of opposing strategies. These are state strategies based on the same deterministic automaton, SYSTEM, and differ only in the output functions used. SYSTEM is designed exactly to carry out the submission-execution mechanism of a computational game as mediated by the pool of submitted but as yet not executed atoms. We get SYSTEM by suppressing the notion of "position" in the description of the pool mechanism in previous paragraph.

Suppose a computational state model of a computational game is given. Let the set S of states of SYSTEM be the set of all pairs $(\pi,\, U)$, π a computational state, U a packet of atoms. Let σ be any symbol from $\sum = \sum^P \cup \sum^C$. Define a deterministic automaton transition function M as follows.

I. $s_{in} = (\pi_0,\, \emptyset)$ (the initial state).
II. Suppose that $\sigma \in \sum^P$. Then

$$M((\pi,\, U),\, \sigma) = (\pi,\, U \cup \sigma)).$$

III. Suppose that $\sigma \in \Sigma^C$. Suppose that σ is an instruction executable at π, and that $\pi' = \sigma\pi$. Then

$$M((\pi,\, U),\, \sigma) = (\pi', U - \{\sigma\}).$$

IV. Suppose that $\sigma \in \Sigma^C$. Suppose that σ is " wait ". Then

$$M((\pi,\, U),\, \sigma) = (\pi,\, U)).$$

V. Suppose that that $\sigma \in \Sigma^C$ is of the form (φ, x), with φ a test predicate such that (φ, x) is executable at π. Then

$$M((\pi, U), \sigma) = (\pi, U - \{(\varphi, t), (\varphi, f)\}).$$

VI. Suppose that that σ is " end ". Then

$$M((\pi, U), \sigma) = (\pi, U).$$

Given a computational game tree, we describe the opposing state strategies whose tree of consistent positions is that computational game tree. SYSTEM is the underlying automaton for both the state strategy for Programmer and that for Computer, but SYSTEM is equipped with different output functions $\check{f}$, $\check{g}$ for the two cases.

- $\check{f}(\pi, U)$ consists of all packets of atoms disjoint from U, plus " end ".

This expresses precisely the restriction on Programmer's choice of move, the symbol chosen has to be either a packet disjoint from the current pool, or " end ".

- $\check{g}(\pi, U)$ consists of all atoms in U which are executable at state π, plus " wait ".

This expresses precisely the restriction on Computer's choice of move, the symbol chosen being either an atom in the pool, or " wait ".

Relativising SYSTEM

Suppose that B is a packet of atoms and that Π' is a set of computational states containing π_0 and closed under the (partial) execution functions for instructions in B. Restrict Programmer to the symbols $\sum^P(B) \subseteq \mathcal{P}(B)$, the set of all packets which are subsets of B. Restrict Computer to the symbols $\sum^C(B) = \{$ wait $\} \cup B$, with $\sum(B) = \sum^P(B) \cup \sum^C(B)$, and restrict computational states to the set $S(B)$ of all pairs (π, U) with $\pi \in \Pi'$ and a packet $U \subseteq B$, and restrict system's transition function similarly. Then SYSTEM restricted to B, is a deterministic automaton with initial state $(\pi_0, \emptyset)$. The desciption of the computation tree corresponding to a pair of opposing strategies also relativizes to B. It turns out also that it is useful to vary the initial state π_0. Then we would write SYSTEM(π_0) to indicate dependency on π_0. We use the same symbolization for relativised versions, letting the context tell which B is intended. Then the tree of positions consistent with the pair of opposing strategies associated with SYSTEM(π_0) is $T(B, \pi_0)$.

In any one application to a program in a programming language, there is a finite packet A of atoms containing all atoms used. Let Π, the set of computation states , be the set of all mappings of the set of all program variables of the program, each into a finite set of allowed values for the program variables; these are just the possible contents of all relevant storage locations. For such a finite computational state model, the relativisation of SYSTEM, allowing only atoms in packet A, is a finite automaton describing the submission-execution mechanism. But this relativisation can have an infinite number of states if any program variable occurring in an atom in A can be legally assigned values from an infinite set, as is often assumed. Real machines don't allow this, but computer scientists conventionally model that way anyway.

Definition. Let r be a run

$$(\pi_0, U_0) \cdot (\pi_1, U_1) \cdot (\pi_2, U_2) \cdots (\pi_{n+1}, U_{n+1})$$

of SYSTEM on position $\alpha = \sigma_0 \cdot \sigma_1 \cdots \sigma_n$.

The corresponding computational state run is

$$\pi_0 \cdot \pi_1 \cdot \pi_2 \cdots \pi_{n+1}, \text{ and,}$$

π_{n+1} is the computational state at position α. The corresponding pool run is

$$r_2 = U_0 \cdot U_1 \cdot U_2 \cdots U_{n+1}, \text{ and}$$

U_{n+1} is the pool at position α.

The pool is the medium of transaction between Programmer and Computer. From the Programmer's point of view, the pool is where the directions for Computer go, from the Computer's point of view, instructions are plucked from the pool, all "passages of control" coming from test predicates chosen from the pool. So the pool runs are interesting as records of transactions and passage of control in a computational state model.

Example. Without precise definitions, we describe two plays consistent with

$$\textbf{cobegin } x := 1;\ x := 2 \textbf{ coend}.$$

as usually conceived. See **Section 6** for definitions. Let the set Π of computational states be $\{< x = 0 >, < x = 1 >, < x = 2 >\}$, the three possible assignments of values to a single program variable. We also assume that $< x = 0 >$ *is* the initial computational state π_0.

Play 1. At move 0, the pool is empty, Programmer submits the packet $\{x := 1,\ x := 2\}$. At move 1, the pool is this set, Computer responds by picking from the pool the atom $x := 1$, and executes it. At move 2, the pool is now $\{x := 2\}$, Programmer submits the empty packet Φ. At move 3, the pool is unchanged, and Computer responds by picking $x := 2$ from the pool and executes it. At move 4, the pool is empty, Programmer submits " end ". The pool is now empty, and the play ends. We write this and all other plays informally in the form

$$\{x := 1,\ x := 2\}, x := 1,\ \text{skip}, x := 2,\ \text{end}.$$

The corresponding SYSTEM run is

$$\begin{aligned}(< x = 0 >, \emptyset) \cdot (< x = 0 >, \{x := 1, x := 2\}) \cdot (< x = 1 >, \{x := 2\})\\ \cdot(< x = 1 >, \{x := 2\}) \cdot (< x = 2 >, \emptyset) \cdot (< x = 2 >, \emptyset).\end{aligned}$$

The corresponding computational state run is

$$< x = 0 > \cdot < x = 0 > \cdot < x = 1 > \cdot < x = 1 > \cdot < x = 2 > \cdot < x = 2 > .$$

The corresponding pool run is

$$\emptyset \cdot \{x := 1, x := 2\} \cdot \{x := 2\} \cdot \{x := 2\} \cdot \emptyset \cdot \emptyset\ .$$

Play 2. At move 0, the pool is empty, and Programmer submits the packet $\{x := 1,\ x := 2\}$. At move 1, the pool is that set, Computer waits. At move 3, the pool is unchanged, and Programmer submits the empty packet " skip ". At move 4, the pool is unchanged, and Computer picks atom $x := 2$ and executes it. At move 5, the pool is $\{x := 1\}$, and Programmer submits the empty packet. At move 6, the pool is unchanged, and Computer picks $x := 1$ and executes it. At move 7, the pool is empty and Programmer submits " end ".

So the play is

$$\{x := 1, x := 2\},\ \text{wait},\ \text{skip}, x := 2,\ \text{skip}, x := 1,\ \text{end}.$$

The corresponding run is

$$(< x = 0 >, \emptyset) \cdot (< x = 0 >, \{x := 1, x := 2\}) \cdot (< x = 0 >, \{x := 1, x := 2\}) \cdot$$
$$(< x = 0 >, \{x := 1, x := 2\}) \cdot (< x = 2 >, \{x := 1\}) \cdot$$
$$(< x = 2 >, \{x := 1\}) \cdot (< x = 1 >, \emptyset) \cdot (< x = 1 >, \emptyset)\ .$$

The corresponding computational state run is

$$< x = 0 > \cdot < x = 0 > \cdot < x = 0 > \cdot < x = 0 >$$
$$\cdot < x = 2 > \cdot < x = 2 > \cdot < x = 1 > \cdot < x = 1 >\ .$$

The corresponding pool run is

$$\emptyset \cdot \{x := 1, x := 2\} \cdot \{x := 1, x := 2\} \cdot \{x := 1, x := 2\} \cdot \{x := 1\} \cdot \{x := 1\} \cdot \emptyset \cdot \emptyset.$$

Computational game trees and CONTROL

Pure computational game trees, like computational state game trees associated with computational state models, can be characterized as the tree of positions consistent with a pair of opposing strategies. Both are unrestricted state strategies based on a deterministic automaton, CONTROL, but using different nondeterministic output functions $\check{f}$, $\check{g}$. CONTROL is obtained by suppressing all reference to states and execution in the definition of SYSTEM.

So suppose given a pure computational tree. CONTROL has as its states all packets of atoms. Its transition function M' satisfies:

I. $S_{in} = \emptyset$ (the initial state).

II. Suppose that $\sigma \in \sum^P$. Then

$$M'(U, \sigma) = (U \cup \sigma).$$

III. Suppose that $\sigma \in \Sigma^C$, and σ is an instruction. Then

$$M'(U, \sigma) = U - \{\sigma\}.$$

IV. Suppose that $\sigma \in \Sigma^C$, and σ is a " wait ". Then

$$M'(U, \sigma) = U.$$

V. Suppose that that $\sigma \in \Sigma^C$ and σ is a (φ, x). Then

$$M'(U, \sigma) = (U - \{(\varphi, t), (\varphi, f)\}).$$

VI. Suppose that that σ is " end ". Then

$$M'(U, \sigma) = U.$$

(The numbering is that of the corresponding clauses for SYSTEM.)
A computational game tree is the tree of consistent positions for the pair of opposing unrestricted state strategies induced by control from output functions $\check{f}$, $\check{g}$, given below.

- $\check{f}(U)$ consists of all packets atoms disjoint from U, plus " end ".
- $\check{g}(U)$ consists of all atoms in U, plus " wait ".

Not surprisingly, these are obtained by eliminating the reference to executability in the output functions for SYSTEM. Finally, just as for SYSTEM we can relativise CONTROL. Note that for every initial computational state, there is a unique computational state subtree of the pure computational tree.

Remark. Instead of using partial execution functions and test predicates for atoms, we could alternately have handled branching atoms by incorporating computational states as part of the internal states of our strategies. But the notion of computational state is implementation-dependent, and there are usually many computational states. By introducing partial execution funtions, we are able to keep the internal states of our strategies small in number and machine-independent.

Part II: STRATEGIES DENOTED BY PROGRAMS

6. Strategy Models

Suppose that we are given a programming language L. Such an L will contain some "atomic programs", and will have some "program construction operations" which produce new programs from old. Then the programs of L will be defined as the least class containing the atomic programs and closed under these operations. Here is what we mean by a strategy model of L.

Definition. A strategy model of L (strategy relational system for L) is an assignment to each atomic program of L of a pure computational tree strategy for Programmer as denotation (called in this context an "atomic strategy"), and an assignment to each program constructor of an operation on pure computational tree strategies, giving an inductive definition of an assignment of a pure computational tree strategy as denotation to every program of L.

In a strategy model, the collection $\mathcal{S}(L)$ of all strategies denoted by programs of L is closed under strategy operations corresponding to program construction operations and can be studied as a calculus. A variety of strategy models can be defined. We can study strategy models of programming languages within any of the classes of strategies discusssed below. Different strategy models are appropriate to address different concerns. State-strategy models are more useful for programming because their internal states carry the information needed to write and verify concurrent programs on a modular basis. They also have "concurrency diagrams", state diagrams which carry much more information than conventional flow diagrams and exactly represent the state strategy semantics of the programs. See the appendix.

Here are some classes of strategies that might be used for strategy models.

(1) We could let programs P denote (positional) strategies (F, T), F a strategy on a computational game tree T. Our strategy operations will build new strategies from old. A strategy operation on strategies (F_0, T_0), (F_1, T_1), will generally yield a strategy (F, T) with possibly different T_0, T_1, T. This has the virtue of being mathematically neat and readable. The underlying categories and functors and natural equivalences governing the subject are then obvious.

(2) We could let programs P denote state-strategies given with an automaton $\mathcal{A}$ guiding the strategy. These state-strategies are triples (F, T, $\mathcal{A}$), with $\mathcal{A} = (\sum^P \cup \sum^C, S, M, , s_{in}, Q, \check{f})$, where $\mathcal{A}$ is deterministic with nondeterministic output function $\check{f}$ and (F,T) is the corresponding "positional" strategy as in 1) above. State strategy operations building F from

F_0 and F_1 will build a new automaton $\mathcal{A}$ for F from old automata $\mathcal{A}_0$ for F_0 and $\mathcal{A}_1$ for F_1.

(3) We could let programs denote arbitrary strategies, but equip each strategy with the intrinsic automaton from Nerode equivalence. Then 1) is a special case of 2).

(4) We could develop deterministic program constructions using only Büchi strategies. Büchi strategies are pairs $(F, \mathcal{B})$, where F is a strategy on the Computer move tree T^C and $\mathcal{B} = (\sum^C, S_B, M_B, s_{inB}, Q_B, \check{f}_B)$. In most present day concurrent programming languages, programs actually written may as well denote Büchi strategies. These give rise to state-strategies and are therefore a special case of 2).

(5) We could limit ourselves to finite state strategies, since these are the only strategies that occur in programming.

(6) We might limit ourselves to strategies $(F, T(A))$ in a fixed pure computational tree $T = T(A)$, with A a set of atoms. A might be the set of all atoms of a programming language L. Such strategies f may be identified with subtrees V of $T(A)$, where V consists of all positions consistent with f (**Section 2**). So all operations on strategies could be given as operations on subtrees of $T(A)$. That is, we could assign as denotation to an atomic program a subtree of T, and define operations on subtrees corresponding to program constructs, and by induction make every program denote a subtree of T. Since these trees induce strategies, this will give a strategy model, entirely with subtrees of $T(A)$. We call this the internal point of view. A disadvantage is that operations on strategies are partial because using only subtrees of $T(A)$ makes collisions of trees inevitable. These collisions are of the same sort as those of trying to form weak direct products of subgroups of a group entirely within the group; the subgroup they generate gives the weak direct product only when their intersection is trivial. But a major advantage of an internal approach is that operations on trees are directly suitable for coding into software environments intended to automate the program development process.

(7) We could choose a computational state model, work within the fixed pure computational tree $T(A)$ associated with the vocabulary A of that model, and associate with each choice of initial state π_0 a computational state subtree $T(A, \pi_0)$ of $T(A)$, with a computational state associated with every position. Each strategy on that computational state subtree is conventionally extended to $T(A)$ by assigning the empty set of moves to posi-

tions in $T(A) - T(A, \pi_0)$. If we wish to allow programs to be used starting from any initial computational state π, the program should be thought of as denoting the family of strategies $\pi \rightarrow F_\pi$, parameterized by computational states π in the role of π_0.

Alternatives 1)-6) meet the criterion for denotational semantics of abstraction from operational elements such as execution sequences of machine or computational states.

We introduce two natural requirements on strategies denoted by programs that are actually used in practice.

(a) Strategies denoted by programs should be perpetual. Perpetuity asserts that the Programmer cannot get stuck before the end of a play due to assigning programmer an empty set of moves with which to respond to Computer.

Remark. We can assign perpetual strategies on pure computational trees to all programs in all standard programming languages. But in program development, non-perpetual strategies are pieced together to get perpetual strategies to satisfy a specification. So we would not wish to use only perpetual strategies as building blocks.

(b) Strategies should respect termination. We say that F respects termination if for all positions p in T for which $p \cdot end$ is consistent with F, the pool at p is empty. If V is the tree of positions consistent with F, this says that every finite play in V is also a finite play in T. We can assure "respecting termination" for all strategies used as denotations, but similarly would not wish to have only these strategies as building blocks for development.

Remark. What "respecting termination" says is that, in modelling execution, a program execution should terminate only if all atoms previously submitted for execution have been executed. How would this fail? If execution stops with yet unexecuted submitted instructions, we would call this "aborting", not "terminating". Usually when the operating system aborts programs, the program itself is not "aware" that it was aborted, the program should have continued executing instead of aborting. We therefore prefer to model aborted executions as infinite plays. Note that since we impose no time constraint for Computer to execute a given atom, at no finite position of a play can Programmer determine whether there has been a problem in executing an atom left in the pool.

7. Operations on Strategies and Their Continuity

First, we informally introduce positional strategies Atom(i) , Atom(φ) corresponding to the atomic programs for instructions and tests, and then we give operations constructing new positional strategies from old corresponding to the program constructs

"**begin** P_0; P_1 **end**",

"**cobegin** P_0; P_1 **coend**",

"**if** φ **then** P_0 **else** P_1",

"**while** φ **do** P_0",

If program P_i denotes $(F_i, T(A_i))$, we need strategy operations " IfThen Else$_\varphi(F_0, F_1)$ ", " WhileDo$_\varphi(F_0)$ ", "$Seq(F_0, F_1)$" for sequential composition, "$Par(F_0, F_1)$" for concurrent connection. After this is done, we sketch corresponding operations for state strategies on pure computational trees, and corresponding operations for internal strategies on a fixed pure computational tree.

Positional strategies

One does not usually think of the empty program which immediately terminates or of programs denoting individual instructions or tests. But we need strategies for each as basic building blocks.

Empty

" Empty " is the Programmer strategy with one and only one move, to submit " end " at the null position e. Its tree of consistent positions consists of the null position and " end ".

Atom(i), Atom(φ)

For instructions i and test predicates φ, there are respective positional strategies Atom(i) and Atom(φ), We describe these simultaneously, by referring to the "smallest packet" for i or φ, respectively defined as $A_i = \{i\}$, $A_\varphi = \{(\varphi, t), (\varphi, f)\}$.

(1) Start at the empty position e by submitting the smallest packet.
(2) Submit " skip " till the pool is empty.
(3) Submit " end " when the pool is empty.

IfThenElse

We need a definition first.

Definition. Let p be a position consistent with Atom(φ) such that Atom(φ)(p) = *end*. Then p is called a positive (resp. negative) branch of Atom(φ) if the last move of Computer in p is (φ, t) (resp. (φ, f)).

Let P_i be a program denoting strategy $(F_i, T(A_i))$, the $T(A_i)$ being pure computational trees. Let φ be a test predicate. Then " **if** φ **then** P_0 **else** P_1" should denote $(F, T(A))$ defined as follows.

Think of Programmer as having three subcontractors: subcontractors $i = 0$, 1, playing according to F_i in $T(A_i)$; and subcontractor ATOM, playing according to Atom(φ). Programmer starts at the empty position by using ATOM's moves in his game for Atom(φ).

(a) If and when a positive branch p of Atom(φ) is reached, Programmer shifts to subcontractor 0 and uses at position $p \cdot x$ the move of subcontractor 0 at position x in the subcontractor 0 game.

(b) If, and when, a negative branch p of Atom(φ) is reached, Programmer shifts to subcontractor 1 and uses at position $p \cdot x$ the move that subcontractor 1 makes at position x in the subcontractor 1 game.

This F is written " IfThenElse$_\varphi(F_0, F_1)$ ". Its pure computational tree $T(A)$ has $A = A_0 \cup A_1 \cup \{(\varphi, t), (\varphi, f)\}$.

WhileDo

Suppose that P_0 denotes a strategy $(F_0, T(A_0))$ in a pure computational tree, and φ is a test predicate. Then " **while** φ **do** P_0" should denote $(F, T(A))$ as defined recursively below. Below, Programmer is playing according to F, the intended "WhileDo$_\varphi(F_0)$" strategy.

Programmer employs one subcontractor, who follows IfThenElse$_\varphi(F_0$, Empty). Programmer starts at the null position, using the moves that the subcontractor would use. If and when he arrives at a position p where " end " would be next move offered by subcontractor, programmer does one of two things.

(a) If p is a negative branch of Atom(φ), Programmer terminates by submitting " end ".

(b) If not, Programmer on $p \cdot x$ now submits what Programmer would have submitted on x. Thus, from the point of view of Programmer, the p reached is forgotten, and the game starts from the null position all over again.

Seq

For programs P_0, P_1 denoting strategies $(F_0, T(A_0))(F_1, T(A_1))$ in pure computational trees, what strategy $(F, T(A))$ should " **begin** P_0; P_1 **end** " denote? Informally, the idea is that first we start following F_0, then if and when F_0 terminates, we follow F_1 till it terminates, if ever. First, the pure computational tree $T(A)$ might as well have $A = A_0 \cup A_1$. Suppose given strategies F_i on A_i are extended to strategies on $T(A)$ by assigning $F_i(p) = \emptyset$ for p in $T(A) - T(A_i)$.

Think of Programmer as having two subcontractors, subcontractor 0 and subcontractor 1, playing against Computer in $T(A_i)$ according to F_i. Informally, Programmmer starts with an empty pool at the empty position. At first, Programmer uses as his submission at positions p the packet in $F_0(p)$ chosen by subcontractor 0 in the F_0-game. He does this forever if " end " is never offered by subcontractor 0. But if subcontractor 0 terminates, that is, offers " end " at p, since we assume that F_0 respects termination, the pool is empty at p. Programmer turns to subcontractor 1 at p. Then Programmer offers, at position $p \cdot x$, the packet that subcontractor 1 would have offered at position x, starting with an empty pool. Note that it is crucial that, in pure computational trees, the end tree consisting of all x with $p \cdot x$ in the original tree is exactly the original tree, so that F_1 can play the game as required in response to all possible future moves.

This describes a (positional) strategy F built from those for F_0, F_1. In the class of strategies, this is a "sequential composition" $Seq(F_0, F_1)$.

Par

If programs P_i denote strategies $(F_i, T(A_i))$, we want to identify a strategy $(F, T(A))$ for "**cobegin** P_0; P_1 **coend**" to denote. Let $Par(F_0, F_1)$ be this F. We will think of Computer as playing a game against two independent subcontractors of Programmer, subcontractors 0 and 1, guided by stategies F_i in $T(A_i)$. Each subcontractor i, until his strategy F_i terminates in its subgame, contributes a packet to Programmer at each Programmer turn. As long as neither of these strategies F_i is terminated in its subgame, Programmer submits the union of these two packets, one received from each subcontractor. If and when F_i terminates in its subgame, subcontractor i goes out of business and, until strategy F_{1-i} terminates, Programmer submits the packet provided by subcontractor $1-i$. When both subcontractors are out of business, Programmer terminates by submitting "end".

Note. We usually assume that F_i satisfies the property that the pool is necessarily empty at termination in the subgame. But even if F_i does not always terminate with an empty pool, when subcontractor i goes out of business, the pool of the F_i subgame has already been included in the pool of the Programmer game. So this pool would influence only Computer moves, and we may dispense with the F_i-subgame at that point of the Programmer game entirely.

Proposition. *All strategies generated from the atomic strategies by the strategy operations given terminate with empty pool.*

Proof. It is easy to see that atomic strategies always terminate with empty pool, and the rest is an easy but detailed induction on the four strategy operations. □

Remark. The " Par " we define will be symmetric in its treatment of subcontractors for the F_i. Whichever subcontractor terminates, the contractor for F then moves over to the play of the other subcontractor, and terminates when the latter has terminated. From the point of view of termination, it will be a parallel "and", which holds at termination because both have terminated.

What is the appropriate game tree $T(A)$? What is the appropriate set of atoms A for that tree? What is the Programmer strategy F for that game? We want to distinguish between atoms from the F_0-alphabet A_0 used for $T(A_0)$ and atoms from the F_1-alphabet $T(A_1)$. They will both be put in a single pool for F, and we do not wish them confused in the pool. So we define $A = A_0 \times \{0\} \cup A_1 \times \{1\}$ and will use $T(A)$ as the tree for F. So (a, i) plays in $T(A)$ the role that a plays in $T(A_i)$. So copies of atoms a in both A_0, A_1 will go into the pool for F disjointly, whether or not A_0, A_1 are themselves disjoint. In the course of the play, Programmer will split the position p in $T(A)$ into two "splinter" positions p_i in $T(A_i)$, passing p_i to subcontractor i. Then subcontractor i will act solely on the basis of splinter p_i, ignoring splinter p_{i-1}. We shall define by simultaneous induction the positions p in $T(A)$ consistent with F and the splinter positions p_i in $T(A_i)$ consistent with F_i.

Case 0. At the beginning, all three players, Programmer and his two subcontractors, are at the empty position e, their pools are empty. So $p = e$, $p_0 = e$, $p_1 = e$.

Case 1. Suppose that p is a Computer position and that $\sigma = (\sigma_i, \text{i})$ is a legitimate Computer's move at p in $T(A)$, other than " wait ". The splinters of $p \cdot \sigma$ in $T(A_i)$ and $T(A_{i+1})$ are respectively $p_i \cdot \sigma_i$ and p_{1-i}·wait.

Case 2. Suppose that p is Computer's position and that $\sigma = wait$. The splinters of $p \cdot \sigma$ in $T(A_0)$ and $T(A_1)$ are respectively p_0·wait and p_1·wait.

Case 3. Suppose that p is a Programmer position. Define $F(p)$ as follows.

Case 3.1. $F(p) = [a \times \{i\} : a \in F_i(p_i)]$ if for one of $i = 0, 1$, p_i is not terminal and p_{1-i} is either terminal or $F_{1-i}(p_{1-i}) = end$. Let $p_{1'-i} = p_{1-i}$ if p_{1-i} is terminal, let $p_{1'-i} = p_{1-i} \cdot end$ if $F_{1-i}(p_{1-i}) = end$. Then the splinters at $p \cdot \sigma$ are respectively $p_i \cdot \sigma$ and $p_{1'-i}$, where $\sigma \in F(p)$.

Case 3.2. $F(p) = \{$ end $\}$ if for each of $i = 0, 1$, p_i is terminal or $F_i(p_i) = \{$ end $\}$. Let $p_i' = p_i$ if p_i is terminal, and let $p_i' = p_i \cdot end$ if $F_i(p_i) = end$. Let $\sigma \in F(p)$. Then the splinters at $p \cdot \sigma$ are respectively p_0' and p_1'.

Case 3.3. $F(p) = [a \times \{0\} \cup b \times \{1\} : a \in F_0(p_0) \wedge b \in F_1(p_1)]$ otherwise. Let $\sigma \in F(p)$. Then the splinters at $p \cdot \sigma$ are respectively $p_0 \cdot \sigma_0$ and $p_1 \cdot \sigma_1$, where

$\sigma_0 = [a : (\text{a}, 0) \in \sigma \cap A_0 \times \{0\}]$,
$\sigma_1 = [a : (\text{a}, 1) \in \sigma \cap A_1 \times \{1\}]$.

The meaning for termination of finite plays we adopt is that the terminal plays of F are those plays for which both the corresponding plays for F_0 and for F_1 have terminated in $T(A_i)$. This is a "concurrent and". This termination rule we have adopted is the same termination rule for F used when a Grandmaster (Computer), plays chess simultaneously on two boards against two opponents using F_0 and F_1. There is no correlation between the F_0, F_1 moves. Each F_i will think it is playing against the Grandmaster alone, viewing moves of the F_{i-1} as " skip ", or waiting for a Computer move.

Note. Such splinter plays do not "communicate" with one another in pure computational game trees. But the same stategies in a computational state game tree can communicate via computational states. This models communication for shared memory programs.

Remark. There is also another variant on " Par " with a different termination condition. This is a parallel "or". There, the play terminates, whenever the play in at least one of the two subgames terminates. With this termination rule, in the chess analogy the Grandmaster would be declared a winner as soon as one opponent is beaten. We omit details.

State strategies

Next, we give the corresponding operations for state strategies guided by automata.

Empty

$\mathcal{A} = (S, M, s_{in}, Q, \check{f})$

(1) $S = Q = \{s_{in}\} = \{$ finish $\}$.

(2) $M = \emptyset$.

(3) $\check{f}($ finish $) = end$

Atom(i)

$\mathcal{A}_i = (S_i, M_i, s_i, Q_i, \check{f}_i)$

(1) $S_i = \{$ output , idle , finish $\}$

(2) $M_i(s, \sigma) = s$ if s is "output" or "idle" and σ is $\{i\}$ or " skip ".

(3) $M_i(s,$ wait $) = idle$ if s is "output" or "idle".

(4) $M_i(s, i) = finish$ if s is "output" or "idle".

(5) $s_i = output$.

(6) $Q_i = \{$ finish $\}$.

(7) $\check{f}_i($ output $) = \{i\}$, $\check{f}_i($ idle $) = skip$, $\check{f}_i($ finish $) = \{$ end $\}$.

Atom(φ)

$\mathcal{A}_\varphi = (S_\varphi, M_\varphi, s_\varphi, Q_\varphi, \check{f}_\varphi)$

(1) $S_\varphi = \{$ output , idle , finish $\}$

(2) $M_\varphi(s, \sigma) = s$ if s is "output" or "idle" and σ is $\{(\varphi, t), (\varphi, f)\}$ or " skip ".

(3) $M_\varphi(s,$ wait $) = idle$ if s is " ouput " or " idle ".

(4) $M_\varphi(s, \sigma) = finish$ if σ is " output " or " idle ", and σ is (φ, t) or (φ, f).

(5) $s_\varphi = output$.

(6) $Q_\varphi = \{$ finish $\}$.

(7) $\check{f}_\varphi($ output $) = \{(\varphi, t), (\varphi, f)\}$, $\check{f}_\varphi($ idle $) = skip$, $\check{f}_\varphi($ finish $) = \{$ end $\}$.

IfThenElse

Assume that Atom(φ) is a state strategy guided by automaton $\mathcal{A}_\varphi$ with three states as above. Let (F_j, T_j) be given with automata $\mathcal{A}_j = (S_j, M_j, s_j, Q_j, \check{f}_j)$. Then define automaton $(S, M, s, Q, \check{f})$ as follows.

(1) $S = (S_\varphi - Q_\varphi) \times \{2\} \cup S_0 \times \{0\} \cup S_1 \times \{1\}$.

(2) *Note.* In (s, x), whether x is 2, 0, or 1 tells whether s is in S_φ, S_0, or S_1.
(3) $M((s, 2), \sigma) = (M_\varphi(s, \sigma), 2)$ if $s \in S_\varphi - Q_\varphi$ and $M_\varphi(s, \sigma) \notin Q_\varphi$.
(4) $M((s, 2), \sigma) = (s_0, 0)$ if $s \in S_\varphi - Q_\varphi$ and σ is (φ, t).
(5) $M((s, 2), \sigma) = (s_1, 1)$ if $s \in S_\varphi - Q_\varphi$ and σ is (φ, f).
(6) $s_{in} = (s_\varphi, 2)$.
(7) $Q = Q_0 \times \{0\} \cup Q_1 \times \{1\}$.
(8) $\check{f}(s, 2) = \check{f}_\varphi(s)$ if $s \in S_\varphi - Q_\varphi$.
(9) $\check{f}(s, x) = \check{f}_x(s)$ if $s \in S_x$ and x is not 2.

Note. The automaton for the "Whiledo" strategy can be obtained from that for the "IfThenElse" strategy by replacing the "end" state by a return to the initial state, as for positional strategies. For variety we give another explicit automaton for this purpose.

WhileDo Let Atom(φ) be a state strategy guided by automaton $\mathcal{A}_\varphi$ as earlier defined. Let (F_0, T_0) be given guided by automaton $\mathcal{A}_0 = (S_0, M_0, s_0, Q_0, \check{f}_0)$. Then define automaton $\mathcal{A} = (S\ , M, s_{in}, Q, \check{f})$ as follows.

(1) $S = S_\varphi \times \{1\} \cup (S_0 - Q_0) \times \{0\}$

(In (s, x), $x = 1$ or 0 as s is in S_φ or in S_0.)

(2) $M((s, 1), \sigma) = (M_\varphi(s, \sigma), 1)$ if $s \in S_\varphi - Q_\varphi$ and $\sigma \neq (\varphi, t)$.
(3) $M((s, 1), \sigma) = (s_0, 0)$ if $s \in S_\varphi - Q_\varphi$ and $\sigma = (\varphi, t)$ and $s_0 \notin Q_0$.
(4) $M((s, 1), \sigma) = (s_\varphi, 1)$ if $s \in S_\varphi - Q_\varphi$ and $\sigma = (\varphi, t)$ and $s_0 \in Q_0$.
(5) $M((s, 0), \sigma) = (M_0(s, \sigma), 0)$ if $s \in S_0 - Q_0$ and $M_0(s, \sigma) \notin Q_0$.
(6) $M((s, 0), \sigma) = (s_\varphi, 1)$ if $s \in S_0 - Q_0$ and $M_0(s, \sigma) \in Q_0$.
(7) $s_{in} = (s_\varphi, 1)$.
(8) $Q = Q_\varphi \times \{1\}$.
(9) $\check{f}(s, 1) = \check{f}_\varphi(s)$ if $s \in S_\varphi$.
(10) $\check{f}(s, 0) = \check{f}_0(s)$ if $s \in S_0$.

Seq

Suppose that the (F_i, T_i) are given with automata $\mathcal{A}_i = (S_i, M_i, s_i, Q_i, \check{f})$. We define automaton $\mathcal{A} = (S, M, s\ , Q, \check{f})$ for F as follows.

(1) $S = (S_0 - Q_0) \times \{0\} \cup S_1 \times \{1\}$

(In (s, x), the $x = 0$ or 1 tells whether s is in S_0 or S_1)

(2) $M((s, 0), \sigma) = ((M_0(s, \sigma), 0)$ if $s \in S_0 - Q_0$ and $M_0(s, \sigma) \notin Q_0$,
(3) $M((s, 0), \sigma) = (s_1, 1)$ if $s \in S_0 - Q_0$ and $M_0(s, \sigma) \in Q_0$,

(4) $M((s, 1), \sigma) = (M_1(s, \sigma), 1)$
(5) $s_{in} = (s_0, 0)$
(6) $Q = Q_1 \times \{1\}$
(7) $\check{f}(s, x) = \check{f}_0(s)$ if $s \in S_0 - Q_0$ and $x = 0$
(8) $\check{f}(s, x) = \check{f}_1(s)$ if $s \in S_1$ and $x = 1$

Par

Each F_i is now equipped with a deterministic automaton $\mathcal{A}_i = (\sum, S_i, M_i, s_i, Q_i, \check{f}_i)$, with nondeterministic output function, guiding F_i. We construct from these an automaton $(\sum, S, M, s, Q, \check{f})$ guiding F. S will be $S_0 \times S_1$. Its initial state is the pair of initial states. Programmer symbols σ are packets from A or " end ". The required transition function is defined by cases corresponding to those in the description of behavior for the strategy above. For giving the definition of the concurrent connector only, we extend $M_i(s, \sigma)$ to be defined also for $s \in Q_i$ and $\sigma =$ " skip " or " wait " by putting $M_i(s, \sigma) = s$.

First, suppose that σ is a Computer's move from $A = A_0 \times \{0\} \cup A_1 \times \{1\}$.

Case 1. If $\sigma = (a, 0)$, then $M((s_0, s_1), \sigma) = (M_0(s_0, \text{a}), M_1(s_1, \text{wait }))$.

Case 2. If $\sigma = (a, 1)$, then $M((s_0, s_1), \sigma) = (M_0(s_0, \text{wait }), M_1(s_1, \text{a}))$.

Case 3. If σ is " wait ", then $M((s_0, s_1), \sigma) = (M_0(s_0, \text{wait }), M_1(s_1, \text{wait }))$.

Second, suppose that σ is a Programmer's move, that is a packet from A. If σ is not " end ", then there are unique packets σ_0 and σ_1 such that σ_i is a (possibly empty) packet from A_i and
$\sigma = \sigma_0 \times \{0\} \cup \sigma_1 \times \{1\}$. Note that here σ_0 and σ_1 are not thought of as moves in the respective splinters. Unlike the situation at the corresponding point in the definition of Par for positional strategies, we are defining the transition function here for all symbols of the alphabet without regard for splinters. The transition function is single valued simply because $A_0 \times \{0\}$ and $A_1 \times \{1\}$ are disjoint. To go on,

$$M((s_0, s_1), \sigma) = (M_0(s_0, \sigma_0), M_1(s_1, \sigma_1))$$

Define $\check{f}$ as follows.

Case 1. $\check{f}(s_0, s_1) = [a \times \{i\} : a \in \check{f}_i(s_i)]$ if there exists an $i = 0, 1$ such that s_i is not in Q_i and s_{1-i} is in Q_{1-i}. (Recall that $f_i(s_i) = \{$ end $\}$ if and only if $s_i \in Q_i$. So, even if we do not use splinters explictly in this definition, it is obvious that $s_{1-i} \in Q_{1-i}$ corresponds to termination for the $1 - i^{th}$ splinter.)

Case 2. $\check{f}(s_0, s_1) = \{ \text{ end } \}$ if S_0 is in Q_0 and s_1 is in Q_1,

Case 3. Otherwise, $\check{f}(s_0, s_1) = [a \times \{0\} \cup b \times \{1\} : a \in \check{f}_0(s_0)] \wedge b \in \check{f}_1(s_1)]$. Finally, put $Q = Q_0 \times Q_1$.

If we "forget" states, up to a natural equivalence, we get the previous " Par " on strategies.

Internal strategies

IfThenElse

We are in a fixed pure tree $T(A)$. Suppose that Atom(φ), F_0, F_1 are Programmer's strategies in $T(A)$ and that V_φ, V_0, V_1 are the corresponding trees of consistent positions. Then define the subtree V of consistent positions for F as follows. Let V consist of all those p in $T(A)$ such that one of (a), (b), (c) below holds.

(a) p is non-terminal in V_φ.

(b) p is an extension of a position q such that $p = q \cdot x$ and q is a positive branch in V_φ and x is in V_0.

(c) p is an extension of a position q such that $p = q \cdot x$ and q is a negative branch in V_φ and x is in V_1.

WhileDo

Suppose that IfThenElse$_\varphi$(F_0, Empty) is defined on a fixed pure tree $T(A)$. (So all atoms submitted by F_0, together with (φ, t), (φ, f) are in A.) Define the subtree V of consistent positions for $F = WhileDo_\varphi(F_0)$ as the smallest subtree of $T(A)$ such that

(1) If p is a non-terminal position consistent with IfThenElse$_\varphi$(F_0, Empty), then $p \in V$.

(2) If p is a negative branch of Atom(φ), then $p \cdot end \in V$.

(3) If $p \cdot end$ is consistent with IfThenElse$_\varphi$(F_0, Empty), and p is not a negative branch of Atom(φ), and $x \in V$, then $p \cdot x \in V$.

Seq

We are within a fixed pure $T(A)$. Suppose that F_0, F_1 are Programmer's strategies in $T(A)$ and V_0, V_1 are the corresponding trees of consistent positions. Then define the subtree V of consistent positions for F as follows. Let V consist of all $p \in T(A)$ such that

- either p is non-terminal and in V_0, or
- p is an extension of a position q in $T(A)$ such that $p = q \cdot x$, $q \cdot end \in V_0$ and $x \in V_1$. Then the internal sequential composition

is defined by $Seq(F_0, F_1) = F$.

Par

Let F_0, F_1 be strategies on pure computational game tree $T(A)$. Before defining the internal Par for "separated" strategies (see below), we define a partial operation $p = \oplus(p_0, p_1)$ on some pairs of positions of $T(A)$ to positions of $T(A)$ by the inductive definition below, starting with the empty word e.

(1) $e \oplus e = e$.

For the rest of the definition, suppose that p, q, $p \cdot \sigma$, $q \cdot \delta$ are positions in T. (They are of the same length in clauses 2 and 3 and may be of different length in clause 4.) Also, to give an inductive definition, assume that $p \oplus q$ is already defined.

(2) Suppose that $\sigma, \delta \in \Sigma^P$, that σ is disjoint from the pool at position $q \cdot \delta$, and δ is disjoint from the pool at position $p \cdot \sigma$. Then

For packets σ, δ, $(p \cdot \sigma) \oplus (q \cdot \delta) = (p \oplus q) \cdot (\sigma \cup \delta)$,

For packet σ, $(p \cdot \sigma) \oplus (q \cdot end) = (p \oplus q) \cdot \sigma$,

For packet δ, $(p \cdot end) \oplus (q \cdot \delta) = (p \oplus q) \cdot \delta$.

So clause 2 deals with the case of combining Programmer's moves in V when positions p, q have extensions in V_0, V_1 respectively (although the definition of operation " $\oplus$ " is independent of the choice of V, V_0, V_1).

(3) Suppose that $\sigma, \delta \in \Sigma^C$. Then

$$(p \cdot \sigma) \oplus (q \cdot \text{wait}) = (p \oplus q) \cdot \sigma,$$

$$(p \cdot \text{wait}) \oplus (q \cdot \delta) = (p \oplus q) \cdot \delta.$$

So clause 3 deals with Computer's moves when computer may wait in V_0 or V_1 or both.

(4) If the length of q does not exceed that of p, and α is a symbol in $\sum$ and $p \cdot \alpha$ is in T, and q ends in " end ", then $(p \cdot \alpha) \oplus q$ is defined, and

$$(p \cdot \alpha) \oplus q = (p \oplus q) \cdot \alpha.$$

If the length of p does not exceed that of q, and α is a symbol in $\sum$, and $q \cdot \alpha$ is in T, and p ends in " end ", then $p \oplus (q \cdot \alpha)$ is defined, and

$$p \oplus (q \cdot \alpha) = (p \oplus q) \cdot \alpha$$

(Clause 4 handles a programmer move at a position where one of the subcontractors positions is terminal (say p in V_0) before the other subcontractor's position (say q in V_1) is terminal. If $(p \oplus q)$ is a position in the V tree, and p is terminal in V_0, and q can be continued in the V_1 tree, to a $q \cdot \alpha$ in V_1, then we can instead continue the position $(p \oplus q)$ in V by putting the string of moves α at the end of that position to get $(p \oplus q) \cdot \alpha$ in V, and define this to be $(p \cdot \alpha) \oplus q$. This says that when we no longer need to follow the moves of subcontractor 0, we then follow only the subsequent moves of subcontractor 1, and regard these as literally the moves of Programmer. So in the clauses, the right side reinterprets the subcontractor move on the left as a Programmer move.)

Definition. Strategies F_0, F_1 (or the corresponding subtrees V_0, V_1) on pure computational tree T are called separated if for all p in V_0, all q in V_1, the pool at position p in V_0 and the pool for position q in V_1 are disjoint. (Of course V_0 and V_1 are separated if V_0 and V_1 involve disjoint sets of atoms.)

Definition. Suppose that strategies F_0, F_1 are separated strategies on the same pure computational game tree T with associated subtrees V_0, V_1. Define $Par(F_0, F_1)$ (or write $Par(V_0, V_1)$) as the set of all positions $p = q_0 \oplus q_1$ such that $q_0 \in V_0$ and $q_1 \in V_1$.

Remark. If $p = p_0 \oplus p_1$, and p_i is in V_i for $i = 1, 2$, we say that p_i is an "internal splinter" of p. These "internal splinters" are analogous to the previous "external splinters" and allow us to extend the chess analogy to the internal Par.

Example. We give two distinct plays μ that are consistent with the strategy denoted by **cobegin** $x := 1$; $x := 2$ **coend** . Both are of the form $\mu = \mu_0 \oplus \mu_1$. Consistency of the "splinter" plays μ_0, μ_1 with the corresponding strategies can easily be checked.

1. Let μ be

   $\{x := 1, x := 2\}$, $x := 1$, skip, $x := 2$, end.

Then

   $\mu_0 = \{x := 1\}$, $x := 1$, end

is consistent with " **begin** $x := 1$ **end** ", and

$\mu_1 = \{x := 2\}$, wait, skip, $x := 2$, end

is consistent with " **begin** $x := 2$ **end** ".

2. Let μ be

$\{x := 1, x := 2\}$, wait, skip, $x := 2$, skip, $x := 1$, end

Then

$\mu_0 = \{x := 1\}$, wait, skip, wait, skip, $x := 1$, end

is consistent with " **begin** $x := 1$ **end** ", and

$\mu_1 = \{x := 2\}$, wait, skip, $x := 2$, end

is consistent with " **begin** $x := 2$ **end** ".

Consistency of internal plays

We give an independent treatment of consistency of plays.

(1) (**Seq**) Play μ is consistent with $Seq(F_0, F_1)$ iff one of 1a), 1b) below hold.

(1a) There is a position p, consistent with F_0, and a play μ_1, consistent with F_1, such that $F_0(p) = end$ and $\mu = p \cdot \mu_1$.

(1b) μ is infinite and consistent with F_0.

2. (**Par**) Play μ is consistent with $Par(F_0, F_1)$ if and only if there exist plays μ_i, consistent with F_i, such that $\mu = \mu_0 \oplus \mu_1$, where "$\oplus$" is extended in the obvious way to plays.

3. (**IfThenElse**) Play μ is consistent with $\text{IfThenElse}_\varphi(F_0, F_1)$ if and only if one of 3a), 3b), 3c) holds.

(3a) There is a positive branch p of $\text{Atom}(\varphi)$ and a play μ_0, consistent with F_0, such that $\mu = p \cdot \mu_0$.

(3b) There is a negative branch q of $\text{Atom}(\varphi)$ and a play μ_0, consistent with F_1, such that $\mu = q \cdot \mu_1$.

(3c) μ is an infinite play consistent with $\text{Atom}(\varphi)$.

Definition. A position p is called a finite cycle of $\text{WhileDo}_\varphi(F_0)$ if there is a positive branch q of $\text{Atom}(\varphi)$ and a position p' consistent with F_0 such that $F_0(p') = end$ and $p = q \cdot p'$. A play μ is called an infinite cycle of $\text{WhileDo}_\varphi(F_0)$ if there is an infinite play μ' consistent with F_0 such that $\mu = q \cdot \mu'$ for some positive branch q of $\text{Atom}(\varphi)$.

4. Play μ is consistent with $\text{WhileDo}_\varphi(F_0)$ if and only if one of 4a)- 4d) holds.

(4a) There is an integer n and finite cycles $p_0, \ldots, p_{n-1}$ of $\text{WhileDo}_\varphi(F_0)$ and a negative branch q of $\text{Atom}(\varphi)$ such that $\mu = p_0 \cdot \ldots \cdot p_{n-1} \cdot q \cdot end$.

(4b) $\mu = p_0 \cdot \ldots \cdot p_{n-1} \cdot \mu'$, where $p_0, \ldots, p_n$ are as above and μ' is an infinite cycle of $\text{WhileDo}_\varphi(F_0)$.

(4c) There is an infinite sequence of finite cycles $p_0, p_1, \ldots$ such that $\mu = p_0 \cdot p_1 \cdot \ldots$.

(4d) $\mu = p_0 \cdot \ldots p_{n-1} \cdot \mu'$, where $p_0, \ldots, p_{n-1}$ is as in 4b) and μ' is an infinite play consistent with $\text{Atom}(\varphi)$.

Computational states revisited

Suppose that we are given a fixed computational state model, which assigns states to positions given the initial assignment of a state to the empty position. Are such assignments naturally carried over by strategy operations? Consider, as an example, $F = Seq(F_0, F_1)$, with F_0, F_1 strategies. Start F and F_0 in the same initial computational state π_0. While F is following F_0, states are assigned to positions for F as they were to positions for F_0. When a terminal position (finite play) $p \cdot end$ is encountered, the play corresponding to p in the F_0-game has a state π_1. When we start following F_1, assign it as a new initial state π_1, at the empty position e of the F_1-play, and assign states at positions $q = p \cdot x$ as the F_1 game would assign states at x. This means that any data in computational state π_1 resulting from a terminating F_0-play is thus passed to F_1 at the termination of the F_0-play through the initial state π_1 of the F_1-play. That is, the F_0-play is in $T(A, \pi_0)$ and the F_1-play is in $T(A, \pi_1)$. Other operations propagate state assignments too, but telling how is more complicated and is omitted.

Remark. Our strategy operations are defined without reference to computational state models, in pure computational game trees. In [Owicki and Gries 1977], [Levin and Gries 1981], [Apt 1981], a computational state model is essential. But properties of strategies on pure game trees yield properties of strategies on computational state game trees, as with Seq above.

Continuity of strategy operations

Each (Programmer) strategy (F, T) is characterized by its subtree $V = Con(F) = Con(F, Vac^C)$ of T, the subtree of positions consistent with V. The strategy (F, T) may as well be written (V, T), or thought of as strategy V on T, and we can introduce the set $\mathcal{S}(T)$ of all strategies V on

T. Each fundamental strategy operation of (say) two variables $op(F_0, F_1)$ maps $\mathcal{S}(T_0) \times \mathcal{S}(T_1)$ to $\mathcal{S}(T)$.

Definition. Sequence V_i of subtrees of T converges to a subtree V of T if for every position p of T, p is in V if and only if

there is an integer n_0 such that for all $n \geq n_0$, p is in V_n.

If we identify subtrees with their characteristic functions as subsets of T, then this is the subspace topology on $\mathcal{S}(T)$ inherited from the product topology on $\{0, 1\}^T$, with $\{0, 1\}$ a discrete space. So this is a topology of pointwise convergence.

Proposition. *Strategy operations are continuous on* $\mathcal{S}(T_0) \times \mathcal{S}(T_1)$ *to* $\mathcal{S}(T)$.

Under strategy operations, perpetual strategies are mapped to perpetual strategies, strategies respecting termination are mapped to strategies respecting termination, so this proposition apply to these subspaces too.

Remark. Programs are terms which can denote strategies, built up by the external strategy operations from atomic strategies. The strategy operations have the feature that, at any position, any instruction in the pool can be traced back through its labelling to reveal the history of passage of control, that is, the history of calls to subprograms which led to the submission of that instruction to the pool. Internal game strategies erase this history completely. It is sometimes useful to have this history, and sometimes not.

Strategy operations such as " Par " and " Seq " combine strategies (F_0, T_1), (F_1, T_1) to obtain new compound strategies (F, T). The T_0, T_1 are naturally embedded as subtrees of T, since atoms and predicates have at worst simply been relabelled so that one can tell which of T_0, T_1 they arose from. By what was said above and composition of embeddings, if P' is a subprogram of P denoting (F', T'), there is a natural embedding of T' as a subtree of T and therefore a natural embedding of the tree V of consistent moves for F' as a subtree of the image of T' in T. So F' induces an isomorphic strategy in a subtree of T. By abuse of notation, we can regard all the games and trees for subprograms of a single program P as occurring in one tree, the tree for F.

When a computational state model is given in which a set of predicates and instructions are interpreted on states, it is understood that in compounding strategies, relabelled predicates and instructions are given the same denotation on states as their unlabelled versions. Suppose we are given a program, denoting a strategy, interpreted in a computational state model. Let p be a position of the tree. By decoding labels, we can construct a unique sequence $\alpha_1, ..., \alpha_n$, of positions in the successively called strategies' trees corresponding to subprograms of P which construct the pool at position p, such that the initial computational state for α_1 is the initial state for position p, and the initial state at which α_i ends is that at which α_{i+1} begins, with the pool at the end of α_i used as the initial pool of α_{i+1}. This duplicates, up to relabelling, the pool run of p. These α_i are the "corresponding plays" of the subprogram games which build the given play p. This actually also sketches the recipe for converting strategies into internal strategies.

8. Park's Example

We introduce the notion of "acceptance of a play". This replaces, in our theory, the somewhat arbitrary notions of justice or weak fairness in other treatments of concurrent programming. The ad hoc assumptions used to make fairness work elsewhere can be replaced here by a natural requirement only because positions fully represent both Programmer and Computer behavior, unlike other current theories of concurrency.

Park's Program

David Park [Park 1981] gave a concurrent program P
" **begin** $x := 0$; $y := 0$; **cobegin** $x := 1$; **while** $x = 0$ **do** $y := y + 1$ **coend** **end** ".
This program has been used as a test problem for many semantics of concurrency. The question is, what does termination for P mean, and how does one prove it? Park states that this program is associated with "unbounded nondeterminism", meaning that even with a termination proof there is no bound computed either on termination time or on the terminal value of y. He gives the opinion that this variety of nondeterminism can be given semantics only by use of monotone non-continuous functions. In fact, our game strategy operations, which operate on atomic strategies as starting-points, are (monotone) continuous functions in the pointwise topology (**Section 7**), and we do get a termination proof. So Park's opinion, on the surface, was hasty. Let us start with the informal, definitionless, "Journeyman Programmer's justification of termination.

Journeyman's justification of termination

Assume that, in the course of execution, the computer eventually executes every instruction ever required to be executed by the program. By this assumption, eventually, "$x := 0$; $y := 0$" will have been executed and terminated and the test predicates $x = 0$ and $y = 0$ are valued true. Then the block " **cobegin** $x := 1$; **while** $x = 0$ **do** $y := y + 1$ **coend** " takes over and starts executing, that is, the block "$x := 1$" is executing in parallel with the block " **while** $x = 0$ **do** $y := y+1$ ". By the assumption, the block "$x := 1$" will eventually be executed, and then the test predicate "$x = 0$" values false. The block " **while** $x = 0$ **do** $y := y + 1$" also terminates after some cycles since the test predicate $x = 0$ eventually values false. So the program terminates altogether, as required.

What does it mean that the computer acts normally? Can it act abnormally? Acting normally is a "deus ex machina", not a definition. In what evolutionary arena of execution is the proof taking place? We say that an appropriate arena is computational games based on a computational state model. We are able to convert the intuition behind the Jouneyman's justification into a real, necessarily more complex, proof, by replacing the usual notion of execution as a sequence of computational or machine states by a stronger notion, the states associated with the positions of the computational game tree of a computational state model. The crucial new feature that we have introduced is the provision for independent moves of Computer, and interaction of Computer with Programmer through the pool. That is, our formulation has the advantage of allowing abnormal as well as normal behaviors of Computer as an active agent in the computational game. Just as in quantum mechanics there have been efforts to introduce "hidden variables" to restore the causality principle, we are by analogy claiming that the Computer's role as opponent in strategies on game trees is a "hidden variable" which make it possible to convert intuitive reasoning about concurrency to at least an outline of a more complex formal correctness proof, using the principle that execution sequences in the conventional approaches are replaced by states of a computational state model on the positions of a computational game tree.

Acceptance

In our games, the assumption that the Computer operates without errors or failure can be enforced by choice of winning condition (see the discussion of Park's example below). Suppose we are given a strategy for Programmer on a computational tree T. We say that a play μ is accepted by Computer if for every atom a, if a is submitted by Programmer at some position $p \leq \mu$ in a packet to the pool, then there is a position q, $p \leq q \leq \mu$, at which Computer moves by choosing a and removing it from the pool (along with its negation if it comes from a test predicate). We say μ is rejected if μ is not accepted. Using the law of the excluded middle, μ is rejected if and only if there exists an atom a, submitted at a position $p \leq \mu$, such that a is never executed at a position q, with $p \leq q \leq \mu$. We also say that a is rejected by play μ.

Rejection of a by μ represents an interaction of Programmer and Computer and can occur for a variety of reasons. One reason for rejection may be that there is no position q with $p \leq q \leq \mu$ and an executable atom a at q. We call this a (Programmer) error. Another reason for rejection may be that there exist q with $p \leq q \leq \mu$ and an executable atom at q, but Computer never chooses a at any of these q. We call this a (Computer) failure of the play μ at atom a.

Example. In the case when every atom is executable at every position at which it is in the pool, play μ rejects atom a if and only if there is a failure at a. We think of failure as representing misbehavior of Computer.

Strategy semantics of Park's program

In a suitable computational game, P denotes F, where

$$F = Seq(G,\, H),$$
$$G = Seq(\text{Atom}(x := 0),\, \text{Atom}(y := 0))$$
$$H = Par(\text{Atom}(x := 1),\, \text{WhileDo}_{x=0}(\text{Atom}(y := y + 1))).$$

We outline, without detail, a computational game tree proof for termination of Park's program. Any of our formulations of strategy semantics will do. The main thing is to have a computational tree associated with a computational state model and F a strategy on that tree, with computational states from the model associated with positions. Each of the subprograms of P corresponds to a substrategy F' of F, with its own computational game and plays. Each position of F induces a corresponding position of each of these games, at least till the plays in these games terminate. The set S of all computation states will consist of all maps $x = a$, $y = b$, with a, b from the non-negative integers ω. We assume that the initial state is given. Each test predicate denotes a predicate $\varphi(s)$ of these states s. Each instruction $x := i$ denotes the map of states to states mapping $x = a$, $y = b$ to $x = i$, $y = b$. (similarly for $y := j$). Also, $y := y + 1$ denotes the map of states to states mapping $x = a$, $y = b$ to $x = a$, $y = b + 1$.

First, to see why precision is needed, examine the play in which Programmer submits $\{x := 0\}$ at the beginning of the game to an empty pool, and afterwards Programmer always submits the empty packet, while Computer never chooses any instruction to execute, so Computer rejects the instruction in the sense defined above. This yields an infinite play

$$\{x := 0\}, \text{wait, skip, wait, skip, ..., wait, skip, ...}$$

This infinite play μ is consistent with " **begin** $x := 0$ **end** ", so by the definition of consistency of a play with $F = Seq(G, H)$, μ is consistent with F. The Computer has rejected the play μ by rejecting the atom $x = 0$, and this is a misbehavior against which the programmer cannot possibly protect himself. The best we can hope to do is to define "P terminates" to mean that if, whenever the Computer accepts a play μ consistent with F, then μ is finite. Let W consist of all rejected plays plus the all finite plays. To prove termination of P is then the same as proving that F is a winning strategy in (T, Ω, W). W is the program specification.

Game-tree termination proof

In our setup, P denotes $F = Seq(G, H)$, where
$G = Seq(\text{Atom}(x := 0), \text{Atom}(y := 0))$
$H = Par(\text{Atom}(x := 1), \text{WhileDo}_{x=0}(\text{Atom}(y := y + 1)))$.
We take a computational state model for granted, so there is a state $x = i(p)$, $y = j(p)$ at each position, with initial state $i(e) = 0$, $j(e) = 0$.

Suppose that μ is a play of the F-game consistent with F and that μ is accepted, that is, every instruction submitted to the pool is later executed. We show that μ is finite. Each instruction among $x := 0$, $y := 0$, $x := 1$ is submitted at most one position, and is executed at at most one later position. So there exists a position p of μ such that if any of $x := 0$, $y := 0$, $x := 1$ is submitted or executed at any position q, then $q < p$. The G-substrategy of $F = Seq(G, H)$ has a corresponding G-game play insuring that $x := 0$, $y := 0$ are submitted, hence executed, in the F-game at a proper prefix of p, thus terminating the play in the G-subgame.

But $F = Seq(G, H)$, so the Atom($x := 1$) in H and the definition of Par assure that that $x := 1$ is submitted, hence executed in the F-play at a proper prefix of p, with the termination of the corresponding play in the Atom($x := 1$)-game. A simple proof shows that, starting from this prefix, the computational state is always $x = 1$, $y =?$. That is, at any further position of the play, the state satisfies $x = 1$. The test on the WhileDo$_{x=0}$(Atom($y := y + 1$))) assures that the corresponding H-game play terminates after p, because Programmer, following Atom($x = 0$) of the WhileDo loop after p, will cause termination of the WhileDo. So, by the definition of termination for Seq, $F = Seq(G, H)$ terminates too.

Thus, the main change in turning the Journeyman's sketch into an actual proof is pinning the execution in the Jouneyman's sketch to the states at positions of a computational tree arising from a computational state model, where the Computer plays the role of opponent.

Remark. For the Park example, think of a "winning set" of plays consisting of all rejected plays plus all finite plays. Due to space limitations, we cannot discuss here how winning sets are used as program specifications, what Gurevich-Harrington winning sets are, why this is one, and how such sets can be used for program development [A. and V. Yakhnis 1990, in prep.].

Consistency of accepted plays

We give a proposition, a variation on the play characterization of operations on strategies, that summarizes systematically basic facts used in program correctness proofs. We apply this to Park's example. This is the sort of computation which one might use in an automated program development environment.

Proposition

(1) (**Seq**) *Accepted play μ is consistent with $Seq(F_0, F_1)$ iff one of* (1a), (1b) *below hold.*
  - (1a) *There is a position p, consistent with F_0, and an accepted play μ_1, consistent with F_1, such that $F_0(p) = end$ and $\mu = p \cdot \mu_1$.*
  - (1b) *μ is infinite and consistent with F_0.*

(2) (**Par**) *Accepted play μ is consistent with $Par(F_0, F_1)$ if and only if there exist accepted plays μ_i, consistent with F_i, such that $\mu = \mu_0 \oplus \mu_1$.*

(3) (**IfThenElse**) *Accepted play μ is consistent with $IfThenElse_\varphi(F_0, F_1)$ if and only if one of* (3a), (3b) *holds.*

(3a) *There is a positive branch p of Atom(φ) and an accepted play μ_0, consistent with F_0, such that $\mu = p \cdot \mu_0$.*

(3b) *There is a negative branch q of Atom(φ) and an accepted play μ_0, consistent with F_1, such that $\mu = q \cdot \mu_1$.*

(4) (**WhileDo**) *Accepted play μ is consistent with WhileDo$_\varphi(F_0)$ if and only if one of* (4a)–(4c) *holds.*

(4a) *There is an integer n and finite cycles p_0, ..., p_{n-1} of WhileDo$_\varphi(F_0)$ and a negative branch q of Atom(φ) such that $\mu = p_0 \cdot ... \cdot p_{n-1} \cdot q \cdot end$.*

(4b) $\mu = p_0 \cdot ... \cdot p_{n-1} \cdot \mu'$, *where p_0, ..., p_n are as above and μ' is an infinite accepted cycle of WhileDo$_\varphi(F_0)$.*

(4c) *There is an infinite sequence of finite cycles p_0, p_1, ... such that $\mu = p_0 \cdot p_1 \cdot ...$.*

(5) *If F is a strategy built up from the atomic operations by strategy operations other than WhileDo's and μ is an accepted play consistent with F, then μ is terminal.*

Formal termination proof

As an illustration, we apply these rules to verify Park's example formally. Let π_0 be any initial state. Let A be the packet

$$\{x := 0,\ y := 0,\ x := 1,\ y := y + 1,\ (x = 0,\ t),\ (x = 0,\ f)\}.$$

We wish to show that every accepted play μ in $T(A, \pi_0)$ consistent with F is terminal. By 5) above, G does not have infinite accepted plays. So, by 1) above, we get $\mu = p \cdot \mu_h$, where p is a position consistent with G, $G(p) = end$, and μ_h is an accepted play consistent with H. By 2) above, $\mu_h = \mu_0 \oplus \mu_1$, where μ_i is accepted for $i = 0$, 1 and μ_0 is consistent with Atom($x := 1$) and μ_1 is consistent with WhileDo$_{x=0}$(Atom($y := y + 1$)). From 5), we conclude that μ_0 must be finite. So $\mu_0 = q \cdot end$, and the last Computer move in q is $x := 1$. Let p' be the prefix of μ_h which has q as splinter. By the definition of "$\oplus$", the last move in p' is also $x := 1$. Since no instruction affecting x is later submitted by WhileDo$_{x=0}$(Atom($y := y{+}1$)), the computational state at every extension of p' in μ_h satisfies $x = 1$. Finally, by 5), WhileDo$_{x=0}$(Atom($y := y{+}1$) does not have infinite accepted cycles. So μ_1 does not satisfy 4b). Can μ_1 satisfy 4c)? If it does, then since the Computer move $(x = 0, t)$ occurs in every finite cycle of WhileDo$_{x=0}$, it occurs infinitely often in μ_1, and therefore it occurs infinitely often in the play $\mu_h = \mu_0 \oplus \mu_1$. Thus the play μ_h has infinitely many positions with computational state satisfying $x = 0$. This contradicts the conclusion of the previous paragraph. So μ_1 has to satisfy 4a), and therefore is terminal.

Recursively bounded nondeterminism and concurrency

Now let us look under the surface of Park's question. We rephrase it in the light of the proof above. What is disquieting to the computer scientist about such a termination proof? We say it is that the proof above has an essential non-constructive use of the law of the excluded middle which, applied to accepted plays, provides a position p by which the three instructions $x := 0$, $y := 0$, $x := 1$ have all been submitted and executed in the play. That is, termination assuming acceptance has been proved in classical, not intuitionistic logic. But with a constructive version of acceptance, that is, a constructive hypothesis that Computer acts normally, one gets constructive termination.

I. Recursively bounded nondeterminism. A Programmer's strategy F has recursively bounded nondeterminism if there is a recursive procedure which, applied to any position p in the tree of consistent positions for F, computes an integer n which is an upper bound for the cardinality of $F(p)$.

II. Recursively bounded concurrency. Computer has recursively bounded concurrency if there is a recursive procedure which acts on positions p and atoms a in the pool at p, to compute an integer m such that for any play μ extending p, if a is withdrawn from the pool by a Computer choice at a position p', $p \leq p' \leq \mu$, then (length p'-length p) $\leq m$. (So if an atom a in the pool at p is not executed after m moves, then it is never executed during the play μ.)

Proposition. *Suppose that Computer has recursively bounded concurrency and that we have a strategy with recursively bounded non-determinism denoted by a program free of WhileDo's, and that μ is accepted. Then, using the supplied procedures for computing m, n as above we can compute an upper bound for the length of μ.*

For Park's example, using the procedures just mentioned we can compute a bound for the length of position such that submission and execution of the three atoms $x := 0$, $y := 0$, $x := 1$ is determined already for all plays μ. This length bounds also the final value of $y = i$ at the termination of μ, for all plays μ, but does not tell us what that value is for any particular μ.

Remark. We have again separated the roles of nondeterminism as a feature of the Programmer and of concurrrency as a feature of the Computer in the games. Viewed by the computer scientist, I, II assert that for accepted plays, one has "recursive responsible" submission-execution behavior. Viewed by the constructivist, I, II assert that for accepted plays, one can construct "bars" for determining submission-execution. Viewed by the classical mathematician, I, II assert that for accepted plays, one has "recursive uniform continuity" for submission-execution. Then again, we could study the case when the procedures supplied are polytime, or linear time, or constant parallel time, etc.

9. Final Remarks

Relation to other semantics

Here is a very brief sketch of the relation of our semantics to the previous semantics mentioned in **Section 1**.

(1) The inference rules of Gries and Owicki are theorems true of our semantics.

(2) Each sequence I of instructions in the denotation of program P in [de Bakker and Zucker 1982] is obtained by suppressing all programmer moves from a play μ consistent positions with a strategy $\mathcal{S}$ denoted by P in a computational state model $\Pi(\pi_0)$ of **Section 6**.

(3) Each sequence C of computational states in the denotation of P in the temporal logic of programs is obtained as the sequence of computational states corresponding to the play μ referred to above.

(4) The sequence of pairs of instructions and computational states in the denotation of P in [Chandy and Misra 1988] are those pairs arising from plays μ as above.

(5) [Alpern and Schneider 1985, 1987] look at infinite sequences of states, each including a computational state and program counters. They use the text of the program to compute which finite segments of such sequences could be a result of execution. Their "proof obligations" are essentially equivalent to the tree of Computer positions consistent with a strategy denoted by the program, with no Programmer positions (see **Section 4**) They use their "proof obligations" in order to verify the program specification, which is a set of paths in the tree of computer positions accepted by a Büchi automaton. But the relation between this tree and the structure of the program, as built up from subprograms, is difficult to grasp. This is why we

separate out the submission-execution mechanism as a computational (two-person) game tree. Alpern and Schneider left the program structure, and instead represented the program as a combination of an initial state and a set of "actions". In contrast, our strategies are represented by terms which preserve the structure of the program. Equipped with internal states, these strategies also resolve rough points. First, in the Alpern-Schneider theory it would be hard to make the value of program counters at a point of execution meaningful. But in computational games this difficulty disappears when internal states of strategies are used as concurrent counters.

(6) [Abadi, Lamport, and Wolper 1989], [Pnueli and Rosner 1989] deal with the question whether program specifications for reactive systems can be realized (implemented) by actual programs. From one point of view, our entire approach, having Programmer and Computer as independent players is a way of considering reactive systems as the basic building blocks for all programming, and therefore covers reactive systems naturally. In our terms, the question of realizability is to determine whether or not there exists at least one winning strategy for Programmer in a computational game with the program specification as winning condition. Most common program specifications can be expressed by Büchi automata or Gurevich-Harrington winning conditions. In this case, the question of realizability can be answered constructively. See [Büchi and Landweber 1969], [Buchi 1981, 1983], [Gurevich and Harrigton 1982], [A. and V. Yakhnis, 1990].

(7) There are close mathematical relations to existing theories of Büchi automata, of Dynamic Logics, and of Branching Temporal Logics. We have included some references.

Program development

Strategy models yield a new paradigm for extracting concurrent programs from specifications.

(1) Reformulate the program specification as a winning condition in a computational game Γ.
(2) Extract a class F of winning strategies for Γ.
(3) Extract a concurrent program in a target programming language L denoting a strategy in F.

See [A. and V. Yakhnis 1989, 90 in prep.]. Our paradigm for program development and verification is based on ordinary mathematical reasoning about strategies winning two-person games.

Program verification

Current formal systems for proving program correctness, and many published informal proofs of correctness, are unsatisfactory in practice. Informal correctness proofs often use fuzzy notions which make unclear assumptions about the computing model, while formal systems often fail to contain natural counterparts of legitimate intuitive notions used in informal correctness proofs. Similar opinions have been expressed in [Abraham and Ben-David 1986], and we have heard similar sentiments from K. Apt. We hope that our model, in which concurrent programs denote strategies for computational games, avoids at least these two pitfalls. Intuitive notions used in informal correctness proofs are easy to introduce, since they are ordinary mathematical notions about games and strategies. To verify that program P is correct we:

(1) Reformulate the specification for P as a winning condition in a computational game Γ.
(2) Convert P into a Programmer strategy on the two-person computational game tree for Γ.
(3) Show that the program P is correct by giving (any kind of) mathematical proof that f wins Γ.

Thus, to show that a program is incorrect amounts to proving there is a play, consistent with f, not in the winning set for Γ.

We followed precisely this scheme in Park's example.

Remark. The winning condition used in Park's example, as well as the ones used in [A. Yakhnis 1990], [V. Yakhnis 1990], are all special cases of a class defined by [Gurevich and Harrington 1982] in their celebrated short determinacy proof of the Rabin S2S decision procedure. This class of sets is known to be the same as the class of sets of infinite strings accepted by Büchi automata. This is an underlying mathematical connection between our work and that of [Alpern and Schneider 1985, 7], [Alpern, 1986]. They showed that the set of execution sequences, of computational states together with program counters, specified by a temporal formula are equally well specified by a Büchi automaton. Given such an automaton, it can be transformed into a Büchi automaton defining a set of plays which correspond perfectly to the set of execution sequences recognized by the first automaton. This is not a straightfoward transformation, since program counters are involved. With our internal states (an implementation-independent model for program counters) this becomes straightforward. It follows from these remarks that Gurevich-Harrington winning conditions for our computational games cover all program specifications in the usual temporal logics of programs.

Other games

At first glance, it is possible to confound our games with other two-person games that are quite different. There are many games connected with computer science, and even with concurrency. We briefly discuss [Abadi, Lamport, Wolper 1989], [Pnueli, Rosner 1989], and [Moschovakis 1989]. Pnueli and Rosner do not give a game semantics for programs and specifications, even though they have reference to Büchi Automata accepting infinite sequences. Abadi-Lamport-Wolper does not distinguish beween programs and specifications. Because these authors do not use strategies as denotations for programs, they cannot always express constraints needed for programming. They also do not cover shared memory models. Moschovakis develops games in which players are processes which play against other processes. In our games, in contrast, the Programmer plays against the Computer as opponent. The operations in Moschovakis are operations on players, not, as here, operations on strategies. In Moschovakis, concurrent program specifications are not treated as winning conditions for games. Concurrent program specifications are there implicitly represented as fixed point conditions on the player; a player wins if and only if the player, and not the opponent, makes the last move, if there is a last move. Proofs of program correctness are not treated as proofs that a strategy satisfies a winning condition. None of the above authors uses winning conditions to extract finite state strategies and programs, or to prove correctness of programs.

Subjects to be reworked

I. There is an active literature in concurrent and distributed programs which should be reworked in computational games, including [Alpern and Scneider 1985, 1987, 1987], [Broy 1986], [Hanna and Daeche 1985], [Hennessy and Milner 1985], [Jonnson 1987], [Lamport and Schneider 1984, 9], [Levin and Gries 1981], [Manna and Pnueli 1984, 1987], [Owicki and Gries 1976], [Peleg 1987], [Sistla 1989].

II. Similarly considerations apply to the theory of reactive systems: [Abadi, Lamport, Wolper 1989], [Pnueli 1986], [Pnueli and Rosner, 1989].

III. In parallel and distributed computing, there are algorithms which should be analyzed more completely in computational games in [Almasi and Gottlieb 1989], [Andrews and Schneider 1985, 1987], [Burns 1985], [Chandy and Misra 1988], [Gibbons and Rytter 1988], [Hwang and Briggs, 1984], [Mikosko and Kotov 1984], [Peterson and Silberschatz, 1985], [Polychronoupoulos 1988], [Raynal 1988].

IV. There are complexity and resource bound questions to be addressed corresponding to those in [Condon 1988] and [Buss and Scott 1990].

V. The "pool model" used for computational games provides a semantics for instruction-based parallelism [Ellis, 1986], [Fisher 1983] such as is used in the RISC architectures, and for trace-based abstract specification [Guttag 1975], [Parnas 1977].

VI. The topological and descriptive set theory methods in [Klarlund 1990] (see also [Arnold, 1986]) can used for computational games as well.

VII. In our paradigm, plays in pure computational games have replaced the usual execution sequences. With this analogy as guide, we can redesign Dynamic Logics or Temporal Logics or other Logics of Programs to have pure computational game semantics. (A roughly appropriate concurrent dynamic logic has been developed in [Nerode and Wijesekera 1991], stemming from prior work of Peleg.) We will carry out some of these logics in later papers.

Fairness

Fairness plays a significant role as a requirement on fair allocation of resources. See, for example, [Park, 1981], [Frances 1986], [Pnueli 1986], [Schneider 1989], [Klarlund 1990]. A typical view is that fairness is a constraint attached to a concurrent program construct. Such a constraint restricts the set of execution sequences produced by a program formed using the construct. What function does fairness play? The answer usually involves discussion of interleavings of execution sequences of component programs joined by the program construct. [Schneider 1989] gives a clear explanation: "A *fairness condition*... allows us, when reasoning, to ignore interleavings that cannot arise in "reasonable" implementations." The three most often discussed fairness conditions are strong fairness, weak fairness, and unconditional fairness. They admit correspondingly larger and larger sets of execution sequences. We can express these conditions very naturally in computational games, due to the independent roles of Programmer and

Computer in the submission-execution mechanism. But we do not express fairness by constraints on concurrent program constructs. Instead, we simply add fairness to the program specification by restricting the "winning set of plays" for the game. That is, a fairness condition is "just another" winning condition on the game.

Due to our separation of the submission from the execution mechanism, all violations of fair distribution of resources can be meaningfully attributed to misbehaviors of the execution mechanism of Computer, while the program constructs are merely operations on the program submission mechanism as expressed in Programmer strategies. But there can also be mistakes of the submission mechanism, or on the Programmer side. This we model by non-executability of a submitted instruction or a test predicate; in this case, the execution mechanism, or Computer, is not at fault. Basically, our treatment of fairness says to Computer "If you misbehave, you lose the game". In Park's example in **Section 8**, the "weakest possible fairness" was modelled by the set of all "accepted plays".

Related research

The research program here was suggested by Nerode in 1986 and results were announced in [A. Yakhnis June 1989, 1990] and [V. Yakhnis June 1989, 1990] and [A. Nerode, A. and V.Yakhnis 1989]. [A. and V. Yakhnis, 1990] have extended the Gurevich-Harrington determinacy theorem [Gurevich and Harrington 1982] to games with restraints. They found a criterion for determining the winning player, and gave an explicit class of winning strategies. The Gurevich-Harrington theorem is then simply a special case with no restraints. Their proof is based on a largest fixpoint argument using a monotonic operator, and yields a new proof of the Büchi-Landweber theorem for games defined by finite automata. Their criteria for the winning player and their extraction and description of winning strategies are much more managable than the corresponding ones originally provided by [Büchi and Landweber 1969]. Winning strategies are represented in a modular form, built up from strategies winning simpler games.

Combined with the semantics for concurrent programs presented in this paper, this leads to a highly modular program development style. We gave examples of programs corresponding to Büchi state-strategies. It is easy to write a finite state strategy with non-deterministic output function corresponding to the following non-deterministic programming construct from [Gries 1983].

if $\varphi_1 \to F_1$
[] $\varphi_2 \to F_2$
...
[] $\varphi_n \to F_n$
fi

"Programs as state-strategies", together with results of A. and V. Yakhnis cited above, lead to concurrent program development. We have investigated several development and correctness problems using state-strategy semantics, including mutual exclusion, and shared-memory and message-passing models. These will appear in later papers.

Concurrency diagrams

In a later paper we will exploit the symmetry between Programmer and Computer as opponents by allowing Computer a strategy other than the unrestricted strategy. This strategy reflects the architecture and the operating system. We get for Computer dual notions of positional, state, internal, Nerode strategies, etc. In the current paper we avoided consideration of arbitrary strategies for Computer by using the results of **Section 2** to eliminate pairs of arbitrary strategies, one for Programmer, one for Computer, in favor of pairs (Programmer, Vac), replacing the original tree of joint positions by the subtree of consistent positions for Computer. This reduces consideration to a Programmer strategy against an unrestricted strategy for Computer. Our concurrency diagrams for describing the behavior of concurrent programs reflect this symmetry and diagram the interaction of the states of the Programmer strategy and the states of the Computer unrestricted strategy. These concurrency diagrams differ from all previous diagrammatic representations of concurrent programs, such as dataflow and flow diagrams, because Computer is fully represented. They are a useful program development and teaching tool, and will be given separate treatment. See the appendix.

ACKNOWLEDGMENTS

Nerode thanks R. Platek for introducing him to program verification questions in the early 1980's, and D. Gries for giving a series of lectures from his work in the Cornell Logic Seminar in 1985. A. and V. Yakhnis thank D. Gries for his 1986 course on "The Science of Programming" and for his friendly criticism of their early work on the mutual exclusion problem. They also wish to thank Y. Gurevich for his encouragement and personal discussion of the Gurevich-Harrington theorem. The authors all thank W. Marek, G. Odifreddi, and J. B. Remmel for helpful readings of the manuscript.

APPENDIX

Concurrency Diagrams for State Strategies

The diagrams below have been drawn obeying the following conventions. In all the diagrams, ovals and overlapping ovals represent internal states of the strategies, terminal states are visually distinguished by using thick borders on the ovals, and thick arrows represent the output function. Figures 1 through 7 represent Büchi strategies. For Büchi strategies, the automaton operates on the tree of Computer moves only, ignoring Programmer moves. In Figures 1 through 7, the thin arrows represent Computer moves. But Figure 8 represents a state strategy in which the automaton operates on the tree of all positions, Computer or Programmer. In Figure 8, thin arrows are used for both Programmer and Computer moves.

Concurrency diagrams are not flowcharts. Flowcharts do not capture concurrency. Even for sequential programs, flowcharts miss essential information which is present in the diagrams. This prevents use of flowcharts for sequential parts of programs from being used as building blocks for concurrent programs. Compare our Figure 2 for the simplest program **begin** a **end** with the corresponding flowchart.

Fig. 1 THE STRATEGY EMPTY DENOTED BY THE NULL PROGRAM **begin end**

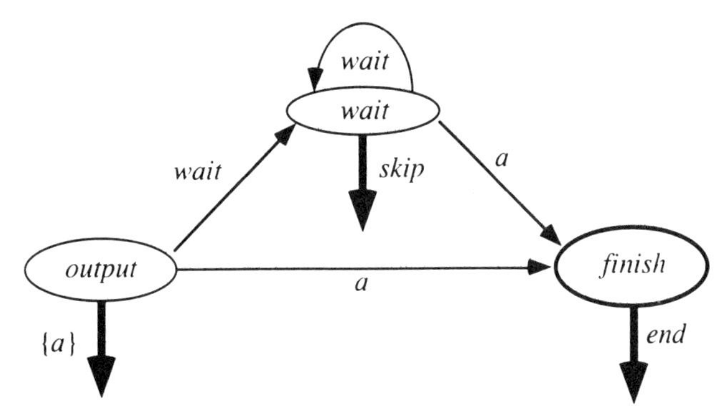

Fig. 2 THE BÜCHI STRATEGY $\text{ATOM}(a)_B$ DENOTED BY THE PROGRAM **begin** a **end**

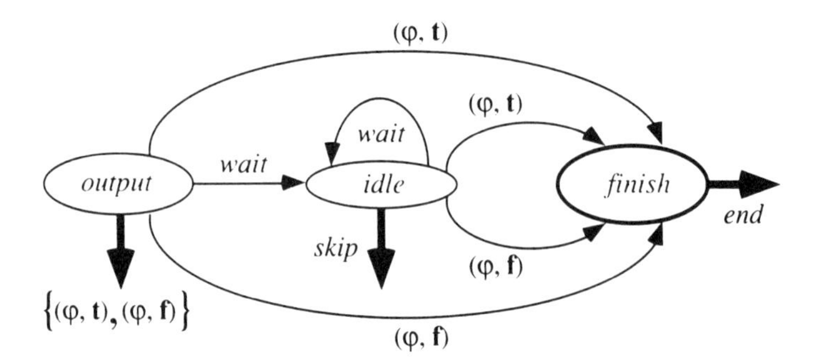

Fig. 3 THE BÜCHI STRATEGY $\text{ATOM}(\varphi)_B$ DENOTED BY THE PROGRAM **if** φ **then begin end**

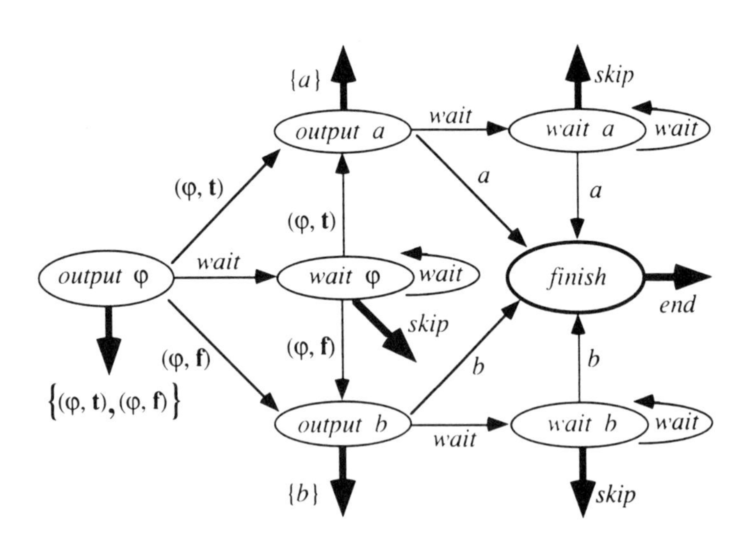

Fig. 4 THE BÜCHI STRATEGY $\text{IFTHENELSE}_\varphi(\text{ATOM}(a), \text{ATOM}(b))_B$ DENOTED BY THE PROGRAM **if** φ **then** a **else** b;

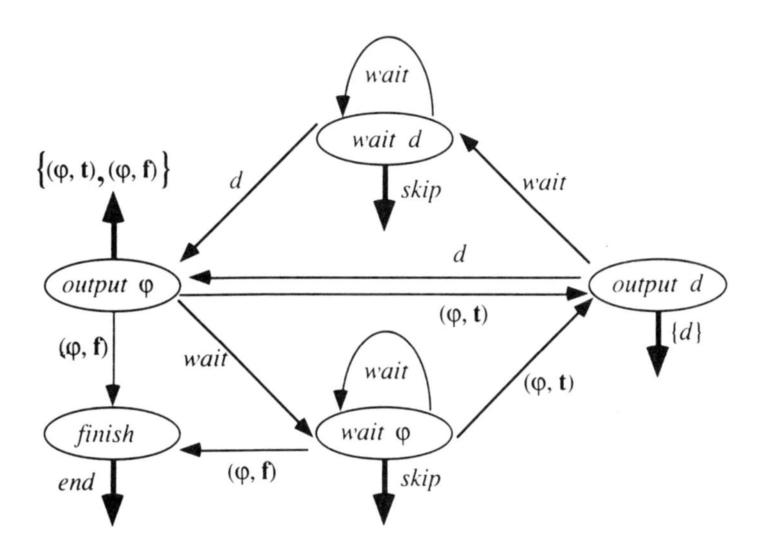

Fig. 5 THE BÜCHI STRATEGY $\text{WHILEDO}_{\varphi}(\text{ATOM}(d))_B$ DENOTED BY THE PROGRAM **While** φ **Do** d;

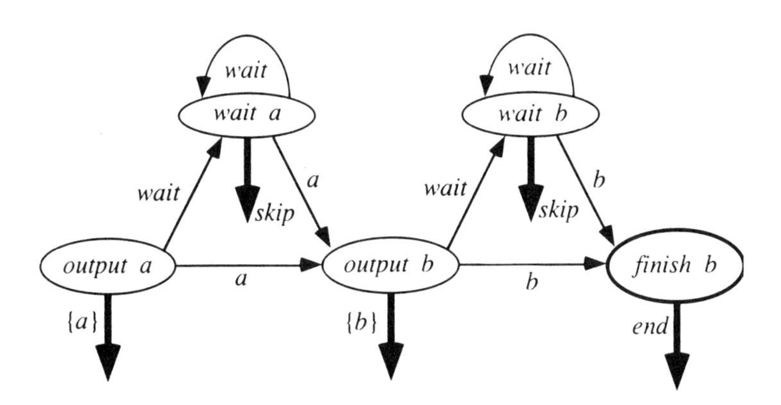

Fig. 6 THE BÜCHI STRATEGY $\text{SEQ}(\text{ATOM}(a), \text{ATOM}(b))_B$ DENOTED BY THE PROGRAM **begin** $a; b$ **end**

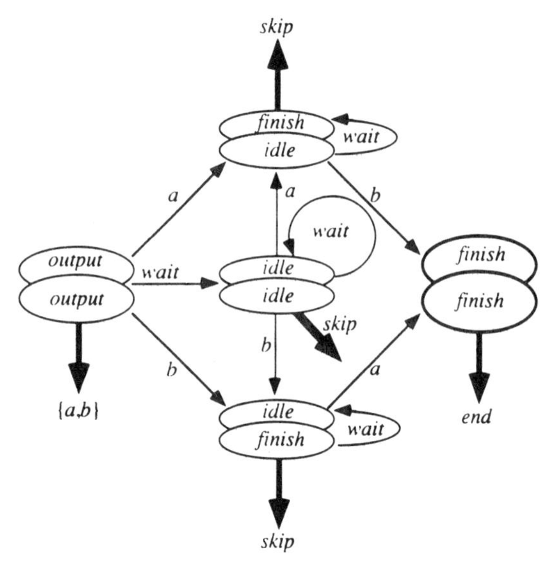

Fig. 7 The Büchi strategy $\text{Par}(\text{Atom}(a), \text{Atom}(b))_B$ denoted by the program **cobegin** $a; b$ **coend**

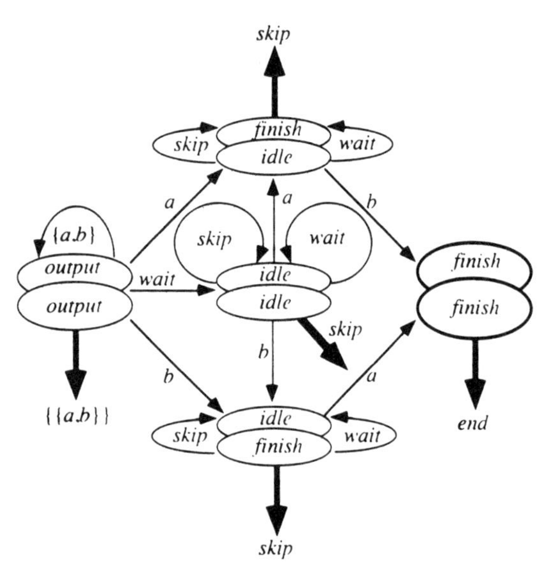

Fig. 8 The strategy $\text{Par}(\text{Atom}(a), \text{Atom}(b))$ denoted by the program **cobegin** $a; b$ **coend**

REFERENCES

M. Abadi, L. Lamport, P. Wolper, *Realizable and Unrealizable Specifications of Reactive Systems*, Proc. ICALP (1989).

U. Abraham and Shai Ben-David, *Informal and Formal Correctness for Programs (For the Critical Section Problem)*, mss., December.

G. S. Almasi and Allen Gottleib, *Highly Parallel Computing*, Benjamin/Cummings, 1989.

B. Alpern, *Proving Temporal Properties of Concurrent Programs: A Non-Temporal Approach*, Ph.D. Diss., Dept. Computer Science, Cornell University, January, 1986.

B. Alpern and F. B. Schneider, *Defining Liveness*, Information Processing Letters **21** (1985), 181–185.

B. Alpern and F. B. Schneider, *Recognizing Safety and Liveness*, Distributed Computing **2** (1987), 117–126..

B. Alpern and F. B. Schneider, *Verifying Temporal Properties without Temporal Logic*, ACM Trans. on Prog. Lang. **11(1)** (1987), 147–167.

G. R. Andrews and F. B. Schneider, *Concepts and Notations for Concurrent Programming*, Computing Surveys **15(1)**, 3–43.

K. R. Apt, *Recursive Assertions and Parallel Programs*, Acta Informatica **15** (1981), 219–232.

K. R. Apt and E.-R. Olderog, *Proof Rules and Transformations Dealing with Fairness*, Science of Computer Programming **3** (1983), 65–100.

K. R. Apt, A. Pnueli, and J. Stavi, *Fair Termination Revisted with Delay*, Science of Computer Programming **3** (1983), 65–100.

A. Arnold, *Topological Characterizations of Infinite Behaviors of Transition Systems*, Proc. 10th Coll. Automata, Languages, and Programming, LNCS 154 (1983), 490–510, Springer-Verlag.

H. Barringer, R. Kuiper, and A. Pnueli, *Now You May Compose Temporal Logic Specifications*, Proc. 16th STOCS, ACM (1984), 51–63.

M. Ben-Ari, Z. Manna, and A. Pnueli, *The Temporal Logic of Branching Time*, 8th POPL, ACM (1981).

M. Ben-Ari, *Principles of Concurrent Programming*, Prentice-Hall, 1982.

M. Broy, *A Theory for Nondeterminism, Parallelism, Communication, and Concurrency*, Theor. Comp. Sci. **45** (1986), 1–61.

A. Burns, *Concurrent Programming in Ada*, Camb. Univ. Press, 1985.

S. Buss and P. J. Scott (eds.), *Feasible Mathematics*, Birkhaüser, 1990.

J. R. Büchi and Lawrence H. Landweber, *Solving Sequential Conditions by Finite State Strategies*, Trans. Amer. Math. Soc. **138** (1969), 295–311.

J. R. Büchi, *Using Determinacy of Games to Eliminate Quantifiers*, Fundamentals of Computation Theory, LNCS 56, 1977, pp. 367–378.

J. R. Büchi, *Winning State-Strategies for Boolean–F_σ Games*, mss. (1981).

J. R. Büchi, *State-Strategies for Games in $F_{\sigma\delta} \cap G_{\delta\sigma}$*, J. Symb. Logic **48** (1983).

S. MacLane and D. Siefkes, *The Collected Works of J. Richard Büchi*, Springer-Verlag, 1990.

J. W. de Bakker and J. I. Zucker, *Processes and the Denotational Semantics of Concurrency*, Inf. and Control **54** (1982), 70–120.

A. Chandra, D. Kozen, and L. Stockmeyer, *Alternation*, JACM **28** (1981), 114–133.

K. Mani Chandy and J. Misra, *Parallel Program Design*, Addison-Wesley, 1988.

Y. Choueka, *Theories of Automata on ω-tapes: A Simplified Approach*, J. Comp. Sys. Sci. **8** (1974), 117–141.

E. M. Clarke, E. Ederson, and A. P. Sistla, *Automatic Verification of Finite State Concurrent Systems Using Temporal Logic Specifications: a Practical Approach*, Proc. 10th POPL, ACM (1983), 117–126.

A. Condon, *Computational Models of Games*, MIT Press, 1989.

John R. Ellis, *"Bulldog": A compiler for VLIW Architectures*, MIT Press, 1986.

E. Emerson and A. Sistla, *Deciding Branching Time Logic*, Proc. 16th ACM STOCS, ACM (1984), 14–24.

E. Emerson and J. Srinivasan, *Branching Time Temporal Logic*, LNCS 354 (1988), 123–172, Springer-Verlag.

E. Emerson and E. Clarke, *Using Branching Time Temporal Logic to Synthesize Synchronization Skeletons*, Science of Computer Programming, 1982, pp. 241–266.

J. A. Fisher, *Very Long Instruction Word Architectures and the ELI–512*, 10th Ann. Inter. Symp. Comp. Arch., IEEE Computer Society and ACM (1983), 140–150.

R. Floyd, *Assigning Meaning to Programs*, Mathematical Aspects of Computer Science, Proc. XIX Symp. Pure Math., Amer. Math. Soc., 1967.

N. Francez, *Fairness*, Springer-Verlag, 1986.

D. Gabbay, A. Pnueli, S. Shelah, and J. Stavi, *The Temporal Analysis of Fairness*, 7th POPL, ACM (1980).

D. Gale and F. M. Stewart, *Infinite Games with Perfect Information*, Contributions to the Theory of Games, Ann. Math. Studies 28, Princeton, 1953, pp. 245–266.

A. Gibbons and Wojciech Rytter, *Efficient Parallel Algorithms*, Cambridge University Press, 1988.

V. M. Glushkov, *Automata Theory and Structural Design Problems of Digital Machines*, Kibernetika **1(1)** (1965), 3–11.

V. M. Glushkov, *Automata Theory and Formal Microprogram Transformations*, Kibernetika **1(5)** (1965), 1–9.

D. Gries, *The Science of Programming*, Springer-Verlag, 1983.

D. Gries and J. Xue, *Generating a Random Cyclic Permutation*, Technical Report TR86–786, Dept. Comp. Science (September, 1986), Cornell University.

Y. Gurevich and L. Harrington, *Trees, Automata, and Games*, STOCS (1982), 60–65.

Y. Gurevich, *Infinite Games*, Bull. European Association for Theoretical Computer Science (June, 1989), 93–100.

J. Guttag, *The Specification and Application to Programming of Abstract Data Types*, Ph.D. diss. CSRG TR 59 (1975), University of Toronto.

Hanna and N. Daeche, *Specification and Verification using Higher Order Logic*, 7th Inter. Conf. on Computer Hardware Design Languages (1985), Tokyo.

D. Harel, *First Order Dynamic Logic*, LNCS 68, Springer-Verlag, 1979.

D. Harel and D. Peleg, *Process Logic with Regular Formulas*, Theor. Comp. Science **38** (1985), 307–322.

M. Hennessy and R. Milner, *Algebraic Laws for Nondeterminism and Concurrency*, JACM **32(1)** (1985), 137–161.

C. A. B. Hoare, *Communicating Sequential Processes*, Prentice-Hall, 1985.

J. Hopcroft and J. D. Ullman, *Introduction to Automata Theory, Language, and Computation*, Addison-Wesley, 1979.

K. Hwang and F. A. Briggs, *Computer Architecture and Parallel Processing*, McGraw-Hill, 1984.

B. Jonnson, *Modular Verification of Asynchronous Networks*, Proc. 6th Symposium on Priinciples of Distributed Computing, ACM (1987), 152–156.

N. Klarlund, *Progress Measures and Finite Arguments for Infinite Computation*, Ph.D. Diss., Cornell University, 1990.

A. H. Lachlan, *On Some Games Which are Relevant to the Theory of Recursively Enumerable Sets*, Ann. Math. **91** (1970), 291–310.

L. Lamport, *Specifying Concurrent Program Modules*, ACM Trans. Prog. Lang. Syst. **5(2)** (1983), 190–222.

L. Lamport and F. B. Schneider, *The Hoare Logic of CSP and All That*, ACM Trans. on Programming Languages and Systems **6(2)** (1984), 281–296.

L. Lamport and F. Schneider, *Pretending Atomicity*, Tech. Rpt. TR 89–1005 (1989), Dept. Computer Science, Cornell University.

L. H. Landweber, *Decision Problems for ω-Automata*, Math. Sys. Theory **3** (1969), 376–384.

G. M. Levin and D. Gries, *A Proof Technique for Communicating Sequential Processes*, Acta Inform. **15** (1981), 281–302.

J. Loeckx, K. Sieber, and R. Stanisfer, *The Foundations of Program Verification*, John Wiley and Sons, 1984.

Z. Manna and A. Pnueli, *The Temporal Framework for Concurrent Programs*, Academic Press (1981), 215–274.

Z. Manna and A. Pnueli, *Adequate Proof Principles for Invariance and Liveness Properties of Concurrent Programs*, Science of Programming **4** (1984), 257–290.

Z. Manna and A. Pnueli, *Specification and Verification of Concurrent Programs by $\forall$-automata*, 14th POPL, ACM (1987), 1–12.

D. A. Martin, *A Purely Inductive Proof of Borel Determinancy*, Recursion Theory (A. Nerode and R. Shore, eds.), Proc. Symp. in Pure Math. **42** (1985), 303–308.

S. Miyano and T. Hayashi, *Alternating Finite Automata on ω-Words*, Theor. Comp. Science **32** (1984), 321–330.

J. Miklosko and V. E. Kotov (Eds.), *Algorithms, Software and Hardware of Parallel Computers*, Springer-Verlag, 1984.

Y. Moschovakis, *A Game-Theoretic Modeling of Concurrency*, Logic in Computer Science, IEEE Computer Society (1989).

Y. Moschovakis, *Computable Processes*, POPL 90, ACM (1990).

Nerode, A., *General Topology and Partial Recursive Functionals*, Summaries of Talks at the AMS Summer Institute in Symbolic Logic, Cornell, 1957.

A. Nerode, *Linear Automaton Transformations*, Proc. AMS **9** (1958), 541–544.

Nerode, A., *Some Stone Spaces and Recursion Theory*, Duke Math. J. **26** (1959), 397–406.

A. Nerode and D. Wijesekera, *Constructive Concurrent Dynamic Logic*, Annals of Mathematics and Artificial Intelligence (1991), (Also Tech. Report of Math. Science Inst., Cornell, 1990)..

M. Nivat and D. Perrin, *Automata on Infinite Words*, LNCS 192, 1985.

A. Nerode, A. Yakhnis, and V. Yakhnis, *Concurrent Computation as Game Playing (mss.)*, presented at the "Logic From Computer Science" Workshop of the Mathematical Sciences Research Institute, Berkeley, California (November, 1989).

A. Nerode, A. Yakhnis, and V. Yakhnis, *Concurrent Programs as Strategies in Games*, Mathematical Sciences Institute Technical Report with appendix (October, 1990), Cornell University.

S. Owicki and D. Gries, *An Axiomatic Proof Technique for Parallel Programs*, Acta. Inf. **6** (1976), 319–340.

D. Parnas, *The Use of Precise Specifications in the development of Software*, Proc. IFIP Congress 1977 (1977), North-Holland.

R. Parikh, *Propositional Game Logic*, 24th FOCS, ACM (1983), 195–200.

D. Park, *On the Semantics of Fair Parallelism*, Abstract Software Specifications (D. Bjorner, ed.) (1979), 504–526, Springer-Verlag.

D. Park, *Concurrency and Automata on Infinite Sequences*, Proc. 5th GI Conference, LNCS 104 (P. Deussen, ed.) (1981), 167–183, Springer-Verlag.

D. Peleg, *Concurrent Program Schemes and Their Logics*, Theor. Comp. Sci. **55** (1987), 1–45.

D. Perrin, *An Introduction to Finite Automata on Infinite Words*, LNCS 192 (1984), 2–17, Springer-Verlag.

J. L. Peterson and A. Silberschatz, *Operating Systems Concepts*, Addison-Wesley, 1985.

A. Pnueli, *Applications of Temporal Logic to the Specifcication and Verification of Reactive Systems: A Survey of Current Trends*, LNCS 224 (1986), 510–584, Springer-Verlag.

A. Pnueli, *Application of Temporal Logic to Specification and Verification of Reactive Systems: Current Trends in Concurrency*, LICS, IEEE Computer Society (1986).

A. Pnueli and R. Rosner, *On the Synthesis of a Reactive Module*, POPL, ACM (1989).

C.D. Polychronopoulos, *Parallel Programming and Compilers*, Kluwer Academic Publishers, 1988.

V. Pratt, *Modelling Concurrency with Partial Orders*, Int. J. Parallel. Programming **15(1)** (1986).

M. O. Rabin, *Decidability of Second Order Theories and Automata on Infinite Trees*, Trans. Amer. Math. Soc. **141** (1969), 1–35.

M. Raynal, *Distributed Algorithms and Protocols*, John Wiley and Sons, 1988.

R. Rosner and A. Pnueli, *A Choppy Logic*, LICS, IEEE Computer Society (1986), 306–313.

S. Safra, *On the Complexity of ω-Automata*, FOCS 1986, ACM, 309–313.

F. B. Schneider, *The State Machine Approach: A Tutorial*, Tech. Rpt. TR86–800, Dept. Comp. Science (1986 (Revised 1987)), Cornell University.

F. B. Schneider, *Concurrent Programming*, (Notes, 160 pp.), Cornell University, 1989.

D. Siefkes, *Decidable Theories I – Büchi's Monadic Second Order Successor Arithmetics*, Lect. Notes in Math. 129, Springer-Verlag, 1970.

A. P. Sistla, *On Verifying that a Concurrent Program Satisfies a Nondeterministic Specification*, Inf. Proc. Letters **32(1)** (July, 1989), 17–24.

G. Slutzki, *Alternating Tree Automata*, Theor. Comp. Science **41** (1985), 305–318.

R. Streett, *Propositional Dynamic Logic of Looping and Converse*, Proc. 13th STOCS, ACM (1981), 375–383.

M. Takahashi, *The Greatest Fixed Points and Rational Omega Tree Languages*, LICS, IEEE Computer Society (1986).

P. Wolper, *Temporal Logic Can be More Expressive*, Inf. and Control **56** (1983), 72–99.

M. Vardi and P. Wolper, *Yet Another Process Logic*, LNCS 164 (1983), 501–512, Springer-Verlag.

P. Wolper, M. Vardi and A. P. Sistla, *Reasoning About Infinite Computation Paths*, Proc. 24th FOCS (1983), 72–99.

M. Vardi and P. Wolper, *An Automata-Theoretic Appoach to Automatic Program Verification*, LICS, IEEE Computer Society (1986).

M. Vardi, *Verification of Concurrent Programs: The Automata-Theoretic Framework*, LICS, IEEE Computer Society (1987).

A. Yakhnis, *Concurrent Specifications and Their Gurevich-Harrington Games and Representation of Programs as Strategies*, 7th Army Conf. on Applied Math. and Comp. (June 1989).

V. Yakhnis, *Extraction of Concurrent Programs from Gurevich-Harrington Games*, 7th Army Conf. on Applied Math. and Comp. (June 1989).

A. Yakhnis,, *Game-Theoretic Semantics for Concurrent Programs and their Specifications*, Ph.D. diss., Cornell University, August, 1990.

V. Yakhnis, *Concurrent Programs, Calculus of State-Strategies, and Gurevich-Harrington Games*, Ph.D. diss., Cornell University, August, 1990.

A. Yakhnis and V. Yakhnis, *Extension of Gurevich-Harrington's Restricted Memory Determinancy Theorem*, Ann. Pure App. Logic **48** (1990), 277–297.

A. Yakhnis and V. Yakhnis, *Gurevich-Harrington's Games Defined by Finite Automata*, Ann. Pure and Applied Logic (to appear).

MATHEMATICAL SCIENCES INSTITUTE
CORNELL UNIVERSITY, ITHACA, NY 14853

RESEARCH SUPPORTED BY THE U.S. ARMY RESEARCH OFFICE THROUGH THE MATHEMATICAL SCIENCES INSTITUTE OF CORNELL UNIVERSITY, CONTRACT $DAAG29-85-C-0018$

E-mail: anil@mssun7.msi.cornell.edu

FINITE AND INFINITE DIALOGUES

ROHIT PARIKH

"But how does he know where and how
he is to look up the word 'red' and
what he is to do with the word 'five'?"
Well, I assume he *acts* as I have described.
Explanations come to an end somewhere.
Ludwig Wittgenstein[1]

ABSTRACT. We consider dialogues over both specific and general Kripke structures where individuals convey and acquire knowledge through statements. We show that conventional 'proofs' of the existence of knowledge actually correspond to *optimal* strategies (which may not occur when real individuals talk). Sometimes these optimal, (and hence all) strategies need to be transfinite in that knowledge can only be acquired at some infinite ordinal. However, the situation changes sharply when we consider dialogues geared not to acquiring knowledge but to taking a justified risk.

1. INTRODUCTION

Imagine the following situation.[2] Two players Ann and Bob are told that the following will happen. Some positive integer n will be chosen and *one* of n, $n+1$ will be written on Ann's forehead, the other on Bob's. Each will be able to see the other's forehead, but not his/her own. After this is done, they will be asked repeatedly, beginning with Ann, if they know what their own number is.

Let us denote the situation where Ann has a and Bob has b as (a, b), and of course $|a-b| = 1$. Consider now the situation (1,2). When Ann is asked if she knows her number, she sees that Bob has a 2 and so her own number must be either 1 or 3. Not knowing which one, she will say, "I don't know". However, if Bob is asked next, he will realise that n must be 1, written on Ann's forehead, with 2 on his, since 0 is not a positive integer.

[1] *Philosophical Investigations* I.1

[2] This sort of problem has been discussed elsewhere, e.g. Littlewood [Li], Emde Boas et al [EGS], etc.

This argument also leads to a solution for the situation (3,2). Ann will respond as before, since her evidence in the beginning is the same as before. However, this time, Bob will also have to say, "I don't know", and so, when Ann is asked a *second* time, she will realise that the situation is not the same as the one just above, and hence that her number cannot be 1. Since 3 is the only other possibility, she will now say, "My number is 3"

Can we continue this argument beyond 3? If the situation is, say, (4,5), then not only must each party say, "I don't know" at the first stage, the other party must already *expect* this response. Thus Bob, seeing a 4, knows that his own number must be either 3 or 5, and in either case, Ann must say, "I don't know". Similarly, Bob must also say "I don't know" when first asked, and Ann must expect this answer. If the answers were already expected, then how can there be any learning, and if there is no learning, how can there be any progress?

Nonetheless, there is a "proof" by induction on n, that the dialogue will always terminate with one or the other player guessing his/her number. In the following, a *stage* will be a single question. A *round* will therefore consist of two stages.

Theorem 1. *In those cases where Ann has the even number, the response at the nth stage will be, "my number is $n+1$", and in the other cases, the response at the $(n+1)$st stage will be "my number is $n+1$". In either case, it will be the person who sees the smaller number, who will respond first.*

Proof. By induction on n, the smaller number. We divide the cases into four categories.
$(A)_n$: n is even, Ann has n.
In this case, Bob sees n and concludes that his own number is $n-1$ or $n+1$. In the first case, we are in case $(B)_{n-1}$ and by induction hypothesis, if Bob's number is $n-1$, then Ann should guess her own number at stage $n-1$. Since she said "I don't know my number", Bob realises that his number is not $n-1$ and hence must be $n+1$, which he will say at the next stage, i.e. at stage n.
$(B)_n$: n is odd, Bob has n.
If n is 1, then at the very first stage, Ann, seeeing a 1, will say, "my number is 2". If $n > 1$, then we revert to the case $(A)_{n-1}$ as above.
$(C)_n$: n is even, Bob has n.
Ann knows that her number is $n-1$ or $n+1$. If it were $n-1$, Bob would say at stage n that his number is n. Hence, when Bob says "I don't know

my number", she realises that she is in case $(C)_n$ rather than in $(D)_{n-1}$ and at the next stage she guesses her number.

$(D)_n$: n is odd, Ann has n.

This case is like the case (B). Note that if n is 1, then the number will be guessed at stage 2, since that is Bob's first chance to speak. □

However, there is a serious defect in the argument in that both Ann's and Bob's reasoning depends heavily on what the other one is thinking, including a consideration of what the other does not know. Ann's reasoning is justified if Bob thinks as she believes he does, and Bob's reasoning is justified if Ann thinks as he believes she does. But there is no guarantee that they do indeed think this way. How do we justify what each thinks and what each does and does not know?

In order to deal with this question we need some apparatus. The following definitions are standard in the literature [HM], [Pa].

Definition 1. A *Kripke model* M for a (two person) knowledge situation consists of a state space W and two equivalence relations $\equiv_1$ and $\equiv_2$. Intuitively $s \equiv_1 t$ means that states s and t are indistinguishable to player 1 (Ann) and $s \equiv_2 t$ means that they are indistinguishable to player 2 (Bob). We shall assume in this paper that W is finite or countable, and we won't need special predicates. (The ones we need are all singletons.)

In the example we are looking at, $W = \{(m,n)|m,n\epsilon N^+ \text{ and } |m-n| = 1\}$. If $s,t\epsilon W$ and $i\epsilon\{1,2\}$, then $s \equiv_i t$ iff $(s)_j = (t)_j$, where $j = 3 - i$, and $(s)_j$ is the j-th component of s. Intuitively, $s \equiv_i t$ means that when the dialogue begins, player i cannot distinguish between s and t, where Ann is player 1 and Bob is player 2.

Definition 2. A subset X of W is *i-closed* if $s\epsilon X$ and $s \equiv_i t$ imply that $t\epsilon X$. X is *closed* if it is both 1-closed and 2-closed. The corresponding topologies will be denoted $\mathcal{T}_1$, $\mathcal{T}_2$ and $\mathcal{T}$. For convenience we will identify a topology with its closed sets.

Example: For example, the set {(2,1),(2,3)} is 2-closed but not 1-closed in the situation we are looking at.

Definition 3. Given Kripke model M, $X \subseteq W$, and $s\epsilon X$, then i *knows* X at s iff for all t, $s \equiv_i t$ implies that $t\epsilon X$, i.e. the i-closure of $\{s\} \subseteq X$. X is *common knowledge* at s iff there is a closed set Y such that $s\epsilon Y \subseteq X$.

For example, at the state (2,3), Bob knows, but Ann does not, that both numbers are less than 4. They have common knowledge that Ann's number is even and Bob's odd, but nothing that is not implied by that.

Definition 4. A *DS* (dialogue system) for M is a map $f : W \times N^+ \to \{\text{"}no\text{"}\} \cup W$ such that for each odd n, $f(s,n)$ (Ann's response at stage n) depends only on the $\equiv_1$ equivalence class of s and on $f(s,m)$ for $m < n$. For each even n, $f(s,n)$ (Bob's response at stage n) depends only on the $\equiv_2$ equivalence class of s and on $f(s,m)$ for $m < n$.

Thus, e.g. if n is odd and $s \equiv_1 t$ and for all $m < n$, $f(s,m) = f(t,m)$ then $f(s,n) = f(t,n)$.

The answer "no" means "I don't know my number", whereas saying one's own number is equivalent to giving the full state. We shall refer to "no" as the trivial response. Any other response will be non-trivial.

Definition 5. The DS f is *sound* if for all s, if $f(s,n) \neq$ "*no*", then $f(s,n) = s$. We define $i_f(s) = \mu_n(f(s,n) \neq \text{"}no\text{"})$ and $p(s) = 1$ if $i_f(s)$ is odd and 2 if $i_f(s)$ is even.

Here μ stands for "least". $i_f(s) = \infty$ if $f(s,n)$ is always "no". We may drop the subscript f from i_f if it is clear from the context. $i(s)$ is the first stage when s is discovered and $p(s)$ is the person who discovers it. Note that knowing the state and knowing one's own number are equivalent and we will often use this equivalence.

Since a DS consists of a *pair* of strategies, one each for Ann and Bob, a strategy is not sound in isolation, but is sound only in conjunction with another strategy.

Note that we allow people to be ignorant even when they should not be, but a sound DS requires that all nontrivial responses be correct. Thus the DS which takes the constant value "no" is sound though it may not be very interesting. The DS used by most non-mathematicians may correspond to the strategy, "if you see a 1, then say 2. If you see a 2 and the other player has already said 'no', then say 3. Otherwise say 'I don't know' ". This strategy is also sound when used by both sides, but not optimal. Without loss of generality we will confine ourselves to functions f where the dialogue after any non-trivial response is constant. I.e. if one person says the state s, then it is s thereafter.

Lemma 1. *Let f be a sound DS. Let s,t be distinct states such that $s \equiv_i t$, $i(s) = k < \infty$ and $p(s) = i$. Then $i(t) < k$ and $p(t) \neq i$.*

Proof. At stage $i(s)$, i has evidence distinguishing between s and t. Since all previous utterances associated with s were "no", some previous utterance associated with t must have been nontrivial. Formally, $f(s, i(s)) = s \neq f(t, i(s))$. But $s \equiv_i t$. Hence $(\exists m < i(s))(f(s, m) \neq f(t, m))$. Since $m < i(s)$, $f(s, m) =$ "*no*" and so $f(t, m) \neq$ "*no*". Thus $i(t) \leq m < i(s)$. Now, if $p(t) = i$, then, by a symmetric argument, we could prove also that $i(s) < i(t)$. But this is absurd. Hence $p(t) \neq i$. $\square$

Corollary 1. *Suppose that $p(s) = i$ and there is a chain $s = s_1 \equiv_1 s_2 \equiv_2 s_3 \equiv_1 ... s_m$ such that no two consecutive elements are equal. Then $i(s) \geq m$.*

Proof. $i(s_m) \geq 1$. Now we can show using induction on k and lemma 1, that $i(s_{m-k}) \geq k + 1$. For we have $i(s_{m-k}) > i(s_{m-k+1}) = i(s_{m-(k-1)}) \geq k$ (by induction hypothesis). Taking $k = m - 1$ we get $i(s_1) \geq m$. $\square$

Corollary 2. *Suppose that there is a chain $s_1 \equiv_1 s_2 \equiv_2 s_3 \equiv_1 ... s_m \equiv_2 s_1$, with $m > 1$. and any two consecutive elements distinct. Then $i(s_k) = \infty$ for all $k \leq m$.*

Proof. Assume that $p(s) = 1$. If, say, $i(s_1) = k < \infty$, we would get $i(s_1) > i(s_2) > ... > i(s_m) > i(s_1)$, a contradiction. If $p(s) = 2$, then use the chain backwards for the same conclusion. $\square$

Remark 1: We now return to a discussion of the proof of theorem 1 above. The theorem is really a proof that the DS f is sound where f is defined by:
Ann's strategy: If you see $2n + 1$, then say n "no"'s and then, if Bob has not said his number, say "$2n + 2$". If you see $2n$, then say n "no"'s and if Bob has not said his number, say "$2n + 1$".
Bob's strategy: If you see $2n + 1$, then say n "no"'s and then, if Ann has not said her number, say "$2n + 2$". If you see $2n$, then say n "no"'s and if Ann has not said her number, say "$2n + 1$".

These strategies yield a DS with the following properties: $i(2n+2, 2n+1) = 2n+1$, $i(2n, 2n+1) = 2n$, $i(2n+1, 2n+2) = 2n+2$ and $i(2n+1, 2n) = 2n+1$. In other words, $i(s)$ is the smaller number if Ann's number is even, and the bigger number if it is odd. This DS is *optimal.* E.g. we have

$$(6, 5) \equiv_1 (4, 5) \equiv_2 (4, 3) \equiv_1 (2, 3) \equiv_2 (2, 1)$$

and hence $i(6, 5)$ has a minimum value of 5, the value achieved by the DS above.

Theorem 2. *The DS implicit in theorem 1 and described in Remark 1 is optimal. I.e. if h is any other sound DS, then $i_f(s) \leq i_h(s)$ for all s.*

Proof. By cases. Suppose, for example, that Ann has an even number and $s = (2n, 2n-1)$. $i_f(s) = 2n-1$. Suppose Bob is the one who first notices the state. Then we have $(2n, 2n-1) \equiv_2 (2n, 2n+1) \equiv_1 (2n+2, 2n+3)$..., and by corollary 1, $i_h(s)$ could not be finite. So Ann *does* first discover s. But then we have $(2n, 2n-1) \equiv_1 (2n-2, 2n-1) \equiv_2 (2n-2, 2n-3)... \equiv_2 (2,1)$ and so, by lemma 1, $i_h(s) \geq 2n-1$. □

Remark: We have not said whether the DS f itself is common knowledge[3] between Ann and Bob. The reason is, it does not matter. If they do act as described in the DS, then their utterances will be correct whenever they are non-trivial. Do Pavlov's dogs *know* that there is food when the bell rings? They act as if they knew, and we could credit them with the knowledge. Whether they *have* the knowledge is really a philosophical issue which we will not take up here.

This is important when we discuss group co-ordination. With Pavlov's dogs, there is at least Pavlov himself who has a plan which he is carrying out. But group activities involving some sort of co-ordination may *evolve* without anyone planning that they evolve just this way. Thus a purely behavioristic analysis of co-ordinated actions, which *look* as if they were guided by knowledge, seems appropriate.

This is especially so in view of the sort of problems raised by Schiffer [Sc], Clark and Marshall [CM] and others regarding common knowledge. Under their view, the correct understanding of a reference requires common knowledge, which consists of infinitely many levels of knowledge of the form: "Ann knows p", "Bob knows p", "Ann knows that Bob knows p", etc. If each level requires some minimal amount of time, say $\epsilon > 0$ to attain, then common knowledge could never be attained.

If, however, we only need the *co-ordinated learning* of some DS, then perhaps this is an attainable goal within the context of current computational learning theory.

2. Infinite Dialogues

In the last section we considered the situation where the pair of numbers is of the form $(n, S(n))$ where $S(n) = n+1$ is the successor function. Now instead of S we use a somewhat more interesting function g defined as follows:

[3]In its intuitive sense rather than the formal sense of Definition 3

$g(n) = 1$ if $n = 2^k$ for some $k > 0$
$g(n) = n + 2$ if n is odd
$g(n) = n - 2$ otherwise, i.e. if n is even, but not a power of 2.

In Figure 1, applying g is equivalent to climbing up the tree one step. Thus 4 is a power of 2 and $g(4)$ is 1. 6 is an even number, but not a power of 2 and so $g(6) = 6 - 2 = 4$. If n is odd, then of course $g(n) = n + 2$. Note that the numbers 2,6,14,... of the form $2^k - 2$ with $k \geq 2$, are not values of g.

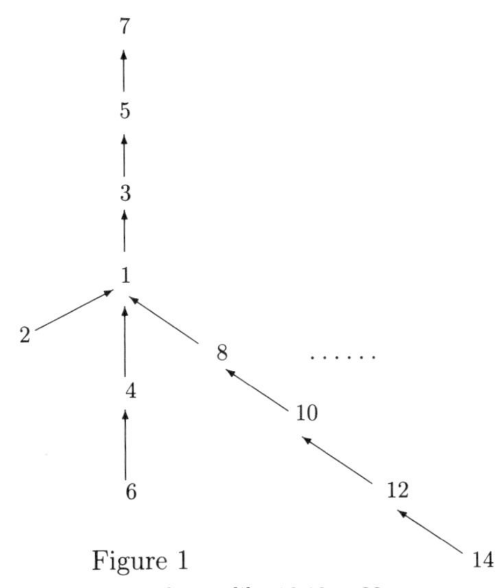

Figure 1

The dots represent numbers not shown, like 16,18,...,32,... etc.

Again the game proceeds by picking a positive integer n, and writing one of $n, g(n)$ on Ann's forehead, the other on Bob's. Figure 2 shows states (a, b), where a is written on Ann's forehead and b on Bob's and either $g(a) = b$ or $g(b) = a$. If two states are connected by a line marked A, then the two values of b are the same, and the states are initially indistiguishable to Ann; if marked by a B, then indistinguishable to Bob.[4]

[4]Figure 2 contains only *half* of all the points and there is of course a *second* tree dual to the one in Figure 2, containing (7,9), (7,5)... etc. where all pairs (a, b) of Figure 2 are replaced by (b, a) and all occurrences of A by B and vice versa. Even though Ann and Bob may not know their numbers, they do know which of the two trees they are in, and in fact the particular tree that they are in is common knowledge.

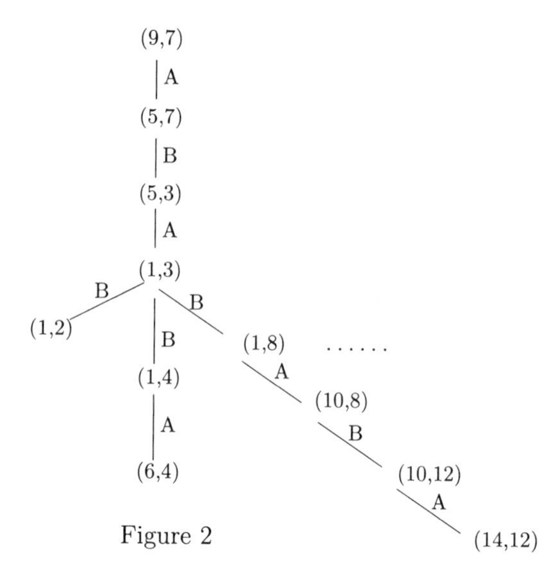

Figure 2

Let us consider now some particular situations. If the state is (6,4), then Ann cannot distinguish it from (1,4) but Bob can. In fact he confuses it with no other state, and when he is asked if he knows his number, he will respond with a "yes". If the state were (1,4) instead, Ann is no better off initially, but when she hears Bob say that he too does not know, then she realises that the state cannot be (6,4) and therefore her number must be 1.

We can now show by an argument analogous to that of theorem 1, that in the state $(1, 2^k)$, Ann realises by stage $2^{k-1} + 1$ that her number must be 1. In fact this will follow from the more general fact that if the state is *below* the point (1,3) in the tree and distance m from a leaf, then the person whose number is of the form $g(n)$ will realise, by stage $m + 2$ at the latest, that his/her number is $g(n)$.

Consider now what happens if the state is (1,3). Bob realises after Ann's first "I don't know", that his number is not 2, for otherwise Ann would have known that her number is 1. After her second "I don't know", he realises that his own number is not 4, for otherwise she would have guessed her own number. More generally, after $2^{k-1} + 1$ stages, he realises that his number is not 2^k.

Thus when ω stages pass, and Ann has *still* not guessed her own number, Bob will realise that his number is not any power of 2, and hence it must be

3. Thus, in the case of the state (1,3), it is at stage $\omega + 1$ that one of the two players realises his number.[5] We can easily see now that if the state is $(5, 3)$, then Ann will realise her own number at stage $\omega + 2$, and so on through all ordinals of the form $\omega + n$.

What is happening is that each time that someone says "I don't know my number", then it becomes common knowledge that they are not in a state where that person *would* have known his/her number. Such a state is always a leaf of the tree, and the "I don't know" removes that leaf, so that a new leaf is created. Thus Ann's saying "I don't know" removes the leaf (1,2), whereas Bob's saying "I don't know" removes the leaves (6,4), (14,12) (and others) creating the new leaves (1,4), (10,12) etc. Then when Ann says "I don't know" a second time, then the leaves (1,4) and (10,12) etc. are removed and (10,8) becomes a leaf.

By stage ω the entire portion of the tree below (1,3) is removed, which then becomes a leaf itself. At this stage, there is no state s left which is equivalent for Bob to (1,3) and on his next turn, at stage $\omega + 1$, he guesses his number to be 3.

This construction is quite similar to that in the Cantor-Bendixson theorem, [Mo], where a closed set is gradually diminished by removing isolated points until, at some countable ordinal, either nothing is left or else a perfect set is left. We now show that the parallel is exact except that we are dealing simultaneously with two topologies on the same space.

We need to correlate the various strands, informal argument, topology, and DS that have turned up so far. We are now dealing with a new Kripke structure M_g.

The space W of M_g equals $graph(g) \cup graph(g)^{-1}$. The equivalence relations $\equiv_i$ are again defined by letting $s \equiv_i t$ iff $(s)_j = (t)_j$ where $j = 3 - i$. We define the notion i-closed as before. We note that closed sets, sets closed in both topologies, are the closed sets of $\mathcal{T} = \mathcal{T}_1 \cap \mathcal{T}_2$. A subset X of W is *weakly closed* if it is closed in the topology $\mathcal{T}^+$, the smallest topology containing both $\mathcal{T}_1$ and $\mathcal{T}_2$. The four topologies $\mathcal{T}_1$, $\mathcal{T}_2$, $\mathcal{T}$ and $\mathcal{T}^+$ correspond, respectively, to Ann's knowledge, Bob's knowledge, commmon knowledge and implicit knowledge. The last represents knowledge *shared* by Ann and Bob, and not necessarily possessed by either individually. Note also that all closed sets in all topologies are in fact clopen.

[5]We assume that it is Ann who has a turn at ω and she will still not know whether her number is 1 or 5.

If X is closed and $p \epsilon X$ then p is *i-isolated* if $X - \{p\}$ is i-closed, and p is isolated if it is i-isolated for *some* i. X is *perfect* if it is nonempty, closed and has no isolated points. "Perfect" is a hybrid notion since the closure is with respect to $\mathcal{T}$ and "isolated" is relative to $\mathcal{T}^+$.

In the general case, if there are more than two individuals rather than just Ann and Bob, then we would define the notions similarly, by quantifying over the individuals i, sometimes existentially and sometimes universally, just as above.

Definition 6. Let $\mathcal{O}$ be the set of countable ordinals, M a Kripke structure. A *TDS* (transfinite dialogue system) for M is a pair of maps $p : \mathcal{O} \rightarrow \{1, 2\}$ and $f : W \times \mathcal{O} \rightarrow \{\text{"no"}\} \cup W$ such that for each s, α, If $j = p(\alpha)$, then $f(s, \alpha)$ depends only on the $\equiv_j$ equivalence class of s and on $f(s, \beta)$ for $\beta < \alpha$. Intuitively, $p(\alpha)$ is the person who responds at stage α and $f(s, \alpha)$ is his response at stage α. Again, "no" stands for "I don't know".

Definition 7. The TDS f, p is *sound* if for all s, if $f(s, \alpha) \neq$ "*no*", then $f(s, \alpha) = s$.

We again define $i_f(s) = \mu_\alpha(s(\alpha) \neq \text{"no"})$. Again, $i_f(s) = \infty$ if $f(s, \alpha)$ is always "no". We think of ∞ as larger than *all* the ordinals, even the infinite ones. By abuse of language, we will write $p(s)$ for $p(i(s))$. This makes our usage consistent with that of the previous section. We assume that it is Ann's turn at all even ordinals (which include 0 and the limit ordinals).

In the situation with the function g, there are two non-trivial closed sets, the set X, the domain of the tree in Figure 2, and its complement. The informal proof at the beginning of the section corresponds to the following TDS.[6]

First define:
$W_0 = W$, $\mathcal{T}_{i,0} = \mathcal{T}_i$, where the topologies $\mathcal{T}_i$ were defined in Definition 2.
$W_{\alpha+1} = W_\alpha -$ the i-isolated points of W_α, where $i = p(\alpha)$.
$\mathcal{T}_{i,\alpha+1} = \mathcal{T}_{i,\alpha} \cap \mathcal{P}(W_{\alpha+1})$
$\mathcal{T}_{j,\alpha+1} = \mathcal{T}_{j,\alpha} \oplus W_{\alpha+1} = \{X \cap W_{\alpha+1} | X \epsilon \mathcal{T}_{j,\alpha}\}$ for $j \neq i$
If λ is a limit ordinal, then

$$W_\lambda = \bigcap_{\alpha < \lambda} W_\alpha$$

$$\mathcal{T}_{i,\lambda} = \{X \cap W_\lambda | \exists \alpha < \lambda, X \epsilon \mathcal{T}_{i,\alpha}\}.$$

[6]Because of its resemblance to the construction in the proof of the Cantor-Bendixson theorem, we shall refer to it as the CB TDS.

Note that the i-isolated points are not j-isolated for $j \neq i$. Thus, in general, $W_{\alpha+1}$ has to be *added* to j's topology. E.g. in Figure 2, the point (6,4) is an isolated point for Bob but not for Ann. When that point is removed, Ann gets more sets in her topology.

Now define the functions p, f by: $p(\alpha) = 1$ if α is even and 2 if α is odd.[7] Let the function f be given by: at stage α, if s is an i-isolated point of W_α and $i = p(\alpha)$ then answer s. If the answer s has ever been given, then answer s. Otherwise answer "no". We show now that this is a sound and optimal DS for all countable Kripke structures M including structures M_g arising from a function g from N^+ to N^+.

Theorem 3. *f is an optimal* [8] *sound TDS and yields,*

$$i(s) = i_f(s) = \mu_\alpha(s \epsilon W_\alpha - W_{\alpha+1}).$$

Proof. f is evidently sound if it is a TDS. To see that it *is* a TDS, suppose, if possible, that there exist s, t, α such that $s \neq t$, $s \equiv_i t$ where $i = p(\alpha)$ and $f(s, \beta) = f(t, \beta)$ for all $\beta < \alpha$, but $f(s, \alpha) \neq f(t, \alpha)$. We may assume that α is the smallest ordinal for which this happens, so that $f(s, \beta) = f(t, \beta) =$ "*no*" for all $\beta < \alpha$. Obviously, at least one of $f(s, \alpha), f(t, \alpha)$, say the first, is different from "no". Now $s, t \epsilon W_\alpha$ (since all previous answers were "no") but s is an i-isolated point of W_α. This contradicts the fact that $s \equiv_i t$.

Suppose now that h is another sound TDS and, there is some s such that $i_h(s) = \alpha < i_f(s)$. I.e., h yields knowledge earlier in some case. Assume α is the smallest ordinal for which h is faster than f. Let $i = p(\alpha)$. Now we have $h(s, \beta) = f(s, \beta) =$ "*no*" for all $\beta < \alpha$ and $f(s, \alpha) =$ "*no*", but $h(s, \alpha) = s$. Since $f(s, \alpha) =$ "*no*", s is not an i-isolated point of W_α. Pick $t \neq s$ such that $s \equiv_i t$ and $t \epsilon W_\alpha$. Then t is not an i-isolated point of W_α, and hence of W_β for any $\beta < \alpha$. Thus we have $f(t, \beta) =$ "*no*" for all $\beta < \alpha$ and by minimality of α, $h(t, \beta) =$ "*no*" for all $\beta < \alpha$. Since h is a TDS, this yields $h(t, \alpha) = h(s, \alpha) = s$. Thus h is not sound. □

Let us consider the problem now over a general Kripke structure with a countable W. Let $W_\infty = \bigcap W_\alpha : \alpha \epsilon \mathcal{O}$.

Definition 8. $< W, \mathcal{T}_1, \mathcal{T}_2 >$ is *scattered* if $W_\infty = \emptyset$.

---

[7] We think of Ann as beginning with the first ordinal, 0, and re-starting the dialogue at each limit ordinal. Thus for instance, she responds at ω, an even ordinal.

[8] It is optimal among all TDS which question Ann at all even ordinals and Bob at all odd ordinals.

Theorem 4. *$< W, \mathcal{T}_1, \mathcal{T}_2 >$ is scattered iff there is a sound TDS for M which always yields a non-trivial answer.*

Proof. If $< W, \mathcal{T}_1, \mathcal{T}_2 >$ is scattered, then the CB TDS always yields an answer. If it is not scattered, then clearly the CB TDS cannot always yield an answer. For there is a perfect core W_∞ which is never removed. However, the CB TDS is optimal. Hence no sound TDS can yield an answer in all cases. □

In a case of this last kind, W splits into a scattered part and a perfect part. An answer will be obtained by the CB TDS in the scattered part, and no TDS can obtain a guaranteed correct answer in the perfect part. Thus, in our original example with the successor function, if we had been dealing with the set Z rather than N^+, then W would be perfect and we could not even get started. See, however, the next section where a kind of probabilistic scatteredness will enter.

Definition 9. g is *well founded* if there is no infinite chain $x_1, x_2, \ldots$ such that $g(x_{n+1}) = x_n$ for all n. g is *finite-one* iff for all n the set $g^{-1}(n) = \{m | g(m) = n\}$ is finite.

Some of the folowing results will depend on the assumption that $g(n) = n$ or $g(g(n)) = n$ never holds and we make this a *blanket assumption* from now on. The reason this condition is relevant is that if $g(g(n)) = n$ or $g(n) = n$, then the point $(n, g(n))$ might be isolated *even though* g is not well founded.

Theorem 5. *(a) The space $< W, \mathcal{T}_1, \mathcal{T}_2 >$ arising from g is scattered iff g is well founded.*
(b) If g is well founded and finite-one, then $W_\omega = \emptyset$, i.e. every state is learned at some finite stage.

Proof. The first part has been proved already. To see the second part, notice that König's lemma applies to the tree of g so that every state has only finitely many states under it. □

Corollary. *g is well founded iff the dialogue between Ann and Bob is guaranteed to terminate (with the CB TDS).*

We remark[9] that for computable well founded functions g, all ordinals less than Church-Kleene ω_1 can arise as ordinals of the corresponding trees. To see this, let R be a computable linear ordering on the positive integers. If R

[9] The reader unfamiliar with hyper-arithmetic sets and Church-Kleene ω_1 might skip the next two paragraphs. They are not needed for the rest of the paper.

is a well -ordering, then it can have any ordinal less than Church-Kleene ω_1. Let us asssume that k is a one-one map from all sequences of N^+ onto N^+. Now consider the game where the master of ceremonies chooses a sequence $(x_1, ..., x_n)$, where for all $i < n$, $x_{i+1} <_R x_i$. Then he writes $k(x_1, ..., x_n)$ on either Ann's or Bob's forehead and he writes $k(x_1, ..., x_{n-1})$ on the other's. We will now get a tree similar to that in Figure 1. The tree is well founded iff R is and if so, then its ordinal is at least as large as the ordinal of R. We can then modify by cutting, to get exactly the required ordinal.

This shows that all ordinals less than ω_1 do arise. It is not hard to see that larger ordinals cannot arise.[10] Also, the tree is well founded iff R is, and the problem of determining whether R is well founded is Π^1_1-complete. Hence the problem of determining whether the tree is well founded is also Π^1_1-complete, [Sp]. However, the tree is well-founded iff the game always terminates with some player guessing his/her number. Hence the problem of game termination is also Π^1_1-complete. A padding argument will show that this is still true if we confine ourselves to g which are polytime computable.

3. The Probabilistic Case

We now show that if we are dealing with *justified risk* rather than knowledge, then the situation of the last section, which required infinite dialogues, improves dramatically.

Suppose that the number n is chosen in accordance with some probability distribution, say $\mu_1(n) = \frac{1}{n(n+1)}$. Thus $\mu_1(1) = 1/2$, $\mu_1(2) = 1/6$, $\mu_1(3) = 1/12$ etc. This μ_1 induces a probability measure μ on W if we assume that the states (a, b) and (b, a) are equally likely.

Now the game is played as follows: each person risks \$1,000 by saying "I know my number, it is ...". If (s)he is right, (s)he receives one dollar. If (s)he is wrong, (s)he loses \$1,000. It is assumed that the parties are rational and that rationality is common knowledge. Thus, for example, if Ann did not guess her number yet, Bob can assume that it was not yet profitable for her, and conversely.

Then it will always make sense to take the risk after a *finite* number of steps. I.e. after a finite number of stages, the expected payoff will be positive for some person.

---

[10]If we do not require R to be well founded, then as in [Sp2], we can find a computable linear order R, whose largest well ordered initial segment has ordinal ω_1. If we use such an R, then we will get ω_1 as the least ordinal α such that $W_\alpha = W_\infty$.

For example, consider the function g of figures I and II, and assume that the state is (1,3). We know that Bob does not know whether his number is of the form 2^k or is 3. Now the probability of state $(1,3)$ is $\frac{1}{4}$. The probability of $(1,2^k) = (1/2)(\frac{1}{2^k \times (2^k+1)}) = \frac{1}{2^{k+1} \times (2^k+1)}$. The series $\sum_k \frac{1}{2^{k+1} \times (2^k+1)}$ is convergent, and hence there is an m ($m = 5$ in fact) such that

$$\sum_{k>m} \frac{1}{2^{k+1} \times (2^k+1)} < \frac{1}{4,000}.$$

Thus once Bob rules out 2^m (he will rule out smaller numbers earlier), he can profitably bet on 3 as his own number.

Theorem 6. *If some function g is well founded, μ is a probability distribution such that $\mu(s)$ is positive for all s, B is some bet with fixed positive payoff for a correct guess, and fixed negative payoff for an incorrect guess, and it is common knowledge that the parties are rational, then after a finite number of rounds, someone will take the risk (and will be justified in taking the risk).*

Proof. If not, then there is some x of lowest rank in the tree of g such that the bet is never profitable for either side. The person who sees x knows that his number is either $g(x)$ or else in $X = \{y | g(y) = x\}$. However, since x has the lowest possible rank as above, all these y, being of lower rank, are finitely bettable, i.e. it is justified to bet on them at some finite stage. Hence, as time passes, as elements of X which *should* have been guessed are *not* guessed, the set X steadily approaches the empty set and its probability approaches 0. Hence after some finite stage, its probability will be as small as needed. At this point it *will* make sense for Bob to take the risk. This contradiction proves the theorem. □

This proof is intuitively correct, but there are too many psychological assumptions. We now carry out a precise development analogous to that in the previous sections.

Definition 10. Let M be a Kripke structure, μ be a probability measure on W and ϵ be a real number > 0. A dialogue system f for M, μ is *ϵ-good* if for all s, there is an n such that $f(s,n) = s$, and if n is the least such, then $\mu(\{s\})/\mu(\{t | f(t,n) = s\}) > 1 - \epsilon$.

In other words, if the answer given at stage n is s, then the probability that it is correct is at least $1-\epsilon$. Moreover, for every s, an answer is obtained sooner or later - but at some *finite* stage.

In the following theorem we are given a probability measure μ_1 on N^+ such that $\mu_1(n) > 0$ for all n. This induces a measure on W given by $\mu(n, g(n)) = \mu(g(n), n) = (1/2) \times \mu_1(n)$. We will call such a μ_1 *computable* if it is a computable function taking values in $\mathcal{Q}$, the rational numbers.[11] There will then be a computable function r such that for all $k \epsilon N^+$, $\mu_1(\{1, 2, ..., r(k)\}) > 1 - (1/k)$.

Theorem 7. *Let M be a Kripke structure arising from a well founded computable g. Suppose that μ_1 is a computable probability measure on N^+ and $\delta > 0$. Then there is a δ-good, computable DS f for M, μ.*

Proof. Let d be an integer such that $1/d < \delta$. Define strategies $h_A(s)$, $h_B(s)$ as follows:
$h_A(s)$: Let $n = (s)_2$. Let k be the least integer greater than $\frac{2d}{\mu_1(n)}$.
Let $X = \{m | m < r(k) \text{ and } g(m) = n\}$.
Then $h_A(s) = 1 + max(h_B(m) : m \epsilon X)$; $h_A(s) = 1$ if X is empty.
$h_B(s)$: Let $n = (s)_1$. Let k be the least integer greater than $\frac{2d}{\mu_1(n))}$.
Let $Y = \{m | m < r(k) \text{ and } g(m) = n\}$.
Then $h_B(s) = 1 + max(h_A(m) : m \epsilon Y)$; $h_B(s) = 2$ if Y is empty.

We claim first that this gives us computable strategies h_A, h_B. The claim follows from the fact that $h_A(s)$ depends only on $(s)_2$ and on $h_B(m)$ for m such that $g(m) = (s)_2$. Similarly or h_B. Since g is well founded, this is a legitimate definition by recursion.

We now combine h_A, h_B into a DS f. If n is odd, $n \geq h_A(s)$ and all previous values $f(s,p)$ have been trivial, then $f(s,n) = (g((s)_2), (s)_2)$. If some previous value has been t then $f(s,n) = t$. Otherwise $f(s,n) =$ "*no*". Similarly with n even, using h_B instead of h_A.

It is easily seen that h_A depends only on information that Ann has, and h_B depends only on information that Bob has. Hence f is a DS.

We now show that this DS is $(1/d)$-good, this will imply that it is δ-good. Given s, let n be the least integer such that $g(s,n) \neq$ "*no*". Assume without loss of generality that n is odd.

If X is empty, then the set $\{m | g(m) = (s)_2\}$ is contained in the set $\{m | m > r(k)\}$ and hence has measure less than $\mu_1(g((s)_2))/d$. Thus the probability that $(s)_1 = g((s)_2)$ is larger than $1 - 1/d$.

If X is not empty, then $n = h_A(s)$. Suppose $(s)_1$ were such that $g((s)_1) = (s)_2$, then if $(s)_1 \epsilon X$, we would already have a non-trivial value earlier. Hence,

[11] This assumption of rational values is only for convenience.

the probability that $g((s)_1) = (s)_2$, given that there have been only trivial answers so far, is less than $\mu_1(g((s)_2) \times (1/d)$. Hence the probability that the state is $((g(s)_2), (s)_2)$ exceeds $1 - (1/d)$. □

It is not necessarily true that f as above is an optimal DS. We made estimates in defining h_A, h_B of how far Ann and Bob have to go before they can discount the rest of N^+ as having too small a probability. But smaller estimates may well serve and the best estimates might not even be computable.

It is possible for δ-good DS to exist even if g is not well founded. For example, we take $g(n) = n + 1$ on the set Z of all integers, then the corresponding W is perfect and the dialogue can make no progress in imparting *knowledge*. But suppose that we take $\mu_1(n) = (\alpha) \times \frac{1}{2^{n^2}}$, where α is some normalising constant. Then the larger numbers will be so improbable that W will be, for all practical purposes, finite. The existence of good DS is then easy to prove.

On the other hand, if g is not well founded, then we can easily construct an example of a μ and a δ such that there are no δ-good DS. If g has a cycle of length greater than 2, then just give each point in the cycle equal weight and take $\delta < 1/2$. E.g. if $g(x_{i+1}) = x_i$ for $i < n$ and $g(x_1) = x_n$, then the states (x_i, x_{i+1}) and (x_i, x_{i-1}) will always seem equally likely to Bob, and states (x_i, x_{i+1}) and (x_{i+2}, x_{i+1}) will always seem equally likely to Ann. Since $\delta < 1/2$, no one will ever bet, and the fact that they did not bet will convey no information. If the chain $x_1, ..., x_n, ..$ as above is infinite, then let the chain as a whole have measure $1/2$, and all other points together have measure $1/2$. *Within* the chain, assign measure $\mu_1(x_n) = \frac{1}{2n(n+1)}$. This yields $\mu(x_n, x_{n+1}) = \frac{1}{4(n+1)(n+2)}$. Now some states *are* more probable than others, but the ratio of probabilities $\mu(x_n, x_{n+1}){:}\mu(x_{n+1}, x_{n+2})$ is never less than 1:2. Hence if $\delta < 1/4$, then 1-equivalent states and 2-equivalent states will be too close in probablity and again the learning process cannot get started - no point in the chain can ever get guessed by a strategy which is .25-good. This yields:

Theorem 8. *g is well founded iff for all μ, δ, there exist δ-good DS for M_g.*

Acknowledgments

We thank Amnon Nissan and Alexey Stolboushkin for comments on an earlier version of the paper.

References

[ChM] M. Chandy and J. Misra, How Processes Learn, *Proceedings of 4th ACM Conference on Principles of Distributed Computing* (1985) pp 204-214.

[CM] H. H. Clark and C. R. Marshall, Definite Reference and Mutual Knowledge, in *Elements of Discourse Understanding*, Ed. Joshi, Webber and Sag, Cambridge U. Press, 1981.

[EGS] P. Emde Boas, J. Groenendijk and M. Stockhof, "The Conway paradox: its solution in an epistemic framework" in J. Groenendijk, T.M.V. Janssen & M. Stokhof (eds.) *Truth, Interpretation and Information, selected papers from the third Amsterdam Colloquium,* Foris Publications, Groningen, Amsterdam studies in semantics (GRASS 2), Dordrecht 1984, pp 159–182. (appeared originally in J.A.G. Groenendijk, T.M.V. Janssen & M.B.J. Stokhof *Formal methods in the study of language* Mathematical Center Tracts 135, Amsterdam 1981, pp.87–111.)

[HM] J. Halpern and Y. Moses, Knowledge and Commmon Knowledge in a Distributed Environment, *Proc. 3rd ACM Symposium on Distributed Computing* 1984 pp. 50-61

[Lew] D. Lewis, *Convention, a Philosophical Study*, Harvard U. Press, 1969.

[Li] J. E. Littlewood, *A Mathematician's Miscellany*, Methuen and company, 1953

[Mo] Y. Moschovakis, *Descriptive Set Theory*, North-Holland 1980.

[Pa] R. Parikh, Logics of Knowledge, Games and Dynamic Logic, *FST-TCS* 1984, Springer LNCS 181, pp. 202-222.

[Pa2] R. Parikh, Levels of Knowledge in Distributed Computing, *IEEE LICS Symposium*, 1986, pp. 314-321.

[PK] R. Parikh and P. Krasucki, "Communication, Consensus and Knowledge", *J. Economic Theory*, **52** (1990), pp. 178-189.

[PR] R. Parikh and R. Ramanujam, Distributed Processes and the Logic of Knowledge, *Logics of Programs*, Springer Lecture Notes in Computer Science, **193**, (1985), pp. 256-268.

[Pl] J. Plaza, "Logics of Public Communications", *Proceedings of the fourth International Symposium on the Methodology of Intelligent Systems*, poster session, Oak Ridge National Laboratory, 1989, pp. 201-216.

[Sc] S. Schiffer, *Meaning*, Oxford U. Press, 1972.

[Sp] C. Spector, "Recursive Well Orderings", *J. Symbolic Logic* 20 (1955), 151-163.

[Sp2] C. Spector, "Hyperarithmetical Quantifiers", *Fundamenta Mathematica* **48** (1959) pp. 313-320

Department of Computer Science, City University Graduate Center, 33 West 42nd Street, New York, NY 10036
Email: RIPBC@CUNYVM.CUNY.EDU

The author is also affiliated with the Department of Computer Science, Brooklyn College, and the Departments of Mathematics and Philosophy, CUNY Graduate Center

Research supported in part by NSF grant CCR-8803409.

SOME RELATIONS BETWEEN SUBSYSTEMS OF ARITHMETIC ND COMPLEXITY OF COMPUTATIONS

PAVEL PUDLÁK

ABSTRACT. We shall introduce a special mode of interactive computations of optimal solutions to optimization problems. A restricted version of such computations was used in [KPT] to show that $T_2^i = S_2^{i+1}$ implies $\Sigma_{i+2}^p = \Pi_{i+2}^p$. Here we shall reduce the question whether $T_2^i = S_2^i$ to a question about interactive computations (in a more general sense) of some optimization problems.

1. INTRODUCTION

We shall consider fragments of bounded arithmetic S_2. This is a first order theory of arithmetic, where the induction schema is restricted to bounded formulae. Our main reason for studying such systems is their close relation to low level computational complexity. We hope that eventually this research will bring new insight into the problems in complexity theory. Ideally, we would like to show some independence results for such theories and sentences stating some unknown relations between complexity classes. In order to be able to prove such results, we have to understand better the mutual relation between the complexity theory and such logical theories. This paper attempts to make another step in this direction.

The first system for bounded arithmetic was proposed and studied by Parikh [Pa]. Nowadays his system is known as $I\Delta_0$; it is Peano Arithmetic with induction restricted to bounded arithmetical formulae. This and related systems have been extensively studied by Paris and Wilkie (see [PW] for a survey paper). The system S_2 and an equivalent system T_2 were introduced by Buss [B1]. These systems are conservative extensions of $I\Delta_0 + \Omega_1$, where Ω_1 is $\forall x \exists y (y = 2^{\lceil \log_2(x+1) \rceil})$. The richer language of S_2 and T_2 enables one to define natural fragments S_2^i and T_2^i, $i = 1, 2, \ldots$. The definition of S_2^i and T_2^i is motivated by Stockmeyer's Polynomial Hierarchy. This hierarchy is a natural extension of the classes $\mathcal{P}$, $\mathcal{NP}$, $co\mathcal{NP}$ in a much similar way as the arithmetical hierarchy is an extension of classes *recursive sets, recursively enumerable sets, complements of recursively enumerable sets*, etc. It is an open problem whether S_2 is finitely axiomatizable. This problem is equivalent to the statement that S_2 collapses to some fragment S_2^i. Similarly it is an open problem whether Polynomial Hierarchy collapses to some level.

In [KPT] we showed that if S_2 is finitely axiomatizable, then Polynomial Hierarchy collapses. This was done by proving

$$T_2^i = S_2^i \Rightarrow \Sigma_{i+2}^p = \Pi_{i+2}^p$$

(where Σ_{i+1}^p and Π_{i+2}^p are levels in the Polynomial Hierarchy), and using the fact that we have the following inclusions

$$T_2^0 \subseteq S_2^1 \subseteq T_2^1 \subseteq S_2^2 \subseteq T_2^2 \subseteq \dots .$$

Hence assuming a plausible conjecture that Polynomial Hierarchy does not collapse, the odd inclusions are strict. Here we try to do a similar thing with the even inclusions. We shall present a conjecture on certain computations which implies that also the odd levels are strict. This is a conjecture about a new concept and, unfortunately, we are not able to reduce it to any problem in complexity theory which has been considered before.

It should be stressed that the motivation for showing $S_2^i \neq T_2^i$ is not only to clear up the remaining problems. A strong conservation result has been proved for pairs T_2^i and S_2^{i+1} in [B2]. If there were some partial conservativity also between pairs S_2^i and T_2^i, we would have some partial conservativity of the whole system S_2 over its fragments S_2^i. This seems rather unlikely. The separation of S_2^i from T_2^i could be the first step toward showing that there is no such conservativity. For related results see also Krajíček's paper [Kr] in these proceedings.

2. Fragments of Bounded Arithmetic

We shall use fragments of systems of bounded arithmetic S_2 and T_2 of Buss [B1]. Each system has a finite set of basic open axioms and an axiom schema of induction. We consider two schemata. Ordinary induction IND:

$$\varphi(0) \mathbin{\&} \forall x(\varphi(x) \rightarrow \varphi(x+1)) \rightarrow \forall x \varphi(x);$$

and *polynomial induction* PIND:

$$\varphi(0) \mathbin{\&} \forall x(\varphi(\lfloor x/2 \rfloor) \rightarrow \varphi(x)) \rightarrow \forall x \varphi(x).$$

S_2 is the system with the basic axioms and the schema PIND for all bounded formulae; T_2 is the system with the basic axioms and the usual schema of induction IND for all bounded formulae.

The particular choice of primitive notions and basic axioms describing the primitive notions is not very important, hence we shall not give a precise definition, instead we shall describe the most important properties that we need:

(1) the language contains a finite number of polynomial time computable functions and predicates;
(2) there are natural classes of formulae defining the sets in the Polynomial Hierarchy.

The reason for extending the usual language of arithmetic is that we want to have naturally defined fragments of these theories. If we did not extend the language, the bottom fragments would not have nice properties. Furthermore, the schemata of induction for bounded formulae are not strong enough to pove the existence of functions that grow faster than the *terms* of the language. Thus it is more convenient to add functions with higher growth rate than extend the axiomatization, which is Π_1^0, by a Π_2^0 axiom. The higher growth rate is needed in order to be able to formalize polynomial time computations. If a sequence s of length k is encoded by a binary expansion of a number n, then n is about 2^k. If we need another sequence of length k^2, then we need a number of size 2^{k^2}, which is about $2^{(\lceil \log_2(n+2)\rceil)^2}$. That is why we need such a function. Note that there is an alternative approach which is based on the sequence as the basic primitive concept instead of the concept of the number. This has been considered by Cook [C], (however his theories PV and $PV1$ are quantifier free).

The classes of bounded formulae Σ_i^b and Π_i^b are defined as follows. First choose a suitable class $\Sigma_0^b(=\Pi_0^b)$ whose formulae define sets in $\mathcal{P}$. Then we define Σ_i^b and respectively Π_i^b as the classes of formulae with the corresponding prefix of *bounded* quantifiers follows by a Σ_0^b formula. The formulae in Σ_0^b are defined using *sharply bounded quantifiers*. These are bounded quantifiers where the bound is always the logarithm of a term. Thus we shall use also the function $\lceil \log_2(x+1)\rceil$, which is denoted by $|x|$. For $i>0$, the formulae in Σ_i^b (respectively in Π_i^b) define just the sets in Σ_i^p (respectively in Π_i^p). Using this relation we shall sometimes identify Σ_i^p sets with their Σ_i^b definitions. Σ_0^b formulae, as they are usually defined, do not define all sets in $\mathcal{P}$. This is rather inconvenient and complicates the statements of the theorems. Therefore in this paper we shall assume that Σ_0^b is a suitable class of formulae that define just the sets in $\mathcal{P}$. Such a class can be constructed by formalizing polynomial time Turing machine computations. This change influences only the weakest fragment that we consider T_2^0: now in T_2^0 we can define all polynomial time computable functions. To get the full symmetry, we shall use another convention which is not quite standard: we shall sometimes denote $\mathcal{P}$ by Σ_0^p and Π_0^p.

Now we can define fragments of bounded arithmetic: S_2^i is S_2 with PIND restricted to Σ_i^b formulae, T_2^i is T_2 with induction restricted to Σ_i^b formulae, $i \geq 0$. Thus, for instance, we can think of T_2^0 as "induction for $\mathcal{P}$", and of S_2^1 as "polynomial induction for $\mathcal{NP}$".

Often it will be more convenient to use the following schema LIND (a form of the least number principle) instead of PIND:

$$\neg\varphi(x,0) \vee \exists t < |x|(\varphi(x,t) \ \& \ \neg\varphi(x,t+1)) \vee \varphi(x,|x|).$$

Fragments S_2^i can be axiomatized by this schema for Σ_i^b formulae. It has been proven in [B1] that

$$T_2^0 \subseteq S_2^1 \subseteq T_2^1 \subseteq S_2^2 \subseteq T_2^2 \subseteq \dots ,$$

hence

$$S_2 = \cup S_2^i \equiv \cup T_2^i = T_2.$$

Many facts about the subsystems $I\Sigma_i^0$ of Peano Arithmetic transfer to fragments S_2^i and T_2^i (e.g. $\mathrm{PIND}\,\Sigma_i^b \equiv \mathrm{PIND}\,\Pi_i^b$), but not all. In particular the proof that Peano Arithmetic is not finitely axiomatizable breaks down in the new context. To prove that Peano Arithmetic is not finitely axiomatizable one shows that

$$I\Sigma_{i+1}^0 \vdash \mathrm{Con}(I\Sigma_i^0)$$

(where Con denotes the consistency), and uses Gödel's Theorem to show that

$$I\Sigma_i^0 \vdash \mathrm{Con}(I\Sigma_i^0)$$

does *not* hold. While Gödel's Theorem holds for fragments of bounded arithmetic (see [B1]), S_2 does not prove even the consistency of basic axioms; (there are stronger results in this direction in [B1], [PW], [Pu], [T]).

3. Optimization Problems

The aim of this paper is to show that there is a close connection between fragments S_2^i and T_2^i on the one hand and certain interactive computations of solutions to optimization problems on the other hand. Here we introduce some basic terminology on optimization problems.

Let $C(x,y)$ be a binary relation, let ρ be a function; both are defined on natural numbers. We shall think of x as an input. For an input x, any y such that $C(x,y)$ holds will be called *a feasible solution to* x, (put otherwise, C is the condition that determines what is feasible for x). Function ρ measures how good a solution is: if y and y' are feasible solutions, then y' is better than y if $\rho(y) < \rho(y')$. If there is no better solution to x than y, then y is called an *optimal solution.* For a pair C, ρ to determine an *optimization problem* we shall assume two purely technical conditions, which will greatly simplify the exposition:

(1) 0 is a feasible solution to any x;
(2) if y is a feasible solution to x, then $y \leq x$.

We shall, of course, assume that all finite objects are encoded as numbers.

From the practical point of view only problems where C is in $\mathcal{P}$ and ρ is polynomial time computable are interesting. Most of the $\mathcal{NP}$-problems come from optimization. Let us consider two examples.

(1) CLIQUE
$C(x,y) \equiv$ "y is a clique in graph x",
$\rho(y) =$ "the size of y",
(0 is assumed to be the code of the empty clique).
(2) TRAVELLING SALESPERSON
$C(x,y) \equiv$ "y is a tour in graph x whose edges are labelled by numbers (or $y = 0$)",
$\rho(y) =$ "sum of all the labels of the graph minus the length (i.e. the sum of the labels) of the tour" (and $\rho(0) = 0$).

It seems that there is some inherent difference between the two problems. In the first case the range of ρ on feasible solutions to x is bounded by a polynomial in the *size* of x, (in fact it is less than the size of x). In the second case the range may be exponentially large in the size of x, since numbers up to $2^n - 1$ have length less than or equal to n in binary notation. For usual deterministic polynomial time computations this makes little difference, since we know that both problems are $\mathcal{NP}$-hard, but if we have some extra information available the first type of a problem is more likely to be solvable.

It is possible to define a hierarchy of optimization problems according to the size of the range of ρ. However, for this paper we need only the distinction between the polynomial size range and the exponential size range. We shall call them *type* 1 and *type* 2 optimization problems respectively. Thus *type* 2 are in fact all optimization problems and we use this term only to stress the difference. For reasons of symmetry which will be apparent later, we define also *type* 0: the problems in which the range of ρ is uniformly bounded by a constant.

Optimization problems with C in $\mathcal{P}$ will be related to fragments T_2^0, S_1^2, T_2^1. For higher fragments we shall need C in Π_{i-1}^p, while ρ will always be polynomial time computable.

4. Computations with Counterexamples

Let an optimization problem $C(x, y), \rho$ be given. Suppose a person called *STUDENT* is to determine an optimal solution to x and suppose he can use the help of another person called *TEACHER*. *STUDENT* has limited ability, (in the simplest case he can perform deterministic polynomial time computations), while *TEACHER* knows everything about the given problem. *STUDENT* can ask questions of the form:

Is y an optimal solution to x?

TEACHER must answer correctly and, moreover, if her answer is negative, she must produce a counterexample which is some better feasible solution to x. The aim of *TEACHER* is to test *STUDENT*, so she can choose counterexamples which convey as little information as possible to *STUDENT*.

Formally we would model *STUDENT* as a mutlitape Turing machine where the queries of *STUDENT* and answers of *TEACHER* will appear on a special tape of the machine. The queries are computed by the machine, while the answers come from outside following an arbitrary strategy. We define that *STUDENT* (i.e. a given Turing machine) *solves* the optimization problem, if for every x and every strategy of *TEACHER*, *STUDENT* computes some optimal solution to x. Thus *TEACHER* does not act as an oracle in the ordinary sense, she is rather a person in a two player game. Further we shall modify this model by allowing *STUDENT* to use an ordinary racle from some class Σ_i^p and by restricting the number of queries posed to *TEACHER*. However we shall always assume that the number of steps in the computation is bounded by a polynomial.

There is a *trivial strategy* for *STUDENT* (which is often used by stupid students) according to which his first conjecture is 0 and then *STUDENT* just repeats the answers of *TEACHER*. Clearly, for optimization problems of type 1, this strategy produces always a solution. We conjecture that there is no strategy for *STUDENT* in general (i.e. for type 2). In fact, if we measure the complexity by the number of queries that *STUDENT* must ask, then it is plausible that there is no better strategy for hard problems than the trivial one.

We define two types of computations with counterexamples:

type 0: the number of queries is bounded by a constant;

type 1: the number of queries is unbounded, (implicitly it is always bounded by a polynomial, since all computations are bounded by a polynomial).

We have already noticed that type 1 computations solve type 1 problems; the same is true for types 0.

Let us consider for a moment computations with counterexamples without an additional Σ_i^p oracle, and suppose we want to find an optimal solution for a predicate $C(x,y)$ which is in $\mathcal{P}$. If we want to use the usual oracle computation instead of the counterexample computation, we can replace *TEACHER* by a Σ_1^p oracle as follows. The query

Is y optimal?

is Π_1^p, and if the answer is *not*, we can ask a Σ_1^p oracle about the bits of a better feasible solution. Hence if optimal solutions can be computed with counterexamples, then they can be computed by computations with a Σ_1^p oracle. However, the computations with Σ_1^p oracles are too strong: it is an easy exercise to show that any (i.e. type 2) problems can be solved using such computations. So it is important that *STUDENT* is allowed to ask only about feasible solutions to x when he is computing an optimal solution to x.

We shall also use computations with counterexamples in a more general situation. Let $B(x,y,z)$ be a ternary predicate, let x be given. Now the aim of *STUDENT* is to find some y such that $\forall z \leq xB(x,y,z)$. Again *STUDENT* can produce a conjecture y and ask *TEACHER* whether $\forall z \leq xB(x,y,z)$ is true. If it is not true, *TEACHER* must give to *STUDENT* some z such that $z \leq x$ & $\neg B(x, y, z)$. Such a z will be called a counterexample. If $C(x, y), \rho(y)$ is an optimization problem, then we define $B(x, y, z)$ by

$$B(x, y, z) \equiv C(x, y) \ \& \ (\rho(y) < \rho(z) \rightarrow \neg C(x, z));$$

thus $\forall z \leq x B(x, y, z)$ expresses that y is an optimal solution to x. Note that if C is in $\Sigma_i^b \cup \Pi_i^b$, then B is in $\Sigma_{i+1}^b \cap \Pi_{i+1}^b$. Now this more general definition of computatios allows *STUDENT* to ask also about elements which are not feasible solutions to C. But this is no real advantage for *STUDENT*, since *STUDENT* can always test himself whether y is a feasible solution, and if it is not, then clearly $z = 0$ is a counterexample and he can use it as a possible answer to *TEACHER*.

5. The Existence of Optimal Solutions in Fragments of Bounded Arithmetic

We consider optimization problems of the form $C(x, y), \rho(y)$ where $C(x, y)$ is Π_i^p and ρ is polynomial time computable. We shall suppose that $C(x, y)$ is defined by a Π_i^b formula, $\rho(y)$ is defined by a Σ_0^b formula, and formulae $C(x, 0)$ and $C(x, y) \rightarrow y \leq x$ are provable in the fragment in question. If $C(x, y), \rho(y)$ is a type 1 optimization problem, we shall require that

$$C(x, y) \rightarrow \rho(y) < |x|$$

is also provable.

Proposition 1. *The following holds modulo the basic axioms for $i \geq 0$.*

1. PIND Σ_{i+1}^b *is equivalent with the schema saying that every Π_i^p optimization problem of type* 1 *has an optimal solution;*
2. IND Σ_{i+1}^b *is equivalent with the schema saying that every Π_i^p optimization problem of type* 2 *has an optimal solution.*

Proof. We shall prove only (1), since (2) is similar. We shall use the equivalent schema LIND Σ_{i+1}^b. Let $C(x, y), \rho(y)$ be given, let $C(x, y)$ be in Π_i^p. Assume that $C(x, y) \rightarrow \rho(y) < |x|$ is provable in the base theory. Take $\varphi(x, t)$ defined by

$$\varphi(x, t) \equiv \exists y \leq x (C(x, y) \ \& \ t \leq \rho(y)).$$

From LIND Σ_{i+1}^b we get

$$\neg\varphi(x, 0) \vee \exists t < |x| (\varphi(x, t) \ \& \ \neg\varphi(x, t + 1)) \vee \varphi(x, |x|),$$

which just expresses the existence of an optimal solution.

Proving the other direction assume that we are given some Σ^b_{i+1} formula $\Psi(x,u)$. We want to derive an instance of PIND for $\Psi(x,u)$, where u is the induction variable and x is a parameter. It is easily seen that we can consider only formulae of the form

$$\Psi(x,u) \equiv \exists v((u,v) \leq x \ \&\ \psi(x,u,v)),$$

where $(-,-)$ is the pairing function and ψ is Π^b_i, since PIND for such formulae proves the full schema PIND Σ^b_{i+1}. Let $C(x,y)$ and $\rho(y)$ be defined by

$$\begin{aligned} C(x,y) &\equiv (y \leq x \ \&\ \psi(x,(y)_0,(y)_1) \ \&\ (y)_0 \leq |x|) \vee y = 0, \\ \rho(0) &= 0, \\ \rho(y) &= (y)_0 \text{ for } y > 0, \end{aligned}$$

where $(y)_0$ and $(y)_1$ are the decoding functions for the pairing function. Suppose we have

$$\Psi(x,0) \text{ and } \forall u < |x|(\Psi(x,u) \rightarrow \Psi(x,u+1)).$$

If the problem C,ρ has an optimal solution y for a given x, then it must be such that $(y)_0 = |x|$, hence we have $\Psi(x,|x|)$. Thus we have shown PIND for Ψ. □

Note that to prove that a type 0 optimization problem has an optimal solution, we do not need *any* induction.

6. Separation of Fragments

It has been noted quite early in the history of proof theory, that if we prove a sentence $\forall x \exists y \varphi(x,y)$ in some theory, then we have some information about how difficult is to find some y for a given x such that $\varphi(x,y)$. In his fundamental paper [Pa] Parikh showed that if $\forall x \exists y \varphi(x,y)$, with φ bounded, is provable in $I\Delta_0$, then y can be bounded by a polynomial in x, consequently it can be computed in linear space. Buss [B1] proved a theorem about S_2 which gives essentially more information. He proved the following theorem.

Theorem 1. *Let $i \geq 0$, let $\varphi(x,y)$ be Σ^b_{i+1}, and suppose*

$$S_2^{i+1} \vdash \forall x \exists y \varphi(x,y).$$

Then there exists a function f computable in polynomial time with a Σ^p_i oracle such that

$$\mathbb{N} \models \forall x \varphi(x, f(x)).$$

(Actually he proved more: under the same assumption $S_2^{i+1} \vdash \forall x \varphi(x, f(x))$.)

We are not able to use this theorem to show that fragments S_2^i are different assuming e.g. that Polynomial Hierarchy does not collapse. (This is however possible for the intuitionistic version of these fragments using the intuitionistic version of Theorem 1 which was proved in [B3].) Therefore we shall use more complex formulae than Σ^b_i, but then also we have to use a stronger mode of computation—this will be just the computations with counterexamples. The concept of counterexamples is also not new in proof theory, it goes back to Kreisel [K]. Recently Jan Krajíček [Kr] noted a close connection between the present concept and the former one, which can be used to give an alternative proof of the following theorems.

Theorem 2 [KPT]. *Suppose that for $i > 0$, and φ in Σ^b_{i+1}*

$$T_2^i \vdash \forall x \exists y \forall z \leq x \varphi(x,y,z).$$

Then, for a given x, one can compute y such that $\forall z \leq x \varphi(x,y,z)$ in polynomial time using a Σ^p_i oracle by a type 0 counterexample computation (i.e. using a constant number of counterexamples).

The following is a new result.

Theorem 3. *Suppose that for $i > 0$, and φ in Σ^b_{i+1}*

$$S_2^{i+1} \vdash \forall x \exists y \forall z \leq x \varphi(x,y,z).$$

Then, for a gien x, one can compute y such that $\forall z \leq x \varphi(x,y,z)$ in polynomial time using a Σ^p_i oracle by a type 1 counterexample computation (i.e. using an unbounded number of counterexamples).

Let C, ρ be an optimization problem with C in Π^p_i and ρ polynomial time computable. We have constructed a Σ^b_{i+1} formula $B(x,y,z)$ such that $\forall z \leq y B(x,y,z)$ expresses that y is an optimal solution to x. Thus we can apply Theorems 2 and 3.

Corollary 1. *For $i \geq 0$ and C in Π_i^p, if T_2^i proves that $C(x,y), \rho(y)$ has an optimal solution for every x, then optimal solutions can be computed using a type* 0 *computation with a Σ_i^p oracle.*

Corollary 2. *For $i \geq 0$ and C in Π_i^p, if S_2^{i+1} proves that $C(x,y), \rho(y)$ has an optimal solution for every x, then optimal solutions can be computed using a type* 1 *computation with a Σ_i^p oracle.*

By Proposition 1 we know that the existence of optimal solutions (for certain type of problems) is equivalent to (certain type of) induction. Thus we get:

Corollary 3 [KPT]. *For $i \geq 0$, if $T_2^i = S_2^{i+1}$, then every type* 1 *optimization problem C, ρ with C in Π_i^p can be computed by a type* 0 *computation with a Σ_i^p oracle.*

Corollary 4. *For $i \geq 0$, if $S_2^{i+1} = T_2^{i+1}$, then every optimization problem C, ρ with C in Π_i^p can be computed by a type* 1 *computation with a Σ_i^p oracle.*

Conclusions in both corollaries seem unlikely which strongly suggests that $T_2^i \neq S_2^{i+1}$ and $S_2^{i+1} \neq T_2^{i+1}$. In [KPT] it has been shown that the conclusion of Corollary 3 implies that $\Sigma_{i+2}^p = \Pi_{i+2}^p$ which is usually conjectured to be false. For the conclusion of Corollary 4 it is an open problem, whether it implies anything like that.

7. Proof of Theorem 3

Here we shall sketch the idea of a proof of Theorem 3. The proof is a modification of Buss' proof of Theorem 1. Theorem 2 was proved using different means. We shall observe that it can be obtained by modifying the proof given below. Hence there is a uniform way to prove all three theorems.

We consider the sequent calculus of Schwichtenberg [Sch]. The sequents are sets of formulae; logical connectives are $\&, \vee, \neg$, where negation is allowed only at atomic formulae (if φ is not atomic, $\neg\varphi$ is an abbreviation for the equivalent formula obtained by applying De Morgan's laws). The system has initial sequents of the form $\Gamma, \varphi, \neg\varphi$, (which is $\Gamma \cup \{\varphi, \neg\varphi\}$), a rule for $\&$, two rules for $\vee$, one rule for each quantifier and a cut rule. The rules which are important for the proof will be explained in the course of the proof. We formalize S_2^{i+1} in this system by allowing initial sequents of the form Γ, φ (which is $\Gamma \cup \{\varphi\}$) for φ a basic axiom and by adding the following rule for each $\psi(x)$ in Σ^b_{i+1}:

$$\frac{\Theta, \neg\psi(\lfloor b/2 \rfloor), \psi(b)}{\Theta, \neg\psi(0), \psi(t)},$$

where b is not free in Θ and t is a term.

Let $\varphi(a, y, z)$ in Σ^b_{i+1} be fixed, let $A(a)$ be defined by

$$A(a) \equiv \exists y \forall z \varphi(a, y, z).$$

In order to simplify notation we shall assume that the bound $z \leq a$ is implicit in $\varphi(a, y, z)$. Suppose a proof of $A(a)$ is given. By cut elimination we can assume that it is free-cut-free, i.e. the cut formulae are only substitution instances of basic axioms or induction formulae. Thus this proof contains only Σ^b_{i+1} and Π^b_{i+1} formulae and substitution instances of subformulae of $A(a)$, which are either $A(a)$ itself, or $\forall z \varphi(a, t, z)$, for some term t, or a Σ^b_{i+1} formulae. Thus the general form of a sequent Γ in the proof is

$$\Pi, \Sigma, \Delta, \Phi, A(a),$$

where

Π are Π^b_{i+1} formulae which are not in $\Pi^b_i \cup \Sigma^b_i$;
Σ are Σ^b_{i+1} formulae which are not in $\Pi^b_i \cup \Sigma^b_i$;
Δ are $\Pi^b_i \cup \Sigma^b_i$ formulae;
Φ is a set of formulae of the form $\forall x \varphi(a, t, z)$;

and we can assume that $A(a)$ is present in each sequent.

Now we are going to define the concept of *witnessing functions* for such a Γ. Recall that we have defined Σ^b_{i+1} (Π^b_{i+1} resp.) formulae so that they consist of a prefix of existential (universal resp.) bounded quantifiers followed by a Π^b_i (Σ^b_i resp.) formula. We shall call these quantifiers *essential.* Let $a, b_1, \ldots, b_k$ be the string of free variables of Γ. We shall denote strings of variables and functions by boldface letters (e.g. $\mathbf{b}$ denotes $b_1, \ldots, b_k$). We choose distinct variables $x_1, \ldots, x_l$ for all (distinct) occurrences of variables at essential quantifiers in Π and, similarly, $y_1, \ldots, y_m$ for variables at essential quantifiers in Σ. A string of functions

$$f_1(a, \mathbf{b}, \mathbf{x}), \ldots, f_m(a, \mathbf{b}, \mathbf{x}), g(a, \mathbf{b}, \mathbf{x}),$$

will be called *witnessing functions* for Γ if the following formula is true for all assignments of natural numbers to $a, \mathbf{b}, \mathbf{x}$:

$$(*) \quad \bigwedge \neg \Pi(x) \rightarrow \bigvee \Sigma(\mathbf{f}(a, \mathbf{b}, \mathbf{x})) \vee \bigvee \Delta \vee A(a) \vee \forall z \varphi(a, g(a, \mathbf{b}, \mathbf{x}), z).$$

Here $\Pi(\mathbf{x})$ denotes Π with essential bounded quanfiers omitted an their variables replaced by $x_1, \ldots, x_l$; similarly in $\Sigma(\mathbf{f}(a, \mathbf{b}, \mathbf{x}))$ essential quantifiers are omitted and their variables are replaced by $f_1(a, \mathbf{b}, \mathbf{x}), \ldots, f_m(a, \mathbf{b}, \mathbf{x})$. Note that no witnessing functions occur in Φ, though the formulae in Φ are not in Π^b_{i+1}.

Using induction on the depth of a sequent in the proof, we shall show that every sequent has witnessing functions computable in the following way. THere is a Turing machine with a Σ^b_i oracle which on an input $a, \mathbf{b}, \mathbf{x}$ produces the values $f_1(a, \mathbf{b}, \mathbf{x}), \ldots, f_m(a, \mathbf{b}, \mathbf{x}), g(a, \mathbf{b}, \mathbf{x})$ in polynomially many steps. During the computation it may ask queries of the form $\forall z \varphi(a, u, z)$? where u is some value produced during the computation. If the answer is negative, it gets a counterexample, i.e. some z_0 such that $\neg\varphi(a, u, z_0)$. If the answer is positive, it puts $g(a, \mathbf{b}, \mathbf{x}) = u$ and some default values for f_i's and stops. In aprticular, we get the type of computation required in the theorem for the end sequent, since it consists of $A(a)$ only. As defined above, we have to consider all possible strategies for the person (*TEACHER*) who answers the queries. Thus it would be more appropriate to talk about *functionals* $f_1, \ldots, f_m$, rather than functions. Or we can say: for any strategy of *TEACHER* the computed functions are witnessing functions for the sequent.

The induction steps are similar to those in Buss' proof, except for those rules where the principalformula is a subformula of $A(a)$.

(1) Consider the following instance of the $\forall$-rule:

$$\frac{\Theta, \varphi(a, t, b_h)}{\Theta, \forall z \varphi(a, t, z)}.$$

Suppose we have witnessing functions $\mathbf{f}(a, \mathbf{b}, \mathbf{x}), g(a, \mathbf{b}, \mathbf{x})$ for the upper sequent. The lower sequent has free variables $a, \mathbf{b}'$, where $\mathbf{b}' = (b_1, \ldots, b_{h-1}, b_{h+1}, \ldots, b_k)$. To get the witnessing functios for the lower sequent, we omit the witnessing functions for $\varphi(a, t, z)$ and change the remaining ones as follows. First we compute the value of the term t, then we ask the query "$\forall z \varphi(a, t, z)$?". If the answer is positive, then the lower sequent is witnessed no matter how we define the functions. If the answer is negative and the counterexample is u, then we define the witnessing functions for the lower sequent by substituting u for b_i. Thus the functions depend only on $a, \mathbf{b}', \mathbf{x}$.

(2) Suppose $\exists$-rule is applied to $\forall x\varphi(a,t,z)$ to obtain $\exists y\forall z\varphi(a,y,z)$. Since this is just $A(a)$, which is present in every sequent, the instance looks like this:

$$\frac{\Theta, \forall z\varphi(a,t,z)}{\Theta}.$$

Suppose we have $\mathbf{f}, g$ for the upper sequent. The computation of $\mathbf{f}', g'$ for the lower sequent will be the following. First compute the value of t, then ask "$forallz\varphi(a,t,z)$?". If the answer is positive, set $g'(a,\mathbf{b},\mathbf{x}) = t$ and the value of $\mathbf{f}'$ is irrelevant; otherwise compute $\mathbf{f}', g'$ as for the upper sequent.

(3) Consider an instance of the PIND Σ^b_{i+1} rule:

$$\frac{\Theta, \forall x_1 \neg\psi(\lfloor b_1/2 \rfloor, x_1), \exists y_1 \psi(b_1, y_1)}{\Theta, \forall x_1 \neg\psi(0, x_1), \exists y_1 \psi(t, y_1)}$$

where ψ is Π^b_i. In order to simplify notation we assume that there is only one essential bounded quantifier and we omit the bound; also we have chosen the indices to be equal to 1. Suppose we have witnessing functions $\mathbf{f}(a,\mathbf{b},\mathbf{x}), g(a,\mathbf{b},\mathbf{x})$ for the upper sequent. We shall assume that $\exists y_1$ is witnessed by f_1. Now we define witnessing functions $\mathbf{f}'(a,\mathbf{b}',\mathbf{x}), g'(a,\mathbf{b}',\mathbf{x})$ for the lower sequent, where $\mathbf{b}' = (b_2, \ldots, b_k)$. First we compute $0 = v_0, \ldots, v_r = t$ such that $\lfloor v_{j+1}/2 \rfloor = v_j$, for $j = 0, \ldots, r-1$. Then we compute $\mathbf{f}^{(s)}, g^{(s)}$ as follows. Set

$$f_1^{(0)} = x_1,$$

and for $s \geq 0$ let

$$f_j^{(s+1)} = f_j(a, v_s, \mathbf{b}', f_1^{(s)}, x_2, \ldots, x_l),$$
$$g^{(s+1)} = g(a, v_s, \mathbf{b}', f_1^{(s)}, x_2, \ldots, x_l),$$

($f_j^{(0)}$ is not defined for $j > 1$, $f_1^{(s)}$ are iterations of f_1). In each step of the iteration $s = 0, 1, \ldots$ we also check whether $\psi(v_s, f_1^{(s)})$ is true and whether Θ is witnessed by $\mathbf{f}^{(s)}, g^{(s)}$. The computation will stop if one of the following four cases occurs:

(i) we get a positive answer to a query "$\forall z\varphi(a,u,z)$?";
(ii) $\psi(0, f_1^{(0)})$ is not true, i.e. $\neg\psi(0, x_1)$ is true;
(iii) if Θ is witnessed by $\mathbf{f}^{(s)}, g^{(s)}$;
(iv) $s = r - 1$.

We define the witnessing functions according to which case occurs:

(i) we set $g'(a, \mathbf{b}, \mathbf{x}) = u$ and $A(a)$ is witnessed;

(ii) the lower sequent is witnessed independently of the values of $\mathbf{f}', g'$;

(iii) if Θ is witnessed by $\mathbf{f}^{(s)}, g^{(s)}$, then we take them as $\mathbf{f}', g'$;

(iv) define $\mathbf{f}', g'$ as $\mathbf{f}^{(r-1)}, g^{(r-1)}$.

We only have to check that the lower sequent is witnessed also in case (iv). If (iv) occurs, then none of the (i)–(iii) has occurred before, in particular $\psi(0, f_1^{(0)})$ is true. Suppose $\psi(v_r, f^{(r-1)})$ is false. Then there is some $s < r$ such that $\psi(v_s, f_1^{(s)})$ is true and $\neg\psi(v_{s+1}, f_1^{(s)})$ is false. Since $\mathbf{f}, g$ witness the upper squent, this is possible only if Θ is witnessed by $\mathbf{f}^{(s)}, g^{(s)}$, which is a contradiction. Thus the lower sequent is witnessed also in case (iv).

We have tacitly assumed that $\exists y_1 \psi(t, y_1)$ does not occur in Θ. If it does, then $f_1^{(s)}$ will not be included in $\mathbf{f}'$ and case (iv) will be subsumed in case (iii).

We hope that this illustrates sufficiently well the changes that must be done in Buss' proof, and we are not going to consider other instances of rules and axioms. To state the main idea briefly: the change is in the possibility that a positive answer to a query "$\forall z \varphi(a, t, z)$?" may occur. In such a case the computation stops, since we have a witness for $A(a)$. □

Now we describe a proof of Theorem 2. The assumptions are similar, except that we have a weaker rule of induction

$$\frac{\Theta, \neg\psi(b_h), \psi(b_n + 1)}{\Theta, \neg\psi(0), \psi(t)}$$

since ψ is only Σ_i^b. We use the same definition of witnessing as in the above proof, hence *no witnessing functions occur in* $\neg\psi(b_h), \psi(b_h+1), \neg\psi(0), \psi(t)$. Let $\mathbf{f}, g$ be witnessing functions for the upper sequent. We shall define witnessing functions $\mathbf{f}', g'$ for the lower sequent. Let $a, \mathbf{b}', \mathbf{x}$ be input, where again $\mathbf{b}' = (b_1, \ldots, b_{h-1}, b_{h+1}, \ldots, b_k)$. First we check whether $\neg\psi(0) \vee \psi(t)$ is true. If it is true, then we take arbitrary values for $\mathbf{f}', g'$. If not, then we use binary search to find some u such that $\psi(u)$ & $\neg\psi(u+1)$, $u < t$. This is possible, since now we can use $\psi(x)$ as an oracle. Then we put

$$\mathbf{f}'(a, \mathbf{b}', \mathbf{x}) = \mathbf{f}(a, b_1, \ldots, b_{h-1}, u, b_{h+1}, \ldots, b_k, \mathbf{x}),$$
$$g'(a, \mathbf{b}', \mathbf{x}) = g(a, b_1, \ldots, b_{h-1}, u, b_{h+1}, \ldots, b_k, \mathbf{x}).$$

In this case we witness Θ. Here no iterations of witnessing functions occur. Hence it holds for *every* rule: if the number of queries is constant (i.e. does not depend on parameters $a, \mathbf{b}, \mathbf{x}$), then it is constant in the lower sequent too. Consequently the number of queries used to compute the witness for the end sequent is constant. (It is not hard to prove a more precise upper bound: the number of queries is bounded by the number of applications of $\forall$-rule to formulae $\varphi(a, t, b_h)$.) □

7. Relativizations

We conjecture that there are optimization problems C, ρ, with C in Π_i^p (of type 2) whose optimal solutiosn cannot be computed in polynomial time using counterexample computations (of type 1) with Σ_i^p oracles. By Corollary 4, this conjecture implies that $S_2^{i+1} \neq T_2^{i+1}$. We shall justify this conjecture for $i = 0$ and $i = 1$ by showing that for suitable oracles the relativized version is true. Clearly it is sufficient to prove it for $i = 1$. These results imply separations of the corresponding fragments of the system obtained from S_2 by adding a new uninterpreted predicate, (see Corollary 6 below).

In the following proof it will be convenient to consider oracles as mappings $A : \mathbb{N}^3 \to \{0, 1\}$.

Theorem 4. *There exists an oracle A such that there is no polynomial time interactive algorithm (type 1) with a $(\Sigma_1^p)^A$ oracle which computes the largest y such that $y \leq x$ and*

$$\forall u \leq x(A(x, y, u) = 0) \vee y = 0.$$

I.e. in the optimization problem $C(x, y)$ is $\forall u \leq x(A(x, y, u) = 0) \vee y = 0$, which is in $(\Pi_1^p)^A$, and $\rho(y) = y$. In the proof of Theorem 5 we shall assume some familiarity with the concept of relativization. As usual we shall use finite approximations to oracle A. The key lemma which enables us to diagonalize at level Σ_1^p is the following.

Lemma 1. *Let $R^\alpha(v)$ be a $(\Sigma_1^p)^\alpha$ predicate, where α is a variable for an oracle. Let a partial mapping A' be given, let v be given. Then there exists an extension A'' of A' such that it has only polynomially more elements than A' (i.e. $|A'' \backslash A'| \leq q(|v|)$, for some polynomial q), and for any two extensions A and B of A''*

$$R^A(v) \equiv R^B(v),$$

i.e. A'' *forces* $R(v)$ *or* $\neg R(v)$.

Proof. First uppose that there exists some $A_0 \supseteq A'$ such that $R^{A_0}(v)$ holds true. Take an accepting computation for v which uses A_0 and add to A' the queries asked by this computation. Any extension which gives the same answer to these queries will allow this accepting computation. If there is no such extension A_0, then put $A'' = A'$. □

Proof of Theorem 5. We construct the oracle in ω steps. In the i-th step we diagonalize the i-th *STUDENT*, which is a polynomially bounded Turing machine and a Σ_1^p oracle. The precise meaning of this statement is that we construct a finite extension A_i of the previous approximation to the oracle, we take an input x_i and define a strategy for *TEACHER* so that for any extension of A_i, *STUDENT* does not compute the optimal solution to x_i in $p(|x_i|)$ steps, where p is the polynomial bound to *STUDENT*. We can always take x_i so large that on x_j with $j < i$, *STUDENT* never asks queries of the form "$A(x_i, r, s) = 0$?".

On input x, *STUDENT* uses Σ_1^p oracle only for inputs whose size is polynomially bounded in $|x|$. Hence there is a polynomial bound $q'(|x|)$ to all possible values $q(|v|)$ for queries v asked during the computation on input x, (q is the polynomial from Lemma 1). Put $p'(|x|) = p(|x|) \cdot q'(|x|)$, (we assume also $q'(|x|) \geq 1$). We choose x_i so large that

$$(p(|x_i|) + 1) \cdot (p'(|x_i|) + 1) < x_i.$$

Each A_i is also constructed in several stages,

$$A_i^0 = A_{i-1}, A_i^1, A_i^2, \dots .$$

These stages will correspond to the queries of *STUDENT*. At the same time we define the strategy for *TEACHER*. We have to define the strategy of *TEACHER* only for $x_1, x_2, \dots,$ since other inputs are not used for the diagonalization. Let i be given. The strategy of *TEACHER* will consist of her answers $y_1, y_2, \dots$. The approximations A_i^k will be constructed using Lemma 1, hence if we take *any* extension of A_i^k and if *TEACHER* uses $y_1, \dots, y_{k-1}$, the computation of *STUDENT* will be the same up to the k-th query. The number of these stages is bounded by the number of queries that *STUDENT* can pose to *TEACHER* and this in turn is bounded by the total running time $p(|x_i|)$ of *STUDENT*.

There are three reasons to extend the current approximation to the oracle A:

(1) to force the computation of *STUDENT*;
(2) to force that the answers of *TEACHER* are correct;
(3) to force that *STUDENT* has asked the wrong question (this can be avoided by taking an enumeration of *STUDENTS* who do not ask wrong questions).

By Lemma 1, we can always add $q'(|x_i|)$ new elements into the approximation of the oracle so that an answer of Σ_1^p oracle is forced. In this way we add at most $q'(|x_i|)$ new elements to the approximation to A in each computation step. Hence for (1) we have to add at most $p'(|x_i|)$ elements. To ensure (3) we need just one element. Once *STUDENT* asked a wrong question (i.e. he asked *TEACHER* whether y is maximal such that $\forall z \leq x_i A_i(x_i, y, z) = 0$ for some y such that $\exists z \leq x_i A_i(x_i, y, z) = 1$), the construction of A_i and *TEACHER*'s strategy is finished. For (2) we have to add all triples (x_i, y_k, z) for each answer y_k of *TEACHER* and each $z \leq x$. We shall construct A_i^k and y_k in such a way that we add new elements into the domain only if we need them because of one of the reasons (1)–(3) and the following condition is satisfied:

> for $k > 0$, y_k is the largest answer of *TEACHER* and $y_k \leq k \cdot (p'(|x_i|) + 1)$.

Suppose the condition is satisfied at stage $k - 1$. We take any extension B of A_i^{k-1} and let *STUDENT* work until he presents a conjecture to *TEACHER*. Let B_i^{k-1} be an approximation which forces this computation of *STUDENT*, $A_i^{k-1} \subseteq B_i^{k-1} \subseteq B$, and let the conjecture of *STUDENT* be y. If $y > y_{k-1}$, then we can extend B_i^{k-1} to A_i^k by adding just one element to it in such a way that *STUDENT*'s answer is wrong (for any extension of A_i^k). This is because, for such a $y \leq x_i$, there can be at most $p'(|x_i|) < x_i$ elements $z \leq x_i$ such that $B_i^{k-1}(x_i, y, z)$ is defined. Otherwise $y \leq y_{k-1}$. There are at most $p'(|x_i|)$ elements $y' > y_{k-1}$ such that $B_i^{k-1}(x_i, y', z)$ is defined. By the assumption that the condition holds for $k - 1$,

$$
\begin{aligned}
y_{k-1} + p'(|x_i|) &\leq (k-1) \cdot (p'(|x_i|) + 1) + p'(|x_i|) \leq \\
&\leq k \cdot (p'(|x_i|) + 1) \leq p(|x_i|) \cdot (p'(|x_i|) + 1) < x_i.
\end{aligned}
$$

Hence we can take y_k such that $y_{k-1} < y_k \leq x_i$, $B_i^{k-1}(x_i, y_k, z)$ is undefined for all z and

$$
y_k \leq y_{k-1} + p'(|x_i|) + 1.
$$

By the assumption that the condition holds for $k-1$,

$$y_k \leq (k-1) \cdot (p'(|x_i|)+1) + p'(|x_i|) + 1 = k \cdot (p'(|x_i|)+1).$$

Thus we can extend B_i^{k-1} to A_i^k by putting

$$A_i^k(x_i, y_k, z) = 0 \text{ for every } z \leq x_i,$$

and the condition will be preserved.

After polynomially many steps *STUDENT* must stop, but we can still extend the oracle so that there exists a larger counterexample, because

$$\begin{aligned} y_k + p'(|x_i|) &\leq k \cdot (p'(|x_i|)+1) + p'(|x_i|) \leq \\ &\leq (k+1) \cdot (p'(|x_i|)+1) \leq (p(|x_i|)+1) \cdot (p'(|x_i|)+1) < x_i. \end{aligned}$$

Hence he is not able to find the optimal solution. □

Remarks. (1) The proof above is essentially the same as for the similar result in [KPT]. (2) We have shown more for this oracle: for every *STUDENT* there exists an input x such that either he asks a wrong question on x or he uses the trivial strategy on x without success.

Let $S_2^i(\alpha)$ (resp. $T_2^i(\alpha)$) be S_2^i (resp. T_2^i) extended by adding a new predicate to the language and extending PIND (resp. IND) to $\Sigma_i^b(\alpha)$ formulae (which are defined as Σ_i^b in the extended language).

Corollary 5. *For $i = 1$ and $i = 2$, $T_2^i(\alpha) \neq S_2^i(\alpha)$.*

Proof. First it is necessary to check that Proposition 1, Theorem 3 and hence also Corollary 4 can be relativized by adding a new uninterpreted predicate α. Take $C(x,y)$ to be

$$\forall u \leq x(\alpha(x,y,u) = 0) \vee y = 0,$$

and $\rho(y) = y$. Then, by relativized Corollary 4, for any interpretation of α as a subset $A \subseteq \mathbb{N}$, we should be able to compute y form x using an interactive computation with oracle A. If we choose the interpretation fo α to be A from Theorem 5, we get a contradiction. In this way, we obtain the result for $i = 2$. For $i = 1$ we take the same A and encode in it some $\mathcal{NP}^A$-complete problem. Thus former Σ_1^p sets become $\mathcal{P}$ sets. Or we can prove a theorem similar to Theorem 5 for this simpler case using a trivial modification of the proof above. □

8. Open Problems

We would like to prove the conjecture that there are optimization problems C, ρ with C in Π_i^p which cannot be computed by (type 1) interactive computations with Σ_i^p oracles, since it implies that $S_2^{i+1} \neq T_2^{i+1}$. As this conjecture implies that $\mathcal{P} \neq \mathcal{NP}$, it is hopeless to try to prove the conjecture directly. In the present situation the following two problems seem to be more feasible:

(1) Reduce the conjecture to the statement that Polynomial Hierarchy is proper, or a similar statement in complexity theory.
(2) Find oracles for each i such that the relativized statements are true.

A proof of (2) would imply $T_2^i(\alpha) \neq S_2^i(\alpha)$. Similar statements for type 1 optimization problems and type 0 interactive computations were proved in [KPT]. Note that there is a different approach to the separation of fragments of Bounded Arithmetic. It is based on proof systems for the propositional calculus [KP]. There we would need to show superpolynomial lower bounds to the length of proofs in certain proof systems for the propositional calculus. This is a weaker question than $\mathcal{NP} \neq co\mathcal{NP}$. Even less we know about the related problem:

(3) Is T_2^i partially conservative over S_2^i, e.g. is T_2^i $\forall\Pi_1^b$-conservative over S_2^i?

Some results on this problem have been recently obtained by Krajíček [Kr]. Also note taht Krajíček and Takeuti [KT] have constructed a consistency statement which is the strongest $\forall\Pi_1^b$-formula provable in T_2^i, hence it is the best candidate for a possible separation of T_2^i from S_2^i.

Acknowledgments

I would like to thank Jan Krajíček and Jiři Sgall for carefully reading the manuscript and suggesting several improvements.

References

[B1] S. R. Buss, *Bounded Arithmetic*, Bibliopolis, Napoli, 1986.

[B2] ———, *Axiomatizations and conservation results for fragments of bounded arithmetic*, Contemporary Mathematics, AMS Proc. of Workshop in Logic and Computation, 1987 **106** (1990), 57–84.

[B3] ———, *The Polynomial Hierarchy and intuitionistic bounded arithmetic*, in "Structure in Complexity Theory", LNCS 223, Springer-Verlag, 1986, pp. 77–103.

[C] S. A. Cook, *Feasibly constructive proofs and the propositional calculus*, Proc. 7-th STOC (1975), 73–89.

[Kr] J. Krajíček, *No counterexample interpretation and interactive computations*, these proceedings.

[KP] J. Krajíček and P. Pudlák, *Quantified propositional calculi and fragments of bounded arithmetic*, Zeitschrift f. Math. Logik **36**(1) (1990), 29–46.

[KPT] J. Krajíček, P. Pudlák and G. Takeuti, *Bounded arithmetic and Polynomial Hierarchy*, Annals of Pure and Applied Logic, to appear.

[KT] J. Krajíček and G. Takeuti, *On induction-free provability*, Discrete Applied Mathematics (to appear).

[K] G. Kreisel, *On the interpretation of non-finitist proofs*, JSL 16 (1951), 241–267.

[PW] J. Paris and A. Wilkie, *On the scheme of induction for bounded arithmetic formulas*, Annals of Pure and Applied Logic **35**(3) (1987), 205–303.

[Pa] R. Parikh, *Existence and feasibility in arithmetic*, JSL 36 (1971), 494–508.

[Pu] P. Pudlák, *A note on bounded arithmetic*, Fundamenta Mathematicae, to appear.

[Sch] H. Schwichtenberg, *Proof Theory: Some applications of cut-elimination*, in "Handbook of Mathematical Logic", J. Barwise ed. (1977), 867–895.

[T] G. Takeuti, *Some relations among systems of bounded arithmetic*, Preprint.

Mathematical Institute ČSAV, Žitná 25, Praha 1, Czechoslovakia

Mathematical Sciences Research Institute, Berkeley CA 94720

A part of this manuscript was prepared while the author was a member of the Mathematical Sciences Research Institute, Berkeley.

Research at MSRI supported in part by NSF Grant DMS-8505550.

LOGICS FOR NEGATION AS FAILURE

J.C. SHEPHERDSON

ABSTRACT. Negation as failure is the version of negation commonly used in logic programming. It decrees that a (ground) goal $\neg A$ succeeds if A fails and fails if A succeeds. It is not sound with respect to the straightforward classical reading of a program. This paper surveys the ways in which different types of logic—classical two-valued, three-valued, intuitionistic, modal, autoepistemic, linear—have been used to provide declarative semantics for negation as failure.

1. INTRODUCTION

1.1. The Problem of Negation as Failure

The usual way of introducing negation into Horn clause logic programming is by "negation as failure": if A is a ground atom

the goal $\neg A$ *succeeds if* A *fails*

the goal $\neg A$ *fails if* A *succeeds.*

This is obviously not justifiable for classical negation, at least not relative to the given program P; the fact that A fails from P does not mean that you can prove $\neg A$ from P, e.g. if P is

$$a \leftarrow \neg b$$

then $? - b$ fails so, using negation as failure, $? - a$ succeeds, but a is not a logical consequence of P.

You could deal with classical negation by using a form of resolution which gave a complete proof procedure for full first order logic. To a logician this would be the natural thing to do. Two reasons are commonly given why this is not done. The first is that it is believed by most, but not all, practitioners, that this would be infeasible because it would lead to a combinatorial explosion, whereas negation as failure does not, since it is not introducing any radically new methods of inference, just turning the old ones round. The second is that, in practical logic programming, negation as failure is often more useful than classical negation. This is the case when the program is a database, e.g. an airline timetable. You list all the flights there are. If there is no listed flight from Zurich to London at 12:31, then you conclude that there is no such flight. The implicit use of negation as failure here saves us the enormous labor of listing all the non-existent flights.

We have just seen that negation as failure is not sound with respect to the official classical semantics for a program according to which a program is the logical statement P obtained by taking the conjunction of its clauses, and a query Q should succeed with answer θ iff $Q\theta$ is a consequence of P and should fail iff $\neg Q$ is a consequence of P. Since negation as failure is useful there have been many attempts to give a convincing declarative semantics with respect to which negation as failure is sound, i.e. modifications to the notion

Q is a consequence of the program P

such that if Q succeeds with answer θ using negation as failure than $Q\theta$ is a consequence of the program P, and if Q fails using negation as failure than $\neg Q$ is a consequence of the program P. These have involved one or more of the following:

(i) replacing the program P by some other sentence or set of sentences obtained by applying some transformation to P, e.g. the Clark completion $comp(P)$ or Reiter's closed world assumption $CWA(P)$,

(ii) interpreting the notion of consequence with respect to some non-classical logic, e.g. three-valued, intuitionistic, modal, autoepistemic or linear,

(iii) interpreting the notion of consequence not as being true in all models of P, but only in all models of a certain kind, e.g. Herbrand, minimal, well-founded, perfect.

The justification for these approaches would seem to be; for case (i), that when you wrote P you did not mean it to be taken literally but as a shorthand or convenient way of expressing $comp(P)$ or $CWA(P)$; for case (ii) that the particular logic was more appropriate than classical two-valued logic for a situation in which you are dealing with constructive proofs, possibly non-terminating procedures, incomplete information, or beliefs; for case (iii) that these were the only models you were really interested in. The reason that there are so many different declarative semantics with respect to which negation s failure is sound is that in general it is not complete for any of them, i.e. there are queries Q which are consequences, in the sense in question, of P which do not succeed using negation as failure. The different semantics give different sets of consequences which all include those which succeed using negation s failure. The situation is rather like theology where there is room for many mutually incompatible religions as long as they agree on the real, physically observable world. When choosing between these different semantics one must ask why one needs a declarative semantics. Negation as failure (which we take here to mean the procedure of SLD-resolution extended by negation as failure to the SLDNF-resolution described in §1.2 below) has a perfectly well-defined procedural semantics. Presumably one reason for looking for a declarative semantics is based on the desire to consider negation as failure as part of logic programming, which is supposed to be about drawing logical consequences, and to apply the techniques of logic to reasoning about negation as failure. Another reason is that the procedural semantics of negation as failure is quite complicated, and a simple declarative semantics could make it easier to reason about programs, e.g. for a programmer to check the correctness of his program.

From this point of view none of the proposed semantics is very satisfactory; in general negation as failure is incomplete for them so they do not correspond exactly to the procedural semantics, and because they involve preliminary transformations, unfamiliar logics, or restricted classes of models, none of them makes it easy to read off the meaning of a program from the set of clauses actually written. This is probably inevitable; negation as failure does not seem to have a simple logical explanation.

The layout of this paper is as follows.

In §1.2 we give a formal definition of the procedure which is usually referred to as negation as failure, i.e. SLDNF-resolution.

In §2 we deal with semantics using classical two-valued logic. These all use modifications of the consequence relation of type (i) above; they achieve soundness by replacing P by some transformed set of sentences. In §2.1 this is the Clark completion $comp(P)$, in §2.2 it is Reiter's closed world assumption $CWA(P)$, (in this case a type (iii) modification is also involved i.e. a restriction to so-called Herbrand or term models), and in §2.3 it is $\bar{T}_\omega(P)$, the union of an iteratively defined sequence of sets of sentences. The Clark completion and the closed world assumption were the first proposed semantics for negation as failure and are the simplest. In general negation as failure is not complete for either of them, but there are large classes of useful types of program and query for which it is complete for *comp*(P), and at present this is the most popular semantics. However when there are many mutually recursive clauses in P the meaning of *comp*(P) may not be easy to read off from P. The closed world assumption is a very strong presumption in favor of negative information and is probably not the meaning the programmer intended, except for database type programs, i.e. definite Horn clause programs without function symbols (like the example at the beginning of §1.1). The semantics based on $\bar{T}_\omega(P)$ in §2.3 is too complex to be regarded as a practically useful one. It is put forward as a technical solution to the problem of finding, using classical logic, a declarative semantics which fits negation as failure as closely as possible, i.e. for which it is not only sound but as complete as possible. In order to do this it is necessary to extend SLDNF-resolution so that negation as failure can also be used on non-ground negative literals.

§3 deals with type (ii) modifications, i.e. replacing the consequence relation of classical logic by that of some non-classical logic. In §3.1 three-valued logic is used. A type (i) modification is also involved, since it is three-valued consequences of *comp*(P) not of P which are considered. This semantics achieves the best fit to SLDNF-resolution, giving completeness for a large class of 'allowed' (see §1.2) programs and queries. In §3.2 it is noted that many of the soundness results of §2 hold also for intuitionistic logic, so that using that instead of classical logic makes these semantics of §2 a closer fit to SLDNF-resolution, i.e. gives less incompleteness. In §3.3 modal logic is used; this also involves a type (i) modification in the form of a translation of the program into modal logic. Good completeness results are obtained for purely propositional programs but when individual variables are present the completeness results apply not to SLDNF-resolution but to a stronger procedure which operates on the infinite propositional program obtained by taking all ground instances of program clauses. In §3.4 the use of autoepistemic logic is surveyed. Semantics based on this also involve a type (i) translation of the pgoram P into autoepistemic logic. The results here do not relate directly to SLDNF-resolution but to the model-theoretic semantics in §4. In §3.5 a recent use of linear logic to provide a semantics for negation as failure is briefly outlined. This also involves a translation of the program P into linear logic. It is unusual in that it aims to provide a sound and complete declarative semantics not for SLDNF-resolution but for standard Prolog with negation as failure, and succeeds in doing this, at least in the case of purely propositional programs. This is a noteworthy success, although its practicality is lessened by the complexity of the translation of the program into linear logic and by the unfamiliarity of linear logic. In §3.6 ad hoc logical calculi are constructed for SLDNF-resolution and other versions of negation as failure such as Prolog. However these cannot be considered as declarative semantics, but as ways of expressing the syntax of these computational procedures in a form closer to the familiar logical systems.

In §4 semantics based on modifications of type (iii), special classes of models, are surveyed. These can also be thought of as involving type (i) modification as well since the models in question are models not only of the program P but also of $comp(P)$. And in some cases type (ii) modifications are used, e.g. three-valued logic. Plausible arguments have been given for each of these semantics. The fact that for a practically important class of 'locally stratified' programs they all coincide, giving a unique model which often appears to be 'the' natural model, adds support to their claim to be chosen as the intended semantics. However SLDNF-resolution will be even more incomplete for them than it is for the semantics based on $comp(P)$ and there is demonstrably no way of extending SLDNF-resolution to give a computable proof procedure which is both sound and complete for them.

1.2. SLDNF-Resolution

Throughout this paper when we speak of negation as failure we mean negation as finite failure formalized as the SLDNF-resolution of Lloyd [1987]. Before describing this we must point out one of the reasons why negation as failure is almost certain to be incomplete with respect to any simple semantics. This is because it cannot deal with non-ground negative literals. This is a price we have to pay for using a quantifier-free system. A query $?-p(x)$ is taken to mean $?-\exists x p(x)$, and a query $?-\neg p(x)$ to mean $?-\exists x \neg p(x)$. It is possible that both of these are true so it would be unsound to fail $?-\exists x \neg p(x)$ just because $?-\exists x p(x)$ succeeded. That is why negation as failure is only allowed for ground negative literals. [Prolog is unsound because it allows negation as failure on any goal.] This means that we cannot deal with queries of the form $?-\neg p(x)$, and in dealing with other queries we may *flounder*, or be unable to proceed because we reach a goal containing only non-ground negative literals.

In SLDNF-resolution program clauses are of the form

$$A \leftarrow L_1, \ldots, L_n$$

and queries are of the form

$$? - L_1, \ldots, L_n$$

where A is an atom and $L_1, \ldots, L_n$ are positive or negative literals. The negation of the query is written as a goal

$$\leftarrow L_1, \ldots, L_n$$

and when a positive literal L_i is selected from the current goal the computation (or derivation) tree proceeds in the same way as the SLD-resolution used for definite Horn clauses; i.e. if L_i unifies with the head of a program clause

$$A \leftarrow M_1, \ldots, M_m$$

with mgu (most general unifier) θ then there is a child goal

$$\leftarrow (L_1, \ldots, L_{i-1}, M_1, \ldots, M_m, L_{i+1}, \ldots, L_n)\theta.$$

When a ground negative literal $\neg B$ (as noted above non-ground negative literals may not be selected) is selected you carry out a subsidiary computation on the goal $\leftarrow B$ before continuing with the main computation. If this results in a finitely failed tree then $\neg B$ succeeds and there is a child goal

$$\leftarrow L_1, \ldots, L_{i-1}, L_{i+1}, \ldots, L_n$$

resulting from its removal. If the goal $\leftarrow B$ succeeds then $\neg B$ fails so the main derivation fails at this point. If neither of these happens the main derivation has a *dead-end* here. This can arise because the derivation tree for $\leftarrow B$ has infinite branches but no successful ones, or if it, or some subsidiary derivation, flounders or has dead-ends. Apart from these dead-ends the goals at the leaves of the computation tree are either empty (successful), or such that the selected literal is either a positive literal which does not unify with the head of any program clause (failed), or consist entirely of non-ground negative literals (flounder).

A query Q *succeeds with answer* θ using a given computation rule (for selecting an admissible literal in a goal) if the computation tree formed using this computation rule has an empty leaf node, and where θ is the restriction to the variables in Q of the product of all the mgu which have been applied on the branch leading to this empty leaf. A query Q *succeeds with answer* θ if it succeeds with answer θ using some computation rule. A query Q *fails* using a given computation rule if the computation tree formed using this computation rule is *finitely failed* (finite and has all its leaves failed). A query Q *fails* if it fails using some computation rule.

(A more detailed description of SLDNF-resolution can be found in Lloyd [1987] in the form of a recursive definition in terms of the depth of nesting of negation as failure calls. We are using a slightly different terminology from Lloyd because we find it more convenient to talk in terms of queries rather than goals. For us a *query* Q is a conjunction $L_1 \wedge \cdots \wedge L_n$ of literals, often written $? - L_1, \ldots, L_n$, and the corresponding *goal* is $\leftarrow L_1, \ldots, L_n$ i.e. the negation $\neg Q$ of the query. When Lloyd says '$P \cup \{\leftarrow Q\}$ *has an* SLDNF-*refutation*' we say 'Q *succeeds from* P *using* SLDNF-*resolution*', and when he says '$P \cup \{\leftarrow Q\}$ *has a finitely failed* SLDNF-*tree*' we say 'Q *fails from* P *using* SLDNF-*resolution*'.)

Kunen [1987a] gives a more succinct definition of these notions as follows. Let P be the program, $\mathbf{R}$ the set of all pairs (Q, θ) such that query Q succeeds with answer θ, and $\mathbf{F}$ the set of all queries which finitely fail. Then $\mathbf{R}, \mathbf{F}$ are defined by simultaneous recursion to be the least sets such that, denoting the identity substitution by 1,

1) $(\mathbf{true}, 1) \in \mathbf{R}$.
2) If Q is $Q_1 \wedge A \wedge Q_2$ where A is a positive literal, if $A' \leftarrow Q'$ is a clause of P, if $\sigma = \text{mgu}(A, A')$ and $((Q_1 \wedge Q' \wedge Q_2)\sigma, \pi) \in \mathbf{R}$ then $(Q, (\sigma\pi) \mid Q) \in \mathbf{R}$.
3) If Q is $Q_1 \wedge \neg A \wedge Q_2$ where A is a positive *ground* literal, if $A \in \mathbf{F}$ and $(Q_1 \wedge Q_2, \sigma) \in \mathbf{R}$ then $(Q, \sigma) \in \mathbf{R}$.
4) Suppose Q is $Q_1 \wedge A \wedge Q_2$ where A is a positive literal. Suppose that for each clause $A' \leftarrow Q'$ of P, if A' is unifiable with A then $(Q_1 \wedge Q' \wedge Q_2)\,\text{mgu}(A, A') \in \mathbf{F}$. Then $Q \in \mathbf{F}$.
5) If Q is $Q_1 \wedge \neg A \wedge Q_2$ where A is a positive *ground* literal and $(A, 1) \in \mathbf{R}$ then $Q \in \mathbf{F}$.

Here, and above, it is assumed that before computing an mgu the query clause and program clause are renamed to have distinct variables. Also $(\sigma\pi) \mid Q$ denotes the restriction of the substitution $\sigma\pi$ to the variables in Q.

One important difference between SLDNF-resolution and SLD-resolution is that, for the latter, any computation rule can be used i.e. if a query succeeds with answer θ using one computation rule, then it does so using any other computation rule. This is no longer true for SLDNF-resolution; e.g. for the program

$$p \leftarrow p, q$$

$$r \leftarrow \neg p$$

the query $? - r$ succeeds if the 'last literal' rule is used but not if the Prolog 'first literal' rule is used. {The reason for this discrepancy is that whether a query *fails* using SLD-resolution may depend on the computation rule.} So it is hard to imagine a feasible way of implementing SLDNF-resolution, since to determine whether a query succeeds requires a search through all possible derivation trees, using all possible selections of literals. It is shown in Shepherdson [1984], [1985] that there are maximal computation rules R_m such that if a query succeeds with answer θ using any computation rule then it does so under R_m, and if it fails using any rule then it fails using R_m, but in Shepherdson [1989] a program is given for which there is no maximal recursive rule. What causes the difficulty is that in SLDNF-resolution once having chosen a ground negative literal $\neg A$ in a goal G you are committed to waiting possibly forever for the result of the query A before proceeding with the main derivation. What you need to do is to keep coming back and trying other choices of literal in G to see whether any of them fail, since when one of these exists it is not always possible to determine it in advance. Although SLDNF-resolution is probably not feasibly implementable it is the system of logic programming usually considered in theoretical studies of negation as failure. There appears to be no chance of finding a declarative semantics sufficiently simple to be of much use for systems actually implemented, like Prolog, which uses a fixed computation rule (and also loses more completeness, even in the absence of negation, because of its depth first search of the derivation tree).

The fact that we cannot deal with non-ground negative literals means that we can only hope to get completeness of SLDNF-resolution, for any semantics, for queries which do not flounder. In general the proglem of deciding whether a query flounders is recursively unsolvable (Börger [1987]), so a strong overall condition on both the program and the query is often used which is sufficient to prevent this. A query is said to be *allowed* if every variable which occurs in it occurs in a positive literal of it; a program clause $A \leftarrow L_1, \ldots, L_n$ is *allowed* if every variable which occurs in it occurs in a positive literal of its body $L_1, \ldots, L_n$, and a program is *allowed* if all of tis clauses are allowed. It is easy to show that if the program and the query are both allowed then the query cannot flounder, because the variables occurring in negative literals are eventually grounded by the positive literals containing them.

Allowedness is a very stringent condition which excludes many common Prolog constructs, such as the definition of equality (equal(X, X)), and both clauses in the standard definition of member(X, L).

2. Semantics Based on Classical Logic

2.1. The Clark Completion, *comp*(P)

The most widely accepted declarative semantics for negation as failure uses the 'completed database' introduced by Clark [1978]. This is now usually called the *completion* or Clark completion of a program P and denoted by *comp*(P). It provides a semantics by a modification of the consequence relation of the type (i) referred to in §1.1, i.e. by replacing P by *comp*(P). It is based on 'the implied iff', the idea that when in a logic program you write

$$\text{even}(0) \leftarrow$$
$$\text{even}(s(s(x))) \leftarrow \text{even}(x),$$

what you usually intend is to give a comprehensive definition of the predicate *even*, that the clauses with *even* in their head are supposed to cover all the cases in which *even* holds, i.e.

$$\text{even}(y) \leftrightarrow (y = 0 \vee \exists x(y = s(s(x)) \wedge \text{even}(x))).$$

In the general case to form *comp*(P) you take each clause

$$p(t_1, \ldots, t_n) \leftarrow L_1, \ldots, L_m$$

of P in which the predicate symbol p appears in the head, rewrite it in *general form*

$$p(x_1, \ldots, x_n) \leftarrow \exists y_1 \ldots \exists y_p(x_1 = t_1 \wedge \cdots \wedge x_n = t_n \wedge L_1 \wedge \cdots \wedge L_m),$$

where $x_1, \ldots, x_n$ are new variables (i.e. not already occurring in any of these clauses) and $y_1, \ldots, y_p$ the variables of the original clause. If the general forms of all these clauses (we assume there are only finitely many) are:

$$p(x_1, \ldots, x_n) \leftarrow E_1$$
$$\ldots$$
$$p(x_1, \ldots, x_n) \leftarrow E_j$$

then the *completed definition* of p is

$$p(x_1, \ldots, x_n) \leftrightarrow E_1 \vee \cdots \vee E_j.$$

The empty disjunction is taken to be false, so if $j = 0$ i.e. if there is no clause with p in its head, then the completed definition of p is

$$\neg p(x_1, \ldots, x_n).$$

The *completion*, *comp*(P) of p is now defined to be the collection of completed definitions of each predicate symbol in P together with the *equality* and *freeness axioms* below, which we shall refer to as CET (Clark's equational theory).

equality axioms

$x = x$

$x = y \rightarrow y = x$

$x = y \wedge y = z \rightarrow x = z$

$x_1 = y_1 \wedge \cdots \wedge x_n = y_n \rightarrow (p(x_1, \ldots, x_n) \leftrightarrow p(y_1, \ldots, y_n))$, *for each predicate symbol* p.

$x_1 = y_1 \wedge \cdots \wedge x_n = y_n \rightarrow (f(x_1, \ldots, x_n) = f(y_1, \ldots, y_n))$, *for each function symbol* f.

freeness axioms

$f(x_1, \ldots, x_n) \neq g(y_1, \ldots, y_m)$, *for each pair of distinct function symbols* f, g,

$(f(x_1, \ldots, x_n) = f(y_1, \ldots, y_n)) \rightarrow x_1 = y_1 \wedge \cdots \wedge x_n = y_n$, *for each function symbol* f,

$t(x) \neq x$, *for each term* $t(x)$ *different from* x *in which* x *occurs.*

The reason why the freeness axioms are needed is that in SLDNF-resolution terms such as $f(x_1, \ldots, x_n), g(y_1, \ldots, y_m)$ are unifiable.

In stating these axioms constants are treated as 0-ary function symbols, and the axioms are stated for all the function and predicate symbols of the language. We assume that this language is given 'in advance of P' so to speak, rather than, as is often assumed in logic programming, being determined by the function and predicate symbols actually occuring in P. Note that, because it contains the freeness axioms, $comp(P)$ depends on the language as well as on P. For example if P is the program $p(a)$ and the language contains no function symbols and just one constant a, then $comp(P)$ consists of the equality axioms and $p(x) \leftrightarrow x = a$; but if the language has another constant b then $comp(P)$ contains the freeness axiom $a \neq b$. In the latter case $comp(P) \models \exists x \neg p(x)$, but not in the former case. Note that $comp(P)$ is an extension of P, *i.e.* $comp(P) \models P$.

The basic result of Clark [1978] is that negation as failure is sound for $comp(P)$ for both success and failure:

if $? - Q$ *succeeds from* P *with answer* θ *using* SLDNF-*resolution then* $comp(P) \models Q\theta$,

if $? - Q$ *fails from* P *using* SLDNF-*resolution then* $comp(P) \models \neg Q$.

For a proof see Lloyd [1987] pp. 92,93. The key step in the proof is to show that $comp(P)$ implies that in an SLDNF-derivation tree a goal is equivalent to the conjunction of its child goals.

This soundness result can be strengthened by replacing the consequence relation $\models$ by $\models_3$ (truth in all 3-valued models—Kunen [1987]; see §3.1) or by $\vdash_I$, the intuitionistic derivability relation (Shepherdson [1985]) or by a relation $\vdash_{3I}$, which admits only rules which are sound both for classical 3-valued and intuitionistic 2-valued logic (Shepherdson [1989a]). This helps to explain why SLDNF-resolution is usually incomplete for $comp(P)$ with respect to the usual 2-valued models of $comp(P)$, it must be true in all 3-valued models, and must be derivable using only intuitionistically acceptable steps. For example, if P is the program $p \leftarrow \neg p$, then $comp(P)$ contains $p \leftrightarrow \neg p$ and is inconsistent, so p is a consequence of it, and if SLDNF-resolution were complete for $comp(P)$ then $? - p$ should succeed. In fact it dead-ends. This can be explained by noting that there is a 3-valued model of $comp(P)$ in which p is not true but undefined. Similarly if

P is the program

$$p \leftarrow q$$
$$p \leftarrow \neg q$$
$$q \leftarrow q,$$

whose completion contains $p \leftrightarrow (q \vee \neg q)$, the fact that $? - p$ does not succeed using SLDNF-resolution can be explained by observing that p is not intuitionistically derivable from $comp(P)$, since the law of the excluded middle does not hold in intuitionistic logic.

Despite these kinds of incompleteness, and those due to floundering, pointed out above, $comp(P)$ is regarded by many logic programmers as the appropriate declarative semantics for the procedure of SLDNF-resolution applied to a program P, i.e. as the meaning he had in mind when he wrote down the program P. This is somewhat removed from the clarity aimed at in ideal logic programming, where the declarative meaning of a program should be apparent from the text of the program as written. Although in simple cases $comp(P)$ may be what most people have in mind when they write P, it is not always easy when writing P to foresee the effect of forming $comp(P)$, particularly when P contains clauses involving mutual recursion and negation. Indeed the problem of deciding whether $comp(P)$ is consistent is recursively unsolvable. It should also be noted that $comp(P)$ does not depend only on the logical content of P but also on the way it is written in clausal form; the completion of $p \leftarrow \neg q$ contains $p \leftrightarrow \neg q$, and $\neg q$, which is equivalent to p and $\neg q$, but the completion of the logically equivalent program $q \leftarrow \neg p$ has these reversed.

There are some people who use $comp(P)$ with an additional type (iii) modification in the form of a restriction to so-called Herbrand (see below for definition) models, i.e. who regard a query $? - Q$ as a request to know whether $\exists Q$ is true in all *Herbrand models* of $comp(P)$. However it should be noted that if this is done it may lead to non-computable semantics, i.e. (Shepherdson [1988a]):

The set of negative ground literals $\neg A$ which are true in all Herbrand models of $comp(P)$ may not be recursively enumerable.

[Unlike the set of sentences which are true in all models of $comp(P)$, which is recursively enumerable by the Gödel completeness theorem.] We have followed the usual convention here of using 'Herbrand model of $comp(P)$', to refer to a term model of *comp*(P) over the language L used to express P, i.e. a model whose domain is the set of ground terms of L (the Herbrand universe of L), with functions given the free interpretation (e.g. the value of the function f applied to the term t is simply $f(t)$). This is a rather misleading notation since it suggests that the Skolem–Herbrand theorem, that if a sentence has a model it has a Herbrand model should be applicable, and that restriction to Herbrand models should make no difference. The reason it does make a difference here is that for this theorem to apply the Herbrand universe must be large enough to contain Skolem functions enabling the sentence to be written in universal form. The appropriate Herbrand universe for *comp*(P) would be one containing Skolem functions allowing the elmination of the existential quantifiers occurring on the right hand sides of the completed definitions of the predicate symbols of P. And even over that universe one could not restrict to free interpretations because *comp*(P) contains the equality predicate. Nevertheless the 'Herbrand models' of *comp*(P) are considered to be of particular interest, presumably because the Herbrand universe of the original language in which P is expressed often does define the domain of individuals one is interested in. There is a very neat fixpoint characterization of these models which we now consider.

First we associate with the program P the operator T_P which maps a subset I of the Herbrand base (set of ground atoms) B_L into the subset $T_P(I)$ comprising all those ground atoms A for which there exists a ground instance

$$A \leftarrow L_1, \ldots, L_m$$

of a clause of P with all of $L_1, \ldots, L_m$ in I. $T_P(I)$ is the set of immediate consequences of I, i.e. those which can be obtained by applying a rule from P once only. [Note that T_P depends not only on the logical content of P but also on the way it is written e.g. adding the tautology $p \leftarrow p$ changes T_P.] Clearly

$$I \textit{ is a model for } P \textit{ iff } T_P(I) \subseteq I.$$

i.e. iff I is a *pre-fixpoint* of T_P. Following Apt, Blair and Walker [1988] let us say an interpretation I is *supported* if for each ground atom A which is true in I there is a ground instance

$$A \leftarrow L_1, \ldots, L_m$$

of a clause of P such that $L_1, \ldots, L_m$ are true in I. The terminology comes from the idea of negation as the default, that ground atoms are assumed false unless they can be supported in some way by the program. Clearly

I is supported iff $T_P(I) \supseteq I$,

I is a supported model of P iff it is a fixpoint T_P, i.e. $T_P(I) = I$.

The notion of being supported is equivalent to satisfying the 'only if' halves of the completed definitions of the predicate symbols in *comp*(P), so we have:

I is a Herbrand model of comp(P) iff it is a supported model of P.

I is a Herbrand model of comp(P) iff it is a fixpoint of T_P.

This fixpoint characterization of Herbrand models of *comp*(P) can be extended to arbitrary models by replacing the Herbrand base by the set of formal expressions $p(d_1, \ldots, d_n)$ where p is a predicate symbol and $d_1, \ldots, d_n$ are elements of the domain; see Lloyd [1987] p. 81.

If P is composed of definite Horn clauses (i.e. with no negative literals) then T_P is continuous and has a least fixpoint $lfp(T_P)$ which is obtainable as $T_P \uparrow \omega$, the result of iterating T_P ω times starting from the empty set. This is the least Herbrand model of P and of *comp*(P) (so *comp*(P) is consistent). In general T_P is not continuous or even monotonic, so it may not have fixpoints, e.g. if P is $p \leftarrow \neg p$.

If P is a definite Horn clause program then *comp*(P) adds no new positive information:

If P is a definite Horn clause program and Q is a positive sentence, then

$$comp(P) \models Q \text{ implies } P \models Q.$$

A positive sentence is one built up using only $\vee$, $\wedge$, $\forall$, $\exists$. This is an immediate consequence of the existence of the least fixpoint model in the case where Q is a ground atom, and the argument is easily extended to cover the case of a positive sentence (Shepherdson [1988]).

If we accept the restriction described in the last section to programs and queries which are both allowed then there are some classes of programs for which SLDNF-resolution is complete for *comp*(P), i.e.

(1) *if comp(P) $\models Q\theta$ then Q succeeds with answer including*[1] θ

(2) *if comp(P) $\models \neg Q$ then Q fails.*

[1] i.p. an answer θ' such that there exists φ with $Q\theta' = Q\theta\varphi$

The simplest is the class of *definite* programs, whose clauses contain no negative literals. For definite programs P, (1) is true for all queries and (2) is true for definite queries, i.e. not containing negation. {For a proof see Lloyd [1987] Ch. 2,3} (2) may not be true for all queries as the example with P as $p \leftarrow p$ and Q as $? - p, \neg p$ shows. Here the use of negation as failure is minimal; the only negative literals involved are those occurring in the query, and negation as failure is only used once on each of these. There is no nested use of negation as failure.

Another class is that of *hierarchical* programs introduced by Clark. These are free of recursion, that is to say the predicate symbols can be assigned to levels so that the predicate symbols occurring in the body of a clause are of lower levels than that occurring in the head. A much larger class has recently been given by Kunen [1987a]. This is the class of programs which are *semi-strict* (or *call-consistent*, as it is more usually called now). Semi-strict means that no predicate symbol depends negatively on itself in the way that p does in the program $p \leftarrow \neg p$, or similarly via any number of intermediate clauses and predicate symbols, e.g. $p \leftarrow \neg q$, $q \leftarrow p$. Formally this is defined as follows:

We say $p \sqsupseteq_{+1} q$ if there is a program clause with p occurring in the head, and q occurring in a *positive* literal in the body. We say $p \sqsupseteq_{-1} q$ iff there is a clause with p occurring in the head, and q occurring in a *negative* literal in the body. Let $\geq_{+1}$ and $\geq_{-1}$ be the least pair of relations on the set of predicate symbols satisfying:

$$p \geq_{+1} p$$

and

$$p \sqsupseteq_i q \ \& \ q \geq_j r \Rightarrow p \geq_{i \cdot j} r.$$

Then the program is *semi-strict* if we never have $p \geq_{-1} p$.

For semi-strict programs the completeness requires also a condition involving the query, that the program is *strict with respect to the query.* This means that there is no predicate symbol p on which the query depends both positively and negatively, as does the query $? - p, \neg p$ in the example above, or the query $? - a, b$ for the program

$$a \leftarrow p$$

$$b \leftarrow \neg p$$

$$p \leftarrow p.$$

Formally, if Q is a query, we say $Q \geq_i p$ iff either $a \geq_i p$ for some a occurring positively in Q, or $a \geq_{-i} p$ for some a occurring negatively in Q. The program is strict with respect to the query Q iff for no predicate symbol p do we have both $Q \geq_{+1} p$ and $Q \geq_{-1} p$.

The semi-strict programs include the *stratified* programs introduced by Apt, Blair and Walker [1988]; these are like the hierarchical programs except that the condition that the level of every predicate symbol in the body be less than the level of the head is maintained for predicate symbols appearing negatively in the body, but for those appearing positively it is relaxed to 'less than or equal to'. The completeness result for stratified programs has been proved independently by Cavedon and Lloyd [1987].

The three kinds of program mentioned above can be conveniently characterized in terms of the *dependency graph* of a program P. This is a directed graph whose nodes are the predicate symbols of P and which has an edge from p to q iff there is a clause in P with p in the head and q in the body. The edge is marked *positive* (resp. *negative*) if q occurs in a *positive* (resp. *negative*) literal in the body (an edge may be both positive and negative). Then a program is *hierarchical* iff its dependency graph contains no cycles, it is *stratified* iff its dependency graph contains no cycles containing a negative edge and it is *call-consistent* iff its dependency graph contains no cycle with an odd number of negative edges.

Kunen's work clarifies the role of the hypotheses of strictness and allowedness. Allowedness gives completeness for the 3-valued semantics based on comop(P), and strictness ensures that the 2-valued and 3-valued semantics coincide.

Summary. The Clark completion *comp*(P) of a program P is one of the simplest declarative semantics which have been offered for negation as failure and is currently the most used. However it is not always easy to read off the meaning of *comp*(P) from P. A further disadvantage is that SLDNF-resolution is not in general complete for it, though there are important classes of programs for which this is the case.

2.2. Reiter's Closed World Assumption $CWA(P)$

The *closed world assumption* is another of the commonly accepted declarative semantics for negation as failure. It is particularly appropriate for database applications, being founded on the idea that the program (database) contains all the positive information about the objects in the domain. Reiter [1978] gave this a precise formulation by saying that any positive ground literal not implied by the program is taken to be false. He axiomatised this by adjoining the negations of these literals to the program P thus obtaining

$$CWA(P) = P \cup \{\neg A : A \text{ is a ground atom and } P \not\models A\}.$$

He also restricted consideration to what in logic programming are usually called *Herbrand models*, i.e. models whose domain of individuals consists of the ground terms. This makes the closed world assumption (where it is consistent) categorical, i.e. if it has an Herbrand model that model must be unique, because a ground atom A must be true in it if it is a consequence of P and false if it is not a consequence of P. Actually in the usual logic programming situation, where P consists of clauses (or, more generally, universal sentences), the query Q is an existential sentence, and neither P nor Q contains $=$, the restriction to Herbrand models is irrelevant, i.e.

$$CWA(P) \models_H Q \text{ iff } CWA(P) \models Q$$

where $T \models_H S$ means 'S is true in all Herbrand models of T'. This is because $CWA(P) \cup \neg Q$ consists of universal sentences not containing equality so, by the usual Herbrand–Skolem argument, if it has a model it has an Herbrand model. In particular if $CWA(P)$ is consistent it has an Herbrand model.

If P is a definite Horn clause program then $CWA(P)$ is consistent, because the Herbrand interpretation in which a ground atom A is true iff $P \models A$, satisfies P (since if $B \leftarrow A_1, \ldots, A_r$ is a clause of P and $P \models A_1, \ldots, P \models A_r$ then $P \models B$); and it clearly satisfies the remaining axioms ($\neg A$ if $P \models A$) of $CWA(P)$. (This is of course the familiar least Herbrand and least fixpoint model of P.)

The closed world assumption is a very strong presumption in favor of negative information and is often inconsistent. This is the case if the program implies indefinite information about ground atoms, e.g. if P is $p \leftarrow \neg q$, which is equivalent to $p \vee q$, then neither p nor q is a consequence of P so both $\neg p$ and $\neg q$ belong to $CWA(P)$, and since $CWA(P)$ also contains P it is inconsistent. This condition is actually necessary and sufficient for the inconsistency of $CWA(P)$:

If P consists of universal sentences and does not contain $=$, then $CWA(P)$ is consistent iff for all ground atoms $A_1, \ldots, A_r$, $P \models (A_1 \vee \cdots \vee A_r)$ implies $P \models A_i$ for some $i = 1, \ldots, r$.

The 'only if' part of this follows as in the example; the 'if' part follows from the compactness theorem, for if $CWA(P)$ is inconsistent so is some finite subset of it, so there exist $\neg A_1, \ldots, \neg A_r$ in $CWA(P)$ such that $P \cup \{\neg A_1, \ldots, \neg A_r\}$ is inconsistent, i.e. $P \models (A_1 \vee \cdots \vee A_r)$.

Indefinite information about non-ground literals need not imply the inconsistency of the closed world assumption, e.g. $CWA(P)$ is consistent for the program P:

$$p(x) \leftarrow \neg q(x)$$
$$p(a)$$
$$q(b).$$

However the consistency of the closed world assumption for certain extensions of P does imply that P is equivalent to a definite Horn clause program, and, by the remark above, conversely:

If P is a set of first order senteces, then P is equivalent to a set of definite Horn clauses iff $CWA(P \cup S)$ is consistent for each set S of ground atoms, possibly involving new constants.

(For a proof see Makowsky [1986] or Shepherdson [1988]; the result holds for first order logic with or without equality.) This means that if you want to be able to apply the closed world assumption consistently to your program, and to any subsequent extension of it by positive facts, possibly involving new constants, then you are confined to definite Horn clause programs. This may be appropriate when the program is a simple kind of database but most logic programmers would consider it too restrictive and would therefore look for some other form of default reasoning for dealing with negation. Note that if the phrase 'possibly involving new constants' is omitted the result above fails, also that the consistency of the closed world assumption depends on the underlying language. The non-Horn program above satisfies the closed world assumption (i.e. $CWA(P)$ is consistent) if the Herbrand universe, i.e. the set of ground terms, is the usual one $\{a, b\}$ determined by the terms appearing in P, and it continues to do so if any more atoms from the corresponding Herbrand base $\{p(a), p(b), q(a), q(b)\}$ are added to the program. But if the Herbrand universe is enlarged to $\{a, b, c\}$ the closed world assumption becomes inconsistent. This ambiguity does not arise if the program is a definite Horn clause program, for the argument above shows that the closed world assumption for such a program is consistent whatever Herbrand universe is used.

Makowsky [1986] observed that the consistency of the closed world assumption is also equivalent to an important model-theoretic property:

If P is a set of first order sentences then a term structure M is a model for $CWA(P)$ iff it is a generic model of P, i.e. for all ground atoms A,

$$M \models A \text{ iff } P \models A.$$

(We use the words 'term structure' rather than 'Herbrand interpretation' here because in this more general setting where the sentences of P may not be all universal, the word 'Herbrand' would be more appropriately used for the language extended to include the Skolem functions needed to express these sentences in universal form.) This notion of a generic model is like that of a free algebraic structure; just as a free group is one in which an equation is true only if it is true in all groups, so a generic model of P is one in which a ground atom is true only if it is true in all models of P, i.e. is a consequence of P. So it is a unique most economical model of P in which a ground atom is true iff it has to be. If we identify a term model in the usual way with the subset of the base (set of ground atoms) which are true in it, then it is literally the smallest term model. It is easy to see that the genericity of a generic term model extends from grond atoms to existential quantifications of conjunctions of ground atoms:

If M is a generic model of P, and Q is a conjunction of atoms, then

$$M \models \exists Q \text{ iff } P \models \exists Q.$$

So a positive query is true in M iff it is a consequence of P, and one can behave almost as though one was dealing with a theory which had a unique model. This is one of the attractive features of definite Horn clause logic programming, which, as the results above show, does not extend much beyond it. But notice that if one is interested in answer substitutions then one cannot restrict consideration to a generic model, e.g. if P is:

$$p(0)$$
$$p(s(x))$$

then the identity substitution is a correct answer to the query $?{-}p(x)$ in the least Herbrand model, but is not a 'correct answer substitution', because $\forall x p(x)$ is not a consequence of P.

For negative queries not only is this genericity property lost, but as Apt, Blair and Walker [1988] have pointed out, the set of queries which are true under the closed world assumption, i.e. true in the generic model, may not even be recursively enumerable, that is to say there may be no computable procedure for generating the set of queries which ought to succeed under the closed world assumption. To show this take a non-recursive recursively enumerable set W and a definite Horn clause program P with constant 0, unary function symbol s and unary predicate symbol p, such that

$$P \models p(s^n(0)) \text{ iff } n \in W.$$

[This is possible since every partial recursive function can be computed by a definite Horn clause program (Lloyd [1987] p. 53).] Now $\neg p(s^n(0))$ is true under the closed world assumption iff $n \notin W$. However this situation does not arise under the conditions under which Reiter originally suggested the use of the closed world assumption, namely when there are no function symbols in the language. Then the Herbrand base is finite and so is the model determined by the closed world assumption, hence the set of true queries is recursive.

Negation as failure is *sound* for the closed world assumption for both success and failure i.e.

if $? - Q$ *succeeds from* P *with answer* θ *using* SLDNF*-resolution then* $CWA(P) \models_H Q\theta$, *if* $? - Q$ *fails from* P *using* SLDNF*-resolution then* $CWA(P) \models_H \neg Q$.

For a proof see Shepherdson [1984]. However SLDNF-resolution is usually incomplete for the closed world assumption. This is bound to be the case when the closed world assumption is inconsistent. Even for definite Horn clause programs, where the closed world assumption will be consistent, and even when there are no function symbols, SLDNF-resolution can be incomplete for the closed world assumption, e.g. for the program $p(a) \leftarrow p(a)$, and the query $? - \neg p(a)$.

{As noted above, the restriction to Herbrand models makes no difference to the first of the soundness statements above, i.e. it remains true when $\models_H$ is replaced by $\models$. This is not true of the second, where we are dealing with the negation of a query. For example if P consists of the single clause $p(a) \leftarrow p(b)$ and Q is $p(x)$ then Q fails from P but $CWA(P) \models \neg Q$ is not true since there are non-Herbrand models of $CWA(P)$ containing elements c with $p(c)$ true.}

The closed world assumption and the completion are superficially similar ways of extending a program. They are both examples of reasoning by default, assuming that if some positive piece of information cannot be proved in a certain way from P then it is not true. But for the closed world assumption the notion of proof involved is that of full first order logic, whereas for the completion it is 'using one of the program clauses whose head matches the given atom'. At first sight this is a narrower notion of proof, so that one would expect that more ground atoms should be false under the completion than under the closed world assumption, i.e. that $CWA(P)$ should be a consequence of $comp(P)$. But it is not so simple because $comp(P)$ adds, in the 'only if' halves of the completed definition of a predicate symbol p, new statements which can be used to prove things about predicates other than p. Also the closed world assumption involves a restriction to Herbrand models, which the completion, a usually defined does not. In fact when they are compatible it must be the closed world assumption which implies the completion, because the former is categorical. Simple examples show that either of the closed world assumption and the completion can be consistent and the other one not, or both can be separately consistent but incompatible. For conditions under which they are compatible, and the relation of $CWA(P) \cup comp(P)$ to $CWA(comp(P))$, see Shepherdson [1988]. The closed world assumption has the merit that $CWA(P)$ depends only on the logical content of P whereas, as we saw in §2.1, $comp(P)$ depend on the way P is expressed in clausal form. Both may be difficult to read off from P; we have seen that the problem even of deciding whether $comp(P)$ is consistent is recursively unsolvable, and for given P the problem of deciding whether a ground literal $\neg A$ belongs to $CWA(P)$ may be recursively unsolvable.

It is possible that one might want to apply the closed world assumption to some predicates but to protect from it other predicates where it was known that the information about them was incomplete. For reference to this notion of protected data see Minker and Perlis [1985], or Jaeger [1988] which studies the model-theoretic aspects of this relativized closed world assumption. For more details of the model-theory of the ordinary closed world asumption see Makowsky [1987]. In §4 we discuss briefly various model-theoretic semantics which can be regarded as weak forms of the closed world assumption, e.g. the generalized closed world assumption of Minker [1982], and the self-referential closed world assumption of Fine

[1989]. However not all of these have an obvious relation to negation as failure.

Summary. The closed world assumption is a natural and simple way of dealing with negation for programs which represent simple databases, i.e. definite Horn clause programs without function symbols. Its use outside this range is limited: as soon as function symbols are introduced there may not be any sound and complete computable proof procedure for it, and if one goes beyond definite Horn clause programs it will be inconsistent, either for the original program or for some extension of it by positive atoms.

2.3. Sound and Complete Semantics for a Version of Negation as Failure

It would be easier to understand the logical meaning of negation as failure, or SLDNF-resolution, if we had a logic-based declarative semantics with respect to which it was both sound and complete, i.e.

$?-Q$ *succeeds from* P *with answer including* θ *using* SLDNF-*resolution iff* $S(P) \models Q\theta$.

We have seen above that $S(P)$ cannot be taken to be either $comp(P)$ or $CWA(P)$, because although it is sound for both, SLDNF-resolution is in general not complete for either of them. Indeed as noted in §1.2 above, we cannot hope to achieve soundness and completeness for all queries for *any* $S(P)$ because of the inability of SLDNF-resolution to deal with non-ground negative literals. For example for the program, $p(x) \leftarrow \neg r(x)$, the query $?-p(a)$ succeeds but $?-p(x)$ does not. The usual way of dealing with this problem is to restrict consideration to programs and queries satisfying the allowedness conditions described above. But that is a very severe restriction so it is worth considering the alternative of extending SLDNF-resolution so that it can avoid flounders. We may extend SLDNF-resolution to allow negation as failure to be applied to non-ground literals by means of a preliminary substitution, i.e.

if $A\theta$ *fails then* $\neg A$ *succeeds with answer* θ.

This is justified as an extension of negation as failure by the fact that

$$\neg\exists A\theta \Rightarrow \forall\neg A\theta$$

is logically valid. In fact if SLDNF-resolution is sound for $S(P)$ for both success and failure then it will remain sound if negation as failure is extended in this way, since if $A\theta$ fails then $S(P) \models \neg A\theta$, so it is sound for $\neg A$

to succeed with answer θ. This extended use of negation as failure can be grafted onto SLDNF-resolution by allowing the selection of a non-ground negative literal $L_i = \neg A$ in a goal $\leftarrow L_1, \ldots, L_n$, if there exists θ such that $A\theta$ fails, to obtain the new goal $\leftarrow (L_1, \ldots, L_{i-1}, L_{i+1}, \ldots, L_n)\theta$. We shall call this SLDNFS-*resolution* (SLD-resolution with negation as failure with substitution). Note that the extended use of negation as failure is only allowed when constructing success branches; the rule for failing $\neg A$ is the same as before:

If A is ground and succeeds then $\neg A$ fails.

[Though this could be extended, as it is in some Prolog systems, e.g. IC-Prolog, to

If A succeeds with answer 1 *then $\neg A$ fails.*

Here 1 denotes the identity substitution, or, more generally, any substitution which maps the variables in A to distinct variables. This is justified by $\forall A \Rightarrow \neg\exists\neg A$.]

SLDNFS-resolution is even less feasibly implementable then SLDNF-resolution because one has to consider all possible substitutions to apply to the negative literals, and if there are function symbols there will be infinitely many substitutions. But the above remarks on soundness show that it is necessary to extend SLDNF-resolution in this way if one wishes to obtain completeness results in the general case. The effect of passing to SLDNFS-resolution is essentially to prevent floundering. To make this precise we have to generalize the notion of flounder to include flounders occurring in the sub-trees of the ground negative literals selected in negation as failure steps. When this notion of *generalized flounder* is defined in the obvious way we have:

If ? − Q succeeds from P using SLDNFS-*resolution then using* SLDNF-*resolution it either succeeds or has a generalized flounder.*

So completeness results for SLDNFS-resolution will give completeness results for SLDNF-resolution for queries which do not flounder.

We now describe the construction of a declarative semantics $\bar{T}_\omega(P)$ for which SLDNFS-resolution is sound and complete. The idea behind the construction is that, until it is selected, a negative literal $\neg p(\underline{t})$ behaves like an instance $\bar{p}(\underline{t})$ of a new predicate $\bar{p}$ unrelated to p. So we start by replacing $\neg p$ by $\bar{p}$. We then arrange for an instance $\bar{p}(\underline{t})$ to be true or false when and only when the falsity or truth of $\neg p(\underline{t})$ is established by a negation as failure step. This is done by an iterative construction paralleling the rank of the SLDNFS-refutation or finitely failed tree. We add $\bar{p}(\underline{t})$ when $\neg p(\underline{t})$ is a consequence of the previous stage, and we add $\neg\bar{p}(\underline{t}_0)$ when $\underline{t}_0$ is ground and $p(\underline{t}_0)$ is a consequence of the previous stage.

The details are as follows. Start by enlarging the underlying language L to a new language $\bar{L}$ by adding, for each predicate p in L, a new predicate $\bar{p}$ of the same arity as p. Then replace all negative literals $\neg p(t)$ in bodies of clauses of P by $\bar{p}(t)$. If we were to form the completion of tis program then all the new predicates p would be everywhere false. So we now add clauses $\bar{p}(x) \leftarrow \bar{p}(x)$ for each new predicate $\bar{p}$. Let $\bar{P}$ be the resulting definite Horn clause program. The new clauses have the effect that the completed definition of $\bar{p}$ in $comp(P)$ is the tautology $\bar{p}(x) \leftrightarrow \bar{p}(x)$. Now define

$$\bar{T}_0(P) = comp(\bar{P})$$
$$\bar{T}_{n+1}(P) = \bar{T}_n(P) \cup \bar{P}_{n+1}(P) \cup \bar{N}_{n+1}(P),$$

where

$$\bar{P}_{n+1}(P) = \{\bar{p}(\underline{t}) : \bar{T}_n(P) \models \neg p(\underline{t})\}$$
$$\bar{N}_{n+1}(P) = \{\neg\bar{p}(\underline{t}_0) : \underline{t}_0 \text{ ground and } \bar{T}_n(P) \models p(\underline{t}_0)\},$$

finally

$$\bar{T}_\omega(P) = \bigcup_{n=0}^{\infty} \bar{T}_n(P).$$

$\bar{T}_\omega(P)$ can be regarded as a weaker version of $comp(P)$ which includes only those instances of the only-if halves of completed definitions of predicates which correspond to uses of negation as failure.

SLDNFS-resolution is sound and complete for $\bar{T}_\omega(P)$ for success, i.e.

? – Q succeeds from P with answer including θ using SLDNFS-*resolution iff* $\bar{T}_\omega(P) \models Q\theta$.

It is also sound for failure, and complete for failure of positive queries, i.e. those which contain only positive literals:

If ? – Q fails from P using SLDNFS-*resolution then* $\bar{T}_\omega(P) \models \neg Q$.

If ? – Q is a positive query and $\bar{T}_\omega(P) \models \neg Q$ *then ? – Q fails from P using* SLDNFS-*resolution.*

This restriction on the completeness results for failure is due to the fact that none of the extensions of SLD-resolution considered is capable of using the law of contradiction, that $p \wedge \neg p$ is false, so that a query $? - p, \neg p$ which should always fail, will only do so if p succeeds or fails. So to get completeness for failure for any (2-valued) semantics we will have to impose some condition on the query. The condition of §2.1 above that the program is strict with respect to the query, seems to be the appropriate one. For Kunen's 3-valued semantics of §3.1 below this is not necessary because the law of contradiction fails in 3-valued logic, since p can be undefined, in which case $p \wedge \neg p$ is not false but undefined.

These soundness results (and, of course, the completeness results) hold also when the classical consequence relation $\models$ is replaced by a weaker relation of provability obtained by using only rules which are sound for both classical 3-valued and intuitionistic 2-valued logic.

(Reynolds [1987], [1987a], [1988], has recently obtained some completeness results for SLDNF-resolution for another default operator $SYNTH(P)$. This has the advantage of consisting of a finite set of sentences, but the completeness results are not as general as those for $\bar{T}_\omega(P)$. In the 'datalog' case, where there are no function symbols he proves ([1987] Theorem 5.6) completeness for success for positive ground queries, subject to a condition like allowedness which prevents floundering. When there are function symbols he imposes a condition on the existence of a 'height function', which is similar to the semi-strictness of §2.1.)

Summary. The semantics based on $\bar{T}_\omega(P)$ is a solution to the problem of finding a declarative semantics for which negation as failure is sound and complete for all queries (at least for success). It involves extending the negation as failure rule of SLDNF-resolution to SLDNFS-resolution, but that is in a sense the minimal extension for which such completeness results are possible, because of the inability of SLDNF-resolution to deal with non-ground negative literals. It can legitimately be described as a 'declarative' rather than a 'procedural' semantics since it does not refer to any specific computational proof procedure but is a set of sentences defined by an iterative procedure involving the relation of logical consequence. However this definition bears some similarlity to the definition of SLDNF-resolution, and the semantics is too complicated to be of any practical use in checking correctness of programs or validity of program transformations, although it might possibly be of some technical use in theoretical studies.

3. Semantics Based on Non-Classical Logics

3.1. Three-Valued Logic

Three-valued logic seems particularly apt for dealing with programs, since they may either succeed or fail or go on forever giving no answer. And it seems particularly apt for discussing database knowledge where we know some things are true, some things are false, but about other things we do not know whether they are true or false. Kleene [1952] introduced such a logic to deal with partial recursive functions and predicates. The three truth values are **t**, true, **f**, false and **u**, undefined or unknown. A connective has the value **t** or **f** if it has that value in ordinary 2-valued logic for all possible replacements of **u**'s by **t** or **f**, otherwise it has the value **u**. For example $p \to q$ gets the truth value **t** if p is **f** or q is **t** but the value **u** if p, q are both **u**. (So $p \to p$ is not a tautology, since it has the value **u** if p has value **u**.) The universal quantifier is treated as an infinite conjunction so $\forall x \phi(x)$ is **t** if $\phi(a)$ is **t** for all a, **f** if $\phi(a)$ is **f** for some a, otherwise **u** (so it is the glb of truth values of the $\phi(a)$). Similarly $\exists x \phi(x)$ is **t** if some $\phi(a)$ is **t**, **f** if all $\phi(a)$ are **f**, otherwise **u**.

This logic has been used in connection with logic programming by Mycroft [1983] and Lassez and Maher [1985]. Recent work of Fitting [1985] and Kunen [1987,1987a] provided an explanation of the incompleteness of negation as failure for 2-valued models of *comp*(P). It turns out that it is also sound for *comp*(P) in 3-valued logic, so it can only derive those consequences that are true not only in all 2-valued models but also in all 3-valued models.

Use of the third truth value avoids the usual difficulty with a non-Horn clause program P that the associated operator T_P which corresponds to one application of ground instances of the clauses regarded as rules, is no longer monotonic. To avoid asymmetric associations of T with true, let us call the corresponding operator for 3-valued logic Φ_P. This operates on pairs (T_0, F_0) of disjoint subsets of B_P, the Herbrand base of P, to produce a new pair $(T_1, F_1) = \Phi_P(T_0, F_0)$, the idea being that if the elements of T_0 are known to be true and the elements of F_0 are known to be false, then one application of the rules of P to ground instances shows that the elements of T_1 are true and the elements of F_1 are false.

Formally, for a ground atom A, we put A in T_1 iff some ground instance of a clause of P has head A and a body made true by (T_0, F_0), and we put A in F_1 iff all ground instances of clauses of P with head A have body made false by (T_0, F_0). In particular if A does not match the head of any clause of P it is put into F_1, which is in accordance with the default reasoning behind negation as failure.

A little care is needed in defining the notion of a 3-valued model and the notion of *comp*(P). A 3-valued model of a set of sentences S is a set of objects together with interpretations of the various function symbols in the same way as in the 2-valued case. The equality relation '=' is interpreted as identity (hence is 2-valued), the sentences in S must all evaluate to $\mathbf{t}$, and so must what Kunen calls CET or 'Clark's Equational Theory', i.e. the equality and freeness axioms listed in §2.1 above as part of *comp*(P). If we were now to write the rest of *comp*(P), the completed definitions of predicates, in the form

$$p(x_1, \ldots, x_n) \leftrightarrow E_1 \vee \cdots \vee E_j$$

using the Kleene truth table for $\leftrightarrow$ then we should be committing ourselves to 2-valued models, for $p \leftrightarrow p$ is not $\mathbf{t}$ but $\mathbf{u}$ when p is $\mathbf{u}$. Kunen therefore replaces this $\leftrightarrow$ with Kleene's weak equivalence $\simeq$ which gives $p \simeq q$ the value $\mathbf{t}$ if p, q have the same truth value, $\mathbf{f}$ otherwise. (Note: Our notation here agrees with Fitting but not with Kunen who uses $\leftrightarrow$ instead of our $\simeq$ and $\equiv$ instead of our $\leftrightarrow$.) This saves *comp*(P) from the inconsistency it can have under 2-valued logic. For example if P is $p \leftarrow \neg p$ then the 2-valued *comp*(P) is $p \leftrightarrow \neg p$, which is inconsistent; what Kunen uses is $p \simeq \neg p$ which has a model with p having the value $\mathbf{u}$. Having done this, if we want *comp*$(P) \models_3 P$ to hold i.e. 3-valued models of *comp*(P) also to be models of P then we must replace the $\rightarrow$ in the clauses of P by $\supset$ where $p \supset q$ is the 2-valued 'if p is true then q is true', which has the value $\mathbf{t}$ except when p is $\mathbf{t}$ and q is $\mathbf{f}$ or $\mathbf{u}$, when it has the value $\mathbf{f}$. (Actually this would still be true with the slightly stronger 2-valued connective which also requires 'and if q is $\mathbf{f}$ then p is $\mathbf{f}$'.) It does not hold with $\rightarrow$ because the program $p \leftarrow p$ has completion $p \simeq p$ which has a model where p is $\mathbf{u}$ which gives $p \leftarrow p$ the value $\mathbf{u}$ not $\mathbf{t}$.

We may identify a pair (T, F) of disjoint subsets of the Herbrand base with the three valued structure which gives all elements of T the value $\mathbf{t}$, all elements of F the value $\mathbf{f}$, and all other elements of the Herbrand base the value $\mathbf{u}$.

We define $\Phi_P \uparrow \alpha$ like we defined $T_P \uparrow \alpha$ i.e. $\Phi_P \uparrow 0 = (\emptyset, \emptyset)$, $\Phi_P \uparrow (\alpha + 1) = \Phi_P(\Phi_P \uparrow \alpha)$, $\Phi_P \uparrow \alpha = \cup_{\beta<\alpha}(\Phi_P \uparrow \beta)$ for α a limit ordinal. We now have the analogue for Φ_P of the 2-valued properties of T_P.

(1) *Φ_P is monotonic.*

(2) *Φ_P has a least fixed point given by $\Phi_P \uparrow \alpha$ for some ordinal α.*

(3) *If T, F are disjoint then (T, F) is a 3-valued Herbrand model of comp(P) iff it is a fixed point of Φ_P.*

In addition:

(4) *comp(P) is always consistent in 3-valued logic.*

For a proof see Fitting [1985]. The monotonicity is obvious and (2) is a well known consequence of that. (3) also follows easily, as in the 2-valued case, and (4) follows from (2) and (3).

However the operator Φ_P is not in general continuous and so the closure ordinal α, i.e. the α such that $\Phi_P \uparrow \alpha$ is the least fixed point may be greater than ω. For example if P is:

$$p(f(x)) \leftarrow p(x)$$
$$q(a) \leftarrow p(x)$$

then it is easy to check that the closure ordinal is $\omega + 1$. Fitting shows that the closure ordinal can be as high as Church–Kleene ω_1, the first non-recursive ordinal. Both Fitting and Kunen show also that a semantics based on this least fixed point as the sole model suffers from the same disadvantage as the closed world assumption, namely that the set of sentences, indeed even the set of ground atoms, that are true in this model may be non-recursively enumerable (as high as Π_1^1 complete in fact). The same is true of a semantics based on all 3-valued Herbrand models, for the programs in the examples constructed by Fitting and Kunen can be taken to have only one 3-valued Herbrand model.

Kunen proposes a very interesting and natural way of avoiding this non-computable syntax: Why should we consider only Herbrand models; why not consider all 3-valued models? In other words, ask whether the query Q is true in all 3-valued models of *comp*(P). He shows that an equivalent way of obtaining the same semantics is by using the operator Φ_P by chopping it off at ω:

A sentence $\emptyset$ has value **t** *in all 3-valued models of comp(P) iff it has value* **t** *in $\Phi_P \uparrow n$ for some finite n.*

There are three peculiar features of this result. First, in determining the truth value of $\emptyset$ in $\Phi_P \uparrow n$, quantifiers are interpreted as ranging over the

Herbrand universe, yet we get equivalence to truth in *all* 3-valued models. This may be partly explained by the fact that *the truth of this result depends on Φ_P and comp(P) being formed using a language L_∞ with infinitely many function symbols.* (Where constants are treated as 0-ary functions. Actually Kunen uses a language with infinitely many function symbols of all arities, but the weaker condition above is sufficient.) For example if P is simply $p(a)$, then the usual language L_p associated with P has just one constant a, the Herbrand base is $\{p(a)\}$, and if Φ_P is evaluated with respect to this, then $\forall x p(x)$ has value $\mathbf{t}$ in $\Phi_P \uparrow 1$. But $\forall x p(x)$ is not true in all 3-valued (or even all 2-valued) models of *comp*(P), and it does not have value $\mathbf{t}$ in any $\Phi_P \uparrow n$ if Φ_P is evaluated with respect to a language with infinitely many function symbols, because if b is a new constant (or the result $f(a, \dots, a)$ of applying a new function symbol to a) then $p(b)$ has value $\mathbf{f}$ in $\Phi_P \uparrow 1$. Second, truth in some $\Phi_P \uparrow n$ is not the same as truth in $\Phi_P \uparrow \omega$, which is not usually a model of *comp*(P). Third the result only holds for sentences built up from the Kleene connectives $\wedge$, $\vee$, $\neg$, $\rightarrow$, $\leftrightarrow$, $\forall$, $\exists$, which have the property that if they have the value $\mathbf{t}$ or $\mathbf{f}$ and one of their arguments changes from $\mathbf{u}$ to $\mathbf{t}$ or $\mathbf{f}$, then their value doesn't change. The weak equivalence $\simeq$ used in the sentences giving the completed definitions of the predicates in *comp*(P) does not have this property, so the sentences of *comp*(P)—which obviously have value $\mathbf{t}$ in all 3-valued models of *comp*(P)—may not evaluate to $\mathbf{t}$ in any $\Phi_P \uparrow n$. For example if P is

$$p(0)$$
$$p(s(x)) \leftarrow p(x)$$

then the completed definition of p in *comp*(P) is

$$p(y) \simeq [y = 0 \vee (\exists x)(y = s(x) \wedge p(x))].$$

This has value $\mathbf{f}$ in each $\Phi_P \uparrow n$ for finite n, because for $y = s^n(0)$ the left hand side is $\mathbf{u}$ but the right hand side is $\mathbf{t}$.

If Φ_P and *comp*(P) are formed using a language L with only finitely many function symbols then (Shepherdson [1988]) the corresponding result is

A sentence $\emptyset$ has value $\mathbf{t}$ *in all* 3*-valued models of comp*$(P) \cup \{DCA\}$ *iff it has value* $\mathbf{t}$ *in* $\Phi_P \uparrow n$ *for some finite* n.

Here DCA is the *domain closure axiom* for L

$$\forall x \bigvee_{f \in L} \exists y_1, \ldots, y_{r_f}(x = f(y_1, \ldots, y_{r_f}))$$

where r_f is the arity of f), which states that every element is the value of a function in L (so it is satisfied in Herbrand models formed using L).

It is worth noting here the way in which $comp(P)$ depends on the language L. As mentioned above this is due to $comp(P)$ containing the freeness axioms for L. Let us denote by $comp_L(P)$ the completion of P formed using the language L, by G_L^k the sentence expressing the fact that there are greater than or equal to k distinct elements which are not values of a function in L, and by G_L^∞ the set of sentences expressing the fact that there are infinitely many such elements. Let $L(P)$ denote the language of symbols occurring in the program. Then (Shepherdson [1988]):

Let $L_1 \supseteq L_1 \supseteq L(P)$. Then $comp_{L_2}(P)$ is a conservative extension of (1) $comp_{L_1}(P)$ *if L_1 has infinitely many function symbols,* (2) $comp_{L_1}(P) \cup G_{L_1}^\infty$ *if L_1 has finitely many function symbols and $L_2 - L_1$ contains a function symbol of positive arity or infinitely many constants,* (3) $comp_{L_1}(P) \cup \{G_{L_1}^k\}$ *if L_1 has finitely many function symbols and $L_2 - L_1$ contains no function symbol of positive arity but exactly k constants.*

These results are true in both 2- and 3-valued logic since only the equality predicate is involved, and this is always taken to be 2-valued.

Kunen shows that whether $\emptyset$ has value $\mathbf{t}$ in $\Phi_P \uparrow n$ is decidable, so when a language with infinitely many function symbols is used:

The set of Q such that $comp(P) \models_3 Q$ is recursively enumerable.

An alternative proof could be obtained by giving a complete and consistent deductive system for 3-valued logic, as Ebbinghaus [1969] has done for a very similar system of 3-valued logic. This alternative proof shows that the result holds for any language.

For definite Horn claus programs 3-valued logic gives results which are in good agreement with those of 2-valued logic.

If P is a definite Horn clause program then

$$\Phi_P \uparrow \alpha = (T_P \uparrow \alpha, \quad B_P - T_P \downarrow \alpha).$$

Since the closure ordinal of $T_P \uparrow$ for a definite Horn clause program P is ω, this shows that the closure ordinal of $\Phi_P \uparrow$ is the same as that of $T_P \downarrow$.

For definite programs and definite queries, where the 2-valued semantics of negation as failure in terms of $comp(P)$ is satisfactory, the 3-valued semantics agrees with it.

If P is a definite Horn clause program and Q is the existential closure of a conjunction of atoms then the following eight statements are equivalent: Q is true in all (2-valued, 3-valued) (models, Herbrand models) of $(comp(P), P)$.

It is to be understood here that, as above, the clauses of P are written with $\rightarrow$ replaced by $\supset$. When this is done all 3-valued models of $comp(P)$ are models of P, so all the classes of models described are contained in the class of 3-valued models of P. Since they all contain the class of 2-valued Herbrand models of $comp(P)$, the result amounts to saying that if Q is true in all 2-valued Herbrand models of $comp(P)$ then it is true in all 3-valued models of P, which follows easily by considering the least fixed point model (Shepherdson [1988]). The result is true whatever language (containing $L(P)$) is used. But if 'existential' is replaced by 'universal' here then even the 2-valued parts of the result fail if the language used is $L(P)$. For example if P is $p(0)$, $p(s(x)) \leftarrow p(x)$ and Q is $\forall x p(x)$ then Q is true in all 2-valued Herbrand models of $comp(P)$, but not in all 2-valued models of $comp(P)$. However

If the language of symbols contains infinitely many constants, or a function symbol not in P, then the above equivalence result holds for all positive sentences.

A positive sentence here means one built up from atomic sentences using only $\wedge$, $\vee$, $\forall$, $\exists$. The reason this holds is that new constants behave like variables; see Shepherdson [1988] for details.

The relevance of 3-valued logic to negation as failure is given by Kunen's soundness and completeness results:

SLDNF-*resolution is sound with respect to $comp(P)$ in 3-valued logic, for all programs P and queries Q i.e.*

if Q succeeds with answer θ then $comp(P) \models_3 Q\theta$,

if Q fails then $comp(P) \models_3 \neg\exists Q$.

SLDNF-*resolution is complete with respect to $comp(P)$ in 3-valued logic for allowed programs P and allowed queries Q i.e.*

if $comp(P) \models_3 Q\theta$ then Q succeeds with answer including θ,

if $comp(P) \models_3 \neg\exists Q$ then Q fails.

Here a program is said to be *allowed* if every variable which occurs in a clause of it occurs in at least one positive literal in the body of that clause, a query is *allowed* if every variable in it occurs in at least one positive literal of it. It is easily verified that if the program and query are both allowed then the query cannot flounder, and this is what permits the completeness result. So for such programs and queries $comp(P)$ and 3-valued logic provide a perfectly satisfactory semantics. However the allowedness condition is very stringent and excludes many common Prolog constructs, such as the definition of equality, $equal(X, X)$, and both clauses in the definition of $member(X, L)$. The first result above is stated in Kunen [1987] (for a full proof see Shepherdson [1989a]); the much deeper completeness result is proved in Kunen [1987a]. Neither of them require the assumption used above that L contains infinitely many function symbols; they are valid for any L containing the function symbols of P (see Shepherdson [1989a,1988a]).

The 2-valued completeness results of Kunen for semi-strict (call-consistent) programs reported in §2.1 follow from the above result and the fact (Kunen [1987a]) that if the program P is semi-strict and the query Q is strict with respect to the program then the 2-valued and 3-valued semantics coincide i.e.

$$comp(P) \models_2 Q\theta \textit{ implies } comp(P) \models_3 Q\theta$$
$$comp(P) \models_2 \neg Q \textit{ implies } comp(P) \models_3 \neg\theta.$$

Other applications of many-valued logic to logic programming are to be found in the papers of Fitting [1986,1987a,1987b], Fitting and Ben-Jacob [1988] and Przymusinski [1988c].

Summary. The three-valued semantics based on $comp(P)$, which is complete for all allowed programs and queries, is elegant, reasonably simple and probably the best fit to SLDNF-resolution one can hope for without applying complicated transformations to P (as in §2.3) or using less familiar logics (as in §3.5, §3.6).

3.2. Intuitionistic Logic

We have already noted in §2.1 that the soundness of SLDNF-resolution for $comp(P)$ holds if the classical consequence relation $\models$ is replaced by the intuitionistic derivability relation $\vdash_I$ or by a relation $\vdash_{3I}$ based on rules which are sound both for classical 3-valued and intuitionistic 2-valued logic.

So use of $\vdash_I$ instead of $\models$ with $comp(P)$ i.e. interpreting

$$Q \textit{ is a consequence of the program } P$$

as meaning

$$comp(P) \vdash_I P \text{ instead of } comp(P) \models Q$$

gives a closer fit to SLDNF-resolution, i.e. better completeness results. Similarly $\vdash_{3I}$ gives a better fit than the $\vdash_3$ of the last section. Since we are dealing with computational proof procedures the use of intuitionistic logic is quite appropriate.

3.3. Modal Logic

Gabbay [1989] presents a view of negation as failure as a modal provability notion.

> '... we use a variation of the modal logic of Solovay, originally introduced to study the properties of the Gödel provability predicate of Peano Arithmetic, and show that $\neg A$ can be read essentially as 'A is not provable from the program'. To be more precise, $\neg A$ is understood as saying 'Either the program is inconsistent, or the program is consistent, in which case $\neg A$ means A is not provable from the program'.
>
> In symbols
>
> $$\neg A = \Box(\text{Program } \to \mathbf{f}) \vee \sim \Box(\text{Program } \to A),$$
>
> where $\neg$ is negation by failure, $\sim$ is classical negation, $\mathbf{f}$ is falsity, and $\Box$ is the modality of Solovay. We provide a modal provability completion for a Prolog program with negation by failure and show that our new completion has none of the difficulties which plague the usual Clark completion.
>
> We begin with an example. Consider the Prolog program $\neg A \to A$, where $\neg$ is negation by failure. This program loops. Its Clark completion is $\neg A \leftrightarrow A$, which is a contradiction, and does not satisfactorily give a logical content to the program.
>
> We regard this program as saying:
>
> $$(\underline{\text{Provable}}(\mathbf{f}) \vee \sim \underline{\text{Provable}}(A)) \leftrightarrow A.$$
>
> In symbols, if x is the program and $\Box$ is the modality of provability, the program x says, $(\Box\mathbf{f} \vee \sim \Box(x \to A)) \leftrightarrow A$ where $\sim$ is classical

negation. In other words, the logical content of the program x is the fixed point solution (which can be proved to always exist) of the equation

$$x = [(\Box \mathbf{f} \vee \sim \Box(x \to A) \leftrightarrow A].$$

This solution turns out to be a consistent sentence of the modal logic of probability to be described later.

We now describe now to get the completion in the general case. Let $\mathbf{P}$ be a Prolog program. Let $\mathbf{P}1$ be its Clark completion. Let x be a new variable. Replace each $\neg A$ in the Clark completion $\mathbf{P}1$ essentially by $(\Box \mathbf{f} \vee \sim \Box(x \to A))$. Thus $\mathbf{P}1$ becomes $\mathbf{P}2(x)$ containing $\Box$, x and classical negation $\sim$. We claim that in the modal logic of probability, the modal completion of the program P is the X such that in the modal logic $\vdash X \leftrightarrow \mathbf{P}2(x)$ holds. We have that a goal A succeeds from $\mathbf{P}$ iff (more or less) the modal completion of $\mathbf{P} \vdash \Box A$. A more precise formulation will be given later. We denote the modal completion of $\mathbf{P}$ by $\mathbf{m}(\mathbf{P})$ '

The modal completion is shown to exist and to be unique up to $\vdash$ equivalence. Defining $GA = A \wedge \Box A$, 'A is true and provable' (in a general provability logic A may be provable but not true), soundness and completeness results are proved:

For each atomic goal Q the substitution θ, $Q\theta$ succeeds from $\mathbf{P}$ *under* SLDNF*-resolution iff* $\mathbf{m}(P) \vdash GQ\theta$ *and $Q\theta$ fails from* $\mathbf{P}$ *iff* $\mathbf{m}(\mathbf{P}) \vdash G\neg Q\theta$.

The completeness result is based on a proof for the propositional case given by Terracini [1988a].

It should be noted that in this result 'succeeds' and 'finite' do not refer to SLDNF-resolution but to another version of negation as failure. This coincides with SLDNF-resolution when P is a propositional program, but when P contains individual variables '$Q\theta$ succeeds from P' means 'there exists η such that $q\theta\eta$ is ground and succeeds from P^g under SLDNF-resolution'; similarly '$Q\theta$ fails from P' means '$Q\theta\eta$ fails from P^g under SLDNF-resolution for all η for which $Q\theta\eta$ is ground'. Here P^g denotes the (possibly infinite) propositional program obtained by taking all ground instances of clauses of P. For example if P is $p \leftarrow \neg q(x)$ then P^g is $p \leftarrow \neg q(a)$ and the goal p is said to succeed, although in SLDNF-resolution from P it would flounder. This device enables the floundering problem to be dealt with and the predicate case reduced to the propositional case.

The modal logic of provability used is defined both semantically, in terms of Kripke type models consisting of finite trees, and also syntactically. Both of these definitions are rather complicated for predicate logic but the propositional form of the syntactic definition consists of the following schemas and rules:

(1) $\vdash A$, if A is a substitution instance of a classical truth functional tautology.

(2) The schemas:
  (a) $\vdash \Box(A \rightarrow B) \rightarrow (\Box A \rightarrow \Box B)$
  (b) $\vdash \Box(\Box(A \rightarrow B) \rightarrow (\Box A \rightarrow \Box B))$

(3) The schemas:
  (a) $\vdash \Box A \rightarrow \Box\Box A$
  (b) $\vdash \Box(\Box A \rightarrow \Box\Box A)$

(4) The schemas:
  (a) $\vdash \Box(\Box A \rightarrow A) \rightarrow \Box A$
  (b) $\vdash \Box[\Box(\Box A \rightarrow A) \rightarrow \Box A]$

(5) For every atom q
  (a) $\vdash \Box(q \vee \Box\mathbf{f}) \rightarrow \Box q$
  (b) $\vdash \Box(\sim q \vee \Box\mathbf{f}) \rightarrow \Box \sim q$
  (c) $\vdash \Diamond\Diamond\mathbf{t}$

(6) $$\frac{\vdash A, \vdash A \rightarrow B}{\vdash B}$$

Here $\Diamond A$ stands for $\neg\Box\neg A$. This is an extension of Solovay's modal logic of provability, which is itself obtained by extending the modal logic K_4 by Löb's axiom schemas 4(a). It is also described in Terracini [1988].

Summary. The soundness and completeness results show that, at least in the propositional case, this is a very satisfactory semantics for negation as failure. Its main disadvantage is that the modal logic used is apparently rather complicated and, at present, not widely known. So it is doubtful whether this semantics would help writers of programs to understand their meaning and check their correctness. However Gabbay argues convincingly that many of the day to day operations of logic programming have a modal meaning and the logic programmers should become more familiar with modal logic.

3.4. Autoepistemic Logic

Gelfond [1987] and Przymusinska [1987] discussed negation as failure in terms of the *autoepistemic logic* of Moore [1985]. This is a propositional calculus augmented by a belief operator L where Lp is to be interpreted as 'p is believed'. Gelfond considers $\neg p$ in a logic program as intended to mean 'p is not believed'.

He defines the autoepistemic translation $I(F)$ of an *objective formula* F (i.e. a propositional formula not containing the belief operator L) to be the result of replacing each negative literal $\neg p$ in F by $\neg Lp$. The autoepistemic translation $I(P)$ of a logic program consists of the set of translations of all ground instances of clauses of P, i.e. the set of all clauses of the form

$$A \leftarrow B_1, \ldots, B_m, \ \neg LC_1, \ldots, \neg LC_n,$$

for all ground instances

$$A \leftarrow B_1, \ldots, B_m, \ \neg C_1, \ldots, \neg C_n \text{ of clauses from } P.$$

Let T be a set of autoepistemic formulae. Moore defined a *stable autoepistemic expansion* of T to be a set $E(T)$ of autoepistemic formulae which satisfies the fixed point condition

$$E(T) = Cn(T \cup \{Lf : f \text{ is in } E(T)\} \cup \{\neg Lf : f \text{ is not in } E(T)\})$$

where $Cn(S)$ denotes the set of autoepistemic logical consequences of S (formulae of the form Lg being treated as atoms). This intuitively represents a set of possible beliefs of an ideally rational agent who believes in all and only those facts which he can conclude from T and from his other beliefs. If this expansion is unique it can be viewed as the set of theorems which follow from T in the autoepistemic logic. The autoepistemic translation $I(P)$ of a logic program P does not always have such a unique expansion. The program $p \leftarrow \neg p$ has no consistent stable autoepistemic expansion because its translation is $p \leftarrow \neg Lp$ and itis easy to verify that both Lp and $\neg Lp$ must belong to such an expansion. On the other hand the program $p \leftarrow \neg q$, $q \leftarrow \neg p$ has two such expansions, one with $Cn(p)$ as its objective part, the other with $Cn(q)$ as its objective part. Gelfond and Przymusinska showed that for stratified propositional programs the autoepistemic and perfect model semantics coincide:

If P is a stratified propositional logic program then $I(P)$ has a unique stable autoepistemic expansion $E(I(P))$ and for every query Q, $\mathrm{PERF}(P) \models Q$ *iff* $E(I(Q)) \models I(Q)$.

Hence PERF(P) denotes the perfect model of P which, as noted in §4.3 below coincides with the minimal supported model of Apt, Blair and Walker, with the well-founded total model and with the unique stable model.

Przymusinska [1988c,1989] obtains results applicable to all logic programs by using a 3-valued autoepistemic logic. Let $M_{WP}(P)$ denote the 3-valued well-founded model defined in §4.4 below. He shows that if P is a logic program then $I(P)$ always has at least one stable autoepistemic expansion, and that the autoepistemic semantics coincides with the well-founded semantics:

For every ground atom A,
A is true in $M_{WP}(P)$ iff A is believed in $I(P)$
A is false in $M_{WP}(P)$ iff A is disbelieved in $I(P)$
A is undefined in $M_{WP}(P)$ iff A is undefined in $I(P)$.

Here A is believed (resp. disbelieved) in $I(P)$ if A is believed (resp. disbelieved) in all stable autoepistemic expansions of $I(P)$, otherwise we say A is undefined in $I(P)$.

Summary. The autoepistemic semantics reflects a natural interpretation of negation as failure. But, like the perfect and well-founded models (see §4) it corresponds less closely to SLDNF-resolution than the semantics based on *comp*(P).

3.5. Linear Logic

Cerrito [1988] has shown that, at least in the propositional case, the *linear logic* of Girard [1987] can be used to give a declarative semantics with respect to which negation as failure as used in Prolog is both sound and complete. We have not space here to explain all the details of linear logic and will just try to convey enough of the essence of this semantics for the reader to decide whether he wishes to pursue the study of this interesting new approach.

Linear logic is a weak logic which lacks the weakening rules

$$\frac{\Gamma \vdash \Delta}{\Gamma, A \vdash \Delta} \qquad \frac{\Gamma \vdash \Delta}{\Gamma \vdash A, \Delta}$$

and contraction rules

$$\frac{\Gamma, A, A \vdash \Delta}{\Gamma, A \vdash \Delta} \qquad \frac{\Gamma \vdash A, A, \Delta}{\Gamma \vdash A, \Delta}$$

of the usual sequent calculi. It has two conjunctions, or *multiplicative* connectives

$$\otimes (\text{read: } \textit{times})$$
$$\& (\text{read: } \textit{with})$$

the two disjunctions, or *additive* connectives

$$(\text{read: } \textit{par})$$
$$\oplus (\text{read: } \textit{plus})$$

which satisfy De Morgan rules with respect to negation, i.e. if $F^{\perp}$ denotes the linear negation of F we have

$$A{\bigstar}B \text{ is logically equivalent to } (A^{\perp} \otimes B^{\perp})^{\perp}$$
$$A \oplus B \text{ is logically equivalent to } (A^{\perp} \& B^{\perp})^{\perp}.$$

Another multiplicative connective is the linear implication which is defined by:

$$A -^{\circ} B = A^{\perp}{\bigstar}B.$$

The intended meaning of a linear sequent

$$G_1, \ldots, G_n \vdash D_1, \ldots, D_n$$

is

$$G_1 \otimes \cdots \otimes G_n -^{\circ} D_1 \ldots D_n.$$

As Cerrito says:

"The philosophy behind linear logic is that a formula A does not necessarily represent a stable situation whose truth, once established, holds forever. A formula A can represent an **action**, which, once performed, produces certain consequences both is also exhausted by its very accomplishment. Thus, for example, the classical inference rule *Modus Ponens*:

$$\frac{A \quad A \Rightarrow B}{B}$$

can formalize a reasoning of this kind: if statement A is true, then its corollary B is true; actually A is true, therefore B is true. On the other hand, the corresponding linear inference:

$$\frac{A \quad A -^{\circ} B}{B}$$

can formalize a reasoning of this kind: If I spend a dollar at the tobacconist, then I can get a pack of cigarettes; I spent a dollar at the tobacconist, therefore now I have a pack of cigarettes (but my dollar is gone!).

The basic difference between the two kinds of reasoning is that in the first case, once A has been used to deduce B, A is still there to be possibly used again, for example, to get another corollary B', whereas in the second case, once having got the cigarettes, I do not have any longer the original dollar to be spent again. In other terms, Linear Logic tries to take in account the fact that there are limited resources that can be used to certain ends and once these resources have been used up, they are not there any longer.

Let us examine the intuitive difference between the two linear conjunctions $\otimes$ and &. Let A, B, C stand for the following sentences:

A is "I spend a dollar"

B is "I get a pack of Marlboro"

C is "I get a pack of Camel".

Then $A -^{\circ} B\&C$ can be read as "If I spend one dollar I get either a pack of Marlboro or a pack of Camels", where I have **both** the possibilities but there is **choice to be made** because one dollar is sufficient just for one pack of cigarettes; however the choice between B and C is **up to me**. The statement $A -^{\circ} B\&C$ is true in the real world. On the other hand, the formula $A -^{\circ} B \otimes C$ has the reading: "If I spend (just) one dollar, I can get a pack of Marlboro and a pack of Camels" where I have **both** the possibilities and I can do B and C **at the same time**. Clearly this last is false in the real world.

It is important to remark that & is not a disjunction as one could think at the first sight: as a matter of fact the implications $A\&B -^{\circ} A$ and $A\&B -^{\circ} B$ are valid. One can also notice that while the statement $A -^{\circ} B \otimes C$ is false in the real world, the statement $A \otimes A -^{\circ} B \otimes C$ is a true one (because for two dollars I can get two packs of cigarettes). The point is that the implication $A -^{\circ} A \otimes A$ is not valid and the converse $A \otimes A -^{\circ} A$ is not valid either (at the level of the formal system, this fact is due to the absence of Weakening and Contraction).

Let us consider the two linear disjunctions. Under the given interpretation, $A -^{\circ} B \oplus C$ means that for one dollar I can get one pack of cigarettes but the choice of the brand (Marlboro or Camel) is **not up to me**. The intuitive meaning of ★ is less evident."

"After this quick description of Linear Logic, we cannow give the definition of the formal system which we use. For the sake of economy, we will write any (two-sided) sequent

$$G_1, \ldots, G_n \vdash D_1, \ldots, D_m$$

as the right-handed sequent

$$\vdash G_1^{\perp}, \ldots, G_n^{\perp}, D_1, \ldots, D_m.$$

By using this trick, we can formulate the calculus writing just the right rules.

(i) The language $\mathbb{L}$

Let $P_1, \ldots, P_i, \ldots$ and $P_1^{\perp}, \ldots, P_i^{\perp}, \ldots$ be infinite sequences of propositional variables. The language $\mathbb{L}$ contains the formulas built out of such propositional variables by using the following binary connectives:

non-commutative multiplicative connectives: $\otimes, \bigstar$

non-commutative additive connectives: $\&, \oplus$.

The linear negation is a defined connective; if $\mathcal{F}$ is a formula of $\mathbb{L}$ its linear negation $\mathcal{F}^{\perp}$ is defined as follows:

for propositional variables, we have two distinct kinds of variables in the syntax, i.e. P_i (positive literal) and $P_i^{\perp}$ (negative literal);

$$\begin{aligned}
&(P_i^{\perp})^{\perp} = P_i;\\
&(\mathcal{F} \otimes \mathcal{F}')^{\perp} = \mathcal{F}^{\perp} \bigstar \mathcal{F}'^{\perp}\\
&(\mathcal{F} \bigstar \mathcal{F}') = \mathcal{F}^{\perp} \otimes \mathcal{F}'^{\perp}\\
&(\mathcal{F} \oplus \mathcal{F}')^{\perp} = \mathcal{F}^{\perp} \& \mathcal{F}'^{\perp}\\
&(\mathcal{F} \& \mathcal{F}')^{\perp} = \mathcal{F}^{\perp} \oplus \mathcal{F}'^{\perp}.
\end{aligned}$$

Also linear implication is a defined symbol:

$$\mathcal{F} -\!\circ\, \mathcal{F}' = \mathcal{F}^{\perp} \bigstar \mathcal{F}' \text{ (read: } \mathcal{F} \textit{ linearly implies } \mathcal{F}'\text{)}.$$

(ii) The system LL

A (right-handed) linear sequent of $\mathbb{L}$ is an expression of the form $\vdash \Gamma$, where Γ is a finite sequence of formulas $G_1, \ldots, G_n$ of $\mathbb{L}$; the implicitly defined meaning of the linear sequent $\vdash \Gamma$ is $G_1 \bigstar G_2 \ldots G_{n-1} \bigstar G_n$.

$$\vdash A, A^{\perp} \qquad (\textit{Logical Axioms})$$

$$\frac{\vdash \Gamma, A \quad \vdash A^{\perp}, \Delta}{\vdash \Gamma, \Delta}(\text{CUT}) \qquad (\textit{Cut Rule})$$

$$\frac{\vdash \Gamma}{\vdash \Gamma'}(\text{EXCH}) \qquad (\textit{Exchange Rule})$$

where Γ' is a permutation of Γ.

$$\frac{\vdash \Gamma, A \quad \vdash \Gamma, B}{\vdash \Gamma, A \& B}(\&) \qquad (\textit{Additive Rules})$$

$$\frac{\vdash \Gamma, A}{\vdash \Gamma, A \oplus B}(\oplus 1)$$

$$\frac{\vdash \Gamma, B}{\vdash \Gamma, A \oplus B}(\oplus 2)$$

$$\frac{\vdash \Gamma, A \quad \vdash B, \Delta}{\vdash \Gamma, A \otimes B, \Delta}(\otimes) \qquad (\textit{Multiplicative Rules})$$

$$\frac{\vdash \Gamma, A, B, \Delta}{\vdash \Gamma, A \quad B, \Delta}(\bigstar)$$

"

Cerrito aims to provide a set of sequents of linear logic which provides a declarative semantics not for SLDNF-resolution but for standard Prolog with its specific search strategy and selection rules: selection of leftmost atom in the body of a goal, first matching clause,

"*Example 1.* Let P_1 be a PROLOG program whose clauses with head C are the following (in their order of appearance in P):

(1) $C : -B, A.$
(2) $C : -E, D.$
(3) $C.$
(4) $C : -F.$

Let C be a goal for P. The situation in which C is successful at the first attempt (namely when the first clause with head C is tried) can be described by the linear formula:

(a) $B \otimes A -^{\circ} C$

because the success of C is caused by the success of B and by the success of A which is successively tested. When C is successful at the second attempt (after failure of the first one) we have that either B has failed or B has been successful but A has failed (failure w.r.t. the first clause) and both E and D have been successful, namely:

(b) $(B^{\perp} \oplus (B \otimes A^{\perp})) \otimes (E \otimes D) -^{\circ} C.$

When C is successful at the third attempt we have:

(c) $(B^{\perp} \oplus (B \otimes A^{\perp})) \otimes (C^{\perp} \oplus (E \otimes D^{\perp})) -^{\circ} C.$

Clearly, PROLOG never tests the last clause.

We call "success set" of the atom C the set:

$$S_C = \{a, b, c\}.$$

The program P_1 will halt with failure of C if all the clauses of P fail; here, such a situation cannot arise (thanks to the presence of the clause 3); thus, the "failure set" of the atom C is empty:

$$F_C = \emptyset.$$

If we apply the distributivity of $\otimes$ with respect to $\oplus$ we can rewrite S_C as the set of formulas whose elements are:

$$B \otimes A -^{\circ} C$$
$$(B \otimes A^{\perp}) \otimes (E \otimes D) -^{\circ} C$$
$$B^{\perp} \otimes (E \otimes D) -^{\circ} C$$
$$B^{\perp} \otimes E^{\perp} -^{\circ} C$$
$$(B \otimes A^{\perp}) \otimes E^{\perp} -^{\circ} C$$
$$B^{\perp} \otimes (E \otimes D^{\perp}) -^{\circ} C$$
$$(B \otimes A^{\perp}) \otimes (E \otimes D^{\perp}) -^{\circ} C.$$

Now, we can express each formula in this set as a right-handed atomic linear sequent, thus getting the set S_C^{seq} whose elements are:

$$\vdash B^{\perp}, A^{\perp}, C$$
$$\vdash B^{\perp}, A, E^{\perp}, D^{\perp}, C$$
$$\vdash B, E^{\perp}, D^{\perp}, C$$
$$\vdash B, E, C$$
$$\vdash B^{\perp}, A, E, C$$
$$\vdash B, E^{\perp}, D, C$$
$$\vdash B^{\perp}, A, E^{\perp}, D, C.$$

Clearly:

$$F_C^{seq} = \emptyset.$$

Example 2. Let P_2 be such that the only clauses with head C are the first two clauses of the previous example:

1) $A, B \vdash C$
2) $D, E \vdash C$

Now the "success set" for C is given by:

$$S_C = \{a, b\}$$

while

$$F_C = \{(B^\perp \oplus (B \otimes A^\perp)) \otimes (E^\perp \oplus (E \otimes D^\perp)) -^\circ C^\perp)\}$$

because P halts with failure of C when all the clauses of P with head C fail. The set S_C^{seq} now has as elements:

$$\vdash B^\perp, A^\perp, C$$
$$\vdash B^\perp A, E^\perp, D^\perp, C$$
$$\vdash B, E^\perp, D^\perp, C$$

while F_C^{seq} has the elements:

$$\vdash B, E, C^\perp$$
$$\vdash B^\perp, A, E, C^\perp$$
$$\vdash B, E^\perp, D, C^\perp$$
$$\vdash B^\perp, A, E^\perp, D, C^\perp.$$

Let P be a PROLOG program without negation whose atoms are all propositional letters, as in the examples above. Let A be one of such atoms and let S_A^{seq} and F_A^{seq} be defined as the above examples suggest (see later on the rigorous definition). We will call "definition of A", for the given program P, the following set of linear sequents:

$$D_A = S_A^{seq} \cup F_A^{seq}.$$

Let $A_1, \ldots, A_n$ be the atoms in P. The linear translation LT_P of the program P will be defined by:

$$LT_P = \bigcup_{i=1\ldots n} D_{A_i}."$$

For technical reasons this translation LT_P needs to be slightly modified:

"One can mark the beginning and the end of each sequent in LT_P by using two special symbols, namely $\mathbf{e}$ (*entrée*) and $\mathbf{s}$ (*sortie*); ...

Syntactically, $\mathbf{e}$ and $\mathbf{s}$ are just two specific propositional letters of $\mathbb{L}$ and, by definition, $\mathbf{e}^{\perp} = \mathbf{s}$, so that $(e \bigstar \mathcal{F})^{\perp} = \mathcal{F} \otimes \mathbf{s}$ for any formula $\mathcal{F}$ of $\mathbb{L}$. A new translation $\mathbf{MLT}_P$ of a program P is now defined by replacing any sequent $\vdash A_1, \ldots, A_n$ of LT_P by the sequent:

$$\vdash \mathbf{e}, A_1, \ldots, A_n \otimes \mathbf{s}.$$

We can now give the correct formulation of our results. Let P be a PROLOG propositional program without negation and A be an atom in P:

Theorem 1.

(a) *If* $\vdash \mathbf{e}$, $A \otimes \mathbf{s}$ *is LL-provable from* $\mathbf{MLT}_P$ *then the goal* A *is successful.*

(b) *If* $\vdash \mathbf{e}$, $A^{\perp} \otimes \mathbf{s}$ *is LL-provable from* $\mathbf{MLT}_P$ *then the goal* A *finitely fails.*

Theorem 2.

(a) *If* A *is successful, then* $\vdash \mathbf{e}$, $A \otimes \mathbf{s}$ *is LL-provable from* $\mathbf{MLT}_P$.

(b) *If* A *finitely fails, then* $\vdash \mathbf{e}$, $A^{\perp} \otimes \mathbf{s}$ *is LL-provable from* $\mathbf{MLT}_P$.

The above theorems may be seen respectively as a completeness result and a soundness result for SLDNF-resolution with respect to the notion of linear logical consequence of the theory $\mathbf{MLT}_P$ which describes the program P (and which plays a role similar to Clark completion)."

Summary. This semantics based on linear logic is remarkable in that, in the propositional case, it provides a sound and complete declarative semantics for standard Prolog with negation as failure, in terms of an already existing logical system. Its disadvantage is that the translation $\mathbf{MLT}_P$ of P into linear logic is rather complicated, and the linear logic itself is unfamiliar so that the intuitive meaning of $\mathbf{MLT}_P$ is not easy to grasp. However it might be useful for proving correctness of programs, particularly if the proofs could be automated.

3.6. Deductive Calculi for Negation as Failure

The standard procedural descriptions of logic programming systems such as Prolog, SLD- and SLDNF-resolution are in terms of trees. This makes proving theorems about them rather awkward because the proofs somehow have to involve the tree structure. So it might be useful to have descriptions in the form of a deductive calculus of the familiar kind, based on axioms and rules of inference, so that proofs can be simply by induction on the length of the derivation. Mints [1986, 1990] gave such a calculus for pure Prolog, and Gabbay and Sergot [1986] implicitly suggested a similar calculus for negation as failure.

The definition of SLDNF-resolution given by Kunen [1987a] which we reproduced in §1.2 can easily be put in the form of a Mints type calculus.

The meaning of the notation is as follows:

Y	A query, i.e. a sequence $L_1, \ldots, L_n$, $n \geq 0$ of literals
$Y\theta$	The result of applying the substitution θ to the goal Y
$(Y;\theta)$	The query Y succeeds with answer substitution θ
(Y)	The query Y succeeds
$\sim (Y)$	The query Y fails finitely
$\sim j(Y)$	The goal $Y = A, X$ (where A is an atom) fails finitely if you consider only the branch starting with the attempt to unify A with the j-th suitable program clause (i.e. whose head contains the same predicate as A does).
$i : A \leftarrow Z$	The clause $A \leftarrow Z$ is (a variant of) the i-th clause of the given program P which is suitable for (i.e. whose head contains the same predicate as) A.

The calculus operates on formulae of the form

$$(Y;\theta), (Y), \sim j(Y), \sim (Y), i : A \leftarrow Z.$$

The axioms will be those formulae

$$i : A \leftarrow Z$$

such that $A \leftarrow Z$ is indeed the i-th clause of P which is suitable for A, together with

(START) $\qquad$ $(\text{true}; 1)$

where 1 denotes the identity substitution. The rules of inference are:

Two rules allowing permutation of literals in a goal

(PERM) $\qquad$ $\dfrac{(Q;\theta)}{(Q^\pi, \theta)}$

($\sim$ PERM) $\qquad$ $\dfrac{\sim (Q)}{\sim (Q^\pi)}$

where Q^π denotes any permutation of the atoms of Q.

A rule corresponding to resolution

(RES) $$\frac{i : A' \leftarrow Z;\ (Z\sigma, X\sigma; \pi)}{(A, X; (\sigma\pi) \mid (A, X))}$$

where A is an atom and $\sigma = mgu(A, A')$. In all these rules $A' \leftarrow Z$ is as usual supposed to be a variant of the program clause which is standardized apart so as to have no variables in common with A, X.

A rule allowing you to pass from 'succeeds with answer θ' to 'succeeds',

(FIN) $$\frac{(Q, \theta)}{(Q)}.$$

Finally five rules for negation as failure,

($\sim_1$) $$\frac{\sim 1(Y); \sim 2(Y); \ldots; \sim k(Y)}{\sim (Y)}$$

where k is the number of suitable clauses for A, the first sub-goal of Y,

($\sim_2$) $$\frac{i : A' \leftarrow Z}{\sim i(A, Z)}$$

provided that A, A' are not unifiable,

($\sim_3$) $$\frac{i : A' \leftarrow Z;\ \sim (Z\sigma, X\sigma)}{\sim i(A, X)}$$

where $\sigma = mgu(A, A')$,

($\neg_1$) $$\frac{\sim (A);\ (Y; \theta)}{(\neg A, Y; \theta)}$$

if A is ground

($\neg_2$) $$\frac{(A)}{\sim (\neg A, X)},$$

if A is ground.

If we take SLDNF-resolution to be defined in the familiar way in terms of trees, as in Lloyd [1987] then the statement that Kunen's definition given in §1.2 above is equivalent to it amounts to the statement:

A goal X succeeds under SLDNF*-resolution from P iff (X) is derivable in this calculus; it succeeds with answer substitution θ iff $(X;\theta)$ is derivable in the calculus; it fails iff $\sim (X)$ is derivable in the calculus.*

The proof is routine, by induction on the length of the derivation for the 'if' halves, and by induction on the number of nodes in the success or failure tree for the 'only if' halves. For a survey of these calculi see Shepherdson [1989b] which also gives the obvious extension of this calculus to deal with the extended negation as failure rules described in §1.6, above:

if $A\theta$ fails then $\neg A$ succeeds with answer θ

if A succeeds with answer 1 then $\neg A$ fails.

Summary. Although we have given a calculus for which SLDNF-resolution is both sound and complete it is not a declarative semantics, but an alternative procedural semantics.

4. Semantics Based on Special Classes of Models

4.1. Minimal Models and the Generalized Closed World Assumption

In these sections of §4 we discuss semantics based on the idea that the meaning of a program P is not a set of sentences (such as P, $comp(P)$ or, $CWA(P)$), but a set $M(P)$ of models of P, and that when we ask a query $?Q$ we want to know whether Q is true in all models of $M(P)$. We have already discussed, in §2.1 the case where P is a definite Horn clause program and $M(P)$ is the singleton consisting of its least Herbrand model, and in §2.2 we have shown its close connection with the semantics based on $CWA(P)$. When P contains negation $CWA(P)$ may be inconsistent and P may not have a least Herbrand model. In attempting to find a weaker assumption which would be consistent for all consistent P, Minker [1982] was led to suggest replacing least Herbrand model by minimal Herbrand model, i.e. one not containing any proper sub model, where as usual we identify a Herbrand model with the subset of the Herbrand base B_P which is true in it. So he considers $M(P)$, the class of intended models of P, to be the class of minimal Herbrand models of P. He shows this semantics is closely related to the one based on his 'Generalised Closed World Assumption', $GCWA(P)$, defined as follows:

$GCWA(P) = P \cup \{\neg A$: *A is a ground atom such that there is no disjunction B of ground atoms such that $P \vdash A \vee B$ but $P \nvdash B$*$\}$.

Indeed the condition on A here is easily shown to be equivalent to:

A is a ground atom which is false in all minimal Herbrand models of P.

Since minimal models of P clearly satisfy $GCWA(P)$ this implies:

If a first order sentence Q is a consequence of $GCWA(P)$ then it is true in all minimal Herbrand models of P.

However the converse is not generally true, for there may be models of $GCWA(P)$ which are not minimal Herbrand models of P. For example if P is $p(a) \leftarrow \neg p(b)$ i.e. $p(a) \vee p(b)$, there are two minimal Herbrand models of P namely $\{p(a)\}$, $\{p(b)\}$ but $GCWA(P)$ is the same as P and has a non-minimal model $\{p(a), p(b)\}$. And the query $\exists x \neg p(x)$ is true in all minimal Herbrand models of P but is not a consequence of $GCWA(P)$. So $GCWA(P)$ is an incomplete attempt to characterize the minimal Herbrand models of P. The converse of the statement displayed above is true if Q is a positive query or, more generally the existential closure of a positive matrix (i.e. a formula built up from atoms using only $\wedge$, $\vee$). But $GCWA(P)$ is not really involved then, because if such a Q is true in all minimal Herbrand models of P it is actually a consequence of P alone. From the displayed statement above it follows that if such a Q is a consequence of $GCWA(P)$ then it is a already a consequence of P so addition of the generalized closed world assumption, like the closed world assumption, does not allow the derivation of any more positive information of this kind (in particular of ground atoms). This is usually thought to be a desirable feature, since closed world assumptions are usually intended as devices for uncovering implicit negative information, thus avoiding the need to state it explicitly, without adding unconsciously to the positive information of the program.

Minker's aim of providing a version of the closed world assumption which is consistent is achieved:

If P is consistent then so is $GCWA(P)$.

This is true not only for normal programs P but for any program P consisting of universal sentences because it is easy to show that such a P has a minimal Herbrand model.

$GCWA(P)$ is a generalization of the $CWA(P)$ in the sense that it agrees with that when P is definite. The generalized closed world assumption is a kind of negation as failure in that $\neg A$ is assumed when A fails to be true in any minimal Herbrand model. However negation as failure as defined here, i.e. as SLDNF-resolution, is not sound with respect to it. This is shown by the program $q \leftarrow \neg p$ where p fails but is not a consequence of $GCWA(P)$. Henschen and Park [1988] discuss computational proof procedures appropriate to the $GCWA(P)$ in the case of principal interest where P is a database, i.e. without function symbols (which Minker's original article restricted itself to). When function symbols are present there may be no sound and complete computational proof procedure for $GCWA(P)$. This follows from our example in §2.2 of a definite clause program for which the set of queries which are consequences of $CWA(P)$ is not recursively enumerable. The same is true for the set of queries true in all minimal Herbrand models of P, since for definite P this does coincide with the set of queries which are consequences of $GCWA(P)$ (i.e. of $CWA(P)$).

For further results on the $GCWA$ and other generalization of the closed world assumption see Gelfond et al. [1986], [1986a], Lifschitz [1988], Shepherdson [1988], Brass and Lipeck [1989], Yahya and Henschen [1989].

4.2. Minimal Supported Models

Apt, Blair and Walker [1988] propose a semantics for negation which combines this last approach with that of the Clark completion, i.e. applies both kinds of default reasoning. They suggest that the models of P which it is reasonable to study are those Herbrand models which are not only minimal but *supported*, i.e. a ground atom A is true only if there is a ground instance of a clause of P with head A and a body which is true. We saw in §2.1 that the supported models are the fixpoints of T_P, and the models of *comp*(P). Apt, Blair and Walker say on p. 100

> '... we are interested here in studying minimal and supported models ... this simply means we are looking for the minimal fixed points of the operator T_P'.

Since they go on to study the minimal fixed points of T_P it looks as though they intended the latter definition i.e. minimal supported models of P i.e. minimal models of *comp*(P). But the first phrase suggests a stronger definition i.e. models of *comp*(P) which are also minimal models of P. To see the difference consider the program $p \leftarrow q$, $q \leftarrow \neg p$, $q \leftarrow q$. The only, and hence the minimal, model of *comp*(P) is $\{p, q\}$; the only minimal model of P is $\{p\}$. There is no supported model of P (i.e. model of *comp*(P)) which is a minimal model of P. However their main concern is with stratified programs, for which they establish the existence of a model satisfying the stronger first definition:

If P is a stratified program then there is a minimal model of P which is also supported (i.e. a model of comp(P), so comp(P) is consistent).

There may be more than one model satisfying these conditions, e.g. if P is the stratified program $p \leftarrow p$, $q \leftarrow \neg p$ there are two such models $\{p\}$ and $\{q\}$. They show that there is one such model M_P which is defined in a natural way and propose that it be taken as defining the semantics for the program P, i.e., that an ideal query evaluation procedure should make a query Q succeed if Q is true in M_P and fail if Q is false in M_P. They give two equivalent ways of defining M_P. A stratified program P can be partitioned

$$P = P_1 \dot{\cup} \ldots \dot{\cup} P_n$$

so that if a predicate occurs positively in the body of a clause in P_i, all clauses where it occurs in the head are in P_j with $j \leq i$, and if a predicate occurs negatively in the body of a clause in P_i, then all clauses where it occurs in the head are in P_j with $j < i$. (So P_i consists of the clauses defining i-th level predicates.) Their first definition of M_P is to start with the empty set, iterate T'_{P_1} ω times when T'_{P_2} ω times,...,T'_{P_n} ω times. (The operator T'_P here is defined by $T'_P(I) = T_P(I) \cup I$ and is more appropriate than T_P when that is nonmonotonic.) The other definition starts by defining $M(P_1)$ as the intersection of all Herbrand models of P_1, then $M(P_2)$ as the intersection of all Herbrand models of P_2 whose intersection with the Herbrand base of P_1 is $M(P_1)$, then $\ldots, M(P_n)$ as the intersection of all Herbrand models of P_n whose intersection with the Herbrand base of P_{n-1} is $M(P_{n-1})$. Finally define $M_P = M(P_n)$. For the program above, this model is $\{q\}$, which does seem to be better in accordance with default reasoning than the other minimal model of *comp*(P), namely $\{p\}$. There is no reason to suppose p is true, so p is taken to be false, hence q to be true. It seems natural to assign truth values to the predicates in the order in which they are defined, which is the essence of the above method. Moreover, a strong point in favor of the model M_P is that they show it does not depend on the actual way a stratifiable program is stratified, i.e., divided into levels.

Notice that like *comp*(P), M_P depends not only on the logical content of P but on the way it is written, for $p \leftarrow \neg q$ and $q \leftarrow \neg p$ give different M_P.

Since SLDNF-resolution is sound for $comp(P)$ and M_P is a model of $comp(P)$, it is certainly sound for M_P but, since more sentences will be true in M_P then in all models of $comp(P)$, SLDNF-resolution will be even more incomplete for M_P then for $comp(P)$. For example, with the program above where q is true in M_P but not in all models of $comp(P)$, there is no chance of proving q by SLDNF-resolution.

In general there may be no sound and complete computational proof procedure for the semantics based on M_P. This is shown by the example in §2.2 which shows this for $CWA(P)$, since for definite programs M_P is the least Herbrand model, so the semantics based on M_P coincides with that based on $CWA(P)$. However Apt, Blair and Walker do give an interpreter which is sound, and which is complete when there are no function symbols.

4.3. Perfect Models

Przymusinski [1988], [1988a] proposes an even more restricted class of models than the minimal, supported models, namely the class of *perfect* models. The argument for a semantics based on this class is that if one writes $p \vee q$, then one intends p, q to be treated equally; but, if one writes $p \leftarrow \neg q$ there is a presupposition, that in the absence of contrary evidence q is false and hence p is true. He allows "disjunctive databases", i.e., clauses with more than one atom in the head, e.g.

$$C_1 \vee \cdots \vee C_p \leftarrow A_1 \wedge \cdots \wedge A_m \wedge \neg B_1 \cdots \wedge \neg B_n,$$

and his basic notion of priority is that the C's here should have lower priority than the B's and no higher priority than the A's. To obtain greater generality, he defines this notion for ground atoms rather than predicates, i.e., if the above clause is a ground instance of a program clause he ways that $C_i < B_j$, $C_i \leq A_k$. Taking the tarnsitive closure of these relations establishes a relation on the ground atoms that is transitive (but may not be asymmetric and irreflexive if the program is not stratified).

His basic philosophy is

> ... if we have a model of DB and if another model N is obtained by possibly adding some ground atoms of M and removing some other ground atoms from M, then we should consider the new model N to be preferable to M only if the addition of a lower priority atom A to N is justified by the simultaneous removal from M of a higher priority atom B (i.e. such that $B > A$). This reflects the general

principle that we are willing to minimize higher priority predicates, even at the cost of enlarging predicates of lower priority, in an attempt to minimize high priority predicates as much as possible. A model M will be considered perfect if there are no models preferable to it. More formally:

[Definition 2.] Suppose that M and N are two different models of a disjunctive database DB. We say that N is preferable to M (briefly, $N < M$) if for every ground atom A in $N - M$ there is a ground atom B in $M - N$, such that $B > A$. We say that a model M of DB is *perfect* if there are no models preferable to M.

He extends the notion of stratifiability to disjunctive databases by requiring that in a clause

$$C_1 \vee \cdots \vee C_p \leftarrow A_1 \wedge \cdots \wedge A_m \wedge \neg B_1 \cdots \wedge \neg B_n$$

the predicates in $C_1, \ldots, C_p$ should all be of the same level i greater than that of the predicates in $B_1, \ldots, B_n$ and greater than or equal to those of the predicates in $A_1, \ldots, A_m$. He then weakens this to *local stratifiability* by applying it to ground atoms and instances of program clauses instead of to predicates and program clauses. (The number of levels is then allowed to be infinite.) It is equivalent to the nonexistence of increasing sequences in the above relation $<$ between ground atoms.

He proves:

Every locally stratified disjunctive database has a perfect model. Moreover every stratified **logic program** *P (i.e. where the head of each clause is a single atom) has exactly one perfect model, and it coincides with the model M_P of Apt, Blair, and Walker.*

He shows that every perfect model is minimal and supported; if the program is positive disjunctive, then a model is perfect iff it is minimal, and a model is perfect if there are no minimal models preferable to it. (Positive disjunctive means the clauses are of the form $C_1 \vee \cdots \vee C_p \leftarrow A_1 \wedge \cdots \wedge A_m$.) He also establishes a relation between perfect models and the concept of prioritized circumscription introduced by McCarthy [1984] and further developed by Lifschitz [1985]:

Let $S_1, \ldots, S_r$ be any decomposition of the set S of all predicates of a database DB into disjoint sets. A model M of DB is called a model of prioritized circumscription of DB with respect to priorities $S_1 > S_2 > \cdots > S_r$, or—briefly—a model of $\mathrm{CIRC}(DB, S_1 > S_2 > \cdots > S_r)$ if for every $i = 1, \ldots, r$ the extension in M of predicates from S_1, is minimal among all models M' of DB in which the extension of predicates from $S_1, S_2, \ldots, S_{i-1}$ coincides with the extension of these predicates in M.

He shows

Suppose that DB is a stratified disjunctive database and $\{S_1, S_2, \ldots, S_r\}$ is a stratification of DB. A model of DB is perfect if and only if it is a model of prioritized circumscription $\mathrm{CIRC}(DB, S_1 > S_2 > \cdots > S_r)$.

Przymusinska and Przymusinski [1988] extend the above results to a wider class of *weakly stratified program* and a corresponding wider class of *weakly perfect models.* The definitions are rather complicated but are based on the idea of removing 'irrelevant' predicate symbols in the dependency graph of a logic program and substituting components of this graph for its vertices in the definitions of stratification and perfect model.

Przymusinski [1988a] observed that the restriction to Herbrand models gives rise to what he calls the *universal query problem.* This is illustrated by the program P consisting of the single clause $p(a)$. Using the language defined by the program this has the sole Herbrand model $\{p(a)\}$ so that $\forall x p(x)$ is true in the least Herbrand model although it is not a consequence of P. So the semantics based on the least Herbrand model implies new positive information, and also prevents standard unification based procedures from being complete with respect to this semantics. One way of avoiding this problem is to consider Herbrand models not with respect to the language defined by the program but with respect to a language containing infinitely many function symbols of all arities, as in §3.1. This seems very cumbersome; given an interpretation for the symbols occurring in the program, to extend this to a model you would have to concoct meanings for all the infinitely many irrelevant constant and function symbols. A simpler way is to consider all models, or, as Przymusinski did, all models satisfying the freeness axioms of §2.1, instead of just Herbrand models. He showed how to extend the notion of perfect model from Herbrand models to all such models, and proposed a semantics based on the class of **all** perfect models. He gave a 'procedural semantics' and showed it to be sound and complete (for non-floundering queries), for stratified programs with respect to this new perfect model semantics. It is an extension of the interpreter given by Apt, Blair and Walker. However it is not a computational procedure. It differs from SLDNF-resolution by considering derivation trees to be failed not only when they are finitely failed, but also when all their branches either end in failure or are infinite. This cannot always be checked in a finite number of steps. Indeed there cannot be any computational procedure which is sound and complete for the perfect model semantics because for definite programs it coincides with the least Herbrand mdoel semantics, and the example in §2.2 shows that the set of ground atoms false in this model may be non-recursively enumerable. Further results on stratified programs and perfect models are found in Apt and Pugin [1987], Apt and Blair [1988].

4.4. Well-founded Models

Van Gelder, Ross and Schlipf [1988] (GRS), building on an idea of Ross and Topor [1987] defined a semantics based on *well-founded models.* These are Herbrand models which are supported in a stronger sense than that defined above. It is explained roughly by the following example. Suppose that $p \leftarrow q$ and $q \leftarrow p$ are the only clauses in the program with p or q in the head. Then p needs q to support it and q needs p to support it so the set $\{p, q\}$ gets no external support and in a well-founded model all its members will be taken to be false. In order to deal with all programs GRS worked with partial interpretations and models. A *partial interpretation* I of a program P is a set of literals which is consistent, i.e. does not contain both p and $\neg p$ for any ground atom (element of the Herbrand base B_p) p. If p belongs to I then p is true in I, if $\neg p$ belongs to I then p is false in I, otherwise p is undefined in I. It is called a *total interpretation* if it contains either p or $\neg p$ for each ground atom p. A total interpretation I is a *total model* of P if every instantiated clause of P is satisfied in I. A *partial model* is a partial interpretation that can be extended to a total model. A subset A of the Herbrand base B_p is an *unfounded set of* P *with respect to the partial interpretation* I if each atom $p \in A$ satisfies the following condition: For each instantiated clause C of P whose head is p, at least one of the following holds:

(1) Some literal in the body of C is false in I
(2) Some positive literal in the body of C is in A.

The well-founded semantics uses conditions (1) and (2) to draw negative conclusions. Essentially it simultaneously infers all atoms in A to be false, on the grounds that there is no one atom in A that can be first established as true by the clauses of P, starting from 'knowing' I, so that if we choose to infer that all atoms in A are false there is no way we would later have to infer one as true. The usual notion of supported uses (1) only. The closed sets of Ross and Topor [1987] use (2) only. It is easily shown that the union of all unfounded sets with respect to I is an unfounded set, the greatest unfounded set of P with respect to I, denoted by $U_P(I)$. Now for each partial interpretation I an extended partial interpretation $W_P(P)$ is obtained by adding to I all those positive literals p such that there is an instantiated clause of P whose body is true in I (this part is like the familiar

T_P operator) and all those negative literals $\neg p$ such that $p \in U_P(I)$. It is routine to show that U_P is monotonic and so has a least fixed point reached after iteration to some countable ordinal. This is denoted by $M_{WP}(P)$ and called the *well-founded partial model of P*. The *well-founded semantics* of P is based on $M_{WP}(P)$. In general $M_{WP}(P)$ will be a partial model, giving rise to a 3-valued semantics. Using the 3-valued logic of §3.1 $M_{WP}(P)$ is a 3-valued model of *comp*(P), but in general it is not the same as the Fitting model defined in §3.1 as the least fixed point of Φp. Since the Fitting model is the least 3-valued model of *comp*(P) it is a subset of $M_{WP}(P)$. For the program $p \leftarrow p$, in the Fitting model p is undefined, but in $M_{WP}(P)$ it is false (since the set $\{p\}$ is unfounded). However for stratified programs this new approach agrees with the two previous ones:

If P is locally stratified then its well-founded model is total and coincides with its unique perfect model i.e. with the model M_P of Apt, Blair and Walker.

Przymusinski [1988] extends this result to weakly stratified programs and weakly perfect models. He also shows how to extend SLS-resolution so that it is sound and complete for all logic programs (for non-floundering queries) with respect to the well-founded semantics. Ross [1988] gives a similar procedure. Przymusinski also shows that if the well-founded model is total the program is in a sense equivalent to a locally stratified program.

4.5. Stable Models

A closely related notion of *stable model* was introduced by Gelfond and Lifschitz [1988]. For a given logic program P they define a *stability transformation S* from total interpretations to total interpretations. Given a total interpretation I its transform $S(I)$ is defined in three stages as follows. Start with the set of all instantiations of clauses of P. Discard those whose bodies contain a negative literal which is false in I. From the bodies of those remaining discard all negative literals. This results in a set of definite clauses. Define $S(I)$ to be its least Herbrand model.

This transformation S is a 'shrinking' transformation, i.e. the set of positive literals true in $S(I)$ is a subset of those true in I. If I is a model of P the interpretation $S(I)$ may not be a model of P; it may shrink too much. However the fixed points of S are always models of P. These are defined to be the *stable models* of P. A stable model is minimal (in terms of the set of positive literals) but not every minimal model is stable. GRS show that for total interpretation being a fixed point of S is the same as being a fixed point of their operator W_P. Since the well-founded model is the least fixed point of W_P it is a subset of every stable model of P. Furthermore

If P has a well-founded total model then that model is the unique stable model.

The converse is not true.

Fine [1989] independently arrived, from a slightly different point of view, at a notion of *felicitous model*, which is equivalent to that of stable model. A felicitous model is one such that the falsehoods of the model serve to generate, via the program, exactly the truths of the model. His idea is that if you make a hypothesis as to which statements are false, and use the program to generate truths from this hypothesis then there are three possible outcomes: some statement is neither a posited falsehood nor a generated truth (a "gap"); some statement is both a posited falsehood and a generated truth (a "glut"); the posited falsehoods are the exact complements of the generated truths (no gap and no glut). A *happy hypothesis* is one which leads to no gaps and no glut, and a felicitous model is one where the hypothesis that the false statements are precisely those which are false in the model, is a happy hypothesis. Fine shows that the restriction to felicitous models can be viewed as a kind of self-referential closed world assumption.

Summary. Plausible arguments have been given for each of the semantics discussed in this section. The minimal models of *comp*(P) of Apt, Blair and Walker, the perfect models of Przymusinski, the well-founded models of van Gelder, Ross and Schlipf, and the stable models of Gelfond and Lifschitz are all models of *comp*(P), so SLDNF-resolution is sound for them. So they all offer plausible semantics for negation as failure in general different from that based on *comp*(P) because they are based on a subset of the models of *comp*(P). The fact that for the important class of locally stratified programs they all coincide, giving a unique model M_P, which often appears to be 'the' natural model, adds support to their claim to be chosen as the intended semantics. However SLDNF-resolution will be even more incomplete for them than it is for the semantics based on *comp*(P) and, as noted above there is, even for the class of definite programs, demonstrably no way of extending SLDNF-resolution to give a computable proof procedure which is both sound and complete for them.

References

1. Apt, K. R. and Blair, H. A. [1988], *Arithmetic classification of perfect models of stratified programs*, Report TR-88-09, University of Texas at Austin.
2. Apt, K. R.; Blair, H. A. and Walker, A. [1988], *Towards a theory of declarative knowledge*, in "Foundations of Deductive Databases and Logic Programming", (J. Minker, Ed.), Morgan Kaufmann, Los Altos, CA, 89–148.
3. Apt. K. R. and Pugin, J.-M. [1987], *Management of stratified databases*, Report TR-87-41, University of Texas at Austin.
4. Apt, K. R. and Emden, M. H. van [1982], *Contributions to the theory of logic programming*, JACM **29**, 841–863.
5. Barbuti, R. and Martelli, M. [1986], *Completeness of* SLDNF*-resolution for structured programs*, submitted to Theoretical Computer Science **21**.
6. Blair, H. A. [1982], *The recursion theoretic complexity of the semantics of predicate logic as a programming language*, Information and Control **54**, 25–47.
7. Blair, H. A. [1986], *Decidability in the Herbrand base*, in "Proceedings Workshop on Foundations of Deductive Databases and Logic Programming", (J. Minker, Ed.), Washington, DC.
8. Börger, E. [1987], *Unsolvable decision problems for* PROLOG *programs*, to appear in Computer Theory and Logic (E. Börger, Ed.), Lecture Notes in Computer Science, Springer-Verlag.
9. Brass, S. and Lipeck, U. W. [1989], *Specifying closed world assumptions for logic databases*, Proc. Second Symposium on Mathematical Fundamentals of Database Systems (MFDBS89).
10. Cavedon, L. [1988], *On the completeness of* SLDNF*-resolution*, Ph.D. Thesis, Melbourne University, 120.
11. Cavedon, L. and Lloyd, J. W. [1989], *A completeness theorem for* SLDNF*-resolution*, J. Logic Programming **7**(3), 177–192.
12. Carvalho, R. L.; de Maibaum, T. S. E.; Pequeno, T. H. C.; Pereda, A. A. and Veloso, P. A. S. [1980], *A model theoretic approach to the theory of abstract data types and structures*, Research Report CS-80-22, Waterloo, Ontario.
13. Cerrito, S., *Negation as failure, a linear axiomatization*, to appear in J. Logic Programming.
14. Chan, D. [1988], *Constructive negation based on the completed database*, Proc. 1988 Conference and Symposium on Logic Programming, Seattle, Washington, September 1988, pp. 111–125.
15. Clark, K. L. [1978], *Negation as failure*, in "Logic and Data Base" (H. Gallaire and J. Minker, Eds.), Plenum, New York, 293–322.
16. Davis, M. [1983], *The prehistory and early history of automated deduction*, in "Automation of Reasoning" (J. Siekmann and G. Wrightson, Eds.), Springer, Berlin, Vol. (1983), 1–28.
17. Ebbinghaus, H. D. [1969], *Über eine prädikaten logik mit partiell definierten prädikaten and funktionen*, Arch. Math. Logik **12**, 39–53.
18. Fine, K. [1989], *The justification of negation as failure*, Logic, Methodology and Philosophy of Science VIII (J. E. Fenstad et al., Eds.) Elsevier Science Publishers B.V.
19. Fitting, M. [1985], *A Kripke–Kleene semantics for general logic programs*, Logic Programming **2**, 295–312.
20. Fitting, M. [1986], *Partial models and logic programming*, to appear in Computer Science.
21. Fitting, M. [1987a], *Pseudo-boolean valued Prolog*, Research Report, H. Lehman College, (CUNY), Bronx, NY.

22. Fitting, M. [1987b], *Logic programming on a topological bilattice*, Research Report, H. Lehman College, (CUNY), Bronx, NY.
23. Fitting, M. [1988b], *Bilattices and the semantics of logic programming*, Research Report, Dept. of Computer Science, CUNY.
24. Fitting, M. and Ben-Jacob, M. [1988], *Stratified and three-valued logic programming semantics*, Research Report, Dept. of Computer Science, CUNY.
25. Gabbay, D. M. [1985], *N-Prolog: An extension of Prolog with hypothetical implication*, II. Logical Foundations, and Negations as Failure, J. Log. Programming **2**(4), 251–283.
26. Gabbay, D. M. [1989], *Modal provability foundations for negation by failure*, Preprint.
27. Gabbay, D. M. and Sergot, M. J. [1986], *Negation as inconsistency*, J. Logic Programming **3**(1), 1–36.
28. Gallier, J. H. and Raatz, S. [1986a], HORNLOG*: A graph based interpreter for general Horn clauses*, Technical Report MS-CIS-86-10, University of Pennsylvania; J. Logic Programming **4** 2, 119–156.
29. Gallier, J. H. and Raatz, S. [1986b], *Extending* SLD*-resolution to equational Horn clauses using E-unification*, J. Logic Programming **6**, 3–44, Short version to appear, 1986 IEEE Symposium on Logic Programming, Salt Lake City, UT..
30. Gelfond, M. [1987], *On stratified autoepistemic theories*, in "Proceedings AAAI-87", 207–211, American Association for Artificial Intelligence, Morgan Kaufmann, Los Altos, CA.
31. Gelfond, M. and Lifschitz, V. [1988], *The stable model semantics for logic programming*, 5th International Conference on Logic Programming, Seattle.
32. Gelfond, M.; Przymusinski, H. and Przymusinski, T. [1986], *The extended closed world assumption and its relationship to parallel circumscription*, Proceedings ACM SIGACT-SIGMOD Symposium on Principles of Database Systems, Cambridge, MA, 133–139.
33. Gelfond, M.; Przymusinski, H. and Przymusinski, T. [1986a], *On the relationship between circumscription and negation as failure*, to appear in Artificial Intelligence.
34. Girard, J. Y. [1987], *Linear logic*, Theoretical Computer Science **50**.
35. Goguen, J. A. and Burstall, R. M. [1984], *Institutions: Abstract model theory for computer science*, Proc. of Logic Programming Workshop (E. Clark and D. Kozen, Eds.), Lecture Notes in Computer Science **164**, Springer-Verlag, 221–256.
36. Haken, A. [1985], *The intractability of resolution*, Theoretical Computer Science **39**, 297–308.
37. Henschen, L. J. and Park, H.-S. [1988], *Compiling the* GCWA *in indefinite databases*, in "Foundations of Deductive Databases and Logic Programming" (J. Minker, Ed.), Morgan Kaufmann Publishers, Los Altos, CA, 395–438.
38. Hodges, W. [1985], *The logical basis of* PROLOG, unpublished text of lecture, 1–10.
39. Jäger, G. [1988], *Annotations on the consistency of the closed world assumption*, Preprint, Computer Science Dept., Technische Hochschule, Zürich (1987).
40. Jaffar, J.; Lassez, J.-L. and Lloyd, J. W. [1983], *Completeness of the negation as failure rule*, IJCAI-83, Karlsruhe, 500–506.
41. Jaffar, J.; Lassez, J.-L. and Maher, M. J. [1984a], *A theory of complete logic programs with equality*, J. Logic Programming **1**(3), 211–223.
42. Jaffar, J.; Lassez, J.-L. and Maher, M. J. [1984b], *A logic programming language scheme*, in "Logic Programming Relations, Functions and Equations" (D. DeGroot and G. Lindstrom, Eds.), Prentice Hall. Also Technical Report TR 84/15, University of Melbourne.
43. Jaffar, J.; Lassez, J.-L. and Maher, M. J. [1986a], *Comments on "General Failure of Logic Programs"*, J. Logic Programming **3**(2), 115–118.

44. Jaffar, J.; Lassez, J.-L. and Maher, M. J. [1986b], *Some issues and trends in the semantics of logic programs*, Proceedings Third International Conference on Logic Programming, Springer, 223–241.
45. Jaffar, J. and Stuckey, P. J. [1986], *Canonical logic programs*, J. Logic Programming **3**, 143–155.
46. Kleene, S. C. [1952], *Introduction to Metamathematics*, van Nostrand, New York.
47. Kowalski, R. [1979], *Logic for Problem Solving*, North Holland, New York.
48. Kunen, K. [1987], *Negation in logic programming*, J. Logic Programming **4**, 289–308.
49. Kunen, K. [1987a], *Signed data dependencies in logic programs*, Computer Sciences Technical Report #719, University of Wisconsin, Madison. Also J. Logic Programming **7**(3), 231–246.
50. Lassez, J.-L. and Maher, M. J. [1984], *Closures and fairness in the semantics of programming logic*, Theoretical Computer Science **29**, 167–184.
51. Lassez, J.-L. and Maher, M. J. [1985], *Optimal fixed points of logic programs*, Theoretical Computer Science **39**, 15–25.
52. Lewis, H. [1978], *Renaming a set of clauses as a Horn set*, JACM 25, 134–135.
53. Lifschitz, V. [1985], *Computing circumscription*, Proceedings IJCAI-85m, 121–127.
54. Lifschitz, V. [1988], *On the declarative semantics of logic programs with negation*, Foundations of Deductive Databases and Logic Programming (J. Minker, Ed.), Morgan Kaufmann Publishers, Los Altos, CA, 177–192.
55. Lloyd, J. W. [1987], *Foundations of Logic Programming*, 2nd edition Springer, Berlin.
56. Lloyd, J. W. and Topor, R. W. [1984], *Making* PROLOG *more expressive*, J. Logic Programming **1**, 225–240.
57. Lloyd, J. W. and Topor, R. W. [1985], *A basis for deductive data base systems,* II, J. Logic Programming **3**, 55–68.
58. Loveland, D. W. [1988], *Near-Horn Prolog*, Proc. ICLP'87, (J.-L. Lassez, Ed.), MIT Press.
59. McCarthy, J. [1984], *Applications of circumscription to formalizing common sense knowledge*, AAAI Workshop on Non-Monotonic Reasoning, 295–323.
60. Maher, M. J. [1987], *Complete axiomatization of the algebras of finite, infinite and rational trees*, Technical Report, IBM T. J. Watson Research Centre, Yorktown Heights, NY.
61. Mahr, B. and Makowsky, J. A. [1983], *Characterizing specification langauges which admit initial semantics*, Proc. 8th CAAP, Lecture Notes in Computer Science **159**, Springer-Verlag, 300–316.
62. Makowsky, J. A. [1986], *Why Horn formulas matter in computer science: Initial structures and generic examples*, Technical Report No. 329, Technion Haifa, 1984 (extended abstract); in Mathematical Foundations of Software Development, Proceedings of the International Joint Conference on Theory and Practice of Software Development (TAPSOFT) (H. Ehrig et al., Eds.), Lecture Notes in Computer Science **185**, Springer (1985), 374–387, and (revised version) May 15, 1986, 1–28, preprint. The references in the text are to this most recent version.
63. Malcev, A. [1971], *Axiomatizable classes of locally free algebras of various types*, in "The Metamathematics of Algebraic Systems: Collected Papers", Chapter 23, 262–281, North Holland, Amsterdam.
64. Meltzer, B. [1983], *Theorem-proving for computers: Some results on resolution and renaming*, in "Automation of Reasoning", **1** (J. Siekmann and G. Wrightson, Eds.), Springer, Berlin, 493–495.
65. Minker, J. [1982], *On indefinite data bases and the closed world assumption*, Proc. 6th Conf. Automated Deduction, Lecture Notes in Computer Science **138**, Springer-Verlag, 292–308.

66. Minker, J. and Perlis, D. [1985], *Computing protected circumscription*, J. Logic Programming **2**, 1–24.
67. Mints, G. [1986], *Complete calculus for pure Prolog (Russian)*, Proc. Acad. Sci. Estonian SSR, **35**, 4, 367–380.
68. Mints, G. [1990], *Several Formal Systems of the Logic Programming*, Computer and Artificial Intelligence **9**, 19–41.
69. Moore, R. C. [1985], *Semantic considerations on non-monotonic logic*, Artificial Intelligence **25**, 75–94.
70. Mycroft, A. [1983], *Logic programs and many-valued logic*, Proc. 1st STACS Conference.
71. Naish, L. [1986], *Negation and quantifiers in NU-Prolog*, Proceedings Third International Conference on Logic Programming, Springer, 624–634.
72. Naqvi, S. A. [1986], *A logic for negation in database systems*, in "Proceedings of Workshop on Foundations of Deductive Databases and Logic Programming", (J. Minker, Ed.), Washington, DC.
73. Plaisted, D. A. [1984], *Complete problems in the first-order predicate calculus*, J. Comp. System Sciences **29**, 8–35.
74. Poole, D. L. and Goebel, R. [1986], *Gracefully adding negation and disjunction to Prolog*, Proceedings Third International Conference on Logic Programming, Springer, 635–641.
75. Przymusinska, H. [1987], *On the relationship between autoepistemic logic and circumscription for stratified deductive databases*, Proceedings of the ACM SIGART International Symposium on Methodologies for Intelligent Systems, Knoxville, Tenn.
76. Przymusinska, H. and Przymusinski, T. [1988], *Weakly perfect model semantics for logic programs*, In R. Kowalski and K. Bowen, Editors, Proceedings of the Fifth Logic Programming Symposium, 1106–1122, Association for Logic Programming, MIT Press, Cambridge, Mass.
77. Przymusinski, T. C. [1988], *On the semantics of stratified deductive databases*, in "Foundations of Deductive Database and Logic Programming" (J. Minker, Ed.), Morgan Kaufmann Publishers, Los Altos, CA, 193–216.
78. Przymusinski, T. C. [1988a], *On the declarative and procedural semantics of logic programs*, J. Automated Reasoning, **4**. In print. (Extended abstract appeared in: Przymusinski, T. C. [1988] *Perfect model semantics*. In R. Kowalski and K. Bowen, Editors, Proceedings of the Fifth Logic Programming Symposium, 1081–1096, Assocaition for Logic Programming, MIT Press, Cambridge, Mass.
79. Przymusinski, T. C. [1988b], *On constructive negation in logic programming*, Technical Report Draft, University of Texas at El Paso.
80. Przymusinski, T. C. [1988c], *Three-valued formalizations of non-monotonic reasoning and logic programming*, Research Report, Dept. of Mathematics, University of Texas at El Paso.
81. Przymusinski, T. C. [1989], *Three-valued stable semantics for normal and disjunctive logic programs*, Technical Report, University of Texas at El Paso.
82. Przymusinski, T. C. [1989], *Three-valued non-monotonic formalisms and logic programming*, in "Proceedings of the First International Conference on Principles of Knowledge Representation and Reasoning (KR'89)", Toronto, Canada. In print.
83. Reiter, R. [1978], *On closed world data bases*, in "Logic and Data Bases" (H. Gallaire and J. Minker, Eds.), Plenum, New York, 55–76.
84. Reynolds, M. [1987], *The expressive power of query languages based on logic programming*, Ph.D. Thesis, University London.
85. Reynolds, M. [1987a], *A completeness result for logic programming*, Manuscript.
86. Reynolds, M. [1988], *Declarative meaning for logic programs with negation as failure*, Manuscript.

87. Ross, K. [1989], *A procedural semantics for well-founded negation in logic programs*, in "Proceedings of the Eighth Symposium on Principles of Database Systems", ACM SIGACT-SIGMOD.
88. Ross, K. and Topor. R. W. [1987], *Inferring negative information from disjunctive databases*, Technical Report 87/1, University of Melbourne.
89. Sakai, K. and Miyachi, T. [1983], *Incorporating naive negation into* PROLOG, ICOT Technical Report: TR-028.
90. Sato, T. [1982], *Negation and semantics of* PROLOG *programs*, Proc. 1st International Conference on Logic Programming, 169–174.
91. Sato, T. [1987], *On the consistency of first order logic programs*, Tech. Report 87-12, Electrotechnical Laboratory, Ibarki, Japan.
92. Sato, T. [1988], *Completed logic programs and their consistency*, Typescript, Electrotechnical Laboratory, Ibaraki, Japan. To appear in J. Logic Programming.
93. Schmitt, P. H. [1986], *Computational aspects of three-valued logic*, Proc. 8th Conf. Automated Deduction, Lecture Notes in Computer Science, **230**, Springer-Verlag, 190–198.
94. Shepherdson, J. C. [1984], *Negation as failure: A comparison of Clark's completed data base and Reiter's closed world assumption*, J. Logic Programming **1**, 51–81.
95. Shepherdson, J. C. [1985], *Negation as failure* II, J. Logic Programming **3**, 185–202.
96. Shepherdson, J. C. [1988], *Negation in logic programming*, in "Foundations of Deductive Databases and Logic Programming" (J. Minker, Ed.), Morgan Kaufmann, Los Altos, CA, 19–88.
97. Shepherdson, J. C. [1988a], *Language and equality theory in logic programming*, Technical Report PM-88-08, Mathematics Dept., Univ. Bristol, 1–41.
98. Shepherdson, J. C. [1988b], SLDNF-*resolution with equality*, Technical Report PM-88-05, Mathematics Dept., Univ. Bristol, 1–15; Journal of Automated Reasoning **6** (1990).
99. Shepherdson, J. C. [1989], *Unsolvable problems for* SLDNF-*resolution*, J. Logic Programming **10**, 19–22.
100. Shepherdson, J. C. [1989a], *A sound and complete semantics for a version of negation as failure*, Theoretical Computer Science **65**, 343–371.
101. Shepherdson, J. C. [1989b], *Mints type deductive calculi for logic programming*, Technical Report, Mathematics Dept., University of Bristol, to appear in special issue of Annals of Pure and Applied Logic in memory of John Myhill.
102. Solovay, R. M. [1976], *Provability interpretation of modal logic*, Israel Journal of Mathematics **25**, 287–304.
103. Stickel, M. E. [1986], *A* PROLOG *technology theorem prover: Implementation by an extended* PROLOG *compiler*, Proceedings Eighth International Conference on Automated Deduction, Springer, 573–587.
104. Terracini, L. [1988], *Modal interpretation for negation by failure*, Atti dell'Academia delle Scienze di Torino **122**, 81–88.
105. Terracini, L. [1988a], *A complete bi-modal system for a class of models*, Atti dell'Academia delle Scienze di Torino **122**, 116–125.
106. Van Gelder, A. [1988], *Negation as failure using tight derivations for general logic programs*, Foundations of Deductive Databases and Logic Programming (J. Minker, Ed.), Morgan Kaufmann Publishers, Los Altos, CA, 149–176; revised version in J. Logic Programming **6**, 109–133, 1989.
107. Van Gelder, A.; Ross, K. and Schlipf [1988], *Unfounded sets and well-founded semantics for general logic programs*, in "Proceedings of the Symposium on Principles of Database Systems", ACM SIGACT-SIGMOD.
108. Voda, P. J. [1986], *Choices in, and limitations of, logic programming*, Proc. 3rd Int. Conf. Logic Programming, Springer, 615–623.

109. Yahya, A. and Henschen, L. [1985], *Deduction in non-Horn databases*, J. Automated Reasoning **1**(2), 141–160.

MATHEMATICS DEPARTMENT, UNIVERSITY OF BRISTOL, ENGLAND
MATHEMATICAL SCIENCES RESEARCH INSTITUTE, BERKELEY, CA 94720

RESEARCH AT MSRI SUPPORTED IN PART BY NSF GRANT DMS-8505550.

NORMAL VARIETIES OF COMBINATORS

RICK STATMAN

1. INTRODUCTION

We adopt for the most part the terminology and notation of [1]. A *combinator* is a term with no free variables. A set of combinators which is both recursively enumerable and closed under β conversion is said to be *Visseral* ([5]). Given combinators F and G, the *variety defined by* $Fx = Gx$ is the set of all combinators M such that $FM = GM$. Such a variety is said to be *normal* if both F and G are normal. In this note we shall be principally concerned with normal varieties.

Example 1 (Böhm and Dezani [2]). The normal variety defined by $x \circ K^n = I$ consists of the combinators $X \underset{\beta}{=} \lambda x.x\mathcal{H}_1 \ldots \mathcal{H}_n$ for some $\mathcal{H}_1 \ldots \mathcal{H}_n$.

Clearly, a variety of combinators is always Visseral. We recall the following theorem from [4].

Theorem 1. *For Σ a set of combinators, the following are equivalent*

(1) *Σ is Visseral*
(2) *Σ is a variety*
(3) *Σ is the variety defined by $Fx = F$, for some combinator F.*

For normal varieties the situation is quite different. For example, the β closure of $\{\Omega\}$ is Visseral but not a normal variety ([1] p. 445). More generally, if F and G are distinct forms then $F\Omega \underset{\beta}{\neq} G\Omega$. Combinators with this property are said to be transcendental. A transcendental is always order zero.

We shall proceed as follows. First, we shall discuss normal varieties of solvable combinators. In this context, the pattern matching equation $Fx = I$ plays a special role. Among the Visseral sets of solvable combinators are the binary languages. We shall characterize those binary languages which are normal varieties. From this characterization follows the result that the collection of normal varieties is Σ_3^0 complete. Next, we shall consider order zero solutions to normal equations. In this context, the fixed point equation $Fx = x$ plays a special role. We shall prove a number of results analogous to theorems from classical algebra and number theory; such as, Hilbert's Nullstellensatz and Lindemann's theorem concerning the transcendence of e. We conclude with several open problems and applications.

It will be useful to have some special terminology for terms. A typical term $\mathcal{H}$ has the form

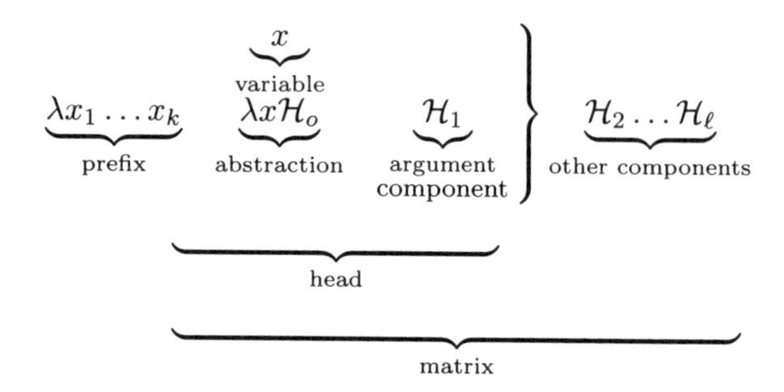

A head normal combinator of order one all whose components are combinators is called a *word*.

$$M \equiv \lambda x.xM_1 \ldots M_m$$

a typical word

A normal combinator of order one is called a *deed*.

$$X \equiv \lambda x.x\mathcal{H}_1 \ldots \mathcal{H}_k$$

a typical deed

Words and deeds with at least one component are said to be *non-trivial*.

2. Varieties of Solvable Combinators

We begin with the following.

Lemma 1. *Suppose Σ is a normal variety. Then one of the following holds.*

(1) $\Sigma = \emptyset$

(2) $K^\infty \in \Sigma$

(3) *There exists a deed F such that Σ is contained in the variety defined by $Fx = I$.*

Proof. Suppose Σ is defined by $Gx = Hx$. The proof is by induction on the structure of G and H, and its routine. □

Remark 1. It is easy to see that the deed F in Lemma 1 be assumed to be non-trivial. Theorem 1 and Lemma 1 yield the following.

Proposition 1. *Suppose that Σ is a set of solvable combinators. Then the following are equivalent.*

(1) *Σ is a normal variety*
(2) *Σ is Visseral and there exists a non-trivial deed F such that Σ is contained in the variety defined by $Fx = I$.*

Proof. (1) $\Rightarrow$ (2) by Lemma 1 and Remark 1. Suppose (2). By Theorem 1 there is a combinator G such that $M \in \Sigma \Leftrightarrow GM \underset{\beta}{=} G$. Let $x\mathcal{F}_1 \ldots \mathcal{F}_k$ be the matrix of F. We "make G normal" by replacing each redex $(\lambda x\mathcal{H})\mathcal{Y}$ by $x\mathcal{F}_1 \ldots \mathcal{F}_k(\lambda x\mathcal{H})\mathcal{Y}$. Set $H_1 \equiv \lambda x[Gx, Fx]$ and $H_2 \equiv \lambda x[G, I]$. Then $M \in \Sigma \Leftrightarrow H_1 M \underset{\beta}{=} H_2 M$. □

Example 2. Suppose k and l are given and Σ is a set of combinators such that whenever $M \in \Sigma$ there exist $m \leq k, n \leq l, i \leq m$, and $\mathcal{H}_1 \ldots \mathcal{H}_n$ satisfying $M \underset{\beta}{=} \lambda x_1 \ldots x_m . x_i \mathcal{H}_i \ldots \mathcal{H}_n$. Then Σ is contained in the variety defined by $Fx = I$ where $F \equiv \lambda x$.

$$x \underbrace{(K^{\ell}I) \ldots (K^{\ell}I)}_{k} \underbrace{I \ldots I}_{\ell}$$

A *language* is a Visseral set of combinators each of which β converts to a word. If Σ is a set of combinators then $\underline{\Sigma}^+$ is the β closure of the set of all non-trivial words with components from Σ. Clearly, Σ is Visseral if and only if Σ^+ is a language.

Proposition 2. *Suppose that Σ is a β closed set of solvable combinators. Then the following are equivalent.*

(1) *Σ is a normal variety*
(2) *Σ^+ is a normal variety*

The proof of Proposition 2 requires two ideas that we shall need to refine for the proof of Theorem 2 below.

First, we consider the symbolic action of a deed on a word under head reduction. Let $\mathcal{U}_n \equiv \lambda x . x u_1 \ldots u_n$. Recall that $\|F\|$ is the number of symbols in F.

Lemma 2. *Suppose that k and n are given with $n > o$. Suppose that F is a normal combinator with at least one component, and*

(i) *F has a prefix of length $k+1$, and*
(ii) *$\|F\| \leq n$.*

Then $F\mathcal{U}_n u_1 \dots u_k$ has a normal form with an empty prefix and some u_i at the head.

Proof. By induction F.

To apply this to Proposition 2, suppose that Σ^+ is $\subseteq$ the variety defined by $Fx = I$ for F a non-trivial deed. Then, for sufficiently large n, $F(\lambda x.\ x\frac{u\dots u}{n})$ has a normal form $\mathcal{U}$ with an empty prefix and u at the head. Thus, for the deed $\lambda u\mathcal{U}$, we have $M \in \Sigma \Rightarrow (\lambda u\mathcal{U})M \underset{\beta}{=} F(\lambda x\ x\frac{M\dots M}{n}) \underset{\beta}{=} I$.

Second we must take into consideration the fact that a combinator computes on a word sequentially from left to right. Memory is added by concatenating on the right. Suppose that F is a deed with $k > o$ components and $M \in \Sigma \Rightarrow FM \underset{\beta}{=} K$. As in [1] 6.1 we can define a combinator G satisfying

$$Gxyz \underset{\beta}{=} x(\lambda u \text{ IF } Fu \text{ THEN } G(\lambda v\ x(Kv)(\overset{k+1}{K\ I}))(\lambda ab\ a(yab))z \text{ ELSE } zxy.$$

Let $\ulcorner n \urcorner$ be the nth Church numeral, and suppose $M \equiv \lambda x.xM_1 \dots M_m \in \Sigma$. We have $GM\ulcorner n \urcorner z \underset{\beta}{=} G(\lambda x.\ xM_2 \dots M_m(\overset{k+1}{K\ I}))\ulcorner n+1 \urcorner zM_2 \dots M_m \underset{\beta}{=}$

$$\frac{\dots}{m-2} \underset{\beta}{=} G(\lambda x.x(K^{k+1}I)\dots(\overset{k+1}{K\ I}))\ulcorner n+m \urcorner z$$

$$\underbrace{\frac{(\overset{k+1}{K\ I})\dots(\overset{k+1}{K\ I})}{m-1} M_m \frac{(\overset{k+1}{K\ I})\dots(\overset{k+1}{K\ I})}{m-2} \dots M_2 \dots M_m}_{m \text{ blocks}} \underset{\beta}{=}$$

$$z(\lambda x\ x\frac{(\overset{k+1}{K\ I})\dots(\overset{k+1}{K\ I})}{m})\ulcorner n+m \urcorner \frac{\dots}{(m+1).(m-1)}.$$

Now it is easy to construct a combinator P satisfying

$$P(\lambda x.\ x\frac{(\overset{k+1}{K\ I})\dots(K^{k+1}I)}{m})\ulcorner n+m \urcorner = \overset{(n+m).(m-1)}{K\qquad I}$$

since the combinators $\lambda x\ x\frac{(\overset{k+1}{K\ I})\dots(\overset{k+1}{K\ I})}{m}$ form an adequate numerical system ([1] 6.4). We have that $GM\ulcorner 1 \urcorner P \underset{\beta}{=} I$, but $Gx\ulcorner 1 \urcorner P$ is not yet normal. We "make $Gy\ulcorner 2 \urcorner P$ normal" as follows. Let $\mathcal{F}$ be the matrix of F

with head variable x. Replace each redex $(\lambda x \mathcal{H})\mathcal{Y}$ by $\mathcal{F}(\lambda x \mathcal{H}) I \mathcal{Y}$. Let the result be $\mathcal{G}$. Finally set $H \equiv \lambda y\ y(\lambda x \mathcal{G})$. For M as above we have $HM \underset{\beta}{=} M(\lambda x[^M|y]\mathcal{G}) \underset{\beta}{=} GM\ulcorner 2 \urcorner P\ M_2 \ldots M_m \underset{\beta}{=} I$. This completes the proof of Proposition 3. □

Thus, for our purposes, for sets of solvable combinators it suffices to study languages.

A language Σ is said to be *binary* if whenever

$$\lambda x.xM_1 \ldots M_m \in \Sigma, \text{ for each } i = 1 \ldots m$$

$$M_i \underset{\beta}{=} \begin{cases} I \\ \text{unsolvable.} \end{cases}$$

Example 3. For each e and k define combinators M_k^e by

$$M_k^e = \begin{cases} I & \text{if } k \in W_e \\ \text{unsolvable} & \text{else} \end{cases}$$

as in [1] p. 179. Let $\Sigma_e =$ the β closure of the set

$$\{\lambda x.\ x \frac{M_k^e \ldots M_k^e}{k} : k = 1, \ldots\}.$$

A binary language Σ is said to be *bounded away from* $\perp$ if there is an infinite recursively enumerable set $\mathcal{J}$ of positive integers such that whenever $s \in \mathcal{J}$, $\lambda x.xM_1 \ldots M_m \in \Sigma$ and $s \leq m$ we have $M_s \underset{\beta}{=} I$.

Theorem 2. *Suppose that Σ is a binary language. Then the following equivalent*

(1) *Σ is a normal variety*
(2) *Σ is bounded away from $\perp$.*

To prove Theorem 2 we need to refine the two ideas in the proof of Proposition 2. First, we refine the "symbolic computation". For this we need a refinement of the standardization theorem ([1] p. 318). A *cut* of $\mathcal{H}$ is a maximal applicative subterm of $\mathcal{H}$ with a redex at its head. The cuts of $\mathcal{H}$ and the redexes of $\mathcal{H}$ are in one to one correspondence, and we shall use notions defined for one freely for the other. The following notions will be used exclusively when $\mathcal{H}$ has a head redex. The *major cut* of $\mathcal{H}$ is the leftmost cut whose abstraction term is in head normal form. The *major variable* of $\mathcal{H}$ is the head variable of the abstraction term of the major cut. The *base* of $\mathcal{H}$ is the matrix of the abstraction term of the major cut. When the major variable of $\mathcal{H}$ is bound in the prefix of the abstraction term of a cut, this cut is called the *minor cut.*

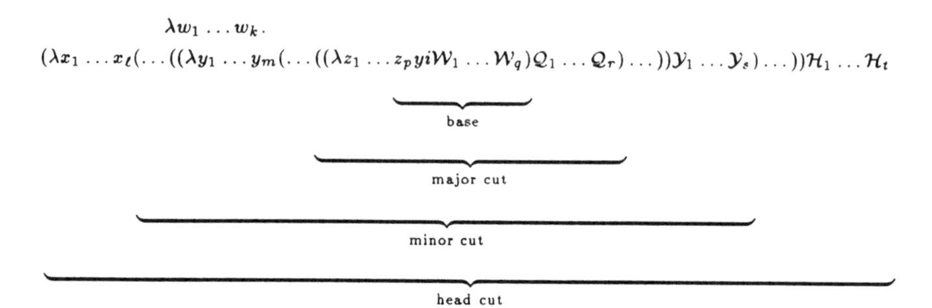

THE MAJOR AND MINOR CUTS OF A TERM

A reduction of $\mathcal{H}$ is said to be *solving* if the redex contracted corresponds to the minor cut if exists and the major cut otherwise.

Lemma 3. *If $\mathcal{H}$ is solvable then the solving reduction sequence beginning with $\mathcal{H}$ achieves a head normal form.*

Proof. Each solving reduction reduces the length of a head reduction to head normal form.

Suppose F is a deed. We symbolically calculate $F\mathcal{U}_m$ as follows. Perform solving reductions until some u_i is the major variable. Next, substitute I for this occurence of u_i and repeat the process. The calculation terminates in I if and only if there is a binary word $M \equiv \lambda x.xM_1 \ldots M_m$ such that $FM \underset{\beta}{=} I$. Let $1 \leq k < m$, and suppose that $\mathcal{U}_m$ is the abstraction term of the minor(= major) cut at some stage, say

$$\mathcal{U}_m \mathcal{H}_1 \mathcal{H}_2 \ldots \mathcal{H}_\ell \to$$

$$\underbrace{\underbrace{\mathcal{H}_1 u_1 \ldots u_k}_{\mathcal{H}_o} u_{k+1} \ldots u_m \mathcal{H}_2 \ldots \mathcal{H}_\ell}_{\mathcal{H}}.$$

From this stage on we trace the decendants ([3] p. 18) of $\mathcal{H}$, so long as they exist and maintain the form $\mathcal{H} \equiv \mathcal{H}_o u_{k+1} \ldots u_m \mathcal{H}_2 \ldots \mathcal{H}_\ell$. Such an $\mathcal{H}$ is either

(a) an initial segment of the base with the major variable at the head of $\mathcal{H}_o$, or

(b) an initial segment of some cut with the abstraction term of the major cut contained in $\mathcal{H}_o$.

Now suppose for some binary word $M \equiv \lambda x.xM_1 \dots M_m$, $FM \underset{\beta}{=} I$. Since $\mathcal{H}$ has no decendant in I, the form of $\mathcal{H}$ must change. Thus there is some stage at which $\mathcal{H}_o$ coincides with the abstraction term of the minor cut if one exists or the major cut, with major variable not one of the u_i, otherwise. At this stage we have the following

(*) If $m - k$ exceeds the length of the prefix of $\mathcal{H}_o$ then some u_i, for $k < i \leq m$, is the major variable at some later stage in the computation.

Since, if $\mathcal{H}_o$ is the abstraction term of the major cut with major variable not one of the u_i, no minor cut exists, and $m - k$ exceeds the number of λ's in the prefix of $\mathcal{H}_o$ then u_m is a component of the last term in the computation. The property (*) is just a refinement of Lemma 2.

To prove (1) $\Leftrightarrow$ (2), suppose that the computation beginning with $F\mathcal{U}_m$ terminates in I. Let k be larger than the number of symbols in the computation. Recall that we can assume that F is non-trivial so $\mathcal{U}_{m+k}$ is the abstraction term of the minor cut at least once in the computation beginning with $F\mathcal{U}_{m+k}$. Consider the first time any one of the $\mathcal{H}$ changes form. By choice of k the hypothesis of (*) is satisfied. Thus if the computation of $F\mathcal{U}_{m+k}$ terminates in I then there is some u_i for $m < i \leq m + k$ which is the major variable at some stage in the computation. Now it is easy to see that (1) $\Leftrightarrow$ (2).

Next, we need to refine the construction in the second part of the proof of Proposition 2. Changes are needed for the following reason. Not every M_i is solvable, so in cycling through them some must be skipped; this is where $\mathcal{J}$ is used. We sketch only the construction of G; the rest is routine. Suppose that $\mathcal{H}$ is a total recursive function which enumerates an infinite subset of $\mathcal{J}$ in increasing order, and H is a combinator which represents ([1] 6.3) $\mathcal{H}$ on the Church numerals. Suppose that P is a combinator which satisfies P

$$(\lambda x \underbrace{xK_* \dots K_*}_{m})\ulcorner n \urcorner \underset{\beta}{=} K^{n.(m-1)}I$$

Now construct G satisfying $Guvwx \underset{\beta}{=}$ IF (Zero$_c x$) THEN $(u(\lambda y$ IF yK^2K_*

THEN $G(\lambda z\ u(Kz)K_*)(S_c^+v)(S_c^+w)(\text{Minus}_c(H(J_c^+w))(Hn))$ ELSE $Puv)$ ELSE $G(\lambda z\ u(Kz)K_*)(S_c^+v)nP_c^-x)$ (see [1] p. 135). To understand the action of G it suffices to understand the function of v, w, and x. v counts the total number of moves made by G, w is the number of the next integer in $\mathcal{J}$, according to h, and x is the number of moves needed to get to $M_{h(w)}$. This completes our sketch of the proof of Theorem 2. □

Corollary 1. *The collection of normal varieties is Σ_3^o complete.*

Proof. Σ_e is a normal variety $\Leftrightarrow W_e$ is confinite. □

Example 3 (continued). Let $\Delta_e =$ the β closure of $\{\lambda x.\ xM_1^e \ldots M_k^e : k = 1 \ldots\}$. Then Δ_e is a normal variety if and only if W_e is infinite.

3. Order Zero Solutions

If F is a deed we write $\check{F}$ for its matrix; we shall always assume that x is the head variable. For deed F, G, and H the relation $F \underset{G}{>\!\!\rightarrow} H$ holds if H is obtained from F by replacing one occurence of x by $\check{G}$. If Δ is a set of deeds the relations $\underset{\Delta}{>\!\!\rightarrow}$, $\underset{\Delta}{>\!\!\rightarrow\!\!>}$, and $\underset{\Delta}{=}$ are defined in the obvious way.

Example 4. $F \underset{G}{>\!\!\rightarrow\!\!>}$ the normal form of $F \circ G$. It is easy to see that $\underset{G}{>\!\!\rightarrow\!\!>}$ is Church-Rosser, upward Church-Rosser, and has unique upward normal forms (see [1] 3.5). If $F \underset{G}{=} H$ and M is any fixed point of G then $FM \underset{\beta}{=} HM$. A set Δ of deeds is called a *behavior* if whenever $F \underset{G}{>\!\!\rightarrow} H$ and two of F, G, and H belongs to Δ then the third of F, G, and H belongs to Δ.

Example 5. The *powers of* F is the set $\{H : F \underset{F}{>\!\!\rightarrow\!\!>} H\} \cup \{I\}$. The powers of F form a behavior.

Suppose $F \equiv \lambda x_1 \ldots x_k.x_i\mathcal{F}_1 \ldots \mathcal{F}_\ell$ and $G \equiv \lambda x_1 \ldots x_m.x_j\mathcal{G}_1 \ldots \mathcal{G}_n$ are normal combinators and there is an order zero combinator α such that $F\alpha \underset{\beta}{=} G\alpha$. Note that $k = m$, $i = j$, and $i \neq q \Rightarrow l = n$. Symmetrically, assume $l \leq n$. We define a set $\delta(F, G)$ of deeds as follows.

$$\delta(F,G) = \begin{cases} \{I\} \cup \bigcup_{p=1}^{\ell} \delta(\lambda x_1 \ldots x_k\mathcal{F}_p, \lambda x_1 \ldots x_m\mathcal{G}_p) & \text{if } i \neq 1 \\ \{\lambda x_1.x_1 \overset{p=1}{\mathcal{G}_1} \ldots \mathcal{G}_{n-\ell}\} \cup \bigcup_{p=1}^{\ell} \delta(\lambda x_1\lambda x_1 \ldots x_k\mathcal{F}_p, \\ \quad \lambda x_1 \ldots x_m\mathcal{G}_{n-\ell+p}) & \text{if } i = 1. \end{cases}$$

Note that $I \in \delta(F,G)$ and whenever $H \in \delta(F,G)$ $\check{H}$ is a subterm of F or G.

Lemma 4. *If F and G are normal and α is an order zero combinator then* $F\alpha = G\alpha \Leftrightarrow \forall H \in \delta(F,G)\alpha \underset{\beta}{=} H\alpha$.

Proof. By induction on F and G. $\square$

Example 6. If a is a set of order zero combinators let $\delta(a) = \{F : \forall\alpha \in a\alpha \underset{\beta}{=} F\alpha$ and F is a deed$\}$. Then $\delta(a)$ is a behavior, since whenever $F, H \in \delta(a)$ and $F \underset{G}{>\!\!\rightarrow} H$ we have $G \in \delta(F,H)$.

A behavior Δ is said to be *non-trivial* if there are order zero α and β such that $\Delta \subseteq \delta(\alpha)$ but $\Delta \not\subseteq \delta(\beta)$.

The following is an analogue of Hilbert's Nullstellensatz.

Proposition 3. *If a is a non empty set of order zero combinators then $\delta(a)$ is the set of all powers of one of its members.*

Proof. Let F be a shortest member of $\delta(a) - \{I\}$. We show by induction on $G \in \delta(a) - \{I\}$, that $F \underset{G}{>\!\!\rightarrow\!\!>} G$. If $F \not\equiv G$ then there is some $H \in \delta(F,G) - \{I\}$. By Lemma 4 $H \in \delta(a)$. Since F is shortest H is a proper subterm of G; so by induction hypothesis $F \underset{F}{>\!\!\rightarrow\!\!>} H$. There is some deed L such that $L \underset{H}{\rightarrow} G$ so $L \in \delta(a) - \{I\}$. Thus by induction hypothesis $F \underset{F}{\rightarrow\!\!>} L$. Hence $F \underset{F}{\rightarrow\!\!>} G$. This completes the proof. $\square$

Corollary 2. *For any normal combinators F and G there exists a normal H such that for all order zero combinators α*

$$F\alpha \underset{\beta}{=} G\alpha \Leftrightarrow \alpha = H\alpha.$$

Proof. Let $a = \{\alpha : F\alpha \underset{\beta}{=} G\alpha$ and α is order zero.$\}$ If $a = \emptyset$ put $H \equiv K$. Otherwise $\delta(a)$ is the set of all powers of some $H \in \delta(a)$. We claim that this is the desired H. For suppose $F\alpha \underset{\beta}{=} G\alpha$. Then $\alpha \in a$ so $\alpha \underset{\beta}{=} H\alpha$. Conversely, if $\alpha \underset{\beta}{=} H\alpha$ then, since $\delta(F,G) \subseteq \delta(a)$, whenever $L \in \delta(F,G)\alpha = L\alpha$. Thus, by Lemma 4, $F\alpha \underset{\beta}{=} G\alpha$. $\square$

Corollary 3. *Every non trivial behavior is contained in the set of powers of some deed.*

Remark 2. We say G is an *atom* if whenever G is a power of F then $F \equiv G$. It can be proved that each deed is a power of a unique atom. This analogous to the fact from formal language theory that whenever $u^n = v^m$ both u and v are power of a common atom. We shall not give the proof here because the result will not be used below.

An order zero combinator α is said to be *algebraic* if there are normal F and G such that $F \not\equiv G$ and $F\alpha \underset{\beta}{=} G\alpha$. Otherwise α is said to be *transcendental.* A set $\{\alpha_1 \ldots \alpha_k\}$ of order zero combinators is said to be *algebraically dependent* if there exist normal combinators F and G with $F \not\equiv G$ such that $F\alpha_1 \ldots \alpha_k \underset{\beta}{=} G\alpha_1 \ldots \alpha_k$.

Lemma 5. *Suppose $\{\alpha_1 \ldots \alpha_k\}$ is a set of algebraically dependent combinators. Then there exists an $i, 1 \leq i \leq k$, and a normal $H \not\equiv J_i^k$ such that $\alpha_i \underset{\beta}{=} H\alpha_1 \ldots \alpha_k$.*

Proof. Suppose F and G are distinct normal combinators such that $F\alpha_1 \ldots \alpha_k \underset{\beta}{=} G\alpha_1 \ldots \alpha_k$. The proof is a routine induction on F and G.

□

Remark 3. In Lemma 5, since each α_i is order zero, H can be put in the form $\lambda x_1 \ldots x_k.x_j(H_1x_1 \ldots x_k) \ldots (H_\ell x_1 \ldots x_k)$ with each H_p normal.

Proposition 4. *Suppose $\{\alpha_1, \ldots, \alpha_k\}$ is a set of algebraically independent combinators. Then there is an order zero combinator α_{k+1} such that $\{\alpha_1, \ldots, \alpha_k, \alpha_{k+1}\}$ is algebraically independent.*

Proof. Consider the equations:

$$
\begin{aligned}
&(1) \quad M \underset{\rho}{=} M(H_1\alpha_1 \ldots \alpha_k M) \ldots (H_\ell\alpha_1 \ldots \alpha_k M), \ell \geq 1 \\
&(2)_i \quad M \underset{\beta}{=} \alpha_i(H_1\alpha_1 \ldots \alpha_k M) \ldots (H_\ell\alpha_1 \ldots \alpha_k M) \\
&(3)_i \quad \alpha_i \underset{\beta}{=} M(H_1\alpha_1 \ldots \alpha_k M) \ldots (H_\ell\alpha_1 \ldots \alpha_k M) \\
&(4)_{ij} \quad \alpha_i \underset{\beta}{=} \alpha_j(H_1\alpha_1 \ldots \alpha_k M) \ldots (H_\ell\alpha_1 \ldots \alpha_k M)
\end{aligned}
$$

where if $i = j$ then $l \geq 1$. For each of the k^2+2k+1 equations $E(H_1, \ldots, H_\ell, M)$ the set $\{M : \exists H_1 \ldots H_\ell$ normal such that $E(H_1, \ldots, H_\ell, M)\}$ is Visseral. For each of these sets the following are not members: $(1)\Omega$, $(2)_i K^\infty$, $(3)_i K^\infty$, $(4)_{ij}\alpha_i$. Thus each of the complements is co-Visseral and non-empty. In addition, the set of order zero combinators is co-Visseral and non-empty. Hence, by [5] 2.5, all these co-Visseral sets intersect, and any members of the interesection is the desired α_{k+1}. This completes the proof. □

Proposition 4 allows the construction of many algebraically independent transcendentals. However, a more direct construction of transcendentals can be achieved using Proposition 3. The following is an analogue of Lindemann's theorem.

Theorem 3. *Suppose α is an order zero algebraic combinator. Then there are only finitely many algebraic combinators β, up to β conversion, of the form $\beta \underset{\beta}{=} F\alpha$, for F a deed.*

Proof. First suppose α is order zero and $\alpha \underset{\beta}{=} \alpha(F_1\alpha)\dots(F_n\alpha), \beta \underset{\beta}{=} \alpha(G_1\alpha)\dots(G_m\alpha)$, and $\beta \underset{\beta}{=} \beta(H_1\beta)\dots(H_k\beta)$, where the F_i, G_j, and H_ℓ are all normal and $n \leq 1$, $m \geq o$, and $k \geq 1$ are as small as possible. Then $\alpha \underset{\beta}{=} \beta(H_1\beta)\dots(H_{k-m}\beta)$, so, setting $G \equiv \lambda x\ x(G_1x)\dots(G_mx)$ and $H^- \equiv \lambda x.\ x(H_1x)\dots(H_{k-m}x)$ we have $\alpha \underset{\beta}{=} (H^- \circ G)\alpha$. Now if F is a deed such that $\delta(\alpha)$ coincides with the powers of F and $F \equiv \lambda x\ x\mathcal{F}_1\dots\mathcal{F}_n$ let $F_i \equiv \lambda x\mathcal{F}_i$. If α is algebraic, we have $n \geq 1$ and n is as small as possible. Let J be the normal form of $H^- \circ G$. If $\beta \underset{\beta}{\neq} \alpha$ then $J \underset{\beta}{\neq} I$ and $F \underset{F}{>\!\!\rightarrow\!\!>} J$. By inspection, there exists some $r \leq n$ such that $\lambda x.\ x\mathcal{F}_1\dots\mathcal{F}_r \underset{F}{>\!\!\rightarrow\!\!>} G$ so $\beta \underset{\beta}{=} (\lambda x.\ x\mathcal{F}_1\dots\mathcal{F}_r)\alpha$. This completes the proof. □

4. Applications and Open Problems

First we solve a problem of Böhm and Dezani ([2] p. 185).

Proposition 5. *It is undecidable whether $Fx = I$ has a (normal) solution, for normal combinators F.*

Proof. For any combinator M, "make M normal" by replacing each redex $(\lambda x\mathcal{H})\mathcal{Y}$ by $x(\lambda x\mathcal{H})\mathcal{Y}$ and replace the result with $M^\# \equiv \lambda x.\ x \circ M \circ x$. Observe that $M^\# I \underset{\beta}{=} M$. Now, if $M^\# N \underset{\beta}{=} I$ then N is both right and left β invertible, so by [1] 21.4.8, $N \underset{\beta}{=} I$. Hence $M_e^\# x \underset{\beta}{=} I$ has a (normal) solution if and only if $e \in W_e$. □

Remark 4. A similar argument works for $\beta\eta$ conversion.

Below we gave a stronger result for general normal equations.

Our next application concerns Hilbert's 10th problem. Suppose d is an adequate numeral system ([1] p. 139) with normal test for zero.

Proposition 6. *Suppose Σ is the β closure of a recursively enumerable set of d_n. Then Σ is a normal variety.*

Proof. It is easy to see that there is deed F such that for all n $Fd_n \underset{\beta}{=} I$. □

The next example shows that the condition of normality on the zero test is necessary.

Example 7. Let $d_n^{\circ} \equiv \lambda x.\ x(K^{n^2+1}\ulcorner n \urcorner)$ and $d_n^{m+1} \equiv \lambda x.\ x d_n^m \underbrace{\Omega \ldots \Omega}_{n}$. We write d_n for d_n^n. Construct F satisfying $F \underset{\beta}{=} (\lambda xy.yx)F$. We have $Fd_n^o \underset{\beta}{=} F(K^{n^2+1}\ulcorner n \urcorner) \underset{\beta}{=} K^{n^2}\ulcorner n \urcorner$. By induction, $Fd_n^m \underset{\beta}{=} K^{n^2-m.n}\ulcorner n \urcorner$, for $m \leq n$. Thus $Fd_n \underset{\beta}{=} \ulcorner n \urcorner$, and d is an adequate numerical system. By the method of the proof of Lemma 2, d is not a normal variety.

Despite Corollary 1 it may be possible to give a coherent solution to the following.

Open Problem. *Characterize the normal varieties.*

References

1. H.P Barendregt, *The Lambda Calculus*, North Holland, 1984.
2. C. Böhm and M. Dezani-Ciancaglini, *Combinatorial problems, combinator equations and normal forms*, Automated Languages and Programming, J. Loeckx, ed., Lecture Notes in Computer Science **14**, Springer-Verlag, 1974, pp. 185–199.
3. J.W. Klop, *Combinatory Reduction Systems*, Math. Centrum Amsterdam, 1980.
4. R. Statman, *On sets of solutions to combinator equations*, T.C.S. **66** (1989), 99–104.
5. A. Visser, *Numerations, λ-calculus and arithmetic*, Essays on Combinatory Logic, Lambda Calculus and Formalism, J.P Seldin and J.R Hindley, eds., Academic Press, 1980.

Department of Mathematics, Carnegie Mellon University, Pittsburgh, PA 15213

NSF CCR–8702699

COMPLEXITY OF PROOFS IN CLASSICAL PROPOSITIONAL LOGIC

ALASDAIR URQUHART

ABSTRACT. This paper surveys the work of the last two decades on the complexity of proofs in classical propositional logic, and its connection with open problems in theoretical computer science. The paper ends with a short list of open problems.

1. INTRODUCTION

The rise of computer science has revealed in a startling fashion several areas of logic which have been neglected by the logical tradition of the last few decades. Nowhere is this neglect more striking than in the area of classical propositional logic. Open questions of computer science such as $\mathcal{P} =?\mathcal{N}P$ put the satisfiability problem of propositional logic in a central position. In contrast to the interest shown by the pioneers of modern logic in such questions (see for example Martin Gardner's interesting little book on the subject [10]), algorithms for the satisfiability problem occupy only a small part of modern logical texts, and are generally dismissed as of no great interest. A similar striking neglect is shown in the area of the simplification of Boolean formulas. Here the general problem takes the form: how many logical gates do we need to represent a given Boolean function? This is surely as simple and central a logical problem as one could hope to find; yet in spite of Quine's early contributions [15], [16] the whole area has been simply abandoned by most logicians, and is apparently thought to be fit only for engineers. As we shall see below, this question has close connections with the topic of our title, and our lack of understanding of the simplification problem retards our progress in the area of the complexity of proofs.

The present paper is concerned with the question: how long does a proof of a tautology have to be, as a function of the size of the tautology? The best upper bound for standard proof systems is exponential in the size of the formula proved. It is conjectured that a similar lower bound holds for such systems, but except for a fairly limited class of systems, no such bound has yet been proved, and the problem of proving such bounds appears very difficult. The present paper gives a brief survey of the progress that has been made to date on the problem, which has been the subject of serious research for just over twenty years. An important topic omitted from this survey is the connection between weak systems of arithmetic and propositional provability. The reader is referred to [1], [2], [4], [5], [7], [13], [14], for some of the results in this area.

2. Proof systems and simulation

This section gives a general definition of a proof system, together with a definition which is intended to formalize the relation which holds between two proof systems when one can simulate the other efficiently. The definitions are adapted from Cook and Reckhow [6].

Let Σ be a finite alphabet; we write Σ^* for the set of all finite strings over Σ.

Definition 1. If Σ_1 and Σ_2 are finite alphabets, a function f from Σ_1^* into Σ_2^* is in $\mathcal{L}$ if it can be computed by a deterministic Turing machine in time bounded by a polynomial in the length of the input.

The class $\mathcal{L}$ of polynomial-time computable functions we take as a way of making precise the vague notion of "feasibly computable function".

Definition 2. If $L \subseteq \Sigma^*$, a *proof system for* L is a function $f : \Sigma_1^* \to L$ for some alphabet Σ_1, where $f \in \mathcal{L}$ and f is onto. A proof system f is *polynomially bounded* if there is a polynomial $p(n)$ such that for all $y \in L$, there is an $x \in \Sigma_1^*$ such that $y = f(x)$ and $|x| \leq p(|y|)$, where $|z|$ is the length of the string z.

The property of proof systems which we are taking to be the defining condition is that, given an alleged proof, there is a feasible method for checking whether or not it really is a proof, and if so, of what it is a proof. A standard axiomatic proof system for the tautologies, for example, can be brought under the definition by associating the following function f with the proof system $\mathcal{F}$: if a string of symbols σ is a legitimate proof in $\mathcal{F}$ of a formula A, then let $f(\sigma) = A$, if it is not a proof in $\mathcal{F}$ then let $f(\sigma) = T$, where T is some standard tautology, say $P \vee \neg P$.

The importance of our main question for theoretical computer science lies in the following result of Cook and Reckhow [6].

Theorem 1. *$\mathcal{N}P = co{-}\mathcal{N}P$ if and only if there is a polynomially-bounded proof system for the classical tautologies.*

The above equivalence underlines the very far-reaching nature of the widely believed conjecture $\mathcal{N}P \neq co{-}\mathcal{N}P$. The conjecture implies that even ZFC, together with any true axioms of infinity that are thought desirable (provided that these axioms of infinity have a sufficiently simple syntactic form) is not a polynomially-bounded proof system for the classical tautologies (where we take a proof of $TAUT(\ulcorner A \urcorner)$ as a proof of the tautology A).

We can say nothing of interest about the complexity of such powerful proof systems as the above (in effect, the strongest we can imagine). We can, however, order proof systems in terms of complexity, and prove some non-trivial separation results for systems low down in the hierarchy.

Definition 3. If $f_1 : \Sigma_1^* \rightarrow L$ and $f_2 : \Sigma_2^* \rightarrow L$ are proof systems for L, then f_2 *p-simulates* f_1 provided that there is a polynomial-time computable function $g : \Sigma_1^* \rightarrow \Sigma_2^*$ and $f_2(g(x)) = f_1(x)$ for all x.

Thus g is a feasible translation function which translates proofs in f_1 into proofs in f_2. We have assumed in the above definition that the language of both proof systems is the same. Reckhow's thesis [17] contains a more general definition of p-simulation which eliminates this restriction. It is easy to see that the p-simulation relation is reflexive and transitive, and also that the following theorem follows from the definitions.

Theorem 2. *If a proof system f_2 for L p-simulates a polynomially bounded proof system f_1, then f_2 is also polynomially bounded.*

The symmetric closure of the p-simulation relation is an equivalence relation; thus we can segregate classes of proof systems into equivalence classes within which the systems are "equally efficient up to a polynomial".

3. A MAP OF PROOF SYSTEMS

The present section is a commentary on the accompanying map showing the relative efficiency of various proof systems. The boxes in the diagram indicate equivalence classes of proof systems under the symmetric closure of the p-simulation relation. Systems below the dotted line have been shown not to be polynomially bounded, while no such lower bounds are known for those which lie above the line. Hence, the dotted line represents the current frontier of research on the main problem. Although systems below the line are no longer candidates for the role of a polynomially bounded proof system, there are still some interesting open problems concerning the relative complexity of such systems. Questions of this sort, although not directly related to such problems as $\mathcal{NP} =? co-\mathcal{NP}$, have some relevance to the more practical problem of constructing efficient automatic theorem provers.

An arrow from one box to the other in the diagram indicates that any proof system in the first box can p-simulate any system in the second box. In the case of cut-free Gentzen systems, this simulation must be understood as referring to a particular language on which both systems are based. An arrow with a slash through it indicates that no p-simulation is possible between any two systems in the classes in question. If a simulation is possible in the reverse direction (as is the case with all the systems in our diagram), then we can say that systems in one class are strictly more powerful than systems in the other (up to a polynomial). The diagram shows that all such questions of relative strength have been settled for systems below the dotted line, with the exception of the case of the relative complexity of resolution and cut-free Gentzen systems where connectives other than the biconditional and negation are involved.

The diagram shows only a selection from the wide variety of proof systems which have been considered in the literature of logic, automatic theorem proving and combinatorics. A more detailed diagram, showing a wider selection of proof systems, is to be found in Reckhow [17].

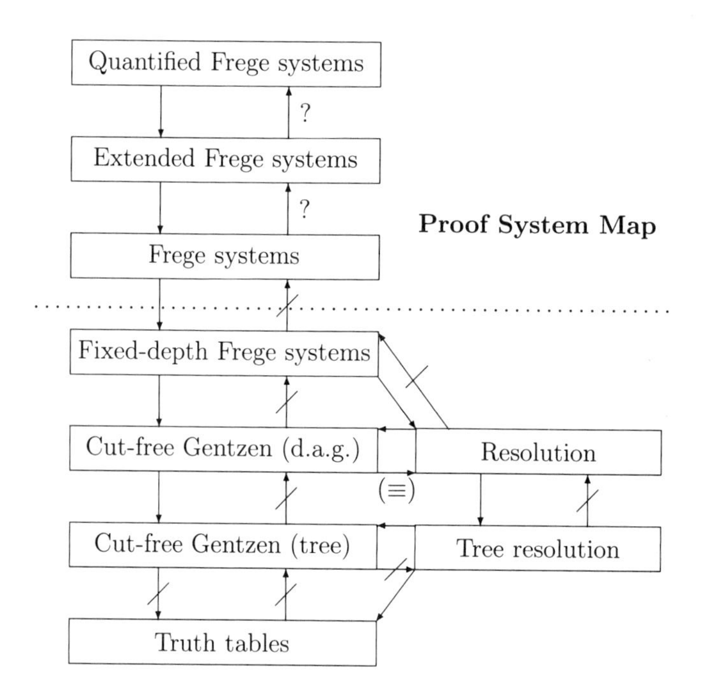

The least efficient of the systems considered is the truth table method. If we consider a formula with n variables, then a truth table, represented as a vector of 0's and 1's, has length 2^n. It should be emphasized, however, that we are only considering asymptotic complexity measures here. In practice, the truth table method may be quite efficient for formulas containing a small number of variables, given a reasonably sophisticated implementation. It is easy, however, to find simple tautologies containing n variables which can be proved quickly by a cut-free Gentzen system, for example the tautologies having the form

$$(P_1 \vee \neg P_1) \wedge \ldots \wedge (P_n \vee \neg P_n).$$

Cut-free Gentzen systems are an improvement on truth tables, at least asymptotically, but in certain cases may even be worse, in the strict sense. In fact, Marcello D'Agostino has pointed out that the family of tautologies consisting of the disjunction of all possible conjunctions of a set of n variables or their negations serves to separate these systems from truth tables. They have been of considerable interest to the automatic theorem proving community, since they lend themselves to goal-directed search strategies, which in favourable cases may result in elegant and attractive proofs. However, they break down badly in certain cases, a fact which is especially plain in the case of biconditional tautologies. Hao Wang wrote one of the earliest theorem provers, basing it on a cut-free Gentzen system; he explicitly noted the inefficient behaviour of his system when given such problems as input [23]. A simple class of difficult problems for such systems is the following family of tautologies. Let U_n be the formula :

$$P_1 \equiv (P_2 \equiv \ldots \equiv (P_n \equiv (P_1 \equiv (P_2 \equiv \ldots \equiv P_n) \ldots).$$

Let G_{Tree} be a cut-free Gentzen system for pure biconditionals, employing sequents with multiple righthand side, where the proofs are represented as trees, having axioms of the form $A \vdash A$. Then it is not hard to prove the following theorem [22].

Theorem 3. *A minimal size proof of the sequent* $\vdash U_n$ *in the system* G_{Tree} *contains* $n.2^n$ *occurrences of axioms.*

By modifying this example, we can show that cut-free Gentzen systems in tree form cannot simulate truth tables, so that the two proof systems are incomparable in the p-simulation ordering; this observation is due to Marcello D'Agostino.

It is normal in the standard literature of proof theory to represent proofs as trees, although it is clear that this representation is inefficient, since a sequent may appear many times in the proof tree. From the usual point of view, where efficiency is disregarded, this is of no consequence. However, in the context of complexity theory, it makes a great deal of difference. This is illustrated in the next result. We shall use the notation G to refer to the Gentzen system for biconditionals, where the proofs are represented as directed acyclic graphs or "d.a.g's" (equivalently, as sequences of sequents).

Theorem 4. *The sequents* $\vdash U_n$ *have derivations in* G *containing* $O(n^2)$ *distinct sequents.*

The above theorem, proved in [22], illustrates the fact that even very simple logical systems may contain surprises, contrary to the opinion of Wittgenstein [24, Proposition 6.1251]. The reader is invited to try to prove the existence of such derivations, which is not obvious. This result provides the second separation result which is indicated in our diagram. An earlier proof of this separation was given by Statman [19], using a different set of examples.

If we represent cut-free Gentzen proofs as d.a.g.'s, or sequences, of sequents, then a significant increase in efficiency results. However, the proofs in this case may not be of the kind which are found by the simple search procedure mentioned above. Even in this case, biconditional problems cause difficulties, though the examples showing this are not of the simple type of the U_n formulas. Let G be a graph (we allow multiple edge in G, but not loops). Associate a distinct propositional variable with each edge in G, and pick a single distinguished vertex v. Then associate with each vertex the biconditional formula $F(v)$:

$$P_1 \equiv P_2 \equiv \ldots \equiv P_k$$

where $P_1 \ldots P_k$ are the variables labelling the edges incident with v. Then the sequent $S(G)$ is defined to be the sequent which has as antecedent the set of biconditionals $\{F(w) : w \neq v\}$, and the consequent $F(v)$. By choosing an appropriate sequence of graphs, we can construct difficult problems for G. The following theorem is proved in [21]:

Theorem 5. *Let* L_n *be the graph on* n *vertices which has exactly two edges joining each pair of vertices. Then any proof in* G *of* $S(L_n)$ *contains at least* $2^{n/16}$ *distinct sequents.*

A proof system which has been of particular interest to those involved in automatic theorem proving is the resolution system. Its simple syntax and single rule makes it particularly suitable for computer implementation. Here the steps in proofs take the form of *clauses*, that is, disjunctions constructed from literals (variables or their negations). The single rule is the resolution rule: from $(A \vee p)$ and $(B \vee \neg p)$ deduce $(A \vee B)$, where p is a propositional variable. We allow the free use of the associative, commutative and contraction rules within disjunctions (that is, a clause is treated as a set of literals).

A *resolution refutation* of a set of clauses is a resolution proof of Λ, where Λ is the empty clause.

Since resolution is designed to deal only with clauses, it is not obvious what is meant by the simulation of cut-free systems by resolution shown in our map. However, there is an efficient method of associating with a given sequent S a set of clauses $Clauses(S)$ so that S is logically valid if and only if $Clauses(S)$ is contradictory; the method was first described by Tseitin [20]. (This paper of Tseitin, which reports on work done in 1966, is a landmark as the first paper giving a deep analysis of the complexity of proofs in propositional logic.) Then a proof of the sequent S by the resolution method is understood to mean a resolution refutation of the set of clauses $Clauses(S)$. With this convention, it is straightforward to prove the simulation claimed, which is due to Tseitin [20].

A converse simulation holds in the case where the only connective employed is $\equiv$. That is to say, if we restrict our attention to sequents involving only this connective, then resolution and G are equally efficient (up to a polynomial) as proof systems for such sequents [22]. This result extends to the case where $\neg$ is included, but it is not known whether it is true when a functionally complete set of connectives is employed.

No such simulation is possible if we use the tree form of representation for both resolution and cut-free Gentzen proofs, since the sequents $\vdash U_n$ have polynomial-size resolution proofs in tree form.

As in the case of cut-free Gentzen systems, the d.a.g. form of proofs is strictly more powerful than the tree form. This result is also due to Tseitin [20]. To describe the examples which witness the separation, consider a graph G with a single distinguished vertex v. Let $\alpha(G)$ be the set of clauses constructed as follows. For a vertex w distinct from v, let $\alpha(w)$ be the clauses whose conjunction is the conjunctive normal form of $F(w)$; let $\alpha(v)$ be the clauses in the conjunctive normal form of $\neg F(v)$; $\alpha(G)$ is the union of all the clauses formed in this way. Let G_n be the graph having vertices $\{1, \dots, 2^n\}$, where the vertex k is joined to the vertex $k+1$ by n edges; take the vertex 1 to be designated. Then the method of Tseitin [20] can be used to show:

Theorem 6. *The clauses $\alpha(G_n)$ have polynomially bounded resolution refutations when the proofs are represented as d.a.g's, but there is no polynomial bound on the size of the refutations when they are represented as trees.*

An *inference rule* is a pair consisting of a sequence of formulas S and a formula A; an inference rule is *sound* if any truth-value assignment making all the formulas in S true also makes A true. If $\langle S, A\rangle$ is a rule, then we say that a formula B is *inferred from a sequence of formulas T by the rule* $\langle S, A\rangle$ if there is a substitution σ of formulas for variables so that $S^\sigma = T$ and $A^\sigma = B$ (where C^σ is the result of performing the substitution σ on the formula C, similarly for sequences of formulas).

A *Frege system* is defined to consist of a finite set of sound rule. If $\mathcal{F}$ is a Frege system, then a *derivation in $\mathcal{F}$ from the set of assumptions S* is a finite sequences of formulas so that every formula in the sequence is either a member of S or is derived from a sequence of earlier formulas by one of the rules of the system (in the case of axioms of $\mathcal{F}$, the sequence in question is empty). A Frege system $\mathcal{F}$ is *implicationally complete* if whenever A is a logical consequence of a set of formulas S, then there is a derivation in $\mathcal{F}$ of A from S.

Frege systems form a class of flexible and powerful systems. The most familiar Frege systems are given by a finite number of axiom schemes, together with a rule such as *modus ponens.* Systems equivalent to Frege systems in complexity are natural deduction systems and Gentzen systems with cut. The details of the simulations involved are to be found in Reckhow's thesis [17].

It is easy to see that there are simple tautologies which require proofs of size $O(n^2)$, where n is the number of symbols in the tautology. No better lower bound is known for an unrestricted Frege system. The strongest result so far obtained in this area was recently proved by Ajtai [1]; it involves placing a restriction on the complexity of the formulas involved in the derivations.

To define the appropriate restriction on formulas, we define a hierarchy which counts the number of alternations of $\vee$'s and $\wedge$'s in a formula. We assume that the language in which we operate is based on the connectives $\{\vee, \wedge, \neg\}$.

Definition 4. The Σ_k- and Π_k-formulas are defined inductively as follows:

(1) A propositional variable is a Σ_0-formula and a Π_0-formula.
(2) If A is a Σ_i-formula (Π_i-formula) then $\neg A$ is a Π_i-formula (Σ_i-formula).
(3) If $A_1, \ldots, A_n$ are Σ_i-formulas (Π_i-formulas), then any conjunction (disjunction) of them is a Π_{i+1}-formula (Σ_{i+1}-formula).

The preceding definition is that of Buss [3]; the hierarchy defined here is closely related to the hierarchy of bounded-depth, unlimited fan-in Boolean circuits employed in the theory of circuit complexity ([9], [12]).

The pigeonhole principle, that is, the assertion that there is no one-one map from the set $\{1, \ldots, n\}$ into $\{1, \ldots, n-1\}$, can be formulated as a sequence of propositional tautologies. For $i \in \{1, \ldots, n\}$, and $j \in \{1, \ldots, n-1\}$, let P_{ij} be a propositional variable. Then PHP_n is the following tautology:

$$\neg[\bigwedge_{i=1}^{n} \bigvee_{j=1}^{n-1} P_{ij} \wedge \bigwedge_{k=1}^{n-1} \bigwedge_{i \neq j} \neg(P_{ik} \wedge P_{jk})].$$

Ajtai's main result [1] is that the pigeonhole tautologies do not have short proofs in a Frege system, if we impose the restriction that the formulas in the proofs are of bounded depth. We define the *size* of a Frege proof to be the number of occurrences of symbols in the proof (note that this measure of complexity in the case of Frege systems is not polynomially related to the *length* of the proof (i.e. the number of lines), as it is in the case of resolution and cut-free Gentzen systems).

Theorem 7. *Let $\mathcal{F}$ be a fixed Frege system, using the primitive connectives $\{\vee, \wedge, \neg\}$. Then for a given u, d, and sufficiently large n, there is no derivation of size n^u or less in $\mathcal{F}$ of PHP_n in which all of the formulas in the derivation are in the class Σ_d.*

Ajtai's proof of this theorem uses a highly ingenious blend of non-standard number theory and combinatorics. Some of the combinatorial lemmas in the proof are adapted from earlier work on computational limitations of bounded-depth circuits. Ajtai's theorem also serves to separate unrestricted Frege systems from bounded-depth Frege systems, since Buss has shown [3] that the pigeonhole tautologies have polynomial-size proofs in a standard Frege system. The existence of these proofs is not at all obvious; in fact Cook and Reckhow conjectured that there were no such proofs in [6]. Buss used the fact that addition of n-bit binary numbers can be computed by circuits with depth $O(\log n)$ by the use of carry-save addition gates. By using this circuit technique in propositional logic, he was able to show directly the existence of relatively short proofs of PHP_n which formalize a counting argument. This illustrates the close connection mentioned above between the complexity of propositional proofs and the theory of formula and circuit complexity.

The remaining separations in our diagram, between bounded-depth Frege systems and the two systems immediately below, follow from results of Buss and Turán [4]. The negation of the pigeonhole tautologies PHP_n can be written as a contradictory set of clauses. Haken [11] proved an exponential lower bound for resolution refutations of these clauses (an earlier exponential lower bound for resolution was proved by Tseitin [20], but only with the restriction that the resolution refutations be *regular*, a condition that we shall not discuss).

It is possible to formalize in the same way a set of clauses whose intuitive reading is that there is a one-to-one map from a set of size $2n$ into a set of size n; we shall denote this set of clauses by PHC_n^{2n}. By generalizing Haken's argument, Buss and Turán [4] were able to show exponential lower bounds for resolution refutations of these sets of clauses.

Theorem 8. *There is a $c > 0$ so that for sufficiently large n, any resolution refutation of PHC_n^{2n} contains at least c^n distinct clauses.*

However, Buss and Turán show that there is a d so that these modified pigeonhole clauses PHC_n^{2n} have refutations of size $O(n^{(\log n)^c})$ for some constant c in a Frege system in which all the formulas in the refutations are required to belong to Σ_d. This demonstrates the last separation results in our diagram, since the Buss/Turán methods also apply to cut-free Gentzen systems.

Of the complexity of the systems above the dotted line, nothing can at present be said, beyond the fact that they present a formidable challenge to the ingenuity of logicians. Part of the problem lies in the fact that at present we do not have techniques for proving super-polynomial lower bounds on the formula size or circuit size complexity of explicity defined Boolean functions [9]. In the case of bounded-depth, unbounded fan-in circuits, such lower bounds are known [9], [12], and hence these techniques could be adapted to show Ajtai's lower bound.

Extended Frege systems allow the use of abbreviative definitions in proofs, in addition to the use of axioms and rules. Thus in the proof of a formula A from a set of assumptions S, we are allowed to introduce as a step in the proof $P \equiv B$, for any formula B, where P is a variable not occurring in A, B, S or in any formula occurring earlier in the proof. This form of abbreviative definition in effect gives us the same logical expressive power as is possessed by Boolean circuits. Extended Frege systems are p-equivalent to Frege systems with the uniform substitution rule. This is a result of Martin Dowd [8], which was later rediscovered by Krajíček and Pudlák [13]. It appears plausible that any advance in proving lower bounds for extended Frege systems must wait on advances in understanding the circuit complexity of Boolean functions. Similar remarks apply to the relation between Frege systems, and the formula complexity of Boolean functions.

Quantified Frege systems allow the use of propositional quantifiers in proving tautologies. Such systems are at least as powerful as extended Frege systems [17], but otherwise little is known about their complexity. It seems for the moment fruitless to consider systems stronger than quantified Frege systems while no progress has been made on weaker systems such as Frege systems, although as we indicated above it is not hard to imagine such methods of proof. It is not known whether or not there is a strongest proof system in the p-simulation ordering; Krajíček and Pudlák discuss this problem and its relation to other open problems in the proof theory of arithmetic and complexity theory in [13].

4. Open problems

This section gives a few open problems. These are all problems which appear to be solvable by appropriate modification of current proof techniques, although some of them may be difficult.

(1) Prove a lower bound on the complexity of Frege systems which is better than quadratic. Quadratic lower bounds were proved by Krapchenko [9] for the size of formulas expressing the parity function over the basis $\{\wedge, \vee, \neg\}$. This suggests that it may be possible to prove cubic bounds for proofs of biconditional tautologies, provided the proof system uses this basis.

(2) Can a cut-free Gentzen system for $\{\vee, \neg\}$ simulate resolution, as a method of refuting sets of disjunctions of literals?

(3) Prove a non-polynomial lower bound on bounded-depth Frege systems, where the definition of depth is modified to allow the formation of arbitrary chains of biconditionals as well as arbitrary conjunctions and disjunctions. That such a bound may be provable is suggested by lower bounds proved by Razborov and Smolensky for bounded-depth circuits which allow parity gates of unbounded fan-in [9].

(4) Are there families of tautologies which separate depth d from depth $d + 1$ Frege systems?

References

1. Miklos Ajtai 'The complexity of the pigeonhole principle', forthcoming. Preliminary version, *29th Annual Symposium on the Foundations of Computer Science* (1988), 346-55.
2. Samuel R. Buss. *Bounded Arithmetic.* Ph.D. dissertation, Princeton University 1985; revised version Bibliopolis Naples, 1986.
3. Samuel R. Buss. 'Polynomial size proofs of the propositional pigeonhole principle', *J. Symbolic Logic*, Vol. 52 (1987), 916-27.
4. Samuel R. Buss and György Turán. 'Resolution proofs of generalized pigeonhole principles', *Theoretical Computer Science*, Vol. 62 (1988), 311-17.
5. Stephen A. Cook. 'Feasibly constructive proofs and the propositional calculus', *Proc. 7th A.C.M. Symposium on the Theory of Computation*, 83-97.

6. Stephen A. Cook and Robert A. Reckhow. 'The relative efficiency of propositional proof systems', *J. Symbolic Logic* Vol. 44 (1979), 36-50.
7. Martin Dowd. *Propositional representation of arithmetic proofs.* Ph.D. dissertation, University of Toronto, 1979.
8. Martin Dowd. 'Model-theoretic aspects of $\mathcal{P} \neq \mathcal{NP}$', preprint, 1985.
9. Paul E. Dunne. *The Complexity of Boolean Networks.* Academic Press, London and San Diego, 1988.
10. Martin Gardner. *Logic machines, diagrams and Boolean algebra.* Dover Publications, Inc. New York 1968.
11. Armin Haken 'The intractability of resolution', *Theoretical Computer Science*, Vol. 39 (1985), 297-308.
12. Johan T. Håstad. *Computational limitations of small-depth circuits.* The MIT Press, Cambridge, Massachusetts, 1987.
13. Jan Krajíček and Pavel Pudlák. 'Propositional proof systems, the consistency of first order theories and the complexity of computations', *J. Symbolic Logic*, Vol. 54 (1989), 1063-79.
14. Jan Krajíček and Pavel Pudlák. 'Quantified propositional calculi and fragments of bounded arithmetic', *Zeitschrift für Mathematische Logik und Grundlagen der Mathematik* (to appear).
15. W.V. Quine. 'Two theorems about truth functions', *Bol. Soc. Math. Mex.* Vol. 10, 64-70.
16. W.V. Quine. 'A way to simplify truth functions', *Am. Math. Monthly* Vol. 62 (1955), 627-631.
17. Robert A. Reckhow *On the lengths of proofs in the propositional calculus.* Ph.D. thesis, Department of Computer Science, University of Toronto, 1976.
18. Jörg Siekmann and Graham Wrightson (eds.) *Automation of Reasoning.* Springer-Verlag, New York, 1983.
19. Richard Statman 'Bounds for proof-search and speed-up in the predicate calculus', *Annals of mathematical logic* Vol. 15, 225-87.
20. G.S. Tseitin 'On the complexity of derivation in propositional calculus', *Studies in Constructive Mathematics and Mathematical Logic Part 2.* Seminars in mathematics, V.A. Steklov Math. Institute 1968. Reprinted in [18].
21. Alasdair Urquhart 'The complexity of Gentzen systems for propositional logic', *Theoretical Computer Science* Vol. 66 (1989), 87-97.
22. Alasdair Urquhart 'The relative complexity of resolution and cut-free Gentzen systems', forthcoming, *Discrete Applied Mathematics.*
23. Hao Wang. 'Towards mechanical mathematics', *IBM Journal for Research and Development* Vol. 4 (1960), 2-22. Reprinted in [18].
24. Ludwig Wittgenstein. *Tractatus Logico-Philosophicus.* Kegan Paul Ltd, London 1922.

Department of Philosophy, University of Toronto

Mathematical Sciences Research Institute Publications

(continued)

Volume 18	Bryant, Chern, Gardner, Goldschmidt, Griffiths: Exterior Differential Systems
Volume 19	Alperin (ed.): Arboreal Group Therapy
Volume 20	Dazord and Weinstein (eds.): Symplectic Geometry, Groupoids, and Integrable Systems
Volume 21	Moschovakis (ed.): Logic from Computer Science
Volume 22	Ratiu (ed.): The Geometry of Hamiltonian Systems
Volume 23	Baumslag and Miller (eds.): Algorithms and Classification in Combinatorial Group Theory
Volume 24	Montgomery and Small (eds.): Noncommutative Rings